VOLUME ONE HUNDRED AND TWENTY ONE

ADVANCES IN CANCER RESEARCH

VOLUME ONE HUNDRED AND TWENTY ONE

ADVANCES IN CANCER RESEARCH

Edited by

KENNETH D. TEW

Professor and Chairman,
Department of Cell and Molecular Pharmacology,
John C. West Chair of Cancer Research,
Medical University of South Carolina,
Charleston, South Carolina, USA

PAUL B. FISHER

Professor and Chair,
Department of Human & Molecular Genetics,
Director, VCU Institute of Molecular Medicine,
Thelma Newmeyer Corman Chair in Cancer Research,
VCU Massey Cancer Center,
Virginia Commonwealth University, School of Medicine,
Richmond, Virginia, USA

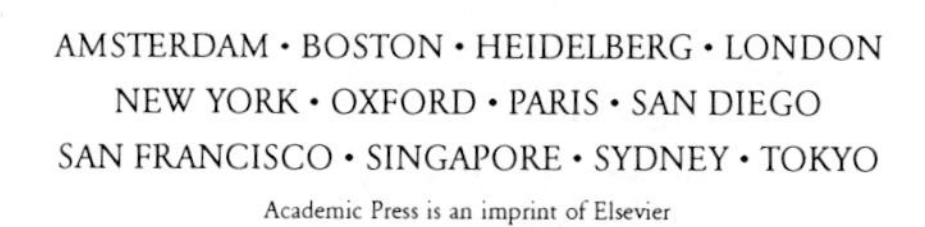

AMSTERDAM • BOSTON • HEIDELBERG • LONDON
NEW YORK • OXFORD • PARIS • SAN DIEGO
SAN FRANCISCO • SINGAPORE • SYDNEY • TOKYO
Academic Press is an imprint of Elsevier

Academic Press is an imprint of Elsevier
525 B Street, Suite 1800, San Diego, CA 92101-4495, USA
225 Wyman Street, Waltham, MA 02451, USA
32 Jamestown Road, London, NW1 7BY, UK
The Boulevard, Langford Lane, Kidlington, Oxford, OX5 1GB, UK
Radarweg 29, PO Box 211, 1000 AE Amsterdam, The Netherlands

First edition 2014

ISBN: 978-0-12-800249-0
ISSN: 0065-230X

For information on all Academic Press publications
visit our website at store.elsevier.com

14 15 16 12 11 10 9 8 7 6 5 4 3 2 1

CONTENTS

CONTRIBUTORS

Aiman Alhazmi
Department of Human and Molecular Genetics, VCU Institute of Molecular Medicine, Massey Cancer Center, Virginia Commonwealth University School of Medicine, Richmond, Virginia, USA

Albert S. Baldwin
Lineberger Comprehensive Cancer Center, University of North Carolina School of Medicine, Chapel Hill, North Carolina, USA

Jennifer W. Bradford
Lineberger Comprehensive Cancer Center, University of North Carolina School of Medicine, Chapel Hill, North Carolina, USA

Amy Lee Bredlau
Department of Pediatrics, and Department of Neurosciences, Medical University of South Carolina, Charleston, South Carolina, USA

Srikumar P. Chellappan
Department of Tumor Biology, H. Lee Moffitt Cancer Center and Research Institute, Tampa, Florida, USA

Swadesh K. Das
Department of Human and Molecular Genetics, and VCU Institute of Molecular Medicine, Virginia Commonwealth University, School of Medicine, Richmond, Virginia, USA

Luni Emdad
Department of Human and Molecular Genetics; VCU Institute of Molecular Medicine, and VCU Massey Cancer Center, Virginia Commonwealth University, School of Medicine, Richmond, Virginia, USA

Benedetto Farsaci
Laboratory of Tumor Immunology and Biology, Center for Cancer Research, National Cancer Institute, National Institutes of Health, Bethesda, Maryland, USA

Paul B. Fisher
Department of Human and Molecular Genetics; VCU Institute of Molecular Medicine, and VCU Massey Cancer Center, Virginia Commonwealth University, School of Medicine, Richmond, Virginia, USA

Daniel J. Foster
Department of Neurology; Department of Neurological Surgery and Brain Tumor Research Center, and Sandler Neurosciences Center, University of California, San Francisco, California, USA

Aaron Frantz
Department of Neurology; Department of Neurological Surgery and Brain Tumor Research Center, and Sandler Neurosciences Center, University of California, San Francisco, California, USA

John W. Greiner
Laboratory of Tumor Immunology and Biology, Center for Cancer Research, National Cancer Institute, National Institutes of Health, Bethesda, Maryland, USA

James L. Gulley
Laboratory of Tumor Immunology and Biology, Center for Cancer Research, National Cancer Institute, National Institutes of Health, Bethesda, Maryland, USA

Christopher R. Heery
Laboratory of Tumor Immunology and Biology, Center for Cancer Research, National Cancer Institute, National Institutes of Health, Bethesda, Maryland, USA

James W. Hodge
Laboratory of Tumor Immunology and Biology, Center for Cancer Research, National Cancer Institute, National Institutes of Health, Bethesda, Maryland, USA

Bin Hu
Department of Human and Molecular Genetics, Virginia Commonwealth University, School of Medicine, Richmond, Virginia, USA

Miller Huang
Department of Neurology, and Helen Diller Family Comprehensive Cancer Center, University of California, San Francisco, California, USA

Shirin Ilkhanizadeh
Department of Neurology, and Helen Diller Family Comprehensive Cancer Center, University of California, San Francisco, California, USA

Caroline Jochems
Laboratory of Tumor Immunology and Biology, Center for Cancer Research, National Cancer Institute, National Institutes of Health, Bethesda, Maryland, USA

Timothy P. Kegelman
Department of Human and Molecular Genetics, Virginia Commonwealth University, School of Medicine, Richmond, Virginia, USA

David N. Korones
Department of Pediatrics, and Department of Palliative Care, University of Rochester, Rochester, New York, USA

Joseph W. Landry
Department of Human and Molecular Genetics, VCU Institute of Molecular Medicine, Massey Cancer Center, Virginia Commonwealth University School of Medicine, Richmond, Virginia, USA

Jasmine Lau
Department of Neurology, and Helen Diller Family Comprehensive Cancer Center, University of California, San Francisco, California, USA

I. Levinger
Avram and Stella Goldstein-Goren Department of Biotechnology Engineering, Ben-Gurion University, Beer-Sheva, Israel

Ravi A. Madan
Laboratory of Tumor Immunology and Biology, Center for Cancer
Research, National Cancer Institute, National Institutes of Health, Bethesda, Maryland, USA

Kimberly Mayes
Department of Human and Molecular Genetics, VCU Institute of Molecular Medicine, Massey Cancer Center, Virginia Commonwealth University School of Medicine, Richmond, Virginia, USA

Mitchell E. Menezes
Department of Human and Molecular Genetics, Virginia Commonwealth University, School of Medicine, Richmond, Virginia, USA

Claudia Palena
Laboratory of Tumor Immunology and Biology, Center for Cancer Research, National Cancer Institute, National Institutes of Health, Bethesda, Maryland, USA

Anders I. Persson
Department of Neurology; Department of Neurological Surgery and Brain Tumor Research Center, and Sandler Neurosciences Center, University of California, San Francisco, California, USA

Smitha Pillai
Department of Tumor Biology, H. Lee Moffitt Cancer Center and Research Institute, Tampa, Florida, USA

Zhijun Qiu
Department of Human and Molecular Genetics, VCU Institute of Molecular Medicine, Massey Cancer Center, Virginia Commonwealth University School of Medicine, Richmond, Virginia, USA

Devanand Sarkar
Department of Human and Molecular Genetics; VCU Institute of Molecular Medicine, and VCU Massey Cancer Center, Virginia Commonwealth University, School of Medicine, Richmond, Virginia, USA

Courtney Schaal
Department of Tumor Biology, H. Lee Moffitt Cancer Center and Research Institute, Tampa, Florida, USA

Jeffrey Schlom
Laboratory of Tumor Immunology and Biology, Center for Cancer Research, National Cancer Institute, National Institutes of Health, Bethesda, Maryland, USA

Kwong-Yok Tsang
Laboratory of Tumor Immunology and Biology, Center for Cancer Research, National Cancer Institute, National Institutes of Health, Bethesda, Maryland, USA

R. Vago
Avram and Stella Goldstein-Goren Department of Biotechnology Engineering, Ben-Gurion University, Beer-Sheva, Israel

Y. Ventura
Avram and Stella Goldstein-Goren Department of Biotechnology Engineering, Ben-Gurion University, Beer-Sheva, Israel

Susan Wang
Department of Neurology; Department of Neurological Surgery and Brain Tumor Research Center, and Sandler Neurosciences Center, University of California, San Francisco, California, USA

Xiang-Yang Wang
Department of Human and Molecular Genetics; VCU Institute of Molecular Medicine, and VCU Massey Cancer Center, Virginia Commonwealth University, School of Medicine, Richmond, Virginia, USA

William A. Weiss
Department of Neurology; Helen Diller Family Comprehensive Cancer Center, and Department of Neurological Surgery and Brain Tumor Research Center, University of California, San Francisco, California, USA

Jolene J. Windle
Department of Human and Molecular Genetics; VCU Institute of Molecular Medicine, and VCU Massey Cancer Center, Virginia Commonwealth University, School of Medicine, Richmond, Virginia, USA

Robyn Wong
Department of Neurology, and Helen Diller Family Comprehensive Cancer Center, University of California, San Francisco, California, USA

CHAPTER ONE

Glial Progenitors as Targets for Transformation in Glioma

Shirin Ilkhanizadeh[*,†], **Jasmine Lau**[*,†,1], **Miller Huang**[*,†,1], **Daniel J. Foster**[*,‡,§,1], **Robyn Wong**[*,†,1], **Aaron Frantz**[*,‡,§], **Susan Wang**[*,‡,§], **William A. Weiss**[*,†,‡,¶], **Anders I. Persson**[*,‡,§,2]

[*]Department of Neurology, University of California, San Francisco, California, USA
[†]Helen Diller Family Comprehensive Cancer Center, University of California, San Francisco, California, USA
[‡]Department of Neurological Surgery and Brain Tumor Research Center, University of California, San Francisco, California, USA
[§]Sandler Neurosciences Center, University of California, San Francisco, California, USA
[¶]Department of Neurology, University of California, San Francisco, California, USA
[1]These authors have contributed equally.
[2]Corresponding author: e-mail address: anders.persson@ucsf.edu

Contents

Advances in Cancer Research, Volume 121
ISSN 0065-230X
http://dx.doi.org/10.1016/B978-0-12-800249-0.00001-9

Abstract

Glioma is the most common primary malignant brain tumor and arises throughout the central nervous system. Recent focus on stem-like glioma cells has implicated neural stem cells (NSCs), a minor precursor population restricted to germinal zones, as a potential source of gliomas. In this review, we focus on the relationship between oligodendrocyte progenitor cells (OPCs), the largest population of cycling glial progenitors in the postnatal brain, and gliomagenesis. OPCs can give rise to gliomas, with signaling pathways associated with NSCs also playing key roles during OPC lineage development. Gliomas can also undergo a switch from progenitor- to stem-like phenotype after therapy, consistent with an OPC-origin even for stem-like gliomas. Future in-depth studies of OPC biology may shed light on the etiology of OPC-derived gliomas and reveal new therapeutic avenues.

1. INTRODUCTION

Gliomas are the most common malignant primary brain tumor and associated with approximately 16,000 cancer-related deaths in United States per year (Louis et al., 2007). Recent advances in the molecular characterization of gliomas have defined subgroups of tumors that are genetically and epigenetically distinct (Noushmehr et al., 2010; Phillips et al., 2006; Sturm et al., 2012; Verhaak et al., 2010). The temporal and regional specificity of genetically distinct gliomas (Sturm et al., 2012), argue that either several discrete populations of precursor cells may be vulnerable to transformation, or that multiple glioma subgroups share a common cell of origin. Glial cells outnumber neurons by 10-fold in the human brain and are composed mainly of terminally differentiated cells and minor discrete precursor populations. Modeling of glioma in mice has demonstrated that cells at various differentiation stages throughout glial and neuronal lineages have the potential to generate gliomas. In this review, we present recent findings suggesting that the most wide-spread population of cycling cells in the pediatric and adult brain of mammalians, the oligodendrocyte progenitor cells (OPCs), represents a likely origin for large cohorts of gliomas. We propose that more in-depth studies of OPC biology will inform novel preventive measures and therapeutic interventions to reverse the fatal outcome of most glioma patients.

Gliomas can grossly be divided into astrocytic, oligodendrocytic, and ependymal phenotypes. Classification by the World Health Organization (WHO) distinguishes malignancy by grade (I–IV).

Based on histological appearance, gliomas of most grades and types are found in children and adults. Recent molecular profiling of grade IV glioblastoma (GBM) exemplifies that subsets of tumors in children, young adults, and adolescents, that are indistinguishable by histology, can be segregated based on genetic alterations, broad-scale gene expression, and methylation patterns. Here, we will present recent experimental advances on the understanding of why humans are diagnosed with a certain type of glioma and where it came from.

Gliomas show profound cellular heterogeneity and influences from the tumor microenvironment; with treatment-resistant tumor cells displaying a high degree of stemness. The failure to target glioma stem cells (GSCs) along with the inability to fully debulk tumors through surgical resection, radiation and chemotherapy, all contribute to poor survival of glioma patients (Huse & Holland, 2010). In this review, we will discuss ways to identify GSCs, their interactions with tumor microenvironment, and therapeutic advances to target GSCs. In 2012, Yanoko Nishiyama and John Gurdon were awarded the Novel Prize in Medicine for identifying factors that can reprogram somatic cells into pluripotent stem cells. Since these factors are also expressed in stem-like cancer cells, it is possible that they arose from more differentiated cells. In fact, viral transduction of oncogenes into mature neurons and astrocytes generate gliomas in mice (Friedmann-Morvinski et al., 2012). Similarly, it is plausible that OPCs also can give rise to more stem-like gliomas.

2. GLIAL CELL LINEAGES

The central nervous system (CNS) represents a mosaic organization of neural stem cells (NSCs) and astrocyte precursors, that generate neurons, astrocytes, and oligodendrocytes with a high degree of regional specificity (Merkle, Mirzadeh, & Alvarez-Buylla, 2007; Tsai et al., 2012). The positional identity is an organizing principle underlying cellular subtype diversification in the brain and is controlled by a homeodomain transcriptional code (Hochstim, Deneen, Lukaszewicz, Zhou, & Anderson, 2008). During embryonic development, expansion and cell fate determination of neural precursors is controlled by gradients of secreted molecules along rostrocaudal and dorsoventral axes. Radial glia and embryonic NSCs generate neurons, glial cells, and ependymal cells during neural development (Rakic, 1990). As a remnant from fetal development, postnatal neurogenesis in mammalians is mainly restricted to the dentate gyrus of the hippocampus and the subventricular zone (SVZ) lining the lateral ventricles (Doetsch, 2003; Eriksson et al., 1998; Sanai et al., 2011), with NSCs also lining the third and fourth ventricles (Weiss et al., 1996; Xu et al., 2005). In the postnatal rodent cerebellum, Bergmann glia express markers associated with NSCs (Koirala & Corfas, 2010; Sottile, Li, & Scotting, 2006). In contrast to rodents, functional SVZ neurogenesis in humans ceases after 18 months, indicating that few SVZ NSCs are present in the aging human brain (Sanai et al., 2011). Given the extensive self-renewal capacity of NSCs, these cells have been suggested as the cell of origin for gliomas (Fig. 1.1). Considering the low abundance of NSCs and the wide distribution of gliomas throughout the human postnatal brain, it is puzzling how such a rare and anatomically restricted cell type could represent the origin of the most common primary malignant brain tumor.

A first wave of oligodendrocyte progenitors arises from the embryonic ventral forebrain, followed by a second wave originating from the lateral and caudal ganglionic eminences, and finally a third wave arises within

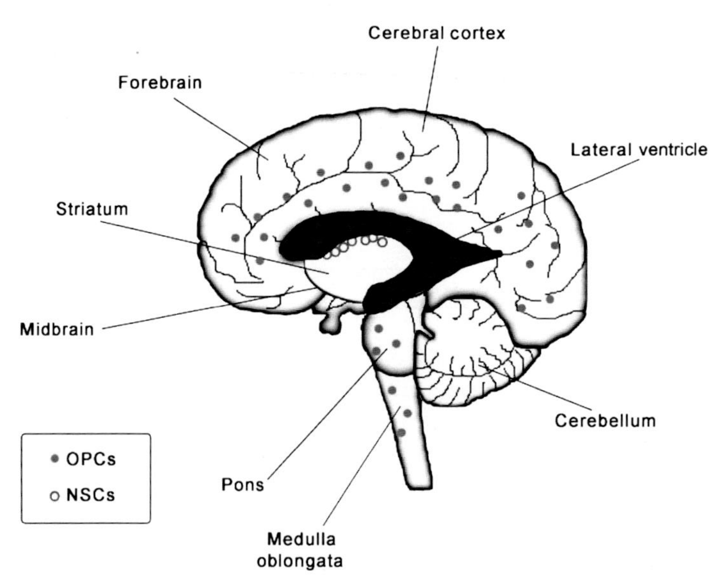

Figure 1.1 Distribution of neural precursor populations in the postnatal brain. OPCs are the most widely distributed population of cycling cells in forebrain and hindbrain regions. In contrast, a discrete population of NSCs is found in the SVZ lining the lateral ventricles. (See the color plate.)

the postnatal cortex (Kessaris et al., 2006). In the developing mouse brain and spinal cord, the first oligodendrocyte-lineage cells appear around embryonic day 12.5 (E12.5) (Zuo & Nishiyama, 2013). The cells are characterized by expression of the basic helix–loop–helix (bHLH) transcription factors *OLIG1*, *OLIG2*, *NKX2.2*, the Sry-related high mobility group box gene (*SOX10*), and platelet-derived growth factor receptor alpha (PDGFRA) (Zuo & Nishiyama, 2013). At E14.5, PDGFRA positive cells also express the chondroitin sulfate proteoglycan neuro-glial 2 (NG2) (in humans CSPG4) in the ventral mouse forebrain (Nishiyama, Lin, Giese, Heldin, & Stallcup, 1996). While *OLIG2* is required for generation of oligodendrocyte specification, the bHLH factor *ASCL1* promotes oligodendrogenesis by repressing *DLX1/2*, a transcriptional repressor of *OLIG2* (Ligon et al., 2006; Petryniak, Potter, Rowitch, & Rubenstein, 2007). Other prerequisites for oligodendrogenesis include the SOXE proteins SOX8, SOX9, and SOX10 (Stolt et al., 2003). In contrast, the SOXD proteins SOX5 and SOX6 inhibit oligodendrocyte specification (Stolt et al., 2006). In addition, the developmentally expressed genes *NOTCH-1*, wingless (*WNT*), and sonic hedgehog (*SHH*), normally associated with NSCs, block differentiation and maintain an undifferentiated state in OPCs (Zuo & Nishiyama, 2013) (Fig. 1.1B).

As the most widely distributed population of cycling cells in the postnatal brain, OPCs, also referred to as polydendrocytes, represent a fourth major type of glia in the CNS (Zuo & Nishiyama, 2013). In fact, approximately 70% of 5-bromo-2′-deoxyuridine (BrdU)-incorporating cells in the adult rat brain co-express NG2 (Dawson, Polito, Levine, & Reynolds, 2003; Lasiene, Matsui, Sawa, Wong, & Horner, 2009). An elegant study showed that OPCs are under homeostatic control to ensure generation of appropriate numbers of myelin-producing oligodendrocytes (Hughes, Kang, Fukaya, & Bergles, 2013). As OPCs are recruited to focal injuries, a proliferative burst of OPCs surrounding the injury restores the cell density. Do all OPCs have the same proliferative capacity or respond to different environment cues? PDGFA acts as a potent mitogen of OPCs expressing *PDGFRA* (Hall, Giese, & Richardson, 1996). Additionally, in white matter, but not gray matter, OPCs proliferate in response to PDGF by activating WNT and phosphatidylinositol 3-kinase (PI3K) (Hill, Patel, Medved, Reiss, & Nishiyama, 2013). Similar to NSCs, the mitogen epidermal growth factor (EGF) induces symmetrical cell division in adult OPCs (Sugiarto et al., 2011) (Fig. 1.2). Treatment of human OPCs with histone deacetylase (HDAC) inhibitors prevents differentiation into oligodendrocytes,

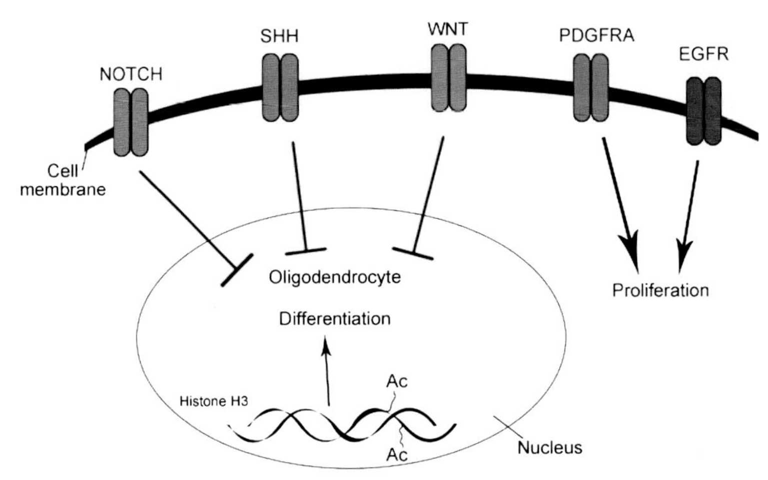

Figure 1.2 Activation of NOTCH, SHH, and WNT signaling inhibits OPC differentiation into oligodendrocytes. On the contrary, acetylation of histone H3 promotes differentiation of OPCs. Growth factor-mediated activation of PDGFRA or EGFR increases proliferation in OPCs. (See the color plate.)

demonstrating the importance of posttranslational modification of histones (Conway, O'Bara, Vedia, Pol, & Sim, 2012). In summary, several drivers of gliomagenesis (*PDGFRA*, *EGFR*, modulation of histone function) also play key roles in OPC development, arguing that proliferating OPCs, abundant throughout life, may transform in response to environmental pressure, and play a key role in gliomagenesis.

3. GLIOMA SUBGROUPS AND CELL OF ORIGIN

Neuropathologists classify gliomas based on grade (I–IV) that includes grade II diffuse gliomas and grade III–IV malignant gliomas. Ependymomas, astrocytomas, oligodendrogliomas, and oligoastrocytomas display fewer mitoses, necrosis, nuclear atypia, and vascular proliferation compared to grade IV GBM (Louis et al., 2007). Microarray expression profiling was able to separate low- versus high-grade gliomas and primary versus recurrent tumors (Godard et al., 2003; Rickman et al., 2001; Van den Boom et al., 2003). New technological advances and the Cancer Genome Atlas network (TCGA) have defined subsets of GBMs based on epigenomic, genomic, and transcriptomal signatures (Noushmehr et al., 2010; Phillips et al., 2006; Verhaak et al., 2010). When stratified based on survival, primary GBM tumors displaying a proneural gene expression signature correlated with improved survival; whereas classical and mesenchymal subsets of tumors correlated with a worse survival (Phillips et al., 2006). Approximately 12% of GBM patients displayed gain of function mutations in isocitrate dehydrogenase (*IDH*) genes and showed increased overall survival (Parsons et al., 2008). Verhaak et al. (2010) later identified a fourth neural GBM subgroup and showed that *IDH* mutations were exclusively found in a subset of proneural GBMs (Verhaak et al., 2010).

A recent study, segregating tumors by methylation rather than gene expression, demonstrated six subgroups of GBMs (Sturm et al., 2012). Childhood GBMs showed a strong correlation with gain of function mutations in histone H3.3 (*H3F3A*) mutations at two critical residues (K27(M)) or G34(R/V), adding two additional GBM subgroups with unique epigenetic and gene expression signatures (Sturm et al., 2012). Interestingly, these *H3F3A* mutations give rise to GBMs in separate anatomic compartments, and in contrast to $IDH1^{R132H}$ mutant tumors, are diagnosed in adolescents (G34) or young children (K27). The authors refer to these six GBM subgroups as; IDH1, K27, G34, RTK I "PDGFRA," mesenchymal, and RTK II "classic," since they are associated with unique genetic alterations (Sturm et al., 2012) (Fig. 1.2). Gene expression profiling of diffuse gliomas shows that approximately 75% and 50% of diffuse oligodendrogliomas and astrocytomas, respectively, belong to the proneural subgroup (Cooper et al., 2010). Interestingly, the fraction of proneural tumors correlates well with the frequency of *IDH* mutations in diffuse gliomas (Yan et al., 2009). Activity among signal transduction pathways at the protein level can also define GBM subclasses (Brennan et al., 2009). Progress in subclassification of gliomas is used to develop biomarkers and gene expression signatures that can be used to stratify patients in clinical trials (Olar & Aldape, 2012). In parallel, magnetic resonance imaging (MRI) predictors that associate with GBM subgroups are in development (Gutman et al., 2013). MRI studies show that proneural GBMs had significantly lower levels of contrast enhancement and $IDH1^{R132H}$ mutant GBMs show accumulation of 2-hydroxyglutarate (2-HG) (Chaumeil et al., 2013; Gutman et al., 2013). However, decision making is still governed by genetic alterations as gliomas show regional heterogeneity for MRI parameters and association to GBM subgroups.

To develop subgroup-specific therapies in glioma, improved preclinical models are needed. Traditionally, researchers have studied therapeutics using cultures or xenografts of human GBM cell lines. Passaging of primary human GBM tumors in immunocompromised mice is an improved model system that better preserves tumor cell heterogeneity (Hodgson et al., 2009).

However, the strong influence of the tumor microenvironment and the importance of blood–brain barrier permeability for drugs have led to generation of several genetically engineered murine models (GEMM) of glioma (Table 1.1). These models are also useful for studies of premalignant events and the cell of origin for different types of gliomas. Although these models have been highly informative, a weakness of previously developed GEMM of glioma is the failure to recapitulate the genetic alterations observed in human tumors. For example, whole-genome sequencing studies have identified *B-RAF*^{V600E}, *FGFR1*, *MYB1*, *MYBL1*, *H3F3A*, and *ATRX* mutations in pediatric gliomas (Zhang et al., 2013). However, recent GEMMs have employed *BRAF*^{V600E} mutation and neurofibromin 1 (*NF1*) loss to generate pilocytic gliomas high-grade GBMs, respectively (Chen et al., 2012; Robinson et al., 2011). Modeling of *H3F3A* and *IDH1*^{R132H} mutations in murine GEMM will be useful tools to develop new therapeutics against larger cohorts of childhood and adult glioma patients.

4. *H3F3A* MUTATIONS DRIVE GLIOMAGENESIS IN SEPARATE BRAIN REGIONS

For NSCs and progenitor cells to achieve production of different types of neurons and glia at appropriate times and places during development, they must integrate cell-intrinsic programs and environmental cues. These developmental dynamics are reflected in changes in gene expression, which is regulated by transcription factors and at the epigenetic level. Methylation and acetylation of histones function as epigenetic modulators of differentiation in NSCs and progenitors (Hu, Wang, & Shen, 2012). However, mutations of epigenetic sites on histones can block differentiation and transformation of cells into cancers. In this section of the review, we will discuss recent findings demonstrating mutations at two distinct residues (K27(M) or G34(R/V)) of *H3F3A* in childhood GBMs. Interestingly, K27 and G34 mutant pediatric GBMs are associated with distinct anatomical locations, have subgroup-specific gene expression signatures, and occur at different ages, suggesting that different precursor cells may represent the cell of origin for K27 and G34 mutant GBMs.

Whole-genome and whole-exome sequencing studies independently identified recurrent mutations in the gene *H3F3A* in pediatric glioma (Schwartzentruber et al., 2012; Wu et al., 2012). Interestingly, Sturm and colleagues found *IDH* mutations that occurred in a distinct group of patient samples from *H3F3A* mutants. The K27 and G34 mutations were mutually

Table 1.1 Murine glioma models

Cancer genes	Cell of Origin	Mechanism	Reference
Astrocytoma models (II–III)			
c-myc	GFAP	Transgenic	Jensen et al. (2003)
K-ras	GFAP	Cre	Abel et al. (2009)
H-Ras	GFAP	Transgenic	Shannon et al. (2005)
H-Ras/Pten$^{loxp/loxp}$	GFAP	Cre	Wei et al. (2006)
EGFR/Ink4a/Arf$^{-/-}$	Nestin	RCAS	Holland, Hively, DePinho, and Varmus (1998)
Oligodendroglioma models			
H-ras/EGFRvlll	GFAP	Transgenic	Ding et al. (2003)
v-ErbB/Ink4a/Arf$^{+/-}$	S100β	Transgenic	Weiss et al. (2003)
v-ErbB/Trp53$^{-/-}$	S100β	Transgenic	Persson et al. (2010)
PDGFB/Ink4a/Arf$^{-/-}$	Nestin	RCAS	Dai et al. (2001)
PDGFB/Akt	Nestin	RCAS	Dai et al. (2005)
PDGFB	CNP	RCAS	Lindberg, Kastemar, Olofsson, Smits, and Uhrbom (2009)
PDGFA$_L$	GFAP	Transgenic	Nazarenko et al. (2011)
Glioblastoma (GBM) models			
PDGFB	Nestin	RCAS	Shih et al. (2004)
PDGFB/Trp53$^{-/-}$	GFAP	Transgenic	Hede et al. (2009)
K-Ras/Akt	Nestin	RCAS	Holland et al. (2000)
K-Ras/Ink4a/Arf$^{-/-}$	GFAP, Nestin	RCAS	Uhrbom et al. (2002)
K-Ras/Akt/Pten$^{loxp/loxp}$	Nestin	RCAS/Cre	Hu et al. (2005)
Nf1$^{loxp/+}$/Trp53$^{+/-}$	GFAP	Cre	Zhu et al. (2005)
Pten$^{loxp/+}$/Trp53$^{loxp/loxp}$	GFAP	Cre	Zheng et al. (2008)
Pten$^{loxp/+}$/Nf1$^{loxp/+}$/Trp53$^{loxp/-}$	GFAP	Cre	Kwon et al. (2008)

Continued

Table 1.1 Murine glioma models—cont'd

Cancer genes	Cell of Origin	Mechanism	Reference
$Pten^{loxp/+}/Nf1^{loxp/+}/Trp53^{loxp/loxp}$	SVZ, Nestin	CreER	Alcantara Llaguno et al. (2009)
$Akt/H\text{-}Ras/Trp53^{+/-}$	SVZ, HC, GFAP	Lentiviral/Cre	Marumoto et al. (2009)
$Pten^{loxp/loxp}/Trp53^{loxp/loxp}/Rb^{loxp/loxp}$	SVZ, GFAP	Adenoviral/Cre	Jacques et al. (2010)
$Nf1^{+/-}/Trp53^{loxp/loxp}$	GFAP	Transgenic/Cre	Wang et al. (2009)
$PDGFB/Pten^{loxp/loxp}/Trp53^{loxp/loxp}$	Subcortical WM	Retroviral/Cre	Lei et al. (2011)
$Pten^{loxp/loxp}/Trp53^{loxp/loxp}/Rb1^{loxp/loxp}$	Astrocyte (GFAP)	CreER	Chow et al. (2011)
$Nf1^{-/-}/Trp53^{-/-}$	OPC (NG2), NSCs (GFAP/Nestin)	MADM/Cre	Liu, Sage, et al. (2011)
$EGFRvIII/Ink4a/Arf^{-/-}$	NSC, Astrocyte	Transplant	Bachoo et al. (2002)

As a useful tool for preclinical studies and the delineation of cell of origins for different types of glioma, GEMMs of glioma generate tumors resembling grade II–IV human counterparts.
NSC, neural stem cell; OPC, oligodendrocyte progenitor cell; SVZ, subventricular zone; WM, white matter; HC, hippocampus.

exclusive, and clustered into their own subgroup when examining gene expression and methylation profiles (Sturm et al., 2012). Anatomically, the K27-mutant tumors arise in pontine and more rostral midline brain structures, whereas the G34-mutant tumors are usually hemispheric and found in the forebrain (Fig. 1.3). K27 mutations occur in children while G34 mutations occur in adolescents; and *IDH* mutations were found in young adults (Khuong-Quang et al., 2012; Sturm et al., 2012). These differences may reflect outcome as patients with K27 mutant tumors typically have shorter overall survival compared to G34 mutant and *H3F3A* wild-type (WT) tumors; while patients with *IDH* mutant tumors have better prognosis compared to patients either *H3F3A* mutation (Sturm et al., 2012). Seventy eight percent of patients with the lethal pediatric tumor, diffuse intrinsic pontine glioma (DIPG), had the K27 mutation, while no DIPG samples had a mutation at G34 (Wu et al., 2012). A complementary study showed

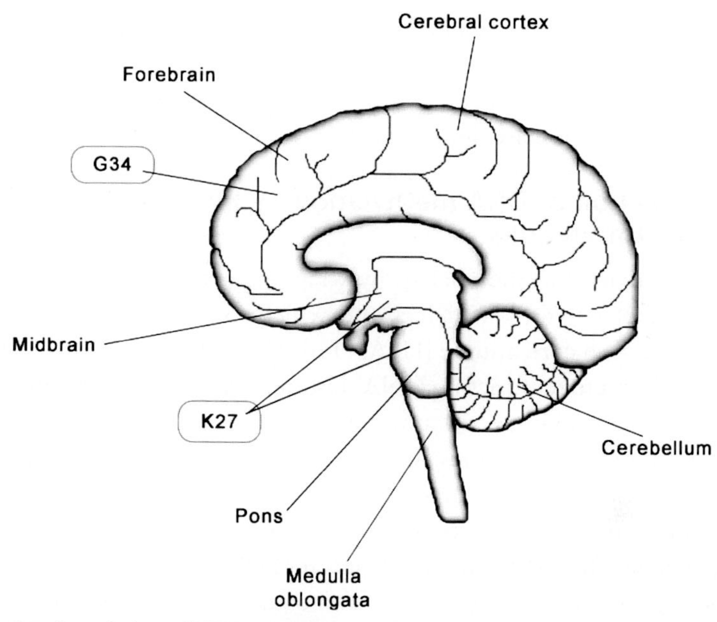

Figure 1.3 Association of *H3F3A* mutations in GBMs to distinct brain regions. GBMs displaying G34 and K27 mutations in the *H3F3A* gene localize to forebrain and hindbrain regions, respectively. (See the color plate.)

K27 mutations in 71% of DIPG samples, while none had G34 or *IDH* mutations (Khuong-Quang et al., 2012). Both K27 and G34 mutations are frequently present with other genetic mutations and associate with the expression of developmentally regulated genes. For example, G34 mutations correlated with hypermethylation and silencing of oligodendrocyte lineage genes such as *OLIG1* and *OLIG2*, while expressing the forebrain marker, *FOXG1*. In contrast, K27 mutant tumors express *OLIG1* and *OLIG2*, but not *FOXG1* (Sturm et al., 2012).

One of the initial studies that identified these recurrent *H3F3A* mutations also found mutations in the chromatin remodeling gene *ATRX* and the tumor suppressor gene *TP53* (Schwartzentruber et al., 2012). *ATRX* mutations were associated with older patients and found in all tumors with G34 mutations (Schwartzentruber et al., 2012; Sturm et al., 2012). *TP53* mutations were also consistently found in tumors with G34 mutations, which may represent an alternate mechanism of p53 inactivation, as *OLIG2*, commonly lost in G34 tumors, has been shown to block p53 function (Mehta et al., 2011). Conversely, this may also explain why K27 mutant tumors (which are OLIG2+) do not correlate as strongly with *TP53* mutations (although ~70% of K27M mutant tumors coexpress mutant *TP53*).

The frequent association of *TP53* and *H3F3A* mutations in pediatric GBM is reminiscent of the strong link between *TP53* and *IDH1*^{R132H} mutations in adult GBM. To summarize, these studies suggest that *H3F3A* mutant GBMs may arise from regionally distinct OLIG2-expressing glial progenitors.

4.1. Regulation of DNA methylation by K27 and G34 *H3F3A* mutations

DNA is condensed in the nucleus by wrapping around histones. Multiple histones assemble to form a nucleosome, which consists of nine histone subunits (two of each core histone [H2A, H2B, H3, H4] and the linker histone H1). In addition to compacting DNA, histones contribute to the epigenetic regulation of the genome via substitution of various histone isoforms and posttranslational modifications, specifically acetylation and methylation of lysine and arginine residues. Although methylation is generally associated with repression of gene expression, methylation can indicate active transcription depending on the residue that is methylated, as well as the number of methyl groups added (mono-, di- and trimethylation). For histone H3, trimethylation of lysine 4 and 36 (K4, K36) is associated with active transcription, while trimethylation of K9 and K27 marks gene silencing (Chi, Allis, & Wang, 2010). While di- and trimethylated K27 mark inhibition of gene expression, monomethylated or acetylated K27 mark highly expressed genes (Creyghton et al., 2010; Steiner, Schulz, Maksimova, Wong, & Gallagher, 2011).

Two DIPG K27 mutant cell lines demonstrated a reduction in trimethylation (me3) and dimethylation (me2) status at K27, while not impacting acetylation of K27 (H3K27ac) compared to normal NSCs (Chan et al., 2013). Similarly, another study found suppressed H3K27me3 levels in 6 GBMs containing K27 mutations (Venneti et al., 2013). Ectopic expression of *H3F3A* K27 in 293T cells, human astrocytes, and mouse embryonic fibroblasts, decreased di- and trimethylation of K27 on endogenous H3F3A proteins and the exogenously expressed mutant, without affecting the methylation status of K36, a posttranslationally modified site near G34 (Chan et al., 2013; Lewis et al., 2013). The mechanism for reduced methylation at K27 may be due to Histone-lysine N-methyltransferase (EZH2), the catalytic subunit of polycomb repressive complex 2 (PRC2), which is a methyltransferase that targets *H3F3A* K27. Since the expression levels of EZH2 is not altered between patient samples expressing either WT or K27 mutant *H3F3A*, the mutation may act in a dominant negative fashion to block *EZH2* function (Venneti et al., 2013). This was supported by the finding that a K27 peptide demonstrated a higher affinity for *EZH2* compared to WT *H3F3A* and inhibited PRC2 activity via interaction with *EZH2* (Chan et al., 2013; Lewis et al., 2013). In K27 mutant DIPG cells, the sites which retained H3K27me3 had increased levels of trimethylation, as well as *EZH2*, and were genes associated with cancer pathways, such as the tumor suppressor *CDKN2A* (Chan et al., 2013). Although overexpression of mutant K27 lowered the methylation status of both the endogenous and exogenous K27, misexpression of an *H3F3A* G34 mutant only decreased trimethylation of K36 on the exogenous peptide, while not affecting the endogenous *H3F3A* K36 (Chan et al., 2013; Lewis et al., 2013). This might suggest that the G34 site needs to be mutated on both alleles to effect a change in methylation (or that methylation is an epiphenomonon in G34 mutant tumors), while the K27 mutation on only one allele subserves this function. The requirement for an additional hit on G34 could explain why these tumors appear later in life than K27 mutant tumors. Sturm et al. found global hypomethlation of genomic DNA in G34 mutant tumors, which contrast the global hypermethylation found in $IDH1^{R132H}$ mutant tumors (Sturm et al., 2012). Hypomethylation in G34 mutant tumors is especially prominent at chromosome ends, which may provide a link with alternative lengthening of telomeres (ALT), a phenomenon commonly observed with ATRX mutations (Schwartzentruber et al., 2012; Sturm et al., 2012).

4.2. Chromosome and *Myc* aberrations in *H3F3A* mutant glioblastoma

Chromosomal aberrations have been investigated in K27 mutant DIPG tumors (Khuong-Quang et al., 2012). As expected due to the role histones play in chromatin remodeling, K27 mutant tumors exhibit disinct chromosomal aberrations, as compared to tumors WT at *H3F3A*. *H3F3A* WT samples exhibited gains of chromosomes 2p and 7p as well as losses in 9p and 12q. K27 tumors harbored losses of 5q, 6q, 17p and 21q and gains in 19p. Focal recurrent gains were also observed in 4q12 (containing *PDGFRA*), and 8q24.21 (*MYC/PVT1* locus). Intriguingly, while *MYC* is amplified in K27 tumors, chromosome 2p24.3 (which includes *MYCN*) is gained in *H3F3A* WT tumors. Although amplification of *MYCN* has not been observed in G34 mutant samples, overexpression of the G34 mutation induces expression of MYCN in normal human astrocytes and fetal glial cells (Bjerke et al., 2013; Khuong-Quang et al., 2012). Thus amplification and overexpression of *MYC* and *MYCN* appear to segregate based on the *H3F3A* mutations. MYC promote apoptosis via activation of *TP53*

(Elson, Deng, Campos-Torres, Donehower, & Leder, 1995) and a number of cancer models utilize overexpression of *MYC* in combination with loss of *TP53* function (Kawauchi et al., 2012; Pei et al., 2012; Schmitt et al., 2002; Yu & Thomas-Tikhonenko, 2002). Since MYC is also known to promote proliferation and block differentiation, the combination of amplified *MYC*, mutant *TP53* and K27 may provide insight to the genetic etiology of DIPG.

4.3. Delineating the cell of origin for K27 and G34 *H3F3A* mutant glioblastoma

Neural stem cells, astrocyte precursors, and OPCs are all abundant at birth in humans and represent possible targets for transformation by K27 mutation. In mice, viral transduction of Nestin-expressing precursors with mutant *TP53* and K27 failed to induce gliomagenesis in the newborn murine brainstem, but did induce proliferating ectopic cell clusters (Lewis et al., 2013). In humans, Nestin and OLIG2 are coexpressed in the ventral pons around 6 years of age, the peak for DIPG occurrence (Monje et al., 2011). In contrast to K27 mutant GBMs, G34 mutant tumors express the forebrain marker *FOXG1* and other forebrain development-related transcription factors such as *ARX*, *DLX5*, *FOXA1*, *NR2E1*, *POU3F2*, and *SP8* (Bjerke et al., 2013). Although NSC are not abundant in the forebrain in adolescents when most G34 mutant GBMs are diagnosed, these tumors express the NSC-associated genes Musashi-1 (*MSI1*), eyes absent homolog 4 (*EYA4*), and *SOX2* under control of the active transcription marker, H3K36me3 (Bjerke et al., 2013). High expression of *MYCN* in G34 mutant GBMs may enable more restricted forebrain precursors to dedifferentiate into a NSC phenotype. Using a mosaic mouse model, and combining loss of *TP53* and *NF1* in Nestin positive NSCs, transformation only occurred after NSCs had differentiated into OPCs (Liu, Sage, et al., 2011). A similar approach could be used to delineate the cell of origin for K27 and G34 mutant GBMs.

5. GLIOMAGENESIS AND MUTATIONS IN ISOCITRATE DEHYDROGENASE GENES

In humans, five genes encode three isoforms of the metabolic enzyme *IDH* involved in the citric acid cycle. IDH1 is found both in the cytoplasm and peroxisomes, while IDH2 and IDH3 are localized to mitochondria. IDH1 and IDH2 are closely related, and catalyze the oxidative decarboxylation of isocitrate to α-ketoglutarate (α-KG) while reducing nicotinamide

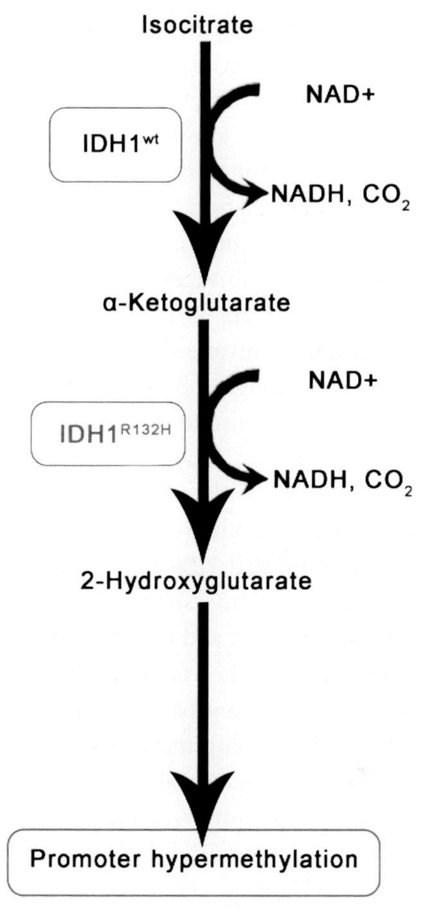

Figure 1.4 IDH1^{R132H} mutant gliomas show reduced levels of α-ketoglutarate and overproduction of 2-hydroxyglutarate (2-HG) that leads to a hypermethylated phenotype.

adenine dinucleotide phosphate (NADP^{+}) to NADPH (Fig. 1.4). The unrelated multisubunit enzyme, IDH3, reduces NAD^{+} to NADH and plays a central role in aerobic energy production in the tricarboxylic acid (TCA) cycle. Arginine 132 (R132), found within the active site of human *IDH1* is critical to the binding of isocitrate, and is evolutionary conserved in the functionally analogous R172 of *IDH2*. The retained WT IDH allele converts isocitrate to α-KG, while the IDH1^{R132H} drives synthesis 2-HG from α-KG. In normal cells, *IDH* genes mediate epigenetic changes that maintain cellular homeostasis. In IDH-mutant cancer cells, 2-HG acts a competitive inhibitor of α-KG-dependent histone demethylases, leading to hypermethylation (Figueroa et al., 2010; Xu et al., 2011). Interestingly,

IDH mutations are generally mutually exclusive and occur uniquely on these arginine residues, which are conserved across species and malignancies.

The first observations of *IDH1/2* mutations were identified in metastatic colon cancer. These mutations are also found in sarcomas, leukemias, and gliomas. Pioneering studies in leukemia demonstrated that *IDH* mutation leads to a block in differentiation in hemapoietic precursor cells (Chaturvedi et al., 2013). The ability of *IDH* mutation to disrupt normal differentiation into defined cell lineages has also been demonstrated in murine forebrain NSCs (Lu et al., 2013). This block in differentiation was attributed to prevention in histone demethylations required for NSCs to terminally differentiate. Although the differentiation program was compromised in NSCs, the authors found no indication of transformation, suggesting that *IDH* mutations may transform glial progenitor or dedifferentiated terminally differentiated glial cells. Parsons et al. (2008) sequenced over 20,000 protein coding genes in order to identify genetic alterations in GBMs and found that 12% of the samples analyzed harbored a recurrent point mutation in the active site of *IDH1/2* (Parsons et al., 2008). Specifically, 5% of primary gliomas and 60–90% of secondary gliomas had *IDH1/2* mutations. Furthermore, they observed that $IDH1^{R132H}$ mutations occurred in a large fraction of young patients and that most patients with secondary GBMs were associated with increased overall survival. Mutations in *IDH2* are much less common and mutually exclusive with $IDH1^{R132H}$ as mentioned before. Virtually all tumors with *IDH* mutations are of the proneural subtype. Other studies found *IDH* mutations in 80% grade II and grade III gliomas as well as in secondary GBMs. Importantly, Yan et al. (2009) reported that $IDH1^{R132H}$ is a favorable prognostic marker for glioma patients (Yan et al., 2009).

Hypoxia-inducible factor 1α (HIF-1α) is a transcription factor that facilitates tumor growth under conditions of low oxygen and its stability is regulated by α-ketoglutarate. Human gliomas with $IDH1^{R132H}$ mutation express higher levels of HIF-1α compared to *IDH* WT tumors. Additionally, expression of $\mathrm{IDH1}^{\mathrm{R132H}}$ in cultured cells reduces α-KG and increases HIF-1α levels (Zhao et al., 2009). These results indicate that $IDH1^{R132H}$ mutations may contribute to tumorigenesis in part through the HIF-1 pathway.

As previously mentioned, mutations occur at a single amino acid residue of the $IDH1^{R132H}$ active site. Interestingly, in tumors, only a single copy of the gene is mutated and the expression of a WT *IDH1* allele is critical for the formation of a heterodimer with the mutated allele. Cancer-associated $IDH1^{R132H}$ mutations have been found to result in the production of

2-HG as a neometabolite. Since levels of 2-HG are elevated in human gliomas harboring *IDH1*^{R132H} mutations, production of the oncometabolite 2-HG in *IDH1*^{R132H} mutated tumors may contribute to glioma initiation and malignant progression.

The accumulation of 2-HG has been associated with increased histone methylation and decreased 5-hydroxylmehtylcytosine (5hmC), resulting in genome-wide histone and DNA methylation alterations (Fig. 1.4). Furthermore, it has been demonstrated that a large number of loci in *IDH1*^{R132H} mutant gliomas display hypermethylation (Noushmehr et al., 2010). Together, these findings suggest that *IDH* mutations may alter the expression of a large number of genes rather than a few specific genes. Reportedly, mutations of *IDH1*^{R132H} occur at an early stage during gliomagenesis (Watanabe, Nobusawa, Kleihues, & Ohgaki, 2009) and may contribute to tumor initiation by globally changing the epigenetic landscape and control cell fate. In fact, glioma samples with *IDH* mutations and increased histone methylation display a gene expression profile enriched for neural progenitor genes. In summary, 2-HG producing *IDH* mutations inhibit histone demethylation and can thereby block cell differentiation.

It is established that acquired alterations in the methylation landscape cause dysregulation and can thereby be involved in oncogenesis (Jones & Baylin, 2007). A distinct subclass of human glioblastomas has the CpG island methylator phenotype (CIMP) and is associated with the proneural subgroups of tumors and *IDH* mutation (Noushmehr et al., 2010). The association of *IDH* mutation and the CIMP phenotype in GBMs raises the question of cause and effect. Turcan et al. (2013) demonstrated *IDH1*^{R132H} mutation as the usual cause of CIMP, establishing the CIMP phenotype by remodeling the epigenome (Turcan et al., 2013). In rare cases however, CIMP also occurs in *IDH* WT tumors (Brennan et al., 2013). Importantly, primary human astrocytes introduced with *IDH1*^{R132H} mutation display epigenomic alterations mirroring low-grade gliomas positive for the CIMP phenotype. These observations further advance our mechanistic understanding of the roles for *IDH* mutation and CIMP phenotype in gliomagenesis, providing targets for development of novel therapies.

5.1. Models of *IDH*-mutant gliomas

Studies of *IDH*-mutant gliomas has been hampered by the lack of glioma-producing GEMMs displaying *IDH1* mutations. Given the effects of *IDH* mutations on tumorigenesis, Sasaki et al. (2012) generated brain-specific

$IDH1^{R132H}$ knock-in (IDH1-KI) mice that were embryonically lethal (Sasaki et al., 2012). Fetuses displayed brain hemorrhage and elevated 2-HG levels but showed decreased reactive oxygen species (ROS). The increased levels of 2-HG stabilized HIF-1α protein and upregulated HIF-induced gene transcription. Moreover, an endoplasmic reticulum stress response was triggered which caused intrinsic cell death. These effects may increase vascular endothelial growth factor (VEGF) levels, leading to aberrant blood vessel formation and ultimately resulting in brain hemorrhage.

Cell lines with $IDH1^{R132H}$ mutation can only be maintained transiently *in vitro*, as the mutation does not persist in cultured cells. Furthermore, primary *IDH*-mutant gliomas from patient tumors are difficult to expand *in vitro* (Piaskowski et al., 2011). In contrast to normal cells, introduction of *IDH* mutations into glioma cells decreases the proliferation rate, which may ultimately cause a selection pressure against cultured glioma cells harboring *IDH* mutations (Bralten et al., 2011). Despite these challenges, at least one group has reported the isolation of a glioma cell line with an endogenous R132H mutation in $IDH1^{R132H}$ (Luchman et al., 2012). The researchers established neurosphere cultures from an $IDH1^{R132H}$ mutant anaplastic oligoastrocytoma sample and confirmed retention of the mutation over passages. These cells were also used in orthotopic xenografts of nonobese diabetic/severe combined immune deficiency (NOD/SCID) mice. Mass spectroscopy experiments were performed to confirm production of 2-HG by glioma cells with endogenous $IDH1^{R132H}$ mutations both *in vivo* and *in vitro*. In summary, there is a great need for glioma models with *IDH* mutations to analyze the roles of IDH mutations during gliomagenesis, since efforts to model IDH mutant gliomas and the ability to grow primary *IDH* mutant tumors have been challenging.

5.2. Glial progenitor-origin for *IDH*-mutant gliomas

Development of *IDH*-mutant sarcoma or leukemia-models has allowed researchers to search for drugable targets up- or down-stream of IDH in a cell-type specific context (Sasaki et al., 2013). To date, most studies of *IDH*-mutations in glioma have relied on overexpression in different glioma cell lines (Rohle et al., 2013). However, genes important for terminal differentiation present in OPCs are lacking in these artificial cell lines, and without those genes expressed, methylation-dependent regulation of gene expression cannot effectively be studied. Since *IDH* mutations are common in oligodendroglioma, development of *IDH*-mutant oligodendroglioma models and the ability to xenograft/culture *IDH*-mutant human oligodendrogliomas will generate tools critical to understand the earliest transformation events and to perform preclinical studies. As *IDH* mutations are found in 12% of all gliomas, efforts are focused on defining the molecular events following *IDH* mutations to find drugable targets. To improve the survival of patients diagnosed with glioma, defining the cell of origin and the first molecular events that initiate transformation will be fundamental for development of new therapies.

The gene expression signature and location of $IDH1^{R132H}$ gliomas (white matter regions of the cerebral hemispheres) suggest that OPCs in these regions may be susceptible to transformation. Is $IDH1^{R132H}$ mutation sufficient to transform OPCs? Oligodendroglioma show distinct histological features including fried-egg morphology (after fixation) that distinguish them from other gliomas. Co-deletion of chromosomal arms 1p and 19q is a well-known prognostic marker found frequently (>60%) in grade II-III oligodendrogliomas. Recent exomic sequencing has identified mutations in tumor suppressor genes: homolog of Drosophila *capicua* (*CIC*) and *far-upstream binding protein 1* (*FUBP1*) in oligodendrogliomas (Bettegowda et al., 2011; Yip et al., 2011). Since *CIC* and *FUBP1* are located on chromosomal arm 19q and 1p, respectively, loss of these regions inactivates the genes and could potentially contribute to oligodendroglioma development. A strong association between $IDH1^{R132H}$ mutation and 1p/19q co-deletion is found in grade II-III oligodendrogliomas, suggesting that these genetic alterations are important for gliomagenesis. While a subfraction of oligodendrogliomas display $IDH1^{R132H}$ and *TP53* mutations, it is more commonly observed in astrocytic tumors.

The reason for the relatively favorable prognosis of $IDH1^{R132H}$ mutant patients is still unknown. Several studies have suggested that $IDH1^{R132H}$ mutation and 2-HG block cell differentiation (Chaturvedi et al., 2013; Figueroa et al., 2010; Lu et al., 2013). The reduced aggressiveness in IDH-mutant tumors may be a result of inhibited differentiation, rather than regulation of cell cycle genes. Alternatively, the better prognosis for *IDH*-mutant patients may reflect an OPC origin. Mutant $IDH1^{R132H}$ accelerates onset of myeloproliferative disease (MPD)-like myeloid leukemia in mice in cooperation with Homeobox protein Hox-A9 (*HOX9A*) (Chaturvedi et al., 2013). Mutant $IDH1^{R132H}$ accelerated cell cycle transition through repression of cyclin-dependent-kinase inhibitors CDKN2A and CDKN2B.

Interestingly, a complex transcriptional regulatory network of *HOX* genes specify neural progenitor cell identity during early development (Gavalas, Trainor, Ariza-McNaughton, & Krumlauf, 2001; Lumsden & Krumlauf, 1996; Trainor & Krumlauf, 2001) Further, sustained activation of MAPK activation in oligodendrocytes stimulates oligodendrocyte progenitor expansion (Ishii, Furusho, & Bansal, 2013). Together, these data may support OPC as candidate cells of origin of $IDH1^{R132H}$ mutant gliomas.

In contrast to classical and mesenchymal GBMs that express gene expression profiles reminiscent of NSCs, *IDH*-mutant gliomas display a proneural phenotype (Verhaak et al., 2010). Interestingly, the transcriptomal signature of proneural GBMs was closely associated with oligodendrocytes rather than astrocytes, neurons, or NSCs (Verhaak et al., 2010). The low activity of NADPH-producing dehydrogenases in rodent versus human brain may suggest that rodents are unsuitable to model gliomas (Atai et al., 2011). However, no studies have demonstrated expression of *IDH* isoforms at high resolution and in defined cell populations, why the cell-specific role of this gene family is still unclear. In the Verhaak study, many genes normally found in OPCs, such as *NKX2.2*, *PDGFRA*, and *OLIG2*, were overexpressed in *IDH*-mutant proneural GBMs (Verhaak et al., 2010). Support from GEMMs show that proneural gliomas can derive from OPCs (Lindberg et al., 2009; Persson et al., 2010). The regional association of *IDH1*-mutant GBMs to frontal cortex and white matter regions argue that OPCs, rather than NSCs, represent a likely origin for *IDH*-mutant gliomas.

6. PRONEURAL-TO-MESENCHYMAL TRANSITION IN GLIOMA

Epithelial-to-mesenchymal transition (EMT) occurs at critical stages during development, such as gastrulation and neural crest formation (Kalluri & Weinberg, 2009; Thiery, 2002). As epithelial cancer cells undergo EMT, they become invasive, metastasize, and become drug-resistant (Craene & Berx, 2013; Singh & Settleman, 2010). In many cancers, including breast cancer, the mesenchymal phenotype is associated with increased activation of transforming growth factor beta (TGFβ) signaling, leading to increased expression of the transcription factor families of *SNAIL*, *SLUG* and *TWIST*, and downregulation of E-cadherin (Craene & Berx, 2013; Mallini, Lennard, Kirby, & Meeson, 2014).

Transcriptomal profiling of gliomas, displaying a neuroepithelial origin, show that the mesenchymal phenotype is associated with stemness, invasiveness, and poor survival (Phillips et al., 2006; Sturm et al., 2012; Verhaak et al., 2010) (Fig. 1.5). Although in a small cohort of patients, recurrent tumors shifted from a proneural to mesenchymal phenotype, reminiscent of EMT (Lai et al., 2011; Phillips et al., 2006). In this section, we will review new exciting data suggesting that expression of a single gene or changes in the tumor microenvironment can shift gliomas between proneural and mesenchymal phenotypes, which we will refer to as a proneural-to-mesenchymal transition (PMT). Since recurrent gliomas tend to have a mesenchymal phenotype (Lai et al., 2011), these findings have implications for therapy and suggest that even more stem-like gliomas can arise from OPCs.

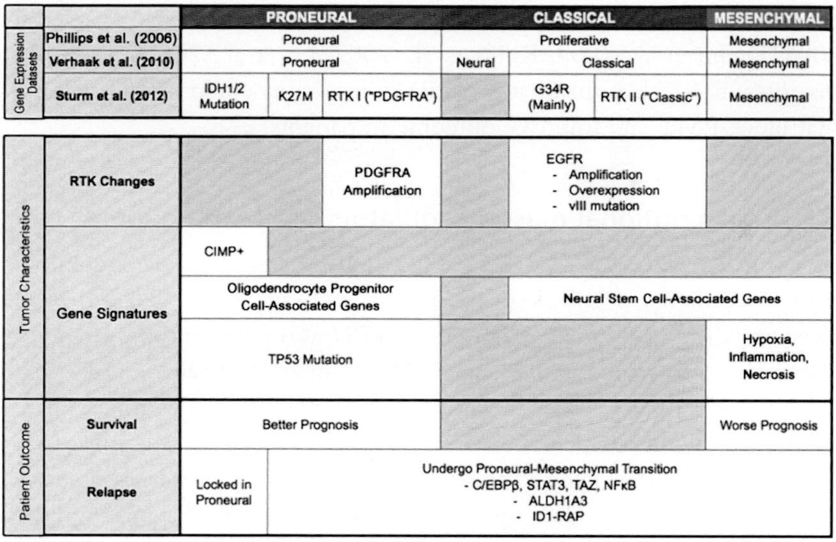

		PRONEURAL			CLASSICAL			MESENCHYMAL
Gene Expression Datasets	Phillips et al. (2006)	Proneural			Proliferative			Mesenchymal
	Verhaak et al. (2010)	Proneural			Neural	Classical		Mesenchymal
	Sturm et al. (2012)	IDH1/2 Mutation	K27M	RTK I ("PDGFRA")		G34R (Mainly)	RTK II ("Classic")	Mesenchymal
Tumor Characteristics	RTK Changes			PDGFRA Amplification		EGFR - Amplification - Overexpression - vIII mutation		
	Gene Signatures	CIMP+						
		Oligodendrocyte Progenitor Cell-Associated Genes				Neural Stem Cell-Associated Genes		
		TP53 Mutation						Hypoxia, Inflammation, Necrosis
Patient Outcome	Survival	Better Prognosis						Worse Prognosis
	Relapse	Locked in Proneural	Undergo Proneural-Mesenchymal Transition - C/EBPβ, STAT3, TAZ, NFκB - ALDH1A3 - ID1-RAP					

Figure 1.5 Transcriptomal profiling of human GBMs identified a proneural subgroup of patients with better overall survival compared to proliferative or mesenchymal subgroups. With no stratification based on survival, Verhaak et al. identified four subgroups and found that approximately 35% of proneural GBMs displayed *IDH1/2* mutations along with a hypermethylated (CIMP+) phenotype. Sturm et al. extended these observations to also include childhood GBMs and found that hotspot mutations in *H3F3A* and *IDH1* defined distinct epigenetic and biological subgroups. While proneural GBMs are associated with OPC-like gene expression signature, more aggressive GBMs express genes known to drive mesenchymal transcriptional networks.

6.1. Mesenchymal phenotype as a function of glioma subgroup

Mesenchymal GBMs are associated with genes expressed in immune cells, vasculature, invasive cells, but also markers expressed in mesenchymal stem cells and NSCs (Phillips et al., 2006; Sturm et al., 2012; Verhaak et al., 2010). In contrast, *IDH* mutant GBMs are strictly proneural (Lai et al., 2011; Sturm et al., 2012). *IDH* mutations are frequent in secondary (60–90%) GBMs and grade II–III gliomas compared to primary (5%) GBMs (Balss et al., 2008; Bleeker et al., 2009; Hartmann et al., 2009; Kang et al., 2009; Sanson et al., 2009; Watanabe et al., 2009; Yan et al., 2009). *IDH* WT, but not mutant, undergo PMT at recurrence (Lai et al., 2011; Phillips et al., 2006). This raises the question of whether methylation of target genes is required for PMT in gliomas. In epithelial cancers, EMT is associated with increased invasion, resistance to therapy, and acquisition of a "cancer stem cell"-like phenotype (Mani et al., 2008; Polyak & Weinberg, 2009). It is increasingly clear that antiangiogenic treatments and radiotherapy enrich for GSCs and generate highly invasive mesenchymal tumors (Bao et al., 2006; Diehn et al., 2009; Kraus et al., 2002; Lu et al., 2012). Therefore, insights into mechanisms that drive PMT in glioma could lead to therapies that block PMT and improve outcome in patients.

6.2. Transcriptional master regulators of PMT in glioma

In epithelial cancers, *SNAIL, TWIST, ZEB1* and *TGFB1* are known master regulators of EMT (Kalluri & Weinberg, 2009; Thiery, 2002). *In vitro* studies suggest that *ZEB1* and *TWIST* also regulate tumor invasion, chemoresistance and stemness in GBM cell lines (Mikheeva et al., 2010; Siebzehnrubl et al., 2013). In an attempt to identify other drivers of the mesenchymal phenotype in GBMs, Carro et al. (2009) used a bioinformatics approach to contrast gene expression signatures between GBM subgroups (Carro et al., 2009). The top six transcription factors that distinguished GBM subgroups (signal transducer and activator of transcription 3 (*STAT3*), *C/EBPβ, bHLH-B2, RUNX1, FOSL2* and *ZNF238*) regulated >74% of the mesenchymal gene expression signature. *C/EBPβ* and *STAT3* were identified as master regulators of mesenchymal transcription networks. During development, *C/EBPβ* and *STAT3* have opposing roles on neurogenesis (Gu et al., 2005; Ménard et al., 2002). While *C/EBPβ* promotes neurogenesis and opposes gliogenesis, *STAT3* promotes astrocytic differentiation and inhibits neurogenesis (Ménard et al., 2002; Nakashima, 1999;

Paquin, 2005). As *C/EBPβ* and *STAT3* were transduced into human fetal NSCs, the authors observed reduced neurogenesis and induction of a program towards a mesenchymal phenotype. In primary human GBM cells, *C/EBPβ* and *STAT3* were essential for the mesenchymal phenotype and tumorigenicity when xenografted into immunocompromised mice. Other studies show that *C/EBPβ* and *STAT3* can individually regulate glioma biology. For example, downregulation of *C/EBPβ* in glioma cells inhibited proliferation, invasion and tumorigenicity in mice (Aguilar-Morante, Cortes-Canteli, Sanz-Sancristobal, Santos, & Perez-Castillo, 2011; Homma et al., 2006), consistent with findings demonstrating that increased mRNA and protein levels of *C/EBPβ* are associated with a worse prognosis (Homma et al., 2006). Similarly, *STAT3* is known to promote tumor growth in glioma. Recent findings suggest that *STAT3* is required for maintenance of GSCs (Garner et al., 2013; Priester et al., 2013; Sherry, Reeves, Wu, & Cochran, 2009).

To identify additional master regulators of the mesenchymal phenotype in glioma, Bhat et al. employed a regulatory network analysis of GBM microarray data sets (Bhat et al., 2011). Compared to mesenchymal GBMs, the authors found lower expression of the *transcriptional coactivator with PDZ-binding motif* (*TAZ*) in proneural GBMs and lower-grade gliomas. Expression of *TAZ* correlated with the degree of CpG island hypermethylation of its promoter. Silencing of *TAZ* in mesenchymal GSCs decreased mesenchymal marker expression, invasion, self-renewal, and tumor formation. Conversely, overexpression of *TAZ* in proneural GSCs as well as murine NSCs induced expression of mesenchymal markers, through a binding with coactivator TEAD. Interestingly, *TAZ* cooperates with PDGF-B to induce high-grade mesenchymal gliomas in the RCAS/Nestin-tv-a model, suggesting that aberrant RAS activation may be a prerequisite for mesenchymal transition by *TAZ* (Bhat et al., 2011). In other cancers, the TAZ–TEAD nuclear complex has previously been shown to play important roles in EMT, cell growth, and organ development (Hong & Yaffe, 2006; Lei et al., 2008; Zhang et al., 2009; Zhao, Li, Lei, & Guan, 2010).

As master regulators of the mesenchymal phenotype, *STAT3*, *C/EBPβ* and *TAZ* represent future therapeutic targets. The ability to inhibit these transcriptional nodes may lead to a collapse of the mesenchymal phenotype, reduced treatment resistance, and improved survival in GBM patients. Pharmacological JAK kinase inhibitors, which lead to decreased STAT activation, are currently being evaluated in clinical trials against solid tumors. A future challenge will be to identify therapeutics that target *C/EBPβ* and *TAZ*. A small molecule inhibitor screen identified the porphyrin family such as verteporfin (a macular degeneration drug) as inhibitors of *YAP/TEAD*-dependent transcription (Liu-Chittenden et al., 2012). As mediators of the Hippo pathway, YAP and *TAZ* are paralogs and display 50% homology (Lei et al., 2008). Furthermore, Rho, ROCK, and WNT inhibitors inhibit *YAP/TAZ* activities (Piccolo, Cordenonsi, & Dupont, 2013). Lastly, since increasing data supports that proneural gliomas arise from OPCs rather than NSCs, future studies should address if introduction of *TAZ* or other mesenchymal master regulators into OPCs can give rise to mesenchymal gliomas.

Interestingly, gene networks regulated by *TAZ* were nonoverlapping compared to those regulated by *STAT3* or *C/EBPβ* (Bhat et al., 2011). Is it possible that an upstream effector gene regulates *TAZ*, *STAT3*, and *C/EBPβ* in gliomas? TCGA analyses revealed that the mesenchymal GBM subgroup were enriched in nuclear factor kappa-light-chain-enhancer of activated B cells (*NFκB*) and tumor necrosis factor (*TNF*) superfamily of genes (Verhaak et al., 2010). Supporting data show that TNFα treatment induced NFκB activation in patient-derived proneural human GBM cells, leading to PMT and increased radioresistance (Bhat et al., 2013). Interestingly, activated NFκB signaling increased the activation of all three master regulators: *STAT3*, *C/EBPβ* and *TAZ*. Inhibition of NFκB activation via a mutant *IκB* blunted the mesenchymal switch, and the authors also showed that minocycline, as an antiinflammatory inhibitor that targets NFκB pathway, effectively reduce tumor proliferation of mesenchymal GBMs. This elegant study suggests that NFκB blockade can block the mesenchymal gene network. Future studies should demonstrate if NFκB inhibition combined with standard of care can improve outcome in human glioma patients.

Pathway enrichment analysis of proneural and mesenchymal GBMs revealed that genes involved in glycolysis and gluconeogenesis were enriched in mesenchymal GBM cells. At the protein level, glycolytic activity was also increased in mesenchymal cells. The authors identified aldehyde dehydrogenase ALDH1A3 as the most highly expressed metabolic enzyme in mesenchymal GBM cells. Blockade of ALDH1 activity reduced proliferation of mesenchymal, but not proneural, GBM cells (Mao et al., 2013). Interestingly, radiation of proneural GBM cells induced an increased expression of mesenchymal markers and a concomitant downregulation of proneural genes, a shift that was reversed by ALDH1A3 inhibition. Results from this study suggest that an upregulation of the glycolysis pathway occurs when glioma cells need to maintain their energy requirement in low nutrient conditions, also observed in mesenchymal tumors in a harsh microenvironment, or in proneural tumors that undergo a stress such as radiotherapy. The importance of glycolysis in tumorigenesis was confirmed by another study that associated expression of glucose transporter, type 3 (GLUT3) with tumorigenicity of GSCs (Flavahan et al., 2013). These studies highlight the adaptation of GSCs to survive in a nutrient-depleted environment and respond to stress following therapy.

Other genes that are associated with the mesenchymal phenotype in gliomas include the inhibitor of differentiation (*ID*) genes *ID1* and *ID2*, two bHLH factors that promote tumorigenicity in mesenchymal high-grade murine gliomas. In this mouse model, murine hippocampal NSCs were transduced with lentiviruses for *H-Ras*^{V12} and short-hairpin RNAs against *TP53* (Niola et al., 2013). RNA interference experiments showed that downregulation of ID proteins reduced stem cell-associated markers such as ITGA6, Nestin and SSEA-1 *in vitro* and the number of perivascular glioma stem-like cells expressing SSEA-1 *in vivo* (Niola et al., 2013). Furthermore, *c-MET* has been linked to the mesenchymal phenotype in epithelial cancers. When overexpressed, *c-MET* induced an EMT-like transition and stemness in GSCs (De Bacco et al., 2012). Similarly, incubation of GBM cells with the c-MET ligand HGF induced a mesenchymal phenotype (Li et al., 2011). It is still unclear if these genes only play a role in existing mesenchymal tumor cells, or alternatively can induce PMT in proneural GBM cells.

6.3. Influence of the tumor microenvironment on the mesenchymal phenotype

A tight relationship exists between tumor microenvironment and EMT during tumor progression (Finger & Giaccia, 2010; Kalluri & Weinberg, 2009; Polyak & Weinberg, 2009). As mentioned, mesenchymal GBMs were found to be enriched in genes of the TNF superfamily and NFκB, reflecting a high level of necrosis and a prominent immune infiltration compared to proneural GBMs (Phillips et al., 2006; Verhaak et al., 2010). This was confirmed with histopathological analyses reflecting high levels of necrosis, hypoxia, and inflammation in mesenchymal specimens compared to other transcriptional subtypes (Cooper et al., 2012; Engler et al., 2012). Furthermore, several mesenchymal network genes identified by Carro et al. (e.g., *C/EBPβ*, *C/EBPδ*, *FOSL2* and *STAT3*) were among the highest ranked genes that correlated with the extent of necrosis; and C/EBPβ and C/EBPδ proteins were expressed in hypoxic perinecrotic regions in GBMs (Cooper et al., 2012). As GBMs, in contrast to lower-grade gliomas, are associated with necrosis and more extensive inflammation, it is plausible that necrosis contributes to the mesenchymal phenotype (Cooper et al., 2010).

In 2009, bevacizumab was approved for use in recurrent GBM patients. However, studies in patients and preclinical models have demonstrated that the beneficial effects of anti-VEGF therapy are transient, as the tumors become highly invasive in nature to circumvent the blockade of the vasculature (Keunen et al., 2011; Lu et al., 2012; Pàez-Ribes et al., 2009). Anti-VEGF therapy and radiotherapy are known to induce HIF1α, leading to increased transcription of VEGF and stromal-cell derived factor 1 (SDF-1), and recruitment of myeloid cells. Ultimately, this cascade of events lead to increased vasculogenesis, and recurrent tumors displaying a high degree of stemness and expression of mesenchymal markers (Kioi et al., 2010; Piao et al., 2012, 2013). Inhibition of C-X-C chemokine receptor 4 (CXCR4) activity or blockade of HIF1α effectively depleted influx of inflammatory myeloid cells following radiation of mice xenografted with human GBM cells (Kioi et al., 2010). Future studies should identify more specific immune targets and employ GEMM of glioma, displaying an intact immune system, to effectively study changes in the tumor microenvironment following radiotherapy or antiangiogenic treatment.

As an important target in GBM therapy, recent studies have identified molecular interactions between VEGF and other RTKs. In a GEMM model of glioma, Lu et al. (2012), found that VEGF recruits tyrosine phosphatase PTP1B to prevent c-MET activation (Lu et al., 2012). Anti-VEGF therapy induced VEGFR2 and c-MET heterodimerization followed by HGF-induced phosphorylation of c-MET. The VEGFR2/c-MET complex produced a switch from T-cadherin to N-cadherin, increased invasion and induction of a mesenchymal phenotype. Surprisingly, these effects were independent on hypoxia, suggesting that recruitment of CXCR4-expressing myeloid cells did not contribute to the mesenchymal phenotype in this mouse model.

Targeting inflammatory cells in the tumor microenvironment has yielded some success in preclinical models. As mentioned previously, NFκB activation induces a PMT in proneural glioma cells. The authors also demonstrated that TNFα from microglia/ macrophages in the tumor microenvironment is a contributing source of NFκB activation, and dual targeting of NFκB in tumors and immune cell activation via minocycline was effective in decreasing tumor growth and radioresistance only in the mesenchymal and not proneural tumors (Bhat et al., 2013). In another recent study, a colony stimulating factor 1 receptor (CSF1R) inhibitor BLZ945 that targets the tumor-associated myeloid cells was effective in causing the regression of tumors in a proneural model of GBM and increased survival of tumor-bearing mice (Pyonteck et al., 2013). Analyses of tumors revealed that myeloid numbers was unchanged, but the expression of tumor-promoting M2 markers was significantly reduced in these cells. This suggests that a reeducation of the immune cells was sufficient to cut off growth support of tumor cells. It will be interesting if CSF1R inhibition will be effective in gliomas of other subclasses, especially mesenchymal tumors that have high inflammatory signature, and in conjunction with standard of care such as irradiation or chemotherapy.

Reactive astrocytes represent another stromal component of the tumor microenvironment in gliomas. Similar to peripheral tumors where fibroblasts respond to tumor growth and injury, astrocytes become reactive in response to pathological conditions. In glioma, the reactive state and the abundance of reactive astrocytes increase with grade. A recent report show that reactive astrocytes is a major component of the tumor microenvironment in a PDGF-driven GEMM of glioma, surrounding the tumor at the periphery and in the perivascular niche, a region suggested to harbor GSCs (Katz et al., 2012). Less is known about the role of reactive astrocytes propagating tumor growth and promoting treatment-resistance in the GSC compartment.

It becomes increasingly clear that radiotherapy and anti-angiogenic treatments effectively target the tumor bulk, but leave behind subpopulations of tumor cells that become treatment-resistant and form invasive recurrent tumors. Additional investigations are needed to fully understand mechanisms underlying PMT in human glioma. As future studies try to prevent PMT or target the mesenchymal phenotype, it is important to appreciate the extensive intratumoral heterogeneity found in human GBMs. Sottoriva et al. (2013) found that multiple regions in an individual GBM can display proneural, classical, and mesenchymal gene expression signatures (Sottoriva et al., 2013). This finding parallels recent progress demonstrating intratumoral heterogeneity for RTK amplifications in human GBMs. Intratumoral GBM heterogeneity in the tumor microenvironment and genetic alterations represent a major challenge for future therapies.

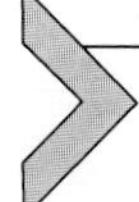

7. RELATIONSHIP BETWEEN GSCs AND GLIAL PROGENITORS

Traditional cancer therapies target tumor cells based on their susceptibility to genotoxic therapies such as radiation and DNA alkylating agents. However, cancer stem cells, also referred to as tumor-initiating or tumor-propagating cells, represent subpopulations that are resistant to genotoxic chemotherapies (Dick, 2008; Reya, Morrison, Clarke, & Weissman, 2001). Considerable research effort is being directed to the identification and characterization of cancer stem cells for determining properties that can be exploited for therapeutic eradication. To date, the identification and isolation of cancer stem cells have relied on differential uptake of cell-permeable dyes or expression of cell-surface antigens.

The first report identifying functional cancer stem cells in glioma was based on expression of glycosylated CD133 protein on the cell surface of GBM cells. CD133-expressing GSCs coexpressed NSC proteins, showed high self-renewal capacity and, in contrast to CD133 negative tumor cells, established tumors in xenografted mice (Singh et al., 2004). Subsequent reports showed that CD133-expressing GBM cells undergo asymmetric cell division (a prerequisite for tumor regrowth), are resistant to treatment with the alkylating agent temozolomide, and display increased radioresistance (Bao et al., 2006; Deleyrolle et al., 2011; Lathia et al., 2011; Liu et al., 2006). Additional, and perhaps more provocative findings suggest that CD133+ GSCs regenerate endothelial cells, as well as pericytes, allowing tumors to reestablish the microenvironmental niche required for regrowth (Ricci-Vitiani et al., 2010; Wang et al., 2010). More recent reports suggest a large number of alternative GSC markers, many of which show partial or no overlap with CD133+ cells (Anido et al., 2010; Bao et al., 2008; He et al., 2011; Kim et al., 2012; Lathia et al., 2010; Li et al., 2011; Mazzoleni et al., 2010; Rasper et al., 2010; Son, Woolard, Nam, Lee, & Fine, 2009; Tchoghandjian et al., 2010) (Table 1.2). Interestingly, CD133, A2B5, CD44, c-Met, PDGFRβ and EGFR are also expressed on OPCs (Bouvier-Labit, Liprandi, Monti, Pellissier, & Figarella-Branger, 2002;

Table 1.2 Examples of cell-surface antigens expressed on GSCs and the corresponding expression in NSC and OPC populations, respectively

GSC marker	Reference	NSC marker	OPC marker
CD133	Singh et al. (2003)	Yes	Yes
A2B5	Tchoghandjian et al. (2010)	No	Yes
CD15/SSEA-1	Son et al. (2009)	Yes	No
ALDH1	Rasper et al. (2010)	Yes	No
CD44	Anido et al. (2010)	Yes	Yes
c-Met	Li et al. (2011)	Yes	Yes

Moransard, Sawitzky, Fontana, & Suter, 2010; Raff, Miller, & Noble, 1983; Verhaak et al., 2010; Wang, O'Bara, Pol, & Sim, 2013).

Is it possible that GSC-rich gliomas can arise from OPCs? In fact, we and other groups have demonstrated that the proneural tumor oligodendroglioma can arise from OPCs (Lindberg et al., 2009; Liu, Sage, et al., 2011; Persson et al., 2010). As oncogenic events were introduced into NSCs and OPCs, mosaic analysis with double markers found that transformation only occurred after NSCs had differentiated into OPCs (Liu, Sage, et al., 2011). As A2B5-expressing GSCs become more mesenchymal with grade (Auvergne et al., 2013), it is possible that cell-surface proteins on GSCs change during the disease progression or following therapy (Bhat et al., 2011, 2013; Lu et al., 2012; Mao et al., 2013). In comparison to lower-grade tumors, GBMs display a transcriptional gene expression signature reminiscent of embryonic stem (ES) cells (Ben-Porath et al., 2008). Higher expression of reprogramming factors in GBM (the same factors used by Yamanaka and colleagues to reprogram somatic cells; Takahashi & Yamanaka, 2006), suggest that dedifferentiation of OPCs or more differentiated cells can produce gliomas with a high degree of stemness. In this section, we will discuss gene families that are normally associated with stem cells, but also play important roles during OPC lineage development (Table 1.2).

7.1. Polycomb gene family

Proneural genes are expressed in a timely and regional fashion during neocortical development (Wilkinson, Dennis, & Schuurmans, 2013). As one example, the Polycomb group (PcG) gene family regulates cell fate during development. PcG proteins function as complexes known as polycomb repressive complexes (PRC) and work by repressing transcription with methylation altering chromatin structure (Simon & Kingston, 2009). Expression of PRCs maintain stemness in ES and repress developmental genes active during differentiation (Lee et al., 2006). The PRCs play a similar role in various cancers (Easwaran et al., 2012). As the catalytic component of PRC, *Enhancer of Zeste homolog 2* (*EZH2*), a histone-lysine *N*-methyltransferase, is downregulated as NSCs differentiate into astrocytes (Sher et al., 2008). Induced overexpression of *EZH2* in astrocytes partially dedifferentiates them back towards NSCs (Sher, Boddeke, & Copray, 2011). Interestingly, when NSCs differentiate into OPCs, *EZH2* expression remains high (Sher, Boddeke, Olah, & Copray, 2012). Downregulation of *EZH2* in OPCs resulted in derangement of the oligodendrocytic phenotype, due to reexpression of neuronal and astrocytic genes, and ultimately apoptosis (Sher et al., 2012). *EZH2* catalyzes the methylations of lysine 27 on histone H3 (H3K27) (Cao et al., 2002). As stated above, the *K27M* mutant has a higher affinity for *EZH3* compared to WT. So it is clear that an oncogenic event such as the *K27M* mutation could dysregulate *EZH2* activity and promote gliomagenesis. Inhibition of *EZH2* with drugs or shRNA has been used as a treatment and has been shown to impair the self-renewal of GBM cancer stem cells (Suvà et al., 2009).

7.2. NOTCH

NOTCH is an important pathway in CNS development and tissue patterning. Moreover, NOTCH activity maintain NSCs in a quiescent state and prevent differentiation. (Ables, Breunig, Eisch, & Rakic, 2011). In addition, NOTCH functions in OPCs as a repressor of differentiation (Sim et al., 2011; Wang et al., 1998) (Fig. 1.2). In glioma, NOTCH-1 and its ligands DELTA-LIKE-1 (DLL1) and JAGGED-1 are overexpressed (Purow et al., 2005; Zhang et al., 2007). Furthermore, NOTCH is expressed in several GBM subgroups (Verhaak et al., 2010), where it forms transcriptional networks with *GLI1*, *Myc*, *BMP2*, and *RUNX2* (Cooper et al., 2010). Several studies suggest that NOTCH-1 activity is essential for survival of GSCs (Ables et al., 2011; Saito et al., 2014; Zhang et al., 2007). For example, expression of the active form of NOTCH-1 in human glioma cell lines increase proliferation and self-renewal (Zhang et al., 2007). RNA interference experiments showed that *DLL1* and *JAGGED-1* knockdown reduce survival of GSCs (Purow et al., 2005). *In vitro* studies suggest cooperative effects of pharmacological inhibition using the γ-secretase inhibitor and γ-irradiation as they target GSCs and non-GSCs, respectively (Ables et al., 2011; Saito et al., 2014). The NOTCH pathway represents an attractive target for future therapies trying to eliminate GSCs.

7.3. Sonic hedgehog

Sonic hedgehog is essential for survival of postnatal NSCs (Machold et al., 2003). In addition, as SHH downstream targets, expression of members of the *GLI* family is associated with proliferative regions during neural development (Dahmane et al., 2001). In GBMs, *GLI-1* is necessary for survival of GSCs and induced by oncogenic drivers (Clement et al., 2007; Dahmane et al., 2001; Santoni et al., 2013). Treatment with the pharmacological SHH inhibitor cyclopamine effectively depleted CD133+ GSCs

(Clement et al., 2007). Interestingly, similar to NOTCH-1 inhibition, cyclopamine and temozolomide cooperates to target GSCs and non-GSCs, respectively (Clement et al., 2007). As a downstream target of *GLI* genes, *NANOG* is known to promote stemness and is expressed in GBMs (Clement et al., 2007; Mitsui et al., 2003). Interestingly, *NANOG* is able to reprogram *TP53*-deficient mouse astrocytes into high-grade gliomas (Moon et al., 2011). SHH also stimulates proliferation of OPCs and is necessary during OPC lineage development (Ferent, Zimmer, Durbec, Ruat, & Traiffort, 2013; Lelievre et al., 2006; Tekki-Kessaris et al., 2001), suggesting that *SHH-GLI1* signaling may also drive tumor growth in more progenitor-like gliomas (Fig. 1.2).

7.4. Wingless

Wingless (WNT) has been extensively studied as a developmental pathway regulating cell fate specification, migration, and proliferation. The WNT pathway also plays a role in self-renewal of adult NSCs (Kalani et al., 2008). A WNT antagonists increased the numbers of immature OPCs in spinal cord explants (Shimizu et al., 2005), demonstrating that endogenous WNT signaling controls oligodendrocyte development. Isolation of OPCs from the human fetal brain based on CD140a (PDGFRA) expression showed that 12/15 WNT target genes were highly expressed in CD140a+ versus CD140a− cell fractions (Sim et al., 2011). The authors concluded that WNT and NOTCH are essential for survival of OPCs. A comparison of development pathways, show that the WNT pathway is more dysregulated than NOTCH or SHH in GSCs compared to human adult NSCs (Sandberg et al., 2013). The authors show that the WNT inhibitor frizzled-related protein 1 (SFRP1) effectively reduced proliferation and self-renewal of GSCs (Sandberg et al., 2013). Furthermore, WNT components are more prominently expressed in A2B5+ cells isolated from GBM versus lower-grade gliomas (Auvergne et al., 2013). During normal development, WNT influence the timing and efficacy of OPC generation in the telencephalon (Langseth et al., 2010). As a target of WNT transcriptional activation, AXIN2 is expressed in OPCs and is essential for normal kinetics of remyelination (Fancy et al., 2011). Interestingly, *SOX17* is a downstream target of the WNT pathway in both OPCs and human oligodendroglioma cells (Chen, Chew, Packer, & Gallo, 2013; Chew et al., 2011), exemplifying how in-depth knowledge of OPCs can increase our understanding of glioma biology (Fig. 1.2).

8. TARGETED THERAPY IN GLIOMA

During development, receptor tyrosine kinase (RTK) signaling mediate effects of growth factors to regulate expansion and differentiation programs in specific neural precursor populations (Forsberg-Nilsson, Behar, Afrakhte, Barker, & McKay, 1998; Hébert & Fishell, 2008; Kilpatrick & Bartlett, 1995). In the brain, members of the fibroblast growth factor receptor (FGFR), epidermal growth factor (EGFR), and platelet-derived growth factor (PDGFR) families are expressed in distinct NSCs and more differentiated progenitor populations. In GBM, amplifications or somatic mutations in *EGFR*, *PDGFRA*, *FGFR1*, or *c-MET* often correlate with transcriptomal subgroups (Verhaak et al., 2010). Approximately 88% of GBMs display genetic amplifications of RTKs and genetic alterations of downstream effector genes. Towards developing a personalized medicine approach and improve outcome for GBM patients, clinical studies increasingly include only subsets of GBM patients that show distinct genetic alterations (Noushmehr et al., 2010; Phillips et al., 2006; Verhaak et al., 2010). Intratumoral heterogeneity of *EGFR*, *PDGFRA*, and *c-MET* in GBMs suggest that future approaches should target several RTKs or common downstream effectors to benefit GBM patients (Snuderl et al., 2011; Stommel, Kimmelman, & Ying, 2007; Szerlip et al., 2012). Furthermore, pharmacological studies show that monotherapy with either *EGFR* or *PDGFR* inhibitors is not sufficient to prevent recurrence in patients (Lo, 2010; Morris & Abrey, 2010). In this section, we will discuss the expression of RTKs in glioma, why inhibition of RTKs or downstream targets has failed in patients, and their roles in OPC lineage development.

8.1. Epidermal growth factor gene family

As the most commonly amplified RTK in GBMs, EGFR is a member of the ErbB family of membrane-bound receptor tyrosine kinases, of which the other members are ErbB2/HER2/Neu, ErbB3/HER3, and ErbB4/HER4 (Mineo et al., 2007; Verhaak et al., 2010). In addition, approximately 50% of *EGFR* amplified GBMs express a constitutively active mutation, called EGFRvIII, lacking the extracellular domain (Gan, Kaye, & Luwor, 2009). Other point mutations in the extracellular ligand binding domain of EGFR have been described (Lee et al., 2006; Vivanco et al., 2012). Over the years, much effort has focused on development of small molecule inhibitors against EGFR inhibitors or vaccines against EGFRvIII (Sampson et al., 2010). To date, EGFR inhibition in patients has largely failed (Table 1.3). Association of poor response in patients displaying *EGFRvIII* mutation and *PTEN* loss may suggest that improved stratification of enrolled patients will improve future response rates (Haas-Kogan et al., 2005; Mellinghoff et al., 2005). Furthermore, development of inhibitors with improved affinity to EGFRvIII or irreversible EGFR inhibitors may more effectively block downstream signaling and produce a better response in patients (Barkovich et al., 2012; Vivanco et al., 2012). However, redundant expression of EGFR family members may also contribute to treatment resistance. For example, proneural GBMs express high levels of ErbB3, functionally active in heterodimers with other ErbB family members (Verhaak et al., 2010).

ErbB3 is highly expressed on both human OPCs and proneural glioma cells expressing A2B5 (Auvergne et al., 2013; Sim et al., 2011; Verhaak et al., 2010). *ErbB3* and *EGFR* regulate survival in OPCs (Flores et al., 2000; Ivkovic, Canoll, & Goldman, 2008) (Fig. 1.2). Introduction of human *EGFRvIII* into OPCs produced hyperplasia in postnatal white matter in mice (Ivkovic et al., 2008). Moreover, we found that a constitutively active avian form of *EGFRvIII*, *v-erbB*, produced oligodendroglioma in mice (Persson et al., 2010). In conclusion, EGFR signaling remains an important therapeutic target as research generates better pharmacological tools and identifies molecular characteristics in glioma subgroups for stratification of patients.

8.2. Targeting the proneural subgroup by PDGFR inhibition

Platelet-derived growth factor isoform α and β form homo- or heterodimers upon ligand binding. Amplification of *PDGFRA* was identified in 11% of GBM patients (The Cancer Genome Atlas Research Network, 2008). Importantly, increased PDGF pathway activity has been reported in up to 33% of adult GBM patients (Brennan et al., 2009). Furthermore, not only focal, but also low-level, amplifications of *PDGFRA* are frequent in pediatric (29%) and adult (20%) GBM patients (Phillips et al., 2013). The authors found that in GBM patients displaying $IDH1^{R132H}$ mutations, *PDGFRA* amplification was a negative prognostic marker (Phillips et al., 2013). Recent studies suggest that PDGFRB is expressed on subsets of GSCs and can cross-talk with EGFR in GBMs (Akhavan et al., 2013; Kim et al., 2012).

Table 1.3 Association of pediatric H3F3A (G24 and K27) mutant high-grade gliomas to distinct brain regions

Subgroup	Target	Inhibitor	Other Targets	Clinical Trials	Reference
Classical					
	EGFR	Gefitinib (Iressa/ZD1839)	–	Yes	Reardon et al. (2006)
		Erlotinib (Tarceva/OSI-779)	–	Yes	Raizer and Abrey (2010)
		Cetuximab (Erbitux, C225)	–	Yes	Eller, Longo, Kyle, and Bassano (2005) and Combs et al. (2006)
		Panitumumab (Vectibix)	–	Yes	Gajadhar, Bogdanovic, Muñoz, and Guha (2012)
		AG1478	–	No	Nagane et al. (2001)
		BIBU-1361	–	No	Ghildiyal, Dixit, and Sen (2013)
		EKI-785	–	No	Rao et al. (2005)
		EtDHC	–	No	Han (1997)
		F90	–	No	Liu et al. (2008)
		NSC56452	–	No	Yang, Yang, Pike, and Marshall (2010)
		Lapantinb (GW572016)	HER2	Yes	Vivanco et al. (2012)
		AEE788	VEGF	Yes	Reardon et al. (2012)
		Vandetanib (Zactima/ZD6474)	VEGFR, RET	Yes	Drappatz et al. (2010) and Shen et al. (2013)
		GW2947	HER2	No	Wang, Wei, et al. (2013)

Proneural					
	PDGFR	α/β: Imatinib (Gleevec, imatinib mesylate, STI571	VEGF, bcr-abl	Yes	Reardon et al. (2009) and Dong et al. (2011)
		α: Sunitinib (Sutent, SU11248, sunitinib malate)	VEGFR2, c-KIT	Yes	Neyns et al. (2011)
		Vatalanib (PTK787, ZK222584)	VEGFR, c-kit	Yes	Gerstner et al. (2011)
		Pazopanib (GW786034)	VEGFR, c-Kit	Yes	Iwamoto (2010)
		β: Dasatinib (Sprycel, BMS-354825)	Src. bcr-abl, c-KIT, EPHA2	Yes	Lu-Emerson et al. (2011)
		β: Sorafenib (Nexavar,)	VEGFR2, Raf	Yes	Lee et al. (2012)
		α/β: Tandutinib (MLN0518)	FLT3, c-KIT	Yes	Boult, Terkelsen, Walker-Samuel, Bradley, and Robinson (2013)
		Leflunomide (SU101)	–	Yes	Vlassenko, Thiessen, Beattie, Malkin, and Blasberg (2000) and Shawver et al. (1997)
		Nintedanib (BIBF-1120)	VEGFR, FGFR	Yes	Muhic, Poulsen, Sorensen, Grunnet, and Lassen (2013)
		CP-673,451		No	Roberts et al. (2005)
	IDH1R132H	AGI-5198	IDH1R132H	No	Rohle et al. (2013)

Continued

Table 1.3 Association of pediatric H3F3A (G24 and K27) mutant high-grade gliomas to distinct brain regions—cont'd

Subgroup	Target	Inhibitor	Other Targets	Clinical Trials	Reference
Mesenchymal					
	VEGF	Bevacizumab	–	Yes	Kreisl, Kim, et al. (2009)
		Cediranib (Recentin, AZD2171)	–	Yes	Batchelor et al. (2010) and Wachsberger et al. (2011)
		CT322 (BMS-844203)	–	Yes	Waters et al. (2012)
		Sunitinib (Sutent, SU11248, sunitinib malate)	PDGFRα, c-KIT	Yes	Neyns et al. (2011)
		Vatalanib (PTK787, ZK222584)	PDGFR, c-Kit	Yes	Gerstner et al. (2011)
		Pazopanib (GW786034)	PDGFR, c-Kit	Yes	Iwamoto (2010)
		Vandetanib (Zactima/ ZD6474)	EGFR, RET	Yes	Broniscer et al. (2013)
		Sorafenib (Nexavar)	PDGFRβ, Raf	Yes	Lee et al. (2012)
	MET	Cabozantinib (XL184)	RET, KIT, VEGFR2	Yes	Yakes et al. (2011) and De Groot et al. (2009)
		SU11274	–	No	Stommel et al. (2007) and Joshi et al. (2012)

Receptor tyrosine kinases have been a major therapeutic target in GBMs. As the most commonly amplified RTK, EGFR represent has been an attractive target in clinical trials. Since subsets of GBMs show *PDGFRA* amplifications and overexpression, Imatinib and other PDGFR inhibitors have been evaluated in patients. Other clinical trials have inhibited VEGF and c-MET signaling in GBMs. For future clinical trials, whole-genome sequencing and transcriptomal studies will be valuable to stratify glioma patients into defined subgroups.

As one of the first successful small molecule kinase inhibitors, Imatinib effectively inhibited kinase activity of the fusion gene *BCR-ABL* (Druker et al., 2001). However, Imatinib also inhibits kinase activity for PDGFRA, tyrosine-protein kinase Kit (c-KIT), and fms-like tyrosine kinase 3 (FLT-3). To inhibit PDGFRA signaling in human GBMs, Imatinib reduced phosphorylated AKT levels in 4/11 patients but had no effect on overall survival benefit (Razis et al., 2009) (Table 1.3). However, since enrolled patients were not stratified based on *PDGFRA* amplification, it is unclear if Imatinib will increase the survival in subsets of GBM patients (Fig. 1.2).

In GBM patients, approximately 35% of proneural tumors display focal amplifications of *PDGFRA* (Verhaak et al., 2010). These proneural GBMs express genes associated with OPCs, including PDGFRA, commonly used to isolate murine or human OPCs (Sim et al., 2011). PDGFB ligand transforms murine OPCs into gliomas (Lindberg et al., 2009). In postnatal mice, PDGF ligand increases proliferation of NG2+ OPCs in white matter, but not gray matter, an effect that was largely dependent on WNT and PI3K signaling (Hill et al., 2013). In summary, *PDGFRA* represents a relevant future therapeutic target in proneural gliomas, in particular *PDGFRA* amplified gliomas.

8.3. Targeting the mesenchymal phenotype through c-MET inhibition

As hepatocyte growth factor (HGF) binds to the c-mesenchymal-epithelial transition factor (c-MET) it can stimulate proliferation of NSCs and OPCs in the postnatal brain (Ohya, Funakoshi, Kurosawa, & Nakamura, 2007; Wang, Zhang, Gyetko, & Parent, 2011). Expression of HGF and c-MET at the boundaries between epithelial and mesenchymal cells regulate normal organogenesis; epithelial cells expressing c-MET, and mesenchymal cells secreting the HGF ligand. In glioma, c-MET and HGF expression levels correlate with grade (Koochekpour et al., 1997). Amplification of *c-MET* is present in approximately 5% of GBMs (Dunn et al., 2012). In a study comparing 19 paired primary and recurrent patient biopsies, overexpression of c-MET was reported in over 30% of newly diagnosed primary GBMs and was much more common (>75%) in relapse cases (Liu, Fu, et al., 2011). Importantly, c-MET overexpression is a key feature of the mesenchymal subclass (Phillips et al., 2006; Verhaak et al., 2010) and associated with worse prognosis across GBM subgroups (Abounader & Laterra, 2005). Mesenchymal GBMs are associated with high expression of immune-related genes and vasculature (Verhaak et al., 2010).

Preclinical and clinical studies show that antiangiogenic therapy in GBMs results in a more invasive phenotype (Gerstner et al., 2010; Lu et al., 2012) (Table 1.3). In an elegant study by Lu et al., HGF-dependent MET phosphorylation and tumor cell migration were suppressed by a direct interaction between the protein tyrosine phosphatase 1B (PTP1B) and a MET/VEGFR2 heterocomplex, implicating VEGF as a negative regulator of tumor cell invasion (Lu et al., 2012). Concordantly, the dual c-MET and VEGFR inhibitor Cabozantinib (XL184) reduced proliferation GBM cells *in vitro* and *in vivo*, potently reducing phosphorylated c-MET, pAKT, and pERK1/2 levels (Navis et al., 2013; Yakes et al., 2011). Cabozantinib is currently used in clinical trials against newly diagnosed or recurrent GBMs (Table 1.3).

Several studies show that c-MET signaling induces a reprogramming network and support the GSC phenotype (Joo et al., 2012; Li et al., 2011). Expression of c-MET has even been suggested as a functional marker of GSCs (De Bacco et al., 2012). Experiments in an orthotopic xenograft model show that c-MET inhibition reduces growth by selectively targeting the GSC compartment in GBMs (Rath et al., 2013). Future studies should characterize the response of rare GSCs expressing c-MET to radiotherapy and temozolomide treatment in glioma patients.

8.4. Treatment-resistance associated with RTK inhibition

8.4.1 RTK cooperativity

Cooperativity between multiple RTKs drive tumor growth in GBMs (Stommel et al., 2007). Combined inhibition of MET, PDGFRA, and EGFR was needed to effectively inhibit proliferation in human primary GBM cultures. This is consistent with findings demonstrating intratumoral GBM heterogeneity of RTKs *in vivo* (Snuderl et al., 2011). Furthermore, significant cross-talk occurs between different RTKs. For example, cross-talk has been observed between c-MET and EGFR (Jo et al., 2000; Reznik et al., 2008), c-MET and VEGFR2 (Lu et al., 2012) IGFR1 and PDGFRA (Bielen et al., 2011), and EGFR and PDGFRB (Akhavan et al., 2013).

8.4.2 Redundant activation of PI3K/mTOR

Ligand binding of the EGFR, PDGFR and c-MET promotes activation of downstream phosphatidylinositol 3-kinase (PI3K) and mammalian target of rapamycin (mTOR) signaling. Activating mutations in PI3K (observed in approximately 15% of GBMs) and missense mutation or deletion (found in >30% of GBMs) effectively uncouple this pathway from upstream RTK control (The Cancer Genome Atlas Research Network, 2008). Monotherapy with a PI3K inhibitor is unlikely to yield any significant clinical benefit as PI3K inhibitors are cytostatic rather than cytotoxic in GBMs (Cheng, Fan, & Weiss, 2009). Given that mTOR kinase is a key effector molecule downstream of PI3K, much effort has focused on development of specific mTOR kinase inhibitors. However, the canonical and allosteric mTOR kinase rapamycin/sirolimus increase phosphorylation of AKT through a feedback loop (Akhavan, Cloughesy, & Mischel, 2010). Furthermore, rapamycin/sirolimus and subsequent "rapalogs", including temsirolimus/CCI-779 and everolimus/RAD-001, showed limited efficacy in clinical trials. Instead, second generation mTOR inhibitors bind the active site of mTOR kinase, block downstream signaling from both mTORC1 and mTORC2, and are currently being evaluated in clinical trials.

Combined inhibition of PI3K and mTOR potently reduce both survival and proliferation of cancer cells. PI-103 is a p110α isoform (catalytic subunit of PI3K) and mTOR kinase dual inhibitor that potently blocks phosphorylation of mTOR targets and induces a robust G_0G_1 arrest (Fan et al., 2006). The clinical compound, NVP-BEZ235 (Maira et al., 2008) demonstrated robust anti-proliferative activity in many cancer types (Baumann, Mandl-Weber, Oduncu, & Schmidmaier, 2009; Cao, Maira, García-Echeverría, & Hedley, 2009; Chapuis et al., 2010; Chiarini et al., 2010; Santiskulvong et al., 2011), including glioma (Liu et al., 2009). However, NVP-BEZ235 fails to cross the BBB and therefore is of limited use in GBM patients. Combined inhibition of EGFR effectively cooperated with PI3K/mTOR inhibitors (Fan et al., 2007). These studies exemplify that combined inhibition of key survival pathways is needed to produce significant survival benefit in glioma patients.

8.4.3 Feedback loops

Blockade of mTOR can paradoxically lead to activation of MAPK (Carracedo et al., 2008) and AKT (O'Reilly et al., 2006; Sun et al., 2005; Zhang et al., 2007). Similarly, blockade of MAPK (MEK-ERK pathway) can increase phosphorylation of EGFR (Li, Huang, Jiang, & Frank, 2008; Prahallad et al., 2012). To avoid the negative effects of feedback loops, researchers use a strategy based on combined inhibition of RTKs and downstream effectors (Rao et al., 2005; Ronellenfitsch, Steinbach, & Wick, 2010; Wang et al., 2006). This strategy has also been evaluated in a clinical setting in GBM patients (Doherty et al., 2006; Kreisl, Lassman, et al., 2009; Nghiemphu, Lai, Green, Reardon, & Cloughesy, 2012; Reardon et al., 2006, 2010). Combined treatment regimens were well tolerated and patients receiving dual inhibitors showed a slight survival benefit (De Witt Hamer, 2010).

8.4.4 Activation of alternative survival pathways

Combined or single treatment with PI3K and mTOR inhibitors induces autophagy in human GBM cells as a survival mechanism (Fan & Weiss, 2010). Interestingly, blockade of autophagy sensitizes GBM cells to radiotherapy, chemotherapy, and RTK inhibition (Kanzawa et al., 2004; Lin et al., 2012; Paglin et al., 2001; Shingu et al., 2009). Preliminary studies in GBM patients suggest that blockade of autophagy in combination with standard of care is a promising strategy (Briceño, Calderon, & Sotelo, 2007; Sotelo, Briceño, & López-González, 2006).

8.5. Therapeutic targeting of IDH-mutant gliomas

Several studies have shown that $IDH1^{R132H}$ mutations in glioma are associated with better survival (Gravendeel et al., 2010; Ichimura et al., 2009; Labussière et al., 2010; Nobusawa, Watanabe, Kleihues, & Ohgaki, 2009; Parsons et al., 2008; Sanson et al., 2009; Zhao et al., 2009). However, there is still controversy over the use of *IDH* mutations as a prognostic indicator. In low-grade glioma patients, Houillier et al. (2010) found that $IDH1^{R132H}$ mutation and 1p-19q codeletion were associated with prolonged overall survival and a better response to temozolomide treatment (Houillier et al., 2010). Unlike 1p-19q codeletion, $IDH1^{R132H}$ mutation was not associated with prolonged progression-free (PFS) survival. The data indicate that $IDH1^{R132H}$ mutation may be a significant prognostic indicator independently of 1p-19q status. Similarly, Hartman and colleagues found that *IDH* status was the most important predictor of progression-free and overall survival only in patients receiving adjuvant therapy (Hartmann et al., 2009). In contrast, van den Bent et al. reported that the presence of $IDH1^{R132H}$ mutation was an indicator of improved prognosis independent of adjuvant therapy (Van den Bent et al., 2010). Taken together, these results emphasize the need for more in-depth studies in order to determine the prognostic versus predictive role of *IDH* status in human glioma.

Reduced expression of $IDH1^{R132H}$ inhibited cell proliferation and clonogenicity *in vitro* (Jin et al., 2012), suggesting that direct targeting of the mutation is a viable strategy. Rohle and colleages identified a selective $IDH1^{R132H}$ inhibitor (AGI-5198) through a high throughput screen that effectively reduced tumor growth in a human GBM xenograft model (Rohle et al., 2013). The inhibitor blocked the ability of *IDH1*^{R132H} mutation to produce 2-HG, demethylation of histone H3K9me3, and changes in global methylation signatures. The SHH pathway has been proposed as a possible target downstream of *IDH1*^{R132H} mutation (Valadez et al., 2013). Since IDH1^{R132H} mutation is associated with a CIMP phenotype, researchers are evaluating hypomethylating agents or epigenetic modifiers (Fathi & Abdel-Wahab, 2012).

In summary, the discovery of *IDH* mutations has been pivotal for furthering our understanding of how tumorigenesis may occur in gliomas. These findings have led scientists to identify molecular targets in glioma patients diagnosed with *IDH*-mutant tumors. Future studies should aim to extend histological identification of *IDH*-mutant tumors to noninvasive imaging correlates in patients.

9. CONCLUDING REMARKS AND FUTURE PERSPECTIVES

The wide-spread distribution and life-long proliferative capacity of OPCs match the temporal and regional occurrence of gliomas, and therefore represent targets for transformation (Fig. 1.6). Genetically distinct gliomas

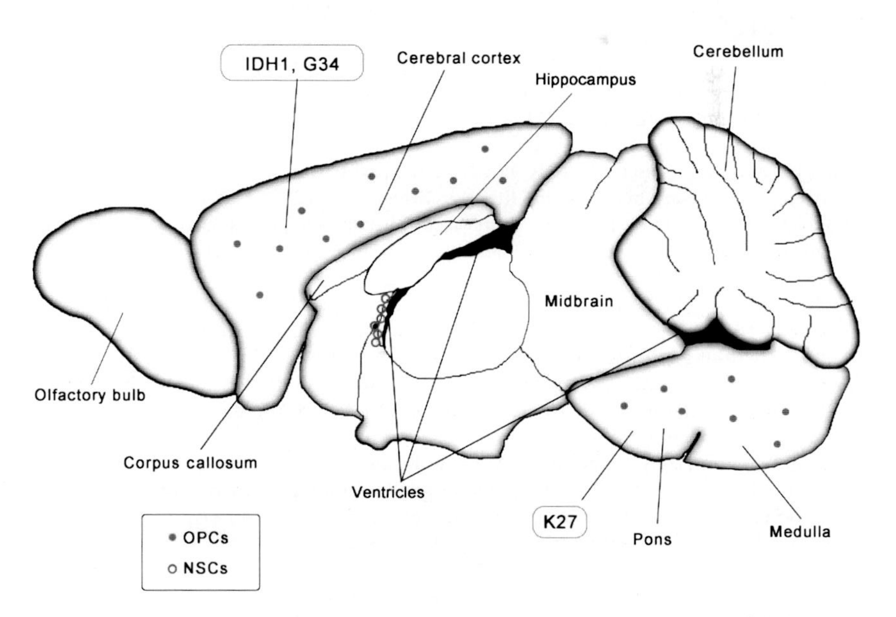

Figure 1.6 OPCs, but not NSCs, are highly abundant in regions where *IDH1*^{R132H} mutant and *H3F3A* K27 mutant tumors arise in GBM patients, implicating their role in gliomagenesis. (See the color plate.)

displaying *PDGFRA* amplifications, *IDH1*^{R132H} mutations, and *H3F3A* mutations express the OPC-related genes OLIG2, NKX2.2, PDGFRα and SOX10, implicating a common cell of origin (Sturm et al., 2012; Verhaak et al., 2010). Even mesenchymal gliomas may arise from OPCs as novel data suggest that defined intrinsic factors and influence from the tumor microenvironment enable gliomas to toggle between proneural and mesenchymal phenotypes (Bhat et al., 2013; Mao et al., 2013). An emerging focus on OPCs in gliomagenesis will provide critical information to researchers studying etiology, tumor microenvironment, and therapeutic intervention in glioma.

ACKNOWLEDGMENTS

A. I. P. is supported by research grants from the TDC Foundation, the Guggenhime Endowment Fund, National Brain Tumor Society, and NIH/U54CA163155. S. I. is supported by the Swedish Childhood Cancer and the Swedish Brain Foundations. M. H. is supported by the Alex's Lemonade Stand Foundation and by PF1329501TBG from the American Cancer Society. The Weiss lab (S. I., J. L., M. H., R. W., W. A. W.) is supported by NIH grants CA176287, CA82103, CA159859, CA133091, CA102321, CA148699, and CA163155, CA81403, and the Alex's Lemonade Stand, CureSearch, Katie Dougherty, National Brain Tumor Society, Pediatric Brain Tumor, St. Baldricks, and Samuel G. Waxman Foundations.

REFERENCES

Abel, T. W., Clark, C., Bierie, B., Chytil, A., Aakre, M., Gorska, A., et al. (2009). GFAP-Cre-mediated activation of oncogenic K-ras results in expansion of the subventricular zone and infiltrating glioma. *Molecular Cancer Research*, *7*, 645–653. Retrieved from, http://www.ncbi.nlm.nih.gov/pubmed/19435821.

Ables, J., Breunig, J., Eisch, A., & Rakic, P. (2011). Not(ch) just development: Notch signalling in the adult brain. *Nature Reviews. Neuroscience*, *12*, 269–283.

Abounader, R., & Laterra, J. (2005). Scatter factor/hepatocyte growth factor in brain tumor growth and angiogenesis. *Neuro-Oncology*, *7*(4), 436–451. http://dx.doi.org/10.1215/S1152851705000050.

Aguilar-Morante, D., Cortes-Canteli, M., Sanz-Sancristobal, M., Santos, A., & Perez-Castillo, A. (2011). Decreased CCAAT/enhancer binding protein β expression inhibits the growth of glioblastoma cells. *Neuroscience*, *176*, 110–119.

Akhavan, D., Cloughesy, T. F., & Mischel, P. S. (2010). mTOR signaling in glioblastoma: Lessons learned from bench to bedside. *Neuro-Oncology*, *12*(8), 882–889. http://dx.doi.org/10.1093/neuonc/noq052.

Akhavan, D., Pourzia, A. L., Nourian, A. A., Williams, K. J., Nathanson, D., Babic, I., et al. (2013). De-repression of PDGFRβ transcription promotes acquired resistance to EGFR tyrosine kinase inhibitors in glioblastoma patients. *Cancer Discovery*, *3*(5), 534–547. http://dx.doi.org/10.1158/2159-8290.CD-12-0502.

Alcantara Llaguno, S., Chen, J., Kwon, C.-H., Jackson, E. L., Li, Y., Burns, D. K., et al. (2009). Malignant astrocytomas originate from neural stem/progenitor cells in a somatic tumor suppressor mouse model. *Cancer Cell*, *15*, 45–56. Retrieved from, http://www.ncbi.nlm.nih.gov/pubmed/19111880.

Anido, J., Sáez-Borderías, A., Gonzàlez-Juncà, A., Rodón, L., Folch, G., Carmona, M. A., et al. (2010). TGF-β receptor inhibitors target the CD44(high)/Id1(high) glioma-initiating cell population in human glioblastoma. *Cancer Cell, 18*(6), 655–668. http://dx.doi.org/10.1016/j.ccr.2010.10.023.

Atai, N. A., Renkema-Mills, N. A., Bosman, J., Schmidt, N., Rijkeboer, D., Tigchelaar, W., et al. (2011). Differential activity of NADPH-producing dehydrogenases renders rodents unsuitable models to study IDH1R132 mutation effects in human glioblastoma. *The Journal of Histochemistry and Cytochemistry: Official Journal of the Histochemistry Society, 59*, 489–503. Retrieved from, http://www.pubmedcentral.nih.gov/articlerender.fcgi?artid=3201175&tool=pmcentrez&rendertype=abstract.

Auvergne, R. M., Sim, F. J., Wang, S., Chandler-Militello, D., Burch, J., Al Fanek, Y., et al. (2013). Transcriptional differences between normal and glioma-derived glial progenitor cells identify a core set of dysregulated genes. *Cell Reports, 3*, 1–15. http://dx.doi.org/10.1016/j.celrep.2013.04.035.

Bachoo, R. M., Maher, E. A., Ligon, K. L., Sharpless, N. E., Chan, S. S., You, M. J., et al. (2002). Epidermal growth factor receptor and Ink4a/Arf: Convergent mechanisms governing terminal differentiation and transformation along the neural stem cell to astrocyte axis. *Cancer Cell, 1*, 269–277. Retrieved from, http://www.ncbi.nlm.nih.gov/pubmed/12086863.

Balss, J., Meyer, J., Mueller, W., Korshunov, A., Hartmann, C., & von Deimling, A. (2008). Analysis of the IDH1 codon 132 mutation in brain tumors. *Acta Neuropathologica, 116*(6), 597–602.

Bao, S., Wu, Q., Li, Z., Sathornsumetee, S., Wang, H., McLendon, R. E., et al. (2008). Targeting cancer stem cells through L1CAM suppresses glioma growth. *Cancer Research, 68*, 6043–6048. Retrieved from, http://www.pubmedcentral.nih.gov/articlerender.fcgi?artid=2739001&tool=pmcentrez&rendertype=abstract.

Bao, S., Wu, Q., McLendon, R. E., Hao, Y., Shi, Q., Hjelmeland, A. B., et al. (2006). Glioma stem cells promote radioresistance by preferential activation of the DNA damage response. *Nature, 444*, 756–760. Retrieved from, http://www.ncbi.nlm.nih.gov/pubmed/17051156.

Barkovich, K. J., Hariono, S., Garske, A. L., Zhang, J., Blair, J. A., Fan, Q.-W., et al. (2012). Kinetics of inhibitor cycling underlie therapeutic disparities between EGFR-driven lung and brain cancers. *Cancer Discovery, 2*(5), 450–457. http://dx.doi.org/10.1158/2159-8290.CD-11-0287.

Batchelor, T. T., Duda, D. G., Di Tomaso, E., Ancukiewicz, M., Plotkin, S. R., Gerstner, E., et al. (2010). Phase II study of cediranib, an oral pan-vascular endothelial growth factor receptor tyrosine kinase inhibitor, in patients with recurrent glioblastoma. *Journal of Clinical Oncology, 28*, 2817–2823. http://dx.doi.org/10.1200/JCO.2009.26.3988.

Baumann, P., Mandl-Weber, S., Oduncu, F., & Schmidmaier, R. (2009). The novel orally bioavailable inhibitor of phosphoinositol-3-kinase and mammalian target of rapamycin, NVP-BEZ235, inhibits growth and proliferation in multiple myeloma. *Experimental Cell Research, 315*(3), 485–497. http://dx.doi.org/10.1016/j.yexcr.2008.11.007.

Ben-Porath, I., Thomson, M. W., Carey, V. J., Ge, R., Bell, G. W., Regev, A., et al. (2008). An embryonic stem cell-like gene expression signature in poorly differentiated aggressive human tumors. *Nature Genetics, 40*, 499–507. Retrieved from, http://www.ncbi.nlm.nih.gov/pubmed/18443585.

Bettegowda, C., Agrawal, N., Jiao, Y., Sausen, M., Wood, L. D., & Hruban, R. H. (2011). Mutations in CIC and FUBP1 contribute to human oligodendroglioma. *Science, 333*(6048), 1453–1455.

Bhat, K. P. L., Balasubramaniyan, V., Vaillant, B., Ezhilarasan, R., Hummelink, K., Hollingsworth, F., et al. (2013). Mesenchymal differentiation mediated by NF-κB promotes radiation resistance in glioblastoma. *Cancer Cell, 24*, 331–346. http://dx.doi.org/10.1016/j.ccr.2013.08.001.

Bhat, K. P. L., Salazar, K. L., Balasubramaniyan, V., Wani, K., Heathcock, L., Hollingsworth, F., et al. (2011). The transcriptional coactivator TAZ regulates mesenchymal differentiation in malignant glioma. *Genes & Development, 25*, 2594–2609. http://dx.doi.org/10.1101/gad.176800.111.

Bielen, A., Perryman, L., Box, G. M., Valenti, M., de Haven Brandon, A., Martins, V., et al. (2011). Enhanced efficacy of IGF1R inhibition in pediatric glioblastoma by combinatorial targeting of PDGFRα/β. *Molecular Cancer Therapeutics, 10*(8), 1407–1418. http://dx.doi.org/10.1158/1535-7163.MCT-11-0205.

Bjerke, L., Mackay, A., Nandhabalan, M., Burford, A., Jury, A., Popov, S., et al. (2013). Histone H3.3 mutations drive pediatric glioblastoma through upregulation of MYCN. *Cancer Discovery, 3*, 512–519.

Bleeker, F. E., Lamba, S., Leenstra, S., Troost, D., Hulsebos, T., Vandertop, W. P., et al. (2009). IDH1mutations at residue p.R132 (IDH1 R132) occur frequently in high-grade gliomas but not in other solid tumors. *Human Mutation, 30*(1), 7–11.

Boult, J. K. R., Terkelsen, J., Walker-Samuel, S., Bradley, D. P., & Robinson, S. P. (2013). A multi-parametric imaging investigation of the response of C6 glioma xenografts to MLN0518 (tandutinib) treatment. *PLoS One, 8*(4), e63024. http://dx.doi.org/10.1371/journal.pone.0063024.

Bouvier-Labit, C., Liprandi, A., Monti, G., Pellissier, J. F., & Figarella-Branger, D. (2002). CD44H is expressed by cells of the oligodendrocyte lineage and by oligodendrogliomas in humans. *Journal of Neuro-Oncology, 60*, 127–134. Retrieved from, http://www.ncbi.nlm.nih.gov/entrez/query.fcgi?cmd=Retrieve&db=PubMed&dopt=Citation&list_uids=12635659.

Bralten, L. B. C., Kloosterhof, N. K., Balvers, R., Sacchetti, A., Lapre, L., Lamfers, M., et al. (2011). IDH1 R132H decreases proliferation of glioma cell lines in vitro and in vivo. *Annals of Neurology, 69*(3), 455–463.

Brennan, C., Momota, H., Hambardzumyan, D., Ozawa, T., Tandon, A., Pedraza, A., et al. (2009). Glioblastoma subclasses can be defined by activity among signal transduction pathways and associated genomic alterations. *PLoS One, 4*(11), e7752. http://dx.doi.org/10.1371/journal.pone.0007752, C. Creighton, Ed.

Brennan, C. W., Verhaak, R. G., McKenna, A., Campos, B., Noushermehr, H., & Salama, S. R. (2013). The somatic genomic landscape of glioblastoma. *Cell, 155*(2), 462–477.

Briceño, E., Calderon, A., & Sotelo, J. (2007). Institutional experience with chloroquine as an adjuvant to the therapy for glioblastoma multiforme. *Surgical Neurology, 67*(4), 388–391. http://dx.doi.org/10.1016/j.surneu.2006.08.080.

Broniscer, A., Baker, S. D., Wetmore, C., Pai Panandiker, A. S., Huang, J., Davidoff, A. M., et al. (2013). Phase I trial, pharmacokinetics, and pharmacodynamics of vandetanib and dasatinib in children with newly diagnosed diffuse intrinsic pontine glioma. *Clinical Cancer Research: An Official Journal of the American Association for Cancer Research, 19*(11), 3050–3058. http://dx.doi.org/10.1158/1078-0432.CCR-13-0306.

Cao, P., Maira, S.-M., García-Echeverría, C., & Hedley, D. W. (2009). Activity of a novel, dual PI3-kinase/mTor inhibitor NVP-BEZ235 against primary human pancreatic cancers grown as orthotopic xenografts. *British Journal of Cancer, 100*(8), 1267–1276. http://dx.doi.org/10.1038/sj.bjc.6604995.

Cao, R., Wang, L., Wang, H., Xia, L., Erdjument-Bromage, H., Tempst, P., et al. (2002). Role of histone H3 lysine 27 methylation in Polycomb-group silencing. *Science, 298*, 1039–1043. Retrieved from, http://www.ncbi.nlm.nih.gov/pubmed/12351676.

Carracedo, A., Ma, L., Teruya-Feldstein, J., Rojo, F., Salmena, L., Alimonti, A., et al. (2008). Inhibition of mTORC1 leads to MAPK pathway activation through a PI3K-dependent feedback loop in human cancer. *The Journal of Clinical Investigation, 118*(9), 3065–3074. http://dx.doi.org/10.1172/JCI34739.

Carro, M. S., Lim, W. K., Alvarez, M. J., Bollo, R. J., Zhao, X., Snyder, E. Y., et al. (2009). The transcriptional network for mesenchymal transformation of brain tumours. *Nature, 463*(7279), 318–325.

Chan, K.-M., Fang, D., Gan, H., Hashizume, R., Yu, C., Schroeder, M., et al. (2013). The histone H3.3K27M mutation in pediatric glioma reprograms H3K27 methylation and gene expression. *Genes & Development, 27*(9), 985–990.

Chapuis, N., Tamburini, J., Green, A. S., Vignon, C., Bardet, V., Neyret, A., et al. (2010). Dual inhibition of PI3K and mTORC1/2 signaling by NVP-BEZ235 as a new therapeutic strategy for acute myeloid leukemia. *Clinical Cancer Research: An Official Journal of the American Association for Cancer Research, 16*(22), 5424–5435. http://dx.doi.org/10.1158/1078-0432.CCR-10-1102.

Chaturvedi, A., Araujo Cruz, M. M., Jyotsana, N., Sharma, A., Yun, H., Görlich, K., et al. (2013). Mutant IDH1 promotes leukemogenesis in vivo and can be specifically targeted in human AML. *Blood, 122*, 2877–2887. http://dx.doi.org/10.1182/blood-2013-03-491571.

Chaumeil, M. M., Larson, P. E. Z., Yoshihara, H. A. I., Danforth, O. M., Vigneron, D. B., Nelson, S. J., et al. (2013). Non-invasive in vivo assessment of IDH1 mutational status in glioma. *Nature Communications, 4*, 2429.

Chen, H.-L., Chew, L.-J., Packer, R. J., & Gallo, V. (2013). Modulation of the Wnt/beta-catenin pathway in human oligodendroglioma cells by Sox17 regulates proliferation and differentiation. *Cancer Letters, 335*(2), 361–371.

Chen, J., Li, Y., Yu, T.-S., McKay, R. M., Burns, D. K., Kernie, S. G., et al. (2012). A restricted cell population propagates glioblastoma growth after chemotherapy. *Nature, 488*(7412), 522–526.

Cheng, C. K., Fan, Q.-W., & Weiss, W. A. (2009). PI3K signaling in glioma–animal models and therapeutic challenges. *Brain Pathology, 19*(1), 112–120. http://dx.doi.org/10.1111/j.1750-3639.2008.00233.x.

Chew, L.-J., Shen, W., Ming, X., Senatorov, V. V., Chen, H.-L., Cheng, Y., et al. (2011). SRY-box containing gene 17 regulates the Wnt/β-catenin signaling pathway in oligodendrocyte progenitor cells. *The Journal of Neuroscience: The Official Journal of the Society for Neuroscience, 31*(39), 13921–13935.

Chi, P., Allis, C. D., & Wang, G. G. (2010). Covalent histone modifications—Miswritten, misinterpreted and mis-erased in human cancers. *Nature Reviews. Cancer, 10*(7), 457–469.

Chiarini, F., Grimaldi, C., Ricci, F., Tazzari, P. L., Evangelisti, C., Ognibene, A., et al. (2010). Activity of the novel dual phosphatidylinositol 3-kinase/mammalian target of rapamycin inhibitor NVP-BEZ235 against T-cell acute lymphoblastic leukemia. *Cancer Research, 70*(20), 8097–8107. http://dx.doi.org/10.1158/0008-5472.CAN-10-1814.

Chow, L. M. L., Endersby, R., Zhu, X., Rankin, S., Qu, C., Zhang, J., et al. (2011). Cooperativity within and among Pten, p53, and Rb pathways induces high-grade astrocytoma in adult brain. *Cancer Cell, 19*, 305–316. Retrieved from, http://www.pubmedcentral.nih.gov/articlerender.fcgi?artid=3060664&tool=pmcentrez&rendertype=abstract.

Clement, V., Sanchez, P., De Tribolet, N., Radovanovic, I., Ruiz, I., & Altaba, A. (2007). HEDGEHOG-GLI1 signaling regulates human glioma growth, cancer stem cell self-renewal, and tumorigenicity. *Current Biology, 17*, 165–172. Retrieved from, http://www.ncbi.nlm.nih.gov/pubmed/17196391.

Combs, S. E., Heeger, S., Haselmann, R., Edler, L., Debus, J., & Schulz-Ertner, D. (2006). Treatment of primary glioblastoma multiforme with cetuximab, radiotherapy and temozolomide (GERT)—Phase I/II trial: Study protocol. *BMC Cancer, 6*, 133. http://dx.doi.org/10.1186/1471-2407-6-133.

Conway, G. D., O'Bara, M. A., Vedia, B. H., Pol, S. U., & Sim, F. J. (2012). Histone deacetylase activity is required for human oligodendrocyte progenitor differentiation. *Glia, 60*, 1944–1953. http://dx.doi.org/10.1002/glia.22410.

Cooper, L. A. D., Gutman, D. A., Chisolm, C., Appin, C., Kong, J., Rong, Y., et al. (2012). The tumor microenvironment strongly impacts master transcriptional regulators and gene expression class of glioblastoma. *American Journal of Pathology, 180*(5), 2108–2119.

Cooper, L. A. D., Gutman, D. A., Long, Q., Johnson, B. A., Cholleti, S. R., Kurc, T., et al. (2010). The proneural molecular signature is enriched in oligodendrogliomas and predicts improved survival among diffuse gliomas. *PloS One, 5*(9), e12548. http://dx.doi.org/10.1371/journal.pone.0012548.

Craene, B. De, & Berx, G. (2013). Regulatory networks defining EMT during cancer initiation and progression. *Nature Reviews Cancer, 13*(2), 97–110.

Creyghton, M. P., Cheng, A. W., Welstead, G. G., Kooistra, T., Carey, B. W., Steine, E. J., et al. (2010). Histone H3K27ac separates active from poised enhancers and predicts developmental state. *Proceedings of the National Academy of Sciences of the United States of America, 107*(50), 21931–21936.

Dahmane, N., Sánchez, P., Gitton, Y., Palma, V., Sun, T., Beyna, M., et al. (2001). The Sonic Hedgehog-Gli pathway regulates dorsal brain growth and tumorigenesis. *Development (Cambridge, England), 128*, 5201–5212. Retrieved from, http://www.ncbi.nlm.nih.gov/pubmed/11748155.

Dai, C., Celestino, J. C., Okada, Y., Louis, D. N., Fuller, G. N., & Holland, E. C. (2001). PDGF autocrine stimulation dedifferentiates cultured astrocytes and induces oligodendrogliomas and oligoastrocytomas from neural progenitors and astrocytes in vivo. *Genes & Development, 15*, 1913–1925. http://dx.doi.org/10.1101/gad.903001.

Dai, C., Lyustikman, Y., Shih, A., Hu, X., Fuller, G. N., Rosenblum, M., et al. (2005). The characteristics of astrocytomas and oligodendrogliomas are caused by two distinct and interchangeable signaling formats. *Neoplasia, 7*, 397–406. http://dx.doi.org/10.1593/neo.04691.

Dawson, M. R. L., Polito, A., Levine, J. M., & Reynolds, R. (2003). NG2-expressing glial progenitor cells: An abundant and widespread population of cycling cells in the adult rat CNS. *Molecular and Cellular Neurosciences, 24*, 476–488. Retrieved from, http://linkinghub.elsevier.com/retrieve/pii/S1044743103002100.

De Bacco, F., Casanova, E., Medico, E., Pellegatta, S., Orzan, F., Albano, R., et al. (2012). The MET oncogene is a functional marker of a glioblastoma stem cell subtype. *Cancer Research, 72*(17), 4537–4550.

De Groot, J. F., Prados, M., Urquhart, T., Robertson, S., Yaron, Y., Sorensen, A. G., et al. (2009). A phase II study of XL184 in patients (pts) with progressive glioblastoma multiforme (GBM) in first or second relapse. *Journal of Clinical Oncology, 27*(15s), 2047. Retrieved from, http://meeting.ascopubs.org/cgi/content/abstract/27/15S/204, ASCO Meeting Abstracts.

Deleyrolle, L. P., Harding, A., Cato, K., Siebzehnrubl, F. A., Rahman, M., Azari, H., et al. (2011). Evidence for label-retaining tumour-initiating cells in human glioblastoma. *Brain: A journal of neurology, 134*, 1331–1343. Retrieved from, http://www.pubmedcentral.nih.gov/articlerender.fcgi?artid=3097894&tool=pmcentrez&rendertype=abstract.

De Witt Hamer, P. (2010). Small molecule kinase inhibitors in glioblastoma: A systematic review of clinical studies. *Neuro-Oncology, 12*(3), 304–316.

Dick, J. E. (2008). Stem cell concepts renew cancer research. *Blood, 112*, 4793–4807. Retrieved from, http://www.ncbi.nlm.nih.gov/pubmed/19064739.

Diehn, M., Cho, R. W., Lobo, N. A., Kalisky, T., Dorie, M. J., Kulp, A. N., et al. (2009). Association of reactive oxygen species levels and radioresistance in cancer stem cells. *Nature, 458*(7239), 780–783.

Ding, H., Shannon, P., Lau, N., Wu, X., Roncari, L., Baldwin, R. L., et al. (2003). Oligodendrogliomas result from the expression of an activated mutant epidermal growth factor receptor in a RAS transgenic mouse astrocytoma model. *Cancer Research, 55*, 1106–1113. Retrieved from, http://www.ncbi.nlm.nih.gov/entrez/query.fcgi?cmd=Retrieve&db=PubMed&dopt=Citation&list_uids=12615729.

Doetsch, F. (2003). The glial identity of neural stem cells. *Nature Neuroscience, 6*, 1127–1134. Retrieved from, http://www.ncbi.nlm.nih.gov/pubmed/14583753.

Doherty, L., Gigas, D. C., Kesari, S., Drappatz, J., Kim, R., Zimmerman, J., et al. (2006). Pilot study of the combination of EGFR and mTOR inhibitors in recurrent malignant gliomas. *Neurology, 67*(1), 156–158. http://dx.doi.org/10.1212/01.wnl.0000223844.77636.29.

Dong, Y., Jia, L., Wang, X., Tan, X., Xu, J., Deng, Z., et al. (2011). Selective inhibition of PDGFR by imatinib elicits the sustained activation of ERK and downstream receptor signaling in malignant glioma cells. *International Journal of Oncology, 38*(2), 555–569. http://dx.doi.org/10.3892/ijo.2010.861.

Drappatz, J., Norden, A. D., Wong, E. T., Doherty, L. M., LaFrankie, D. C., Ciampa, A., et al. (2010). Phase I study of vandetanib with radiotherapy and temozolomide for newly diagnosed glioblastoma. *International Journal of Radiation Oncology, Biology, Physics, 78*(1), 85–90. Retrieved from, http://www.sciencedirect.com/science/article/pii/S0360301609029319.

Druker, B. J., Talpaz, M., Resta, D. J., Peng, B., Buchdunger, E., Ford, J. M., et al. (2001). Efficacy and safety of a specific inhibitor of the BCR-ABL tyrosine kinase in chronic myeloid leukemia. *The New England Journal of Medicine, 344*(14), 1031–1037.

Dunn, G. P., Rinne, M. L., Wykosky, J., Genovese, G., Quayle, S. N., Dunn, I. F., et al. (2012). Emerging insights into the molecular and cellular basis of glioblastoma. *Genes & Development, 26*(8), 756–784. http://dx.doi.org/10.1101/gad.187922.112.

Easwaran, H., Johnstone, S., Vanneste, L., Ohm, J., Mosbruger, T., Wang, Q., et al. (2012). A DNA hypermethylation module for the stem/progenitor cell signature of cancer. *Genome Research, 22*, 837–849. http://dx.doi.org/10.1101/gr.131169.111.

Eller, J. L., Longo, S. L., Kyle, M. M., & Bassano, D. (2005). Anti-epidermal growth factor receptor monoclonal antibody cetuximab augments radiation effects in glioblastoma multiforme in vitro and in vivo. *Neurosurgery, 56*(1), 155–162. http://dx.doi.org/10.1227/01.NEU.0000145865.25689.55.

Elson, A., Deng, C., Campos-Torres, J., Donehower, L. A., & Leder, P. (1995). The MMTV/c-myc transgene and p53 null alleles collaborate to induce T-cell lymphomas, but not mammary carcinomas in transgenic mice. *Oncogene, 11*(1), 181–190.

Engler, J. R., Robinson, A. E., Smirnov, I., Hodgson, J. G., Berger, M. S., Gupta, N., et al. (2012). Increased microglia/macrophage gene expression in a subset of adult and pediatric astrocytomas. *PloS One*, 7(8), e43339. http://dx.doi.org/10.1371/journal.pone.0043339.

Eriksson, P. S., Perfilieva, E., Bjork-Eriksson, T., Alborn, A. M., Nordborg, C., Peterson, D. A., et al. (1998). Neurogenesis in the adult human hippocampus. *Nature Medicine, 4*, 1313–1317. Retrieved from, http://www.ncbi.nlm.nih.gov/entrez/query.fcgi?cmd=Retrieve&db=PubMed&dopt=Citation&list_uids=9809557.

Fan, Q.-W., Cheng, C. K., Nicolaides, T. P., Hackett, C. S., Knight, Z. A., Shokat, K. M., et al. (2007). A dual phosphoinositide-3-kinase alpha/mTOR inhibitor cooperates with blockade of epidermal growth factor receptor in PTEN-mutant glioma. *Cancer Research, 67*(17), 7960–7965. http://dx.doi.org/10.1158/0008-5472.CAN-07-2154.

Fan, Q.-W., Knight, Z. A., Goldenberg, D. D., Yu, W., Mostov, K. E., Stokoe, D., et al. (2006). A dual PI3 kinase/mTOR inhibitor reveals emergent efficacy in glioma. *Cancer cell, 9*(5), 341–349. http://dx.doi.org/10.1016/j.ccr.2006.03.029.

Fan, Q.-W., & Weiss, W. A. (2010). Targeting the RTK-PI3K-mTOR axis in malignant glioma: Overcoming resistance. *Current Topics in Microbiology and Immunology*, *347*, 279–296. http://dx.doi.org/10.1007/82_2010_67.

Fancy, S. P. J., Harrington, E. P., Yuen, T. J., Silbereis, J. C., Zhao, C., Baranzini, S. E., et al. (2011). Axin2 as regulatory and therapeutic target in newborn brain injury and remyelination. *Nature Neuroscience*, *14*, 1009–1016. http://dx.doi.org/10.1038/nn.2855.

Fathi, A. T., & Abdel-Wahab, O. (2012). Mutations in epigenetic modifiers in myeloid malignancies and the prospect of novel epigenetic-targeted therapy. *Advances in Hematology*, *2012*(12), 1–12.

Ferent, J., Zimmer, C., Durbec, P., Ruat, M., & Traiffort, E. (2013). Sonic Hedgehog signaling is a positive oligodendrocyte regulator during demyelination. *The Journal of Neuroscience: The Official Journal of the Society for Neuroscience*, *33*, 1759–1772. http://dx.doi.org/10.1523/JNEUROSCI.3334-12.2013.

Figueroa, M. E., Abdel-Wahab, O., Lu, C., Ward, P. S., Patel, J., Shih, A., et al. (2010). Leukemic IDH1 and IDH2 mutations result in a hypermethylation phenotype, disrupt TET2 function, and impair hematopoietic differentiation. *Cancer Cell*, *18*(6), 553–567. http://dx.doi.org/10.1016/j.ccr.2010.11.015.

Finger, E. C., & Giaccia, A. J. (2010). Hypoxia, inflammation, and the tumor microenvironment in metastatic disease. *Cancer and Metastasis Reviews*, *29*(2), 285–293.

Flavahan, W. A., Wu, Q., Hitomi, M., Rahim, N., Kim, Y., Sloan, A. E., et al. (2013). Brain tumor initiating cells adapt to restricted nutrition through preferential glucose uptake. *Nature Neuroscience*, *16*, 1373–1382.

Flores, A. I., Mallon, B. S., Matsui, T., Ogawa, W., Rosenzweig, A., Okamoto, T., et al. (2000). Akt-mediated survival of oligodendrocytes induced by neuregulins. *The Journal of Neuroscience: The Official Journal of the Society for Neuroscience*, *20*(20), 7622–7630. Retrieved from, http://www.ncbi.nlm.nih.gov/pubmed/11027222.

Forsberg-Nilsson, K., Behar, T. N., Afrakhte, M., Barker, J. L., & McKay, R. D. (1998). Platelet-derived growth factor induces chemotaxis of neuroepithelial stem cells. *Journal of Neuroscience Research*, *53*, 521–530. Retrieved from, http://www.ncbi.nlm.nih.gov/pubmed/9726423.

Friedmann-Morvinski, D., Bushong, E. A., Ke, E., Soda, Y., Marumoto, T., Singer, O., et al. (2012). Dedifferentiation of neurons and astrocytes by oncogenes can induce gliomas in mice. *Science*, *338*, 1080–1084. http://dx.doi.org/10.1126/science.1226929.

Gajadhar, A. S., Bogdanovic, E., Muñoz, D. M., & Guha, A. (2012). In situ analysis of mutant EGFRs prevalent in glioblastoma multiforme reveals aberrant dimerization, activation, and differential response to anti-EGFR targeted therapy. *Molecular Cancer Research: MCR*, *10*(3), 428–440. http://dx.doi.org/10.1158/1541-7786.MCR-11-0531.

Gan, H. K., Kaye, A. H., & Luwor, R. B. (2009). The EGFRvIII variant in glioblastoma multiforme. *Journal of Clinical Neuroscience: Official Journal of the Neurosurgical Society of Australasia*, *16*(6), 748–754. http://dx.doi.org/10.1016/j.jocn.2008.12.005.

Garner, J. M., Fan, M., Yang, C. H., Du, Z., Sims, M., Davidoff, A. M., et al. (2013). Constitutive activation of signal transducer and activator of transcription 3 (STAT3) and nuclear factor B signaling in glioblastoma cancer stem cells regulates the notch pathway. *Journal of Biological Chemistry*, *288*(36), 26167–26176.

Gavalas, A., Trainor, P., Ariza-McNaughton, L., & Krumlauf, R. (2001). Synergy between Hoxa1 and Hoxb1: The relationship between arch patterning and the generation of cranial neural crest. *Development (Cambridge, England)*, *128*(15), 3017–3027. Retrieved from, http://www.ncbi.nlm.nih.gov/pubmed/11532923.

Gerstner, E. R., Chen, P.-J., Wen, P. Y., Jain, R. K., Batchelor, T. T., & Sorensen, G. (2010). Infiltrative patterns of glioblastoma spread detected via diffusion MRI after treatment with cediranib. *Neuro-Oncology*, *12*(5), 466–472. http://dx.doi.org/10.1093/neuonc/nop051.

Gerstner, E. R., Eichler, A. F., Plotkin, S. R., Drappatz, J., Doyle, C. L., Xu, L., et al. (2011). Phase I trial with biomarker studies of vatalanib (PTK787) in patients with newly diagnosed glioblastoma treated with enzyme inducing anti-epileptic drugs and standard radiation and temozolomide. *Journal of Neuro-Oncology*, *103*(2), 325–332. http://dx.doi.org/10.1007/s11060-010-0390-7.

Ghildiyal, R., Dixit, D., & Sen, E. (2013). EGFR inhibitor BIBU induces apoptosis and defective autophagy in glioma cells. *Molecular Carcinogenesis*, *52*, 970–982. http://dx.doi.org/10.1002/mc.21938.

Godard, S., Getz, G., Delorenzi, M., Farmer, P., Kobayashi, H., Desbaillets, I., et al. (2003). Classification of human astrocytic gliomas on the basis of gene expression: A correlated group of genes with angiogenic activity emerges as a strong predictor of subtypes. *Cancer Research*, *63*(20), 6613–6625.

Gravendeel, L. A. M., Kloosterhof, N. K., Bralten, L. B. C., van Marion, R., Dubbink, H. J., Dinjens, W., et al. (2010). Segregation of non-p.R132H mutations in IDH1 in distinct molecular subtypes of glioma. *Human Mutation*, *31*(3), E1186–E1199.

Gu, F., Hata, R., Ma, Y.-J., Tanaka, J., Mitsuda, N., Kumon, Y., et al. (2005). Suppression of Stat3 promotes neurogenesis in cultured neural stem cells. *Journal of Neuroscience Research*, *81*(2), 163–171.

Gutman, D. A., Cooper, L. A. D., Hwang, S. N., Holder, C. A., Gao, J., Aurora, T. D., et al. (2013). MR imaging predictors of molecular profile and survival: Multi-institutional study of the TCGA glioblastoma data set. *Radiology*, *267*(2), 560–569.

Haas-Kogan, D. A., Prados, M. D., Tihan, T., Eberhard, D. A., Jelluma, N., Arvold, N. D., et al. (2005). Epidermal growth factor receptor, protein kinase B/Akt, and glioma response to erlotinib. *Journal of the National Cancer Institute*, *97*(12), 880–887. http://dx.doi.org/10.1093/jnci/dji161.

Hall, A., Giese, N. A., & Richardson, W. D. (1996). Spinal cord oligodendrocytes develop from ventrally derived progenitor cells that express PDGF alpha-receptors. *Development (Cambridge, England)*, *122*, 4085–4094. Retrieved from, http://discovery.ucl.ac.uk/109515/.

Han, Y., Caday, C. G., Umezawa, K., & Nanda, A. (1997). Preferential inhibition of glioblastoma cells with wild-type epidermal growth factor receptors by a novel tyrosine kinase inhibitor ethyl-2,5-dihydroxycinnamate. *Oncology Research*, *9*, 581–587.

Hartmann, C., Meyer, J., Balss, J., Capper, D., Mueller, W., Christians, A., et al. (2009). Type and frequency of IDH1 and IDH2 mutations are related to astrocytic and oligodendroglial differentiation and age: A study of 1,010 diffuse gliomas. *Acta Neuropathologica*, *118*(4), 469–474.

He, J., Liu, Y., Zhu, T., Zhu, J., Dimeco, F., Vescovi, A. L., et al. (2011). CD90 is identified as a marker for cancer stem cells in primary high-grade gliomas using tissue microarrays. *Molecular Cellular Proteomics: MCP*, *11*, 1–24. http://dx.doi.org/10.1074/mcp.M111.010744.

Hébert, J. M., & Fishell, G. (2008). The genetics of early telencephalon patterning: Some assembly required. *Nature Reviews Neuroscience*, *9*, 678–685. Retrieved from, http://www.pubmedcentral.nih.gov/articlerender.fcgi?artid=2669317&tool=pmcentrez&rendertype=abstract.

Hede, S.-M., Hansson, I., Afink, G. B., Eriksson, A., Nazarenko, I., Andrae, J., et al. (2009). GFAP promoter driven transgenic expression of PDGFB in the mouse brain leads to glioblastoma in a Trp53 null background. *Glia*, *57*, 1143–1153. Retrieved from, http://www.ncbi.nlm.nih.gov/pubmed/19115382.

Hill, R. A., Patel, K. D., Medved, J., Reiss, A. M., & Nishiyama, A. (2013). NG2 cells in white matter but not gray matter proliferate in response to PDGF. *The Journal of Neuroscience: The Official Journal of the Society for Neuroscience*, *33*(36), 14558–14566. http://dx.doi.org/10.1523/JNEUROSCI.2001-12.2013.

Hochstim, C., Deneen, B., Lukaszewicz, A., Zhou, Q., & Anderson, D. J. (2008). Identification of positionally distinct astrocyte subtypes whose identities are specified by a homeodomain code. *Cell*, *133*, 510–522. Retrieved from, http://www.ncbi.nlm.nih.gov/pubmed/18455991.
Hodgson, J. G., Yeh, R.-F., Ray, A., Wang, N. J., Smirnov, I., Yu, M., et al. (2009). Comparative analyses of gene copy number and mRNA expression in glioblastoma multiforme tumors and xenografts. *Neuro-Oncology*, *11*(5), 477–487.
Holland, E. C., Celestino, J., Dai, C., Schaefer, L., Sawaya, R. E., & Fuller, G. N. (2000). Combined activation of Ras and Akt in neural progenitors induces glioblastoma formation in mice. *Nature Genetics*, *25*, 55–57. Retrieved from, http://www.ncbi.nlm.nih.gov/pubmed/10802656.
Holland, Eric C., Hively, W. P., DePinho, R. A., & Varmus, H. E. (1998). A constitutively active epidermal growth factor receptor cooperates with disruption of G1 cell-cycle arrest pathways to induce glioma-like lesions in mice. *Genes & Development*, *12*, 3675–3685. Retrieved from, http://www.genesdev.org/cgi/doi/10.1101/gad.12.23.3675.
Homma, J., Yamanaka, R., Yajima, N., Tsuchiya, N., Genkai, N., Sano, M., et al. (2006). Increased expression of CCAAT/enhancer binding protein β correlates with prognosis in glioma patients. *Oncology Reports*, *15*(3), 595–601.
Hong, J.-H., & Yaffe, M. B. (2006). TAZ: A beta-catenin-like molecule that regulates mesenchymal stem cell differentiation. *Cell Cycle*, *5*(2), 176–179.
Houillier, C., Wang, X., Kaloshi, G., Mokhtari, K., Guillevin, R., Laffaire, J., et al. (2010). IDH1 or IDH2 mutations predict longer survival and response to temozolomide in low-grade gliomas. *Neurology*, *75*(17), 1560–1566.
Hu, X., Pandolfi, P. P., Li, Y., Koutcher, J. A., Rosenblum, M., & Holland, E. C. (2005). mTOR promotes survival and astrocytic characteristics induced by Pten/Akt signaling in glioblastoma1. *Neoplasia*, *7*, 356–368. Retrieved from, http://www.ingentaselect.com/rpsv/cgi-bin/cgi?ini=xref&body=linker&reqdoi=10.1593/neo.04595.
Hu, X.-L., Wang, Y., & Shen, Q. (2012). Epigenetic control on cell fate choice in neural stem cells. *Protein & Cell*, *3*(4), 278–290.
Hughes, E. G., Kang, S. H., Fukaya, M., & Bergles, D. E. (2013). Oligodendrocyte progenitors balance growth with self-repulsion to achieve homeostasis in the adult brain. *Nature Neuroscience*, *16*, 668–676. Retrieved from, http://www.nature.com/doifinder/10.1038/nn.3390.
Huse, J. T., & Holland, E. C. (2010). Targeting brain cancer: Advances in the molecular pathology of malignant glioma and medulloblastoma. *Nature Reviews Cancer*, *10*, 319–331. Retrieved from, http://www.ncbi.nlm.nih.gov/pubmed/20414201.
Ichimura, K., Pearson, D. M., Kocialkowski, S., Bäcklund, L. M., Chan, R., Jones, D. T. W., et al. (2009). IDH1 mutations are present in the majority of common adult gliomas but rare in primary glioblastomas. *Neuro-Oncology*, *11*(4), 341–347.
Ishii, A., Furusho, M., & Bansal, R. (2013). Sustained activation of ERK1/2 MAPK in oligodendrocytes and schwann cells enhances myelin growth and stimulates oligodendrocyte progenitor expansion. *The Journal of Neuroscience: The Official Journal of the Society for Neuroscience*, *33*(1), 175–186. http://dx.doi.org/10.1523/JNEUROSCI.4403-12.2013.
Ivkovic, S., Canoll, P., & Goldman, J. E. (2008). Constitutive EGFR signaling in oligodendrocyte progenitors leads to diffuse hyperplasia in postnatal white matter. *The Journal of Neuroscience: The Official Journal of the Society for Neuroscience*, *28*(4), 914–922. http://dx.doi.org/10.1523/JNEUROSCI.4327-07.2008.
Iwamoto, F. M. (2010). trial of pazopanib (GW786034), an oral multi-targeted angiogenesis inhibitor, for adults with recurrent glioblastoma (North American Brain Tumor Consortium Study 06). *Neuro-Oncology*, *12*(8), 855–861. Retrieved from, http://neuro-oncology.oxfordjournals.org/content/12/8/855.short.

Jacques, T. S., Swales, A., Brzozowski, M. J., Henriquez, N. V., Linehan, J. M., Mirzadeh, Z., et al. (2010). Combinations of genetic mutations in the adult neural stem cell compartment determine brain tumour phenotypes. *The European Molecular Biology Organization Journal*, *29*, 222–235. http://dx.doi.org/10.1038/emboj.2009.327.

Jensen, N. A., Pedersen, K. M., Lihme, F., Rask, L., Nielsen, J. V., Rasmussen, T. E., et al. (2003). Astroglial c-Myc overexpression predisposes mice to primary malignant gliomas. *The Journal of Biological Chemistry*, *278*, 8300–8308. Retrieved from, http://www.ncbi.nlm.nih.gov/pubmed/12501251.

Jin, G., Pirozzi, C. J., Chen, L. H., Lopez, G. Y., Duncan, C. G., Feng, J., et al. (2012). Mutant IDH1 is required for IDH1 mutated tumor cell growth. *Oncotarget*, *3*(8), 774–782.

Jo, M., Stolz, D. B., Esplen, J. E., Dorko, K., Michalopoulos, G. K., & Strom, S. C. (2000). Cross-talk between epidermal growth factor receptor and c-Met signal pathways in transformed cells. *The Journal of Biological Chemistry*, *275*(12), 8806–8811.

Jones, P. A., & Baylin, S. B. (2007). The epigenomics of cancer. *Cell*, *128*(4), 683–692. http://dx.doi.org/10.1016/j.cell.2007.01.029.

Joo, K. M., Jin, J., Kim, E., Ho Kim, K., Kim, Y., Gu Kang, B., et al. (2012). MET signaling regulates glioblastoma stem cells. *Cancer Research*, *72*, 3828–3838. http://dx.doi.org/10.1158/0008-5472.CAN-11-3760.

Joshi, A. D., Loilome, W., Siu, I.-M., Tyler, B., Gallia, G. L., & Riggins, G. J. (2012). Evaluation of tyrosine kinase inhibitor combinations for glioblastoma therapy. *PloS One*, 7(10), e44372. http://dx.doi.org/10.1371/journal.pone.0044372.

Kalani, M. Y. S., Cheshier, S. H., Cord, B. J., Bababeygy, S. R., Vogel, H., Weissman, I. L., et al. (2008). Wnt-mediated self-renewal of neural stem/progenitor cells. *Proceedings of the National Academy of Sciences of the United States of America*, *105*, 16970–16975. http://dx.doi.org/10.1073/pnas.0808616105.

Kalluri, R., & Weinberg, R. A. (2009). The basics of epithelial-mesenchymal transition. *The Journal of Clinical Investigation*, *119*(6), 1420–1428.

Kang, M. R., Kim, M. S., Oh, J. E., Kim, Y. R., Song, S. Y., Seo, S. I., et al. (2009). Mutational analysis of IDH1 codon 132 in glioblastomas and other common cancers. *International Journal of Cancer*, *125*(2), 353–355.

Kanzawa, T., Germano, I. M., Komata, T., Ito, H., Kondo, Y., & Kondo, S. (2004). Role of autophagy in temozolomide-induced cytotoxicity for malignant glioma cells. *Cell Death and Differentiation*, *11*(4), 448–457. http://dx.doi.org/10.1038/sj.cdd.4401359.

Katz, A. M., Amankulor, N. M., Pitter, K., Helmy, K., Squatrito, M., & Holland, E. C. (2012). Astrocyte-specific expression patterns associated with the PDGF-induced glioma microenvironment. *PloS One*, 7(2), e32453. http://dx.doi.org/10.1371/journal.pone.0032453.

Kawauchi, D., Robinson, G., Uziel, T., Gibson, P., Rehg, J., Gao, C., et al. (2012). A mouse model of the most aggressive subgroup of human medulloblastoma. *Cancer cell*, *21*(2), 168–180.

Kessaris, N., Fogarty, M., Iannarelli, P., Grist, M., Wegner, M., & Richardson, W. D. (2006). Competing waves of oligodendrocytes in the forebrain and postnatal elimination of an embryonic lineage. *Nature Neuroscience*, *9*, 173–179. http://dx.doi.org/10.1038/nn1620.

Keunen, O., Johansson, M., Oudin, A., Sanzey, M., Rahim, S. A. A., Fack, F., et al. (2011). Anti-VEGF treatment reduces blood supply and increases tumor cell invasion in glioblastoma. *Proceedings of the National Academy of Sciences of the United States of America*, *108*, 3749–3754. http://dx.doi.org/10.1073/pnas.1014480108.

Khuong-Quang, D.-A., Buczkowicz, P., Rakopoulos, P., Liu, X.-Y., Fontebasso, A. M., Bouffet, E., et al. (2012). K27M mutation in histone H3.3 defines clinically and biologically distinct subgroups of pediatric diffuse intrinsic pontine gliomas. *Acta Neuropathologica*, *124*(3), 439–447.

Kilpatrick, T. J., & Bartlett, P. F. (1995). Cloned multipotential precursors from the mouse cerebrum require FGF-2, whereas glial restricted precursors are stimulated with either FGF-2 or EGF. *Journal of Neuroscience*, *15*, 3653–3661. Retrieved from, http://www.ncbi.nlm.nih.gov/pubmed/7751935.

Kim, Y., Kim, E., Wu, Q., Guryanova, O., Hitomi, M., Lathia, J. D., et al. (2012). Platelet-derived growth factor receptors differentially inform intertumoral and intratumoral heterogeneity. *Genes & Development*, *26*, 1247–1262. http://dx.doi.org/10.1101/gad.193565.112.

Kioi, M., Vogel, H., Schultz, G., Hoffman, R. M., Harsh, G. R., & Brown, J. M. (2010). Inhibition of vasculogenesis, but not angiogenesis, prevents the recurrence of glioblastoma after irradiation in mice. *The Journal of Clinical Investigation*, *120*(3), 694–705.

Koirala, S., & Corfas, G. (2010). Identification of novel glial genes by single-cell transcriptional profiling of bergmann glial cells from mouse cerebellum. *PLoS ONE*, *5*, 15. http://dx.doi.org/10.1371/journal.pone.0009198.

Koochekpour, S., Jeffers, M., Rulong, S., Taylor, G., Klineberg, E., Hudson, E. A., et al. (1997). Met and hepatocyte growth factor/scatter factor expression in human gliomas. *Cancer Research*, *57*(23), 5391–5398.

Kraus, A. C., Ferber, I., Bachmann, S.-O., Specht, H., Wimmel, A., Gross, M. W., et al. (2002). In vitro chemo- and radio-resistance in small cell lung cancer correlates with cell adhesion and constitutive activation of AKT and MAP kinase pathways. *Oncogene*, *21*(57), 8683–8695.

Kreisl, T. N., Kim, L., Moore, K., Duic, P., Royce, C., Stroud, I., et al. (2009). Phase II trial of single-agent bevacizumab followed by bevacizumab plus irinotecan at tumor progression in recurrent glioblastoma. *Journal of Clinical Oncology: Official Journal of the American Society of Clinical Oncology*, *27*(5), 740–745. http://dx.doi.org/10.1200/JCO.2008.16.3055.

Kreisl, T. N., Lassman, A. B., Mischel, P. S., Rosen, N., Scher, H. I., Teruya-Feldstein, J., et al. (2009). A pilot study of everolimus and gefitinib in the treatment of recurrent glioblastoma (GBM). *Journal of Neuro-Oncology*, *92*(1), 99–105. http://dx.doi.org/10.1007/s11060-008-9741-z.

Kwon, C.-H., Zhao, D., Chen, J., Alcantara, S., Li, Y., Burns, D. K., et al. (2008). Pten haploinsufficiency accelerates formation of high-grade astrocytomas. *Cancer Research*, *68*, 3286–3294. Retrieved from, http://www.pubmedcentral.nih.gov/articlerender.fcgi?artid=2760841&tool=pmcentrez&rendertype=abstract.

Labussière, M., Idbaih, A., Wang, X.-W., Marie, Y., Boisselier, B., Falet, C., et al. (2010). All the 1p19q codeleted gliomas are mutated on IDH1 or IDH2. *Neurology*, *74*(23), 1886–1890.

Lai, A., Kharbanda, S., Pope, W. B., Tran, A., Solis, O. E., Peale, F., et al. (2011). Evidence for sequenced molecular evolution of IDH1 mutant glioblastoma from a distinct cell of origin. *Journal of Clinical Oncology: Official Journal of the American Society of Clinical Oncology*, *29*(34), 4482–4490. http://dx.doi.org/10.1200/JCO.2010.33.8715.

Langseth, A. J., Munji, R. N., Choe, Y., Huynh, T., Pozniak, C. D., & Pleasure, S. J. (2010). Wnts influence the timing and efficiency of oligodendrocyte precursor cell generation in the telencephalon. *Journal of Neuroscience*, *30*, 13367–13372. Retrieved from, http://www.jneurosci.org/cgi/doi/10.1523/JNEUROSCI.1934-10.2010.

Lasiene, J., Matsui, A., Sawa, Y., Wong, F., & Horner, P. J. (2009). Age-related myelin dynamics revealed by increased oligodendrogenesis and short internodes. *Aging Cell*, *8*, 201–213. Retrieved from, http://www.ncbi.nlm.nih.gov/pmc/articles/PMC2703583/?tool=pubmed.

Lathia, Justin D., Gallagher, J., Heddleston, J. M., Wang, J., Eyler, C. E., Macswords, J., et al. (2010). Integrin alpha 6 regulates glioblastoma stem cells. *Cell Stem Cell*, *6*, 421–432.

Retrieved from, http://www.pubmedcentral.nih.gov/articlerender.fcgi?artid=2884275&tool=pmcentrez&rendertype=abstract.

Lathia, J. D., Hitomi, M., Gallagher, J., Gadani, S. P., Adkins, J., Vasanji, A., et al. (2011). Distribution of CD133 reveals glioma stem cells self-renew through symmetric and asymmetric cell divisions. *Cell Death Disease, 2*, e200. http://dx.doi.org/10.1038/cddis.2011.80.

Lee, T. I., Jenner, R. G., Boyer, L. A., Guenther, M. G., Levine, S. S., Kumar, R. M., et al. (2006). Control of developmental regulators by Polycomb in human embryonic stem cells. *Cell, 125*, 301–313. Retrieved from, http://discovery.ucl.ac.uk/7048/.

Lee, E. Q., Kuhn, J., Lamborn, K. R., Abrey, L., DeAngelis, L. M., Lieberman, F., et al. (2012). Phase I/II study of sorafenib in combination with temsirolimus for recurrent glioblastoma or gliosarcoma: North American Brain Tumor Consortium study 05-02. *Neuro-Oncology, 14*(12), 1511–1518. http://dx.doi.org/10.1093/neuonc/nos264.

Lee, J. C., Vivanco, I., Beroukhim, R., Huang, J. H. Y., Feng, W. L., DeBiasi, R. M., et al. (2006). Epidermal growth factor receptor activation in glioblastoma through novel missense mutations in the extracellular domain. *PLoS Medicine, 3*(12), e485, A. Lassman, Ed.

Lei, L., Sonabend, A. M., Guarnieri, P., Soderquist, C., Ludwig, T., Rosenfeld, S., et al. (2011). Glioblastoma models reveal the connection between adult glial progenitors and the proneural phenotype. *PLoS ONE, 6*, e20041. Retrieved from, http://www.pubmedcentral.nih.gov/articlerender.fcgi?artid=3100315&tool=pmcentrez&rendertype=abstract.

Lei, Q.-Y., Zhang, H., Zhao, B., Zha, Z.-Y., Bai, F., Pei, X.-H., et al. (2008). TAZ promotes cell proliferation and epithelial-mesenchymal transition and is inhibited by the hippo pathway. *Molecular and Cellular Biology, 28*(7), 2426–2436.

Lelievre, V., Ghiani, C. A., Seksenyan, A., Gressens, P., De Vellis, J., & Waschek, J. A. (2006). Growth factor-dependent actions of PACAP on oligodendrocyte progenitor proliferation. *Regulatory Peptides, 137*, 58–66. Retrieved from, http://www.ncbi.nlm.nih.gov/entrez/query.fcgi?cmd=Retrieve&db=PubMed&dopt=Citation&list_uids=16989910.

Lewis, P. W., Müller, M. M., Koletsky, M. S., Cordero, F., Lin, S., Banaszynski, L. A., et al. (2013). Inhibition of PRC2 activity by a gain-of-function H3 mutation found in pediatric glioblastoma. *Science, 340*(6134), 857–861.

Li, X., Huang, Y., Jiang, J., & Frank, S. J. (2008). ERK-dependent threonine phosphorylation of EGF receptor modulates receptor downregulation and signaling. *Cellular Signalling, 20*, 2145–2155. Retrieved from, http://www.pubmedcentral.nih.gov/articlerender.fcgi?artid=2613789&tool=pmcentrez&rendertype=abstract.

Li, Y., Li, A., Glas, M., Lal, B., Ying, M., Sang, Y., et al. (2011). c-Met signaling induces a reprogramming network and supports the glioblastoma stem-like phenotype. *Proceedings of the National Academy of Sciences of the United States of America, 108*, 9951–9956. Retrieved from, http://www.pubmedcentral.nih.gov/articlerender.fcgi?artid=3116406&tool=pmcentrez&rendertype=abstract.

Ligon, K. L., Kesari, S., Kitada, M., Sun, T., Arnett, H. A., Alberta, J. A., et al. (2006). Development of NG2 neural progenitor cells requires Olig gene function. *Proceedings of the National Academy of Sciences of the United States of America, 103*, 7853–7858. http://dx.doi.org/10.1073/pnas.0511001103.

Lin, C.-J., Lee, C.-C., Shih, Y.-L., Lin, C.-H., Wang, S.-H., Chen, T.-H., et al. (2012). Inhibition of mitochondria- and endoplasmic reticulum stress-mediated autophagy augments temozolomide-induced apoptosis in glioma cells. *PloS One*, 7(6), e38706. http://dx.doi.org/10.1371/journal.pone.0038706.

Lindberg, N., Kastemar, M., Olofsson, T., Smits, A., & Uhrbom, L. (2009). Oligodendrocyte progenitor cells can act as cell of origin for experimental glioma. *Oncogene, 28*, 2266–2275. Retrieved from, http://www.ncbi.nlm.nih.gov/pubmed/19421151.

Liu, W., Fu, Y., Xu, S., Ding, F., Zhao, G., Zhang, K., et al. (2011). c-Met expression is associated with time to recurrence in patients with glioblastoma multiforme. *Journal of Clinical Neuroscience: Official Journal of the Neurosurgical Society of Australasia, 18*(1), 119–121. http://dx.doi.org/10.1016/j.jocn.2010.05.010.

Liu, F.-J., Gui, S.-B., Li, C.-Z., Sun, Z.-L., Zhang, Y.-Z., & Academy, C. (2008). Antitumor activity of F90, an epidermal growth factor receptor tyrosine kinase inhibitor, on glioblastoma cell line SHG-44. *Chinese Medical Journal, 121*(17), 1702–1706.

Liu, T.-J., Koul, D., LaFortune, T., Tiao, N., Shen, R. J., Maira, S.-M., et al. (2009). NVP-BEZ235, a novel dual phosphatidylinositol 3-kinase/mammalian target of rapamycin inhibitor, elicits multifaceted antitumor activities in human gliomas. *Molecular Cancer Therapeutics, 8*(8), 2204–2210. http://dx.doi.org/10.1158/1535-7163.MCT-09-0160.

Liu, C., Sage, J. C., Miller, M. R., Verhaak, R. G. W., Hippenmeyer, S., Vogel, H., et al. (2011). Mosaic analysis with double markers reveals tumor cell of origin in glioma. *Cell, 146*(2), 209–221. http://dx.doi.org/10.1016/j.cell.2011.06.014.

Liu, G., Yuan, X., Zeng, Z., Tunici, P., Ng, H., Abdulkadir, I. R., et al. (2006). Analysis of gene expression and chemoresistance of CD133+ cancer stem cells in glioblastoma. *Molecular Cancer, 5*, 67. http://dx.doi.org/10.1186/1476-4598-5-67.

Liu-Chittenden, Y., Huang, B., Shim, J. S., Chen, Q., Lee, S.-J., Anders, R. A., et al. (2012). Genetic and pharmacological disruption of the TEAD-YAP complex suppresses the oncogenic activity of YAP. *Genes & Development, 26*(12), 1300–1305.

Lo, H.-W. (2010). EGFR-targeted therapy in malignant glioma: Novel aspects and mechanisms of drug resistance. *Current Molecular Pharmacology, 3*(1), 37–52.

Louis, D. N., Ohgaki, H., Wiestler, O. D., Cavenee, W. K., Burger, P. C., Jouvet, A., et al. (2007). The 2007 WHO classification of tumours of the central nervous system. *Acta Neuropathologica, 114*(2), 97–109. http://dx.doi.org/10.1007/s00401-007-0243-4.

Lu, K. V., Chang, J. P., Parachoniak, C. A., Pandika, M. M., Aghi, M. K., Meyronet, D., et al. (2012). VEGF inhibits tumor cell invasion and mesenchymal transition through a MET/VEGFR2 complex. *Cancer Cell, 22*(1), 21–35. http://dx.doi.org/10.1016/j.ccr.2012.05.037.

Lu, C., Ward, P. S., Kapoor, G. S., Rohle, D., Turcan, S., Abdel-Wahab, O., et al. (2013). IDH mutation impairs histone demethylation and results in a block to cell differentiation. *Nature, 483*(7390), 474–478.

Luchman, H. A., Stechishin, O. D., Dang, N. H., Blough, M. D., Chesnelong, C., Kelly, J. J., et al. (2012). An in vivo patient-derived model of endogenous IDH1-mutant glioma. *Neuro-Oncology, 14*(2), 184–191.

Lu-Emerson, C., Norden, A. D., Drappatz, J., Quant, E. C., Beroukhim, R., Ciampa, A. S., et al. (2011). Retrospective study of dasatinib for recurrent glioblastoma after bevacizumab failure. *Journal of Neuro-Oncology, 104*(1), 287–291. http://dx.doi.org/10.1007/s11060-010-0489-x.

Lumsden, A., & Krumlauf, R. (1996). Patterning the vertebrate neuraxis. *Science, 274*(5290), 1109–1115. Retrieved from, http://www.ncbi.nlm.nih.gov/pubmed/8895453.

Machold, R., Hayashi, S., Rutlin, M., Muzumdar, M. D., Nery, S., Corbin, J. G., et al. (2003). Sonic hedgehog is required for progenitor cell telencephalic stem cell niches. *Neuron, 39*, 937–950.

Maira, S.-M., Stauffer, F., Brueggen, J., Furet, P., Schnell, C., Fritsch, C., et al. (2008). Identification and characterization of NVP-BEZ235, a new orally available dual phosphatidylinositol 3-kinase/mammalian target of rapamycin inhibitor with potent in vivo antitumor activity. *Molecular Cancer Therapeutics*, 7(7), 1851–1863. http://dx.doi.org/10.1158/1535-7163.MCT-08-0017.

Mallini, P., Lennard, T., Kirby, J., & Meeson, A. (2014). Epithelial-to-mesenchymal transition: What is the impact on breast cancer stem cells and drug resistance. *Cancer Treatment Reviews, 40*, 341–348.

Mani, S. A., Guo, W., Liao, M.-J., Eaton, E. N., Ayyanan, A., Zhou, A. Y., et al. (2008). The epithelial-mesenchymal transition generates cells with properties of stem cells. *Cell, 133*(4), 704–715.

Mao, P., Joshi, K., Li, J., Kim, S.-H., Li, P., Santana-Santos, L., et al. (2013). Mesenchymal glioma stem cells are maintained by activated glycolytic metabolism involving aldehyde dehydrogenase 1A3. *Proceedings of the National Academy of Sciences of the United States of America, 110*(21), 8644–8649. http://dx.doi.org/10.1073/pnas.1221478110.

Marumoto, T., Tashiro, A., Friedmann-Morvinski, D., Scadeng, M., Soda, Y., Gage, F. H., et al. (2009). Development of a novel mouse glioma model using lentiviral vectors. *Nature Medicine, 15*, 110–116. Retrieved from, http://www.ncbi.nlm.nih.gov/pubmed/19122659.

Mazzoleni, S., Politi, L. S., Pala, M., Cominelli, M., Franzin, A., Sergi Sergi, L., et al. (2010). Epidermal growth factor receptor expression identifies functionally and molecularly distinct tumor-initiating cells in human glioblastoma multiforme and is required for gliomagenesis. *Cancer Research, 70*(19), 7500–7513. http://dx.doi.org/10.1158/0008-5472.CAN-10-2353.

Mehta, S., Huillard, E., Kesari, S., Maire, C. L., Golebiowski, D., Harrington, E. P., et al. (2011). The central nervous system-restricted transcription factor Olig2 opposes p53 responses to genotoxic damage in neural progenitors and malignant glioma. *Cancer Cell, 19*(3), 359–371.

Mellinghoff, I. K., Wang, M. Y., Vivanco, I., Haas-Kogan, D. A., Zhu, S., Dia, E. Q., et al. (2005). Molecular determinants of the response of glioblastomas to EGFR kinase inhibitors. *The New England Journal of Medicine, 353*(19), 2012–2024. http://dx.doi.org/10.1056/NEJMoa051918.

Ménard, C., Hein, P., Paquin, A., Savelson, A., Yang, X. M., Lederfein, D., et al. (2002). An essential role for a MEK-C/EBP pathway during growth factor-regulated cortical neurogenesis. *Neuron, 36*(4), 597–610.

Merkle, F. T., Mirzadeh, Z., & Alvarez-Buylla, A. (2007). Mosaic organization of neural stem cells in the adult brain. *Science, 317*, 381–384. Retrieved from, http://www.ncbi.nlm.nih.gov/pubmed/17615304.

Mikheeva, S. A., Mikheev, A. M., Petit, A., Beyer, R., Oxford, R. G., Khorasani, L., et al. (2010). TWIST1 promotes invasion through mesenchymal change in human glioblastoma. *Molecular Cancer, 9*, 194.

Mineo, J.-F., Bordron, A., Baroncini, M., Maurage, C.-A., Ramirez, C., Siminski, R.-M., et al. (2007). Low HER2-expressing glioblastomas are more often secondary to anaplastic transformation of low-grade glioma. *Journal of Neuro-Oncology, 85*(3), 281–287. http://dx.doi.org/10.1007/s11060-007-9424-1.

Mitsui, K., Tokuzawa, Y., Itoh, H., Segawa, K., Murakami, M., Takahashi, K., et al. (2003). The homeoprotein Nanog is required for maintenance of pluripotency in mouse epiblast and ES cells. *Cell, 113*, 631–642. Retrieved from, http://www.ncbi.nlm.nih.gov/pubmed/12787504.

Monje, M., Mitra, S. S., Freret, M. E., Raveh, T. B., Kim, J., Masek, M., et al. (2011). Hedgehog-responsive candidate cell of origin for diffuse intrinsic pontine glioma. *Proceedings of the National Academy of Sciences of the United States of America, 108*(11), 4453–4458.

Moon, J.-H., Kwon, S., Jun, E. K., Kim, A., Whang, K. Y., Kim, H., et al. (2011). Nanog-induced dedifferentiation of p53-deficient mouse astrocytes into brain cancer stem-like cells. *Biochemical and Biophysical Research Communications, 412*, 175–181. http://dx.doi.org/10.1016/j.bbrc.2011.07.070.

Moransard, M., Sawitzky, M., Fontana, A., & Suter, T. (2010). Expression of the HGF receptor c-met by macrophages in experimental autoimmune encephalomyelitis. *Glia, 58*, 559–571. Retrieved from, http://www.ncbi.nlm.nih.gov/pubmed/19941340.

Morris, P. G., & Abrey, L. E. (2010). Novel targeted agents for platelet-derived growth factor receptor and c-KIT in malignant gliomas. *Targeted Oncology*, *5*(3), 193–200. http://dx.doi.org/10.1007/s11523-010-0160-7.

Muhic, A., Poulsen, H. S., Sorensen, M., Grunnet, K., & Lassen, U. (2013). Phase II open-label study of nintedanib in patients with recurrent glioblastoma multiforme. *Journal of Neuro-Oncology*, *111*(2), 205–212. http://dx.doi.org/10.1007/s11060-012-1009-y.

Nagane, M., Narita, Y., Mishima, K., Levitzki, A., Burgess, A. W., Cavenee, W. K., et al. (2001). Human glioblastoma xenografts overexpressing a tumor-specific mutant epidermal growth factor receptor sensitized to cisplatin by the AG1478 tyrosine kinase inhibitor. *Journal of Neurosurgery*, *95*(3), 472–479. http://dx.doi.org/10.3171/jns.2001.95.3.0472.

Nakashima, K. (1999). Synergistic signaling in fetal brain by STAT3-Smad1 complex bridged by p300. *Science*, *284*(5413), 479–482.

Navis, A. C., Bourgonje, A., Wesseling, P., Wright, A., Hendriks, W., Verrijp, K., et al. (2013). Effects of dual targeting of tumor cells and stroma in human glioblastoma xenografts with a tyrosine kinase inhibitor against c-MET and VEGFR2. *PloS One*, *8*(3), e58262. http://dx.doi.org/10.1371/journal.pone.0058262.

Nazarenko, I., Hedrén, A., Sjödin, H., Orrego, A., Andrae, J., Afink, G. B., et al. (2011). Brain abnormalities and glioma-like lesions in mice overexpressing the long isoform of PDGF-A in astrocytic cells. *PLoS ONE*, *6*, 14. http://dx.doi.org/10.1371/journal.pone.0018303.

Neyns, B., Sadones, J., Chaskis, C., Dujardin, M., Everaert, H., Lv, S., et al. (2011). Phase II study of sunitinib malate in patients with recurrent high-grade glioma. *Journal of Neuro-Oncology*, *103*(3), 491–501. http://dx.doi.org/10.1007/s11060-010-0402-7.

Nghiemphu, P. L., Lai, A., Green, R. M., Reardon, D. A., & Cloughesy, T. (2012). A dose escalation trial for the combination of erlotinib and sirolimus for recurrent malignant gliomas. *Journal of Neuro Oncology*, *110*(2), 245–250. http://dx.doi.org/10.1007/s11060-012-0960-y.

Niola, F., Zhao, X., Singh, D., Sullivan, R., Castano, A., Verrico, A., et al. (2013). Mesenchymal high-grade glioma is maintained by the ID-RAP1 axis. *The Journal of Clinical Investigation*, *123*(1), 405–417. http://dx.doi.org/10.1172/JCI63811DS1.

Nishiyama, A., Lin, X. H., Giese, N., Heldin, C. H., & Stallcup, W. B. (1996). Co-localization of NG2 proteoglycan and PDGF alpha-receptor on O2A progenitor cells in the developing rat brain. *Journal of Neuroscience Research*, *43*, 299–314. http://dx.doi.org/10.1002/(SICI)1097-4547(19960201)43:3<299::AID-JNR5>3.0.CO;2-E.

Nobusawa, S., Watanabe, T., Kleihues, P., & Ohgaki, H. (2009). IDH1 mutations as molecular signature and predictive factor of secondary glioblastomas. *Clinical Cancer Research: An Official Journal of the American Association for Cancer Research*, *15*(19), 6002–6007.

Noushmehr, H., Weisenberger, D. J., Diefes, K., Phillips, H. S., Pujara, K., Berman, B. P., et al. (2010). Identification of a CpG island methylator phenotype that defines a distinct subgroup of glioma. *Cancer Cell*, *17*(5), 510–522.

O'Reilly, K. E., Rojo, F., She, Q.-B., Solit, D., Mills, G. B., Smith, D., et al. (2006). mTOR inhibition induces upstream receptor tyrosine kinase signaling and activates Akt. *Cancer Research*, *66*(3), 1500–1508. http://dx.doi.org/10.1158/0008-5472.CAN-05-2925.

Ohya, W., Funakoshi, H., Kurosawa, T., & Nakamura, T. (2007). Hepatocyte growth factor (HGF) promotes oligodendrocyte progenitor cell proliferation and inhibits its differentiation during postnatal development in the rat. *Brain Research*, *1147*, 51–65. Retrieved from, http://www.ncbi.nlm.nih.gov/pubmed/17382307.

Olar, A., & Aldape, K. D. (2012). Biomarkers classification and therapeutic decision-making for malignant gliomas. *Current Treatment Options in Oncology*, *13*, 1–20. http://dx.doi.org/10.1007/s11864-012-0210-8.

Pàez-Ribes, M., Allen, E., Hudock, J., Takeda, T., Okuyama, H., Viñals, F., et al. (2009). Antiangiogenic therapy elicits malignant progression of tumors to increased local invasion and distant metastasis. *Cancer Cell, 15*(3), 220–231.
Paglin, S., Hollister, T., Delohery, T., Hackett, N., McMahill, M., Sphicas, E., et al. (2001). A novel response of cancer cells to radiation involves autophagy and formation of acidic vesicles. *Cancer Research, 61*(2), 439–444.
Paquin, A. (2005). CCAAT/enhancer-binding protein phosphorylation biases cortical precursors to generate neurons rather than astrocytes in vivo. *The Journal of Neuroscience: The Official Journal of the Society for Neuroscience, 25*(46), 10747–10758.
Parsons, D. W., Jones, S., Zhang, X., Lin, J. C.-H., Leary, R. J., Angenendt, P., et al. (2008). An integrated genomic analysis of human glioblastoma multiforme. *Science, 321*, 1807–1812. Retrieved from, http://www.ncbi.nlm.nih.gov/pubmed/18772396.
Pei, Y., Moore, C. E., Wang, J., Tewari, A. K., Eroshkin, A., Cho, Y.-J., et al. (2012). An animal model of MYC-driven medulloblastoma. *Cancer Cell, 21*(2), 155–167.
Persson, A. I., Petritsch, C., Swartling, F. J., Itsara, M., Sim, F. J., Auvergne, R., et al. (2010). Non-stem cell origin for oligodendroglioma. *Cancer Cell, 18*, 669–682. Retrieved from, http://www.pubmedcentral.nih.gov/articlerender.fcgi?artid=3031116&tool=pmcentrez&rendertype=abstract.
Petryniak, M. A., Potter, G. B., Rowitch, D. H., & Rubenstein, J. L. R. (2007). Dlx1 and Dlx2 control neuronal versus oligodendroglial cell fate acquisition in the developing forebrain. *Neuron, 55*, 417–433. Retrieved from, http://www.pubmedcentral.nih.gov/articlerender.fcgi?artid=2039927&tool=pmcentrez&rendertype=abstract.
Phillips, J. J., Aranda, D., Ellison, D. W., Judkins, A. R., Croul, S. E., Brat, D. J., et al. (2013). PDGFRA amplification is common in pediatric and adult high-grade astrocytomas and identifies a poor prognostic group in IDH1 mutant glioblastoma. *Brain Pathology, 23*(5), 565–573.
Phillips, H. S., Kharbanda, S., Chen, R., Forrest, W. F., Soriano, R. H., Wu, T. D., et al. (2006). Molecular subclasses of high-grade glioma predict prognosis, delineate a pattern of disease progression, and resemble stages in neurogenesis. *Cancer Cell, 9*(3), 157–173. http://dx.doi.org/10.1016/j.ccr.2006.02.019.
Piao, Y., Liang, J., Holmes, L., Henry, V., Sulman, E., & de Groot, J. F. (2013). Acquired resistance to anti-VEGF therapy in glioblastoma is associated with a mesenchymal transition. *Clinical Cancer Research: An Official Journal of the American Association for Cancer Research, 19*(16), 4392–4403. http://dx.doi.org/10.1158/1078-0432.CCR-12-1557.
Piao, Y., Liang, J., Holmes, L., Zurita, A. J., Henry, V., Heymach, J. V., et al. (2012). Glioblastoma resistance to anti-VEGF therapy is associated with myeloid cell infiltration, stem cell accumulation, and a mesenchymal phenotype. *Neuro-Oncology, 14*(11), 1379–1392.
Piaskowski, S., Bienkowski, M., Stoczynska-Fidelus, E., Stawski, R., Sieruta, M., Szybka, M., et al. (2011). Glioma cells showing IDH1 mutation cannot be propagated in standard cell culture conditions. *British Journal of Cancer, 104*(6), 968–970.
Piccolo, S., Cordenonsi, M., & Dupont, S. (2013). Molecular pathways: YAP and TAZ take center stage in organ growth and tumorigenesis. *Clinical Cancer Research: An Official Journal of the American Association for Cancer Research, 19*(18), 4925–4930.
Polyak, K., & Weinberg, R. A. (2009). Transitions between epithelial and mesenchymal states: Acquisition of malignant and stem cell traits. *Nature Reviews. Cancer, 9*(4), 265–273.
Prahallad, A., Sun, C., Huang, S., Di Nicolantonio, F., Salazar, R., Zecchin, D., et al. (2012). Unresponsiveness of colon cancer to BRAF(V600E) inhibition through feedback activation of EGFR. *Nature, 483*(7387), 100–103. http://dx.doi.org/10.1038/nature10868.
Priester, M., Copanaki, E., Vafaizadeh, V., Hensel, S., Bernreuther, C., Glatzel, M., et al. (2013). STAT3 silencing inhibits glioma single cell infiltration and tumor growth. *Neuro-Oncology, 15*(7), 840–852.

Purow, B. W., Haque, R. M., Noel, M. W., Su, Q., Burdick, M. J., Lee, J., et al. (2005). Expression of Notch-1 and its ligands, Delta-like-1 and Jagged-1, is critical for glioma cell survival and proliferation. *Cancer Research*, *65*, 2353–2363. Retrieved from, http://www.ncbi.nlm.nih.gov/pubmed/15781650.

Pyonteck, S. M., Akkari, L., Schuhmacher, A. J., Bowman, R. L., Sevenich, L., Quail, D. F., et al. (2013). CSF-1R inhibition alters macrophage polarization and blocks glioma progression. *Nature Medicine*, *19*, 1–12. http://dx.doi.org/10.1038/nm.3337.

Raff, M. C., Miller, R. H., & Noble, M. (1983). A glial progenitor cell that develops in vitro into an astrocyte or an oligodendrocyte depending on culture medium. *Nature*, *303*, 390–396. Retrieved from, http://www.ncbi.nlm.nih.gov/pubmed/6304520.

Raizer, J. J., & Abrey, L. E. (2010). A phase II trial of erlotinib in patients with recurrent malignant gliomas and nonprogressive glioblastoma multiforme postradiation therapy. *Neuro-Oncology*, *12*(1), 95–103. Retrieved from, http://neuro-oncology.oxfordjournals.org/content/12/1/95.short.

Rakic, P. (1990). Principles of neural cell migration. *Experientia*, *46*(9), 882–891.

Rao, R. D., Mladek, A. C., Lamont, J. D., Goble, J. M., Erlichman, C., James, C. D., et al. (2005). Disruption of parallel and converging signaling pathways contributes to the synergistic antitumor effects of simultaneous mTOR and EGFR inhibition in GBM cells. *Neoplasia*, 7(10), 921–929. http://dx.doi.org/10.1593/neo.05361.

Rasper, M., Schäfer, A., Piontek, G., Teufel, J., Brockhoff, G., Ringel, F., et al. (2010). Aldehyde dehydrogenase 1 positive glioblastoma cells show brain tumor stem cell capacity. *Neuro-Oncology*, *12*, 1024–1033. Retrieved from, http://www.pubmedcentral.nih.gov/articlerender.fcgi?artid=3018920&tool=pmcentrez&rendertype=abstract.

Rath, P., Lal, B., Ajala, O., Li, Y., Xia, S., Kim, J., et al. (2013). In vivo c-Met pathway inhibition depletes human glioma xenografts of tumor-propagating stem-like cells. *Translational Oncology*, *6*(2), 104–111.

Razis, E., Selviaridis, P., Labropoulos, S., Norris, J. L., Zhu, M.-J., Song, D. D., et al. (2009). Phase II study of neoadjuvant imatinib in glioblastoma: Evaluation of clinical and molecular effects of the treatment. *Clinical Cancer Research: An Official Journal of the American Association for Cancer Research*, *15*(19), 6258–6266. http://dx.doi.org/10.1158/1078-0432.CCR-08-1867.

Reardon, David A., Conrad, C. A., Cloughesy, T., Prados, M. D., Friedman, H. S., Aldape, K. D., et al. (2012). Phase I study of AEE788, a novel multitarget inhibitor of ErbB- and VEGF-receptor-family tyrosine kinases, in recurrent glioblastoma patients. *Cancer Chemotherapy and Pharmacology*, *69*(6), 1507–1518. http://dx.doi.org/10.1007/s00280-012-1854-6.

Reardon, David A., Desjardins, A., Vredenburgh, J. J., Gururangan, S., Friedman, A. H., Herndon, J. E., et al. (2010). Phase 2 trial of erlotinib plus sirolimus in adults with recurrent glioblastoma. *Journal of Neuro-Oncology*, *96*(2), 219–230. http://dx.doi.org/10.1007/s11060-009-9950-0.

Reardon, D. A., Dresemann, G., Taillibert, S., Campone, M., van den Bent, M., Clement, P., et al. (2009). Multicentre phase II studies evaluating imatinib plus hydroxyurea in patients with progressive glioblastoma. *British Journal of Cancer*, *101*(12), 1995–2004. http://dx.doi.org/10.1038/sj.bjc.6605411.

Reardon, David A., Quinn, J. A., Vredenburgh, J. J., Gururangan, S., Friedman, A. H., Desjardins, A., et al. (2006). Phase 1 trial of gefitinib plus sirolimus in adults with recurrent malignant glioma. *Clinical Cancer Research: An Official Journal of the American Association for Cancer Research*, *12*(3 Pt 1), 860–868. http://dx.doi.org/10.1158/1078-0432.CCR-05-2215.

Reya, T., Morrison, S. J., Clarke, M. F., & Weissman, I. L. (2001). Stem cells, cancer, and cancer stem cells. *Nature*, *414*, 105–111. Retrieved from, http://www.ncbi.nlm.nih.gov/pubmed/11689955.

Reznik, T. E., Sang, Y., Ma, Y., Abounader, R., Rosen, E. M., Xia, S., et al. (2008). Transcription-dependent epidermal growth factor receptor activation by hepatocyte growth factor. *Molecular Cancer Research: MCR*, *6*(1), 139–150. http://dx.doi.org/10.1158/1541-7786.MCR-07-0236.

Ricci-Vitiani, L., Pallini, R., Biffoni, M., Todaro, M., Invernici, G., Cenci, T., et al. (2010). Tumour vascularization via endothelial differentiation of glioblastoma stem-like cells. *Nature*, *468*, 824–828. Retrieved from, http://www.ncbi.nlm.nih.gov/pubmed/21102434.

Rickman, D. S., Bobek, M. P., Misek, D. E., Kuick, R., Blaivas, M., Kurnit, D. M., et al. (2001). Distinctive molecular profiles of high-grade and low-grade gliomas based on oligonucleotide microarray analysis. *Cancer Research*, *61*(18), 6885–6891.

Roberts, W. G., Whalen, P. M., Soderstrom, E., Moraski, G., Lyssikatos, J. P., Wang, H.-F., et al. (2005). Antiangiogenic and antitumor activity of a selective PDGFR tyrosine kinase inhibitor, CP-673,451. *Cancer Research*, *65*(3), 957–966. Retrieved from, http://www.ncbi.nlm.nih.gov/pubmed/15705896.

Robinson, J. P., Vanbrocklin, M. W., Lastwika, K. J., McKinney, A. J., Brandner, S., & Holmen, S. L. (2011). Activated MEK cooperates with Ink4a/Arf loss or Akt activation to induce gliomas in vivo. *Oncogene*, *30*, 1341–1350. Retrieved from, http://discovery.ucl.ac.uk/441249/.

Rohle, D., Popovici-Muller, J., Palaskas, N., Turcan, S., Grommes, C., Campos, C., et al. (2013). An inhibitor of mutant IDH1 delays growth and promotes differentiation of glioma cells. *Science*, *340*(6132), 626–630.

Ronellenfitsch, M. W., Steinbach, J. P., & Wick, W. (2010). Epidermal growth factor receptor and mammalian target of rapamycin as therapeutic targets in malignant glioma: Current clinical status and perspectives. *Targeted Oncology*, *5*(3), 183–191. http://dx.doi.org/10.1007/s11523-010-0154-5.

Saito, N., Fu, J., Zheng, S., Yao, J., Wang, S., Liu, D. D., et al. (2014). A high Notch pathway activation predicts response to γ secretase inhibitors in proneural subtype of glioma tumor initiating cells. *Stem Cells*, *32*, 301–312.

Sampson, J. H., Heimberger, A. B., Archer, G. E., Aldape, K. D., Friedman, A. H., Friedman, H. S., et al. (2010). Immunologic escape after prolonged progression-free survival with epidermal growth factor receptor variant III peptide vaccination in patients with newly diagnosed glioblastoma. *Journal of Clinical Oncology: Official Journal of the American Society of Clinical Oncology*, *28*(31), 4722–4729. http://dx.doi.org/10.1200/JCO.2010.28.6963.

Sanai, N., Nguyen, T., Ihrie, R. A., Mirzadeh, Z., Tsai, H.-H., Wong, M., et al. (2011). Corridors of migrating neurons in the human brain and their decline during infancy. *Nature*, *478*, 1–6. http://dx.doi.org/10.1038/nature10487.

Sandberg, C. J., Altschuler, G., Jeong, J., Strømme, K. K., Stangeland, B., Murrell, W., et al. (2013). Comparison of glioma stem cells to neural stem cells from the adult human brain identifies dysregulated Wnt- signaling and a fingerprint associated with clinical outcome. *Experimental Cell Research*, *319*, 1–14. http://dx.doi.org/10.1016/j.yexcr.2013.06.004.

Sanson, M., Marie, Y., Paris, S., Idbaih, A., Laffaire, J., Ducray, F., et al. (2009). Isocitrate dehydrogenase 1 codon 132 mutation is an important prognostic biomarker in gliomas. *Journal of Clinical Oncology: Official Journal of the American Society of Clinical Oncology*, *27*(25), 4150–4154.

Santiskulvong, C., Konecny, G. E., Fekete, M., Chen, K.-Y. M., Karam, A., Mulholland, D., et al. (2011). Dual targeting of phosphoinositide 3-kinase and mammalian target of rapamycin using NVP-BEZ235 as a novel therapeutic approach in human ovarian carcinoma. *Clinical Cancer Research: An Official Journal of the American Association for Cancer Research*, *17*(8), 2373–2384. http://dx.doi.org/10.1158/1078-0432.CCR-10-2289.

Santoni, M., Burattini, L., Nabissi, M., Beatrice Morelli, M., Berardi, R., Santoni, G., et al. (2013). Essential role of gli proteins in glioblastoma multiforme. *Current Protein and Peptide Science, 14*, 8. Retrieved from, http://www.ingentaconnect.com.sci-hub.org/content/ben/cpps/2013/00000014/00000002/art00005.

Sasaki, M., Knobbe, C. B., Itsumi, M., Elia, A. J., Harris, I. S., Chio, I. I. C., et al. (2012). D-2-hydroxyglutarate produced by mutant IDH1 perturbs collagen maturation and basement membrane function. *Genes & Development, 26*(18), 2038–2049.

Sasaki, M., Knobbe, C. B., Munger, J. C., Lind, E. F., Brenner, D., Brüstle, A., et al. (2013). IDH1(R132H) mutation increases murine haematopoietic progenitors and alters epigenetics. *Nature, 488*(7413), 656–659.

Schmitt, C. A., Fridman, J. S., Yang, M., Baranov, E., Hoffman, R. M., & Lowe, S. W. (2002). Dissecting p53 tumor suppressor functions in vivo. *Cancer Cell, 1*(3), 289–298.

Schwartzentruber, J., Korshunov, A., Liu, X.-Y., Jones, D. T. W., Pfaff, E., Jacob, K., et al. (2012). Driver mutations in histone H3.3 and chromatin remodelling genes in paediatric glioblastoma. *Nature, 482*(7384), 226–231.

Shannon, P., Sabha, N., Lau, N., Kamnasaran, D., Gutmann, D. H., & Guha, A. (2005). Pathological and molecular progression of astrocytomas in a GFAP:12 V-Ha-Ras mouse astrocytoma model. *The American Journal of Pathology, 167*, 859–867. Retrieved from, http://www.pubmedcentral.nih.gov/articlerender.fcgi?artid=1698742&tool=pmcentrez&rendertype=abstract.

Shawver, L. K., Schwartz, D. P., Mann, E., Chen, H., Tsai, J., Chu, L., et al. (1997). Inhibition of platelet-derived growth factor-mediated signal transduction and tumor growth by N-[4-(trifluoromethyl)-phenyl]5-methylisoxazole-4-carboxamide. *Clinical Cancer Research: An Official Journal of the American Association for Cancer Research, 3*(7), 1167–1177. Retrieved from, http://www.ncbi.nlm.nih.gov/pubmed/9815796.

Shen, J., Zheng, H., Ruan, J., Fang, W., Li, A., Tian, G., et al. (2013). Autophagy inhibition induces enhanced proapoptotic effects of ZD6474 in glioblastoma. *British Journal of Cancer, 109*(1), 164–171. http://dx.doi.org/10.1038/bjc.2013.306.

Sher, F., Boddeke, E., & Copray, S. (2011). Ezh2 expression in astrocytes induces their dedifferentiation toward neural stem cells. *Cellular Reprogramming, 13*, 1–6. Retrieved from, http://www.ncbi.nlm.nih.gov/pubmed/20979531.

Sher, F., Boddeke, E., Olah, M., & Copray, S. (2012). Dynamic changes in Ezh2 gene occupancy underlie its involvement in neural stem cell self-renewal and differentiation towards oligodendrocytes. *PLoS ONE*, 7, e40399.

Sher, F., Rössler, R., Brouwer, N., Balasubramaniyan, V., Boddeke, E., & Copray, S. (2008). Differentiation of neural stem cells into oligodendrocytes: Involvement of the polycomb group protein Ezh2. *Stem Cells, 26*, 2875–2883. Retrieved from, http://www.ncbi.nlm.nih.gov/pubmed/18687996.

Sherry, M. M., Reeves, A., Wu, J. K., & Cochran, B. H. (2009). STAT3 is required for proliferation and maintenance of multipotency in glioblastoma stem cells. *Stem Cells, 27*(10), 2383–2392.

Shih, A. H., Dai, C., Hu, X., Shih, A. H., Dai, C., Hu, X., et al. (2004). Dose-dependent effects of platelet-derived growth factor-B on glial tumorigenesis dose-dependent effects of platelet-derived growth factor-B on glial tumorigenesis. *Cancer Research, 64*, 4783–4789.

Shimizu, T., Kagawa, T., Wada, T., Muroyama, Y., Takada, S., & Ikenaka, K. (2005). Wnt signaling controls the timing of oligodendrocyte development in the spinal cord. *Developmental Biology, 282*, 397–410. Retrieved from, http://www.ncbi.nlm.nih.gov/pubmed/15950605.

Shingu, T., Fujiwara, K., Bögler, O., Akiyama, Y., Moritake, K., Shinojima, N., et al. (2009). Inhibition of autophagy at a late stage enhances imatinib-induced cytotoxicity in human malignant glioma cells. *International Journal of Cancer, 124*(5), 1060–1071. http://dx.doi.org/10.1002/ijc.24030.

Siebzehnrubl, F. A., Silver, D. J., Tugertimur, B., Deleyrolle, L. P., Siebzehnrubl, D., Sarkisian, M. R., et al. (2013). The ZEB1 pathway links glioblastoma initiation, invasion and chemoresistance. *EMBO Molecular Medicine*, *5*(8), 1196–1212.

Sim, F. J., McClain, C. R., Schanz, S. J., Protack, T. L., Windrem, M. S., & Goldman, S. A. (2011). CD140a identifies a population of highly myelinogenic, migration-competent and efficiently engrafting human oligodendrocyte progenitor cells. *Nature Biotechnology*, *29*, 934–941. http://dx.doi.org/10.1038/nbt.1972.

Simon, J. A., & Kingston, R. E. (2009). Mechanisms of polycomb gene silencing: Knowns and unknowns. *Nature Reviews Molecular Cell Biology*, *10*, 697–708. Retrieved from, http://www.ncbi.nlm.nih.gov/pubmed/19738629.

Singh, S. K., Clarke, I. D., Terasaki, M., Bonn, V. E., Hawkins, C., Squire, J., et al. (2003). Identification of a cancer stem cell in human brain tumors. *Cancer Research*, *63*, 5821–5828.

Singh, S. K., Hawkins, C., Clarke, I. D., Squire, J. A., Bayani, J., Hide, T., et al. (2004). Identification of human brain tumour initiating cells. *Nature*, *432*(7015), 396–401.

Singh, A., & Settleman, J. (2010). EMT, cancer stem cells and drug resistance: An emerging axis of evil in the war on cancer. *Oncogene*, *29*(34), 4741–4751.

Snuderl, M., Fazlollahi, L., Le, L. P., Nitta, M., Zhelyazkova, B. H., Davidson, C. J., et al. (2011). Mosaic amplification of multiple receptor tyrosine kinase genes in glioblastoma. *Cancer Cell*, *20*(6), 810–817.

Son, M. J., Woolard, K., Nam, D.-H., Lee, J., & Fine, H. A. (2009). SSEA-1 is an enrichment marker for tumor-initiating cells in human glioblastoma. *Cell Stem Cell*, *4*, 440–452. Retrieved from, http://www.ncbi.nlm.nih.gov/pubmed/19427293.

Sotelo, J., Briceño, E., & López-González, M. A. (2006). Adding chloroquine to conventional treatment for glioblastoma multiforme: A randomized, double-blind, placebo-controlled trial. *Annals of Internal Medicine*, *144*(5), 337–343.

Sottile, V., Li, M., & Scotting, P. J. (2006). Stem cell marker expression in the Bergmann glia population of the adult mouse brain. *Brain Research*, *1099*(1), 8–17.

Sottoriva, A., Spiteri, I., Piccirillo, S. G. M., Touloumis, A., Collins, V. P., Marioni, J. C., et al. (2013). Intratumor heterogeneity in human glioblastoma reflects cancer evolutionary dynamics. *Proceedings of the National Academy of Sciences of the United States of America*, *110*(10), 4009–4014. http://dx.doi.org/10.1073/pnas.1219747110.

Steiner, L. A., Schulz, V. P., Maksimova, Y., Wong, C., & Gallagher, P. G. (2011). Patterns of histone H3 lysine 27 monomethylation and erythroid cell type-specific gene expression. *The Journal of Biological Chemistry*, *286*(45), 39457–39465.

Stolt, C. C., Lommes, P., Sock, E., Chaboissier, M.-C., Schedl, A., & Wegner, M. (2003). The Sox9 transcription factor determines glial fate choice in the developing spinal cord. *Genes & Development*, *17*, 1677–1689. http://dx.doi.org/10.1101/gad.259003.

Stolt, C., Schlierf, A., Lommes, P., Hillgartner, S., Werner, T., Kosian, T., et al. (2006). SoxD proteins influence multiple stages of oligodendrocyte development and modulate SoxE protein function. *Developmental Cell*, *11*, 697–709. Retrieved from, http://discovery.ucl.ac.uk/143376/.

Stommel, J., Kimmelman, A., & Ying, H. (2007). Coactivation of receptor tyrosine kinases affects the response of tumor cells to targeted therapies. *Science*. *287*(2007). http://dx.doi.org/10.1126/science.1142946.

Sturm, D., Witt, H., Hovestadt, V., Khuong-Quang, D.-A., Jones, D. T. W., Konermann, C., et al. (2012). Hotspot mutations in H3F3A and IDH1 define distinct epigenetic and biological subgroups of glioblastoma. *Cancer Cell*, *22*(4), 425–437.

Sugiarto, S., Persson, A. I., Munoz, E. G., Waldhuber, M., Lamagna, C., Andor, N., et al. (2011). Asymmetry-defective oligodendrocyte progenitors are glioma precursors. *Cancer Cell*, *20*, 328–340. Retrieved from, http://linkinghub.elsevier.com/retrieve/pii/S1535610811003084.

Sun, S. Y., Rosenberg, L. M., Wang, X., Zhou, Z., Yue, P., Fu, H., et al. (2005). Activation of Akt and eIF4E survival pathways by rapamycin mediated mammalian target of rapamycin inhibition. *Cancer Research, 65*, 7052–7058.

Suvà, M.-L., Riggi, N., Janiszewska, M., Radovanovic, I., Provero, P., Stehle, J.-C., et al. (2009). EZH2 is essential for glioblastoma cancer stem cell maintenance. *Cancer Research, 69*, 9211–9218. Retrieved from, http://www.ncbi.nlm.nih.gov/pubmed/19934320.

Szerlip, N. J., Pedraza, A., Chakravarty, D., Azim, M., McGuire, J., Fang, Y., et al. (2012). Intratumoral heterogeneity of receptor tyrosine kinases EGFR and PDGFRA amplification in glioblastoma defines subpopulations with distinct growth factor response. *Proceedings of the National Academy of Sciences of the United States of America, 109*(8), 3041–3046. http://dx.doi.org/10.1073/pnas.1114033109.

Takahashi, K., & Yamanaka, S. (2006). Induction of pluripotent stem cells from mouse embryonic and adult fibroblast cultures by defined factors. *Cell, 126*(4), 663–676. http://dx.doi.org/10.1016/j.cell.2006.07.024.

Tchoghandjian, A., Baeza, N., Colin, C., Cayre, M., Metellus, P., Beclin, C., et al. (2010). A2B5 cells from human glioblastoma have cancer stem cell properties. *Brain Pathology, 20*(1), 211–221.

Tekki-Kessaris, N., Woodruff, R., Hall, A. C., Gaffield, W., Kimura, S., Stiles, C. D., et al. (2001). Hedgehog-dependent oligodendrocyte lineage specification in the telencephalon. *Development (Cambridge, England), 128*, 2545–2554. Retrieved from, http://discovery.ucl.ac.uk/10570/.

The Cancer Genome Atlas Research Network (2008). Comprehensive genomic characterization defines human glioblastoma genes and core pathways. *Nature, 455*(7216), 1061–1068. http://dx.doi.org/10.1038/nature07385.

Thiery, J. P. (2002). Epithelial—Mesenchymal transitions in tumour progression. *Nature Reviews Cancer, 2*(6), 442–454.

Trainor, P. A., & Krumlauf, R. (2001). Hox genes, neural crest cells and branchial arch patterning. *Current Opinion in Cell Biology, 13*(6), 698–705. Retrieved from, http://www.ncbi.nlm.nih.gov/pubmed/11698185.

Tsai, H.-H., Li, H., Fuentealba, L. C., Molofsky, A. V., Taveira-Marques, R., Zhuang, H., et al. (2012). Regional astrocyte allocation regulates CNS synaptogenesis and repair. *Science, 358*, 1–15. http://dx.doi.org/10.1126/science.1222381.

Turcan, S., Rohle, D., Goenka, A., Walsh, L. A., Fang, F., Yilmaz, E., et al. (2013). IDH1 mutation is sufficient to establish the glioma hypermethylator phenotype. *Nature, 483*(7390), 479–483.

Uhrbom, L., Dai, C., Celestino, J. C., Rosenblum, M. K., Fuller, G. N., & Holland, E. C. (2002). Ink4a-Arf loss cooperates with KRas activation in astrocytes and neural progenitors to generate glioblastomas of various morphologies depending on activated Akt. *Cancer Research, 62*, 5551–5558. Retrieved from, http://www.ncbi.nlm.nih.gov/entrez/query.fcgi?cmd=Retrieve&db=PubMed&dopt=Citation&list_uids=12359767.

Valadez, J. G., Grover, V. K., Carter, M. D., Calcutt, M. W., Abiria, S. A., Lundberg, C. J., et al. (2013). Identification of Hedgehog pathway responsive glioblastomas by isocitrate dehydrogenase mutation. *Cancer Letters, 328*(2), 297–306.

Van den Bent, M. J., Dubbink, H. J., Marie, Y., Brandes, A. A., Taphoorn, M. J. B., Wesseling, P., et al. (2010). IDH1 and IDH2 mutations are prognostic but not predictive for outcome in anaplastic oligodendroglial tumors: A report of the European Organization for Research and Treatment of Cancer Brain Tumor Group. *Clinical Cancer Research: An Official Journal of the American Association for Cancer Research, 16*(5), 1597–1604. http://dx.doi.org/10.1158/1078-0432.CCR-09-2902.

Van den Boom, J., Wolter, M., Kuick, R., Misek, D. E., Youkilis, A. S., Wechsler, D. S., et al. (2003). Characterization of gene expression profiles associated with glioma progression using oligonucleotide-based microarray analysis and real-time reverse transcription-polymerase chain reaction. *The American Journal of Pathology, 163*(3), 1033–1043.

Venneti, S., Garimella, M. T., Sullivan, L. M., Martinez, D., Huse, J. T., Heguy, A., et al. (2013). Evaluation of histone 3 lysine 27 trimethylation (H3K27me3) and enhancer of zest 2 (EZH2) in pediatric glial and glioneuronal tumors shows decreased H3K27me3 in H3F3A K27M mutant glioblastomas. *Brain Pathology, 23*, 558–564.

Verhaak, R. G. W., Hoadley, K. A., Purdom, E., Wang, V., Qi, Y., Wilkerson, M. D., et al. (2010). Integrated genomic analysis identifies clinically relevant subtypes of glioblastoma characterized by abnormalities in PDGFRA, IDH1, EGFR, and NF1. *Cancer Cell, 17*(1), 98–110. http://dx.doi.org/10.1016/j.ccr.2009.12.020.

Vivanco, I., Robins, H. I., Rohle, D., Campos, C., Grommes, C., Nghiemphu, P. L., et al. (2012). Differential sensitivity of glioma- versus lung cancer-specific EGFR mutations to EGFR kinase inhibitors. *Cancer Discovery, 2*(5), 458–471. http://dx.doi.org/10.1158/2159-8290.CD-11-0284.

Vlassenko, A. G., Thiessen, B., Beattie, B. J., Malkin, M. G., & Blasberg, R. G. (2000). Evaluation of early response to SU101 target-based therapy in patients with recurrent supratentorial malignant gliomas using FDG PET and Gd-DTPA MRI. *Journal of Neuro-Oncology, 46*(3), 249–259. Retrieved from, http://www.ncbi.nlm.nih.gov/pubmed/10902856.

Wachsberger, P. R., Lawrence, R. Y., Liu, Y., Xia, X., Andersen, B., & Dicker, A. P. (2011). Cediranib enhances control of wild type EGFR and EGFRvIII-expressing gliomas through potentiating temozolomide, but not through radiosensitization: Implications for the clinic. *Journal of Neuro-Oncology, 105*(2), 181–190. http://dx.doi.org/10.1007/s11060-011-0580-y.

Wang, R., Chadalavada, K., Wilshire, J., Kowalik, U., Hovinga, K. E., Geber, A., et al. (2010). Glioblastoma stem-like cells give rise to tumour endothelium. *Nature, 468*, 829–833. http://dx.doi.org/10.1038/nature09624.

Wang, M. Y., Lu, K. V., Zhu, S., Dia, E. Q., Vivanco, I., Shackleford, G. M., et al. (2006). Mammalian target of rapamycin inhibition promotes response to epidermal growth factor receptor kinase inhibitors in PTEN-deficient and PTEN-intact glioblastoma cells. *Cancer Research, 66*(16), 7864–7869. http://dx.doi.org/10.1158/0008-5472.CAN-04-4392.

Wang, J., O'Bara, M. A., Pol, S. U., & Sim, F. J. (2013). CD133/CD140a-based isolation of distinct human multipotent neural progenitor cells and oligodendrocyte progenitor cells. *Stem Cells and Development, 22*(15), 2121–2131. http://dx.doi.org/10.1089/scd.2013.0003.

Wang, S., Sdrulla, A. D., diSibio, G., Bush, G., Nofziger, D., Hicks, C., et al. (1998). Notch receptor activation inhibits oligodendrocyte differentiation. *Neuron, 21*, 63–75. Retrieved from, http://www.ncbi.nlm.nih.gov/entrez/query.fcgi?cmd=Retrieve&db=PubMed&dopt=Citation&list_uids=9697852.

Wang, Q., Wei, F., Li, C., Lv, G., Wang, G., Liu, T., et al. (2013). Combination of mTOR and EGFR kinase inhibitors blocks mTORC1 and mTORC2 kinase activity and suppresses the progression of colorectal carcinoma. *PloS One, 8*(8), e73175. http://dx.doi.org/10.1371/journal.pone.0073175.

Wang, Y., Yang, J., Zheng, H., Tomasek, G. J., Zhang, P., McKeever, P. E., et al. (2009). Expression of mutant p53 proteins implicates a lineage relationship between neural stem cells and malignant astrocytic glioma in a murine model. *Cancer Cell, 15*, 514–526. Retrieved from, http://www.ncbi.nlm.nih.gov/pubmed/19477430.

Wang, T.-W., Zhang, H., Gyetko, M. R., & Parent, J. M. (2011). Hepatocyte growth factor acts as a mitogen and chemoattractant for postnatal subventricular zone-olfactory bulb neurogenesis. *Molecular and Cellular Neurosciences, 48*, 38–50. Retrieved from, http://www.pubmedcentral.nih.gov/articlerender.fcgi?artid=3160177&tool=pmcentrez&rendertype=abstract.

Watanabe, T., Nobusawa, S., Kleihues, P., & Ohgaki, H. (2009). IDH1 mutations are early events in the development of astrocytomas and oligodendrogliomas. *The American Journal of Pathology*, *174*(4), 1149–1153.

Waters, J. D., Sanchez, C., Sahin, A., Futalan, D., Gonda, D. D., Scheer, J. K., et al. (2012). CT322, a VEGFR-2 antagonist, demonstrates anti-glioma efficacy in orthotopic brain tumor model as a single agent or in combination with temozolomide and radiation therapy. *Journal of Neuro-Oncology*, *110*(1), 37–48. http://dx.doi.org/10.1007/s11060-012-0948-7.

Wei, Q., Clarke, L., Scheidenhelm, D. K., Qian, B., Tong, A., Sabha, N., et al. (2006). High-grade glioma formation results from postnatal pten loss or mutant epidermal growth factor receptor expression in a transgenic mouse glioma model. *Cancer Research*, *66*, 7429–7437. Retrieved from, http://www.ncbi.nlm.nih.gov/pubmed/16885338.

Weiss, W. A., Burns, M. J., Hackett, C., Aldape, K., Hill, J. R., Kuriyama, H., et al. (2003). Genetic determinants of malignancy in a mouse model for oligodendroglioma. *Cancer Research*, *63*, 1589–1595.http://www.ncbi.nlm.nih.gov/pubmed/12670909.

Weiss, S., Dunne, C., Hewson, J., Wohl, C., Wheatley, M., Peterson, A. C., et al. (1996). Multipotent CNS stem cells are present in the adult mammalian spinal cord and ventricular neuroaxis. *The Journal of Neuroscience: The Official Journal of the Society for Neuroscience*, *16*(23), 7599–7609.

Wilkinson, G., Dennis, D., & Schuurmans, C. (2013). Proneural genes in neocortical development. *Neuroscience*, *253C*, 256–273.

Wu, G., Broniscer, A., McEachron, T. A., Lu, C., Paugh, B. S., Becksfort, J., et al. (2012). Somatic histone H3 alterations in pediatric diffuse intrinsic pontine gliomas and non-brainstem glioblastomas. *Nature Genetics*, *44*(3), 251–253.

Xu, Y., Tamamaki, N., Noda, T., Kimura, K., Itokazu, Y., Matsumoto, N., et al. (2005). Neurogenesis in the ependymal layer of the adult rat 3rd ventricle. *Experimental Neurology*, *192*(2), 251–264.

Xu, W., Yang, H., Liu, Y., Yang, Y., Wang, P., Kim, S. H., et al. (2011). Oncometabolite 2-hydroxyglutarate is a competitive inhibitor of α-ketoglutarate-dependent dioxygenases. *Cancer Cell*, *219*(1), 17–30.

Yakes, F. M., Chen, J., Tan, J., Yamaguchi, K., Shi, Y., Yu, P., et al. (2011). Cabozantinib (XL184), a novel MET and VEGFR2 inhibitor, simultaneously suppresses metastasis, angiogenesis, and tumor growth. *Molecular Cancer Therapeutics*, *10*(12), 2298–2308. http://dx.doi.org/10.1158/1535-7163.MCT-11-0264.

Yan, H., Parsons, D. W., Jin, G., McLendon, R., Rasheed, B. A., Yuan, W., et al. (2009). IDH1 and IDH2 mutations in gliomas. *The New England Journal of Medicine*, *360*(8), 765–773.

Yang, R. Y. C., Yang, K. S., Pike, L. J., & Marshall, G. R. (2010). Targeting the dimerization of epidermal growth factor receptors with small-molecule inhibitors. *Chemical Biology & Drug Design*, *76*(1), 1–9. http://dx.doi.org/10.1111/j.1747-0285.2010.00986.x.

Yip, S., Butterfield, Y. S., Morozova, O., Chittaranjan, S., Blough, M. D., An, J., et al. (2011). Concurrent CIC mutations, IDH mutations, and 1p/19q loss distinguish oligodendrogliomas from other cancers. *The Journal of Pathology*, *226*(1), 7–16.

Yu, D., & Thomas-Tikhonenko, A. (2002). A non-transgenic mouse model for B-cell lymphoma: In vivo infection of p53-null bone marrow progenitors by a Myc retrovirus is sufficient for tumorigenesis. *Oncogene*, *21*(12), 1922–1927.

Zhang, Hongbing, Bajraszewski, N., Wu, E., Wang, H., Moseman, A. P., Dabora, S. L., et al. (2007). PDGFRs are critical for PI3K/Akt activation and negatively regulated by mTOR. *The Journal of Clinical Investigation*, *117*(3), 730–738. http://dx.doi.org/10.1172/JCI28984.

Zhang, H., Liu, C.-Y., Zha, Z.-Y., Zhao, B., Yao, J., Zhao, S., et al. (2009). TEAD transcription factors mediate the function of TAZ in cell growth and epithelial-mesenchymal transition. *The Journal of Biological Chemistry*, *284*(20), 13355–13362.

Zhang, J., Wu, G., Miller, C. P., Tatevossian, R. G., Dalton, J. D., Tang, B., et al. (2013). Whole-genome sequencing identifies genetic alterations in pediatric low-grade gliomas. *Nature Genetics*, *45*(6), 602–612.

Zhao, B., Li, L., Lei, Q., & Guan, K.-L. (2010). The Hippo-YAP pathway in organ size control and tumorigenesis: An updated version. *Genes & Development*, *24*(9), 862–874.

Zhao, S., Lin, Y., Xu, W., Jiang, W., Zha, Z., Wang, P., et al. (2009). Glioma-derived mutations in IDH1 dominantly inhibit IDH1 catalytic activity and induce HIF-1alpha. *Science*, *324*(5924), 261–265.

Zheng, H., Ying, H., Yan, H., Kimmelman, A. C., Hiller, D. J., Chen, A. J., et al. (2008). *Nature*, *455*(7216), 1129–1133. http://dx.doi.org/10.1038/nature07443.

Zhu, Y., Guignard, F., Zhao, D., Liu, L., Burns, D. K., Mason, R. P., et al. (2005). Early inactivation of p53 tumor suppressor gene cooperating with NF1 loss induces malignant astrocytoma. *Cancer Cell*, *8*, 119–130. Retrieved from, http://www.pubmedcentral.nih.gov/articlerender.fcgi?artid=3024718&tool=pmcentrez&rendertype=abstract.

Zuo, H., & Nishiyama, A. (2013). Polydendrocytes in development and myelin repair. *Neuroscience Bulletin. 1–12.* http://dx.doi.org/10.1007/s12264-013-1320-4.

FURTHER READING

Chesler, L., Goldenberg, D. D., Collins, R., Grimmer, M., Kim, G. E., Tihan, T., et al. (2008). Chemotherapy-induced apoptosis in a transgenic model of neuroblastoma proceeds through p53 induction. *Neoplasia*, *10*(11), 1268–1274.

CHAPTER TWO

Therapeutic Cancer Vaccines

Jeffrey Schlom[1], James W. Hodge, Claudia Palena, Kwong-Yok Tsang, Caroline Jochems, John W. Greiner, Benedetto Farsaci, Ravi A. Madan, Christopher R. Heery, James L. Gulley

Laboratory of Tumor Immunology and Biology, Center for Cancer Research, National Cancer Institute, National Institutes of Health, Bethesda, Maryland, USA

[1]Corresponding author: e-mail address: js141c@nih.gov

Contents

Advances in Cancer Research, Volume 121
ISSN 0065-230X
http://dx.doi.org/10.1016/B978-0-12-800249-0.00002-0

Abstract

Therapeutic cancer vaccines have the potential of being integrated in the therapy of numerous cancer types and stages. The wide spectrum of vaccine platforms and vaccine targets is reviewed along with the potential for development of vaccines to target cancer cell "stemness," the epithelial-to-mesenchymal transition (EMT) phenotype, and drug-resistant populations. Preclinical and recent clinical studies are now revealing how vaccines can optimally be used with other immune-based therapies such as checkpoint inhibitors, and so-called nonimmune-based therapeutics, radiation, hormonal therapy, and certain small molecule targeted therapies; it is now being revealed that many of these traditional therapies can lyse tumor cells in a manner as to further potentiate the host immune response, alter the phenotype of nonlysed tumor cells to render them more susceptible to T-cell lysis, and/or shift the balance of effector:regulatory cells in a manner to enhance vaccine efficacy. The importance of the tumor microenvironment, the appropriate patient population, and clinical trial endpoints is also discussed in the context of optimizing patient benefit from vaccine-mediated therapy.

1. INTRODUCTION

This chapter encompasses the numerous factors involved in the design, development, and clinical application of therapeutic cancer vaccines both as a monotherapy and in combination with other forms of immunotherapy, as well as with nonimmune-based therapies. Among the topics discussed are (a) the wide spectrum of cancer vaccine targets; (b) the pros and cons of different vaccine platforms; (c) how animal models can be used, and should not be used, in vaccine development; (d) the influence of the tumor microenvironment and regulatory entities on vaccine efficacy; (e) how vaccine combination therapies with certain chemotherapeutic agents, radiation, hormone therapy, and small molecule targeted therapies and other immune therapeutics can potentially be used to enhance vaccine efficacy; (f) the appropriate patient populations and trial endpoints in vaccine clinical studies and the importance of tumor growth kinetics; and (g) the potential of new vaccines that can target cancer cell "stemness," the epithelial-to-mesenchymal transition (EMT) phenotype, and drug resistance. While this chapter presents an overview of several aspects of therapeutic cancer vaccine development, many of the examples given are based on preclinical and clinical studies carried out at the National Cancer Institute, National Institutes of Health. Much of the information has been presented in previous review articles (Palena & Schlom, 2013; Schlom, 2012; Schlom, Hodge, et al., 2013; Schlom, Palena, et al., 2013) and/or in peer-reviewed publications as cited.

2. CANCER VACCINE TARGETS

The validity of a target for a therapeutic cancer vaccine will depend on the ability of a tumor cell to process the tumor-associated antigen (TAA) expressed by the vaccine in the context of a peptide–major histocompatibility complex (MHC) for T-cell recognition or on the surface of the tumor cell for B-cell recognition. The level of expression of the TAA in the tumor, the relative specificity of the TAA for tumor versus normal adult tissue, and the degree of inherent "tolerance" to the given TAA (Cheever et al., 2009; Gulley, Arlen, Hodge, & Schlom, 2010) are also of extreme importance. Common targets include oncofetal antigens, oncoproteins, differentiation-associated proteins, and viral proteins, among others (Table 2.1). The potential ideal target is a somatic point mutation that initiates and/or drives the neoplastic process. Clinical trials are underway to evaluate vaccines that target the various *ras* mutations found in colorectal and pancreatic cancer. However, large numbers of tumor-associated mutations among the various exons of the p53 suppressor gene, for example, make generating the large number of possible mutant p53 vaccines somewhat prohibitive. Similarly, it is also logistically difficult to develop vaccines to target the wide array of frameshift mutations and unique mutations that occur in individual tumors, which may differ among different tumor masses of the same patient. On the other hand, nonmutated oncoproteins can more easily be developed as targets; they include overexpressed HER2/neu (ERBB2), p53, and the C-terminal transmembrane subunit of mucin-1 (MUC-1), that is, MUC1-C (Kufe, 2009; Raina et al., 2011).

Numerous trials have targeted "tissue lineage" antigens that are overexpressed in tumors and normally expressed in a nonvital organ, such as prostatic acid phosphatase (PAP), prostate-specific antigen (PSA), and the melanoma-associated antigens glycoprotein 100 (gp100) and tyrosinase. Numerous vaccine trials have also targeted a class of antigens categorized as oncofetal antigens, such as carcinoembryonic antigen (CEA), underglycosylated MUC-1, tumor-associated glycopeptides (Gilewski et al., 2007; Marshall et al., 2000; Ragupathi et al., 2009), and "cancer–testis" antigens defined by serological expression cloning (SEREX) immunodetection such as melanoma-associated antigen (MAGE-A3) and B melanoma antigen (BAGE) (Gnjatic, Old, & Chen, 2009; Gnjatic et al., 2010; Gnjatic, Wheeler, et al., 2009; Karbach et al., 2011). These antigens are overexpressed in many tumor types and to a lesser extent in some normal adult

Table 2.1 Spectrum of current and potential therapeutic cancer vaccine targets

Target type	Examples	References
Oncoprotein	Point-mutated: ras, B-raf, frameshift mutations, undefined unique tumor mutations; HER2/neu, MUC-1 C-terminus, p53	Disis (2009), Salazar et al. (2009), Brichard and Lejeune (2008), Kufe (2009), Raina et al. (2011)
Stem cell/EMT	Brachyury, SOX-2, OCT-4, TERT, $CD44^{high}/CD24^{lo}$, $CD133^{+}$	Polyak and Weinberg (2009), Dhodapkar et al. (2010), Dhodapkar and Dhodapkar (2011), Hua et al. (2011), Mine et al. (2009), Spisek et al. (2007), Fernando et al. (2010), Palena et al. (2007), Fernando et al. (2010)
Oncofetal antigen	CEA, MUC-1, MUC1-C	Butts et al. (2005), Pejawar-Gaddy et al. (2010), Finn et al. (2011), Jochems et al. (2014)
Cancer–testis	MAGE-A3, BAGE, SEREX-defined, NY-ESO	Karbach et al. (2011), Gnjatic et al. (2010), Gnjatic, Wheeler, et al. (2009), Hofmann et al. (2008)
Tissue lineage	PAP, PSA, gp100, tyrosinase, glioma antigen	Schwartzentruber et al. (2011), Sosman et al. (2008), Kantoff, Schuetz, et al. (2010), Gulley, Arlen, Madan, et al. (2010), Kaufman et al. (2004), Kantoff, Higano, et al. (2010), Okada et al. (2011), Wheeler and Black (2011)
Viral	HPV, HCV	Kemp et al. (2008, 2011)
B-cell lymphoma	Anti-id	Schuster et al. (2011), Bendandi (2009), Inoges et al. (2006), Freedman et al. (2009)
Antiangiogenic	VEGF-R	Kaplan et al. (2006), Xiang, Luo, Niethammer, and Reisfeld (2008), Frazer, Lowy, and Schiller (2007)
Glycopeptides	STn-KLH	Gilewski et al. (2007), Ragupathi et al. (2009)

BAGE, B melanoma antigen; CEA, carcinoembryonic antigen; EMT, epithelial–mesenchymal transition; gp100, glycoprotein 100; HCV, hepatitis C virus; HPV, human papillomavirus; MAGE-A3, melanoma-associated antigen-A3; MUC-1, mucin 1; NY-ESO, New York esophageal carcinoma antigen 1; OCT-4, octamer-binding transcription factor 4; PAP, prostatic acid phosphatase; PSA, prostate-specific antigen; SOX-2, (sex-determining region Y)-box-2; STn-KLH, sialyl-Tn-keyhole limpet hemocyanin; TERT, telomerase reverse transcriptase; VEGF-R, vascular endothelial growth factor receptor.

tissues. The recent approval by the U.S. Food and Drug Administration (FDA) of the Gardasil vaccine targeting the human papillomavirus (HPV) for the prevention of cervical cancer also renders HPV an attractive target for cervical cancer therapy, as does targeting the hepatitis C virus for liver cancer therapy. Preclinical studies have also shown the potential of vaccines that target molecules involved in tumor angiogenesis, such as the vascular endothelial growth factor receptor (VEGF-R) (Kaplan et al., 2006; Xiang et al., 2008).

The most provocative group of potential targets for vaccine therapy are molecules that are associated with cancer "stem cells" and/or the EMT process, both of which are associated with drug resistance (Table 2.1). Transcription factors that are drivers of EMT are also associated with tumor cell dissemination to the metastatic site. Recent studies have described the plasticity of so-called "cancer stem cells" and the similarities between human carcinoma cells undergoing EMT and the acquisition of "stem-like" characteristics (Polyak & Weinberg, 2009).

EMT and cancer "stemness" are associated with the proteins (sex-determining region Y)-box-2 (SOX-2) and octamer-binding transcription factor 4 (OCT-4) and with carcinoma cells that are $CD44^{high}$ and $CD24^{low}$ and/or are $CD133^{+}$ (Dhodapkar & Dhodapkar, 2011; Dhodapkar et al., 2010; Hua et al., 2011; Mine et al., 2009; Spisek et al., 2007). Each of these gene products is currently being evaluated for immunogenicity in terms of generating human T-cell responses *in vitro*, but some also have a relatively broad range of expression in normal adult tissues. The T-box transcription factor Brachyury has recently been identified as a major driver of EMT (Fernando et al., 2010). It has been shown to be selectively expressed on both primary and metastatic lesions of several carcinoma types. T-cell epitopes have been identified on the Brachyury protein that have the ability to generate human T cells capable of lysing a range of human carcinoma cells (Palena et al., 2007). This will be discussed in detail below.

A potent means to enhance immunogenicity of a self-antigen is by altering specific amino acids of TAAs to develop enhancer agonist epitopes, which are designed to enhance binding either to the MHC or to the T-cell receptor, resulting in higher levels of T-cell responses and/or higher avidity T cells (Dzutsev, Belyakov, Isakov, Margulies, & Berzofsky, 2007; Hodge, Chakraborty, Kudo-Saito, Garnett, & Schlom, 2005). For example, the gp100 melanoma vaccine contains an enhancer agonist epitope, the PROSTVAC vaccine contains a PSA enhancer agonist epitope, and the PANVAC vaccine contains enhancer agonist epitopes for both CEA and

MUC-1. These vaccines will be discussed in more detail below. It is important to note, however, that the T cells generated by agonist epitopes must be shown to lyse human tumor cells that express the corresponding endogenous native epitope.

3. SPECTRUM OF CURRENT THERAPEUTIC CANCER VACCINE PLATFORMS

Each of the 14 platforms in Table 2.2 has strengths and weaknesses; clinical benefit with the use of the platform can be influenced by the particular TAA that is targeted, the disease and disease stage that are targeted, the clinical trial endpoint, and whether the vaccine is evaluated in combination with an immune stimulant, an inhibitor of immune suppression, or another mode of cancer therapy. Many diverse vaccine platforms have now been evaluated in Phase II and/or Phase III clinical trials, including the use of peptides or proteins in adjuvant, recombinant viruses, bacteria or yeast, whole tumor cells, or delivery of protein- or peptide-activated dendritic cells (DCs) (Table 2.2).

DCs are the most potent antigen-presenting cells (APCs) (Banchereau et al., 2001). Numerous Phase II studies have now evaluated the use of peptide- or protein-pulsed, or viral vector-infected DCs to treat patients with prostate cancer, colorectal cancer, melanoma, glioma, and other cancers (Table 2.2), (Okada et al., 2011; Wheeler & Black, 2011). The Sipuleucel-T vaccine (Kantoff, Higano, et al., 2010), which was recently approved by the FDA for the therapy of minimally symptomatic metastatic castrate-resistant prostate cancer (mCRPC), consists of APCs from peripheral blood mononuclear cells (PBMCs) that have been incubated with the prostate antigen PAP fused to granulocyte macrophage colony-stimulating factor (GM-CSF). This vaccine regimen consists of leukaphereses to purify PBMCs from the patient and processing in a central facility where the PAP fusion protein is added to the APCs; these cells are then reinfused to the patient for the purpose of conferring immunity; this process is repeated three times at biweekly intervals. One drawback of DC and/or APC platforms is that they require leukapheresis(es) and cell culture processing of PBMCs, and thus a limited number of vaccinations can be used.

Vaccines based on peptides from TAAs, which are usually administered in an adjuvant and/or with an immune modulator, are generally cost-effective and have the advantage that the investigator knows exactly which epitope to evaluate in terms of patients' immune responses (Disis, 2009).

Table 2.2 Spectrum of current vaccine platforms in Phase II/III clinical studies

Vaccine platform	Example	Cancer type	References
Dendritic cells/APCs			
APC–protein	Sipuleucel-T (PAP-GM-CSF)	Prostate	Kantoff, Higano, et al. (2010), Higano et al. (2009)
Dendritic cell–peptide	Glioma peptides	Glioma, melanoma	Banchereau and Steinman (1998), Okada et al. (2011), Banchereau et al. (2001)
Dendritic cells–vector infected	rV, rF-CEA–MUC1–TRICOM (Panvac-DC)	Colorectal	Lyerly et al. (2011) Morse et al. (2011), Morse, Chaudhry, et al. (2013), Morse, Niedzwiecki, et al. (2013)
Tumor cells			
Autologous	Adeno-CD40L, colon (BCG)	CLL, colon, melanoma	Okur et al. (2011), Hoover et al. (1993); Luiten et al. (2005)
Dendritic cell/autologous tumor cell fusions	Myeloma/DC fusion	Myeloma	Avigan et al. (2004)
Allogeneic	GVAX (+GM-CSF)	Pancreatic	Laheru et al. (2008), Lutz et al. (2011), Emens et al. (2009)
Recombinant vectors			
Poxvirus	rV, rF-PSA–TRICOM (Prostvac)	Prostate	Moss (1996), Hodge, Higgins, and Schlom (2009), Marshall et al. (2000), Hodge, Grosenbach, Aarts, Poole, and Schlom (2003), Hodge, Poole, et al. (2003), von Mehren et al. (2001, 2000), Madan, Mohebtash, Schlom, and Gulley (2010), Kantoff, Schuetz, et al. (2010), Gulley, Madan, and Schlom (2011), Marshall et al. (2005), Kaufman et al. (2004), Arlen et al. (2006), Sanda et al. (1999)

Continued

Table 2.2 Spectrum of current vaccine platforms in Phase II/III clinical studies—cont'd

Vaccine platform	Example	Cancer type	References
Saccharomyces cerevisiae (yeast)	Yeast ras	Pancreatic	Remondo et al. (2009), Wansley et al. (2008)
Listeria	Listeria mesothelin	Pancreatic	Singh and Paterson (2006)
Alpha- and adenoviruses	Adeno-CEA, alpha-CEA	Carcinoma	MacDonald and Johnston (2000)
Peptides/proteins			
Peptide	gp100 (modified), MUC-1 (Stimuvax), HER2/neu	Melanoma, lung	Disis (2009), Butts et al. (2005), Pejawar-Gaddy et al. (2010), Finn et al. (2011), Schwartzentruber et al. (2011), Sosman et al. (2008), Salazar et al. (2009), Brichard and Lejeune (2008), Disis (2011)
Protein	MAGE-A3, NY-ESO	Melanoma	Karbach et al. (2011)
Antibody	Anti-idiotype	Lymphoma	Schuster et al. (2011), Bendandi (2009), Inoges et al. (2006), Freedman et al. (2009)
Glycoproteins	sTn-KLH	Melanoma	Gilewski et al. (2007), Ragupathi et al. (2009)

APC, antigen-presenting cell; BCG, Bacillus Calmette-Guerin; CD40L, CD40 ligand; CEA, carcinoembryonic antigen; CLL, chronic lymphocytic leukemia; DC, dendritic cell; gp100, glycoprotein 100; GM-CSF, granulocyte macrophage colony-stimulating factor; MAGE-A3, melanoma-associated antigen 3; MUC-1, mucin 1; NY-ESO, New York esophageal carcinoma antigen 1; PAP, prostatic acid phosphatase; PSA, prostate-specific antigen; rF, recombinant fowlpox; rV, recombinant vaccinia; STn-KLH, sialyl-Tn-keyhole limpet hemocyanin.

However, they also have a potential drawback because they target only one epitope or a few epitopes of the TAA. It is generally believed that for a cancer vaccine to be optimally efficacious, it must induce antigen-specific $CD8^+$ cytotoxic T cells (CTLs), which are responsible for tumor cell lysis, and antigen-specific $CD4^+$ "helper" T cells, which provide cytokines to enhance CTL activity. Protein-based vaccines are more costly than peptide-based vaccines, but they usually also contain both CD4 and CD8

Table 2.3 Properties of recombinant poxviral vaccine vectors

- Vectors
  Vaccinia (rV-) elicits a strong immune response
    - host induced immunity limits its continuous use
    - MVA (replication defective)

  Avipox (fowlpox rF-, ALVAC)
    - derived from avian species
    - safe; does not replicate
    - can be used repeatedly with little if any host neutralizing immunity
- Can insert multiple transgenes
- Do not integrate into host DNA
- Efficiently infect antigen presenting cells including dendritic cells

epitopes. Many peptide- and protein-based vaccines are used as part of a DC vaccine platform.

Numerous clinical trials are ongoing involving recombinant poxvirus vaccines (Table 2.3). They include the use of recombinant vaccinia, modified vaccinia strain Ankara (MVA), and the avipoxviruses (fowlpox and canarypox). Poxviruses have the ability to accept large inserts of foreign DNA and thus can accommodate multiple genes each on individual poxviral promoters. Intracellular expression of the transgene(s) allows for processing of the tumor antigen by both the class I and class II MHC pathways (Moss, 1996). Because poxvirus replication and transcription are restricted to the cytoplasm, there is minimal risk to the patient (or host) of insertional mutagenesis. It has been shown in preclinical studies that when the transgene for a TAA is inserted in vaccinia or MVA, it becomes more immunogenic, most likely because of the Toll-like receptors and other properties of the virus that induce a local inflammatory response. This same property of vaccinia, however, leads to virus neutralization by the host immune response and limits its use to one, or at most two, vaccination(s). Recombinant avipoxviruses can be used multiple times; they will induce antiviral immune responses, but they are not neutralizing because their "late" viral coat proteins are not produced in mammalian cells (Hodge et al., 2009; Hodge, Poole, et al., 2003; Marshall et al., 2000).

Other viruses are also being evaluated as vectors for TAAs. Alphaviruses such as venezuelan equine encephalitis virus are attractive vectors because, once they have infected the host, they replicate RNA in the cytoplasm and express high levels of a transgene (MacDonald & Johnston, 2000). Recombinant adenovirus vectors are easy to engineer and have shown utility as vaccines and gene therapy agents (Okur et al., 2011), but clinical evaluation has been hindered by high levels of preexisting antiviral immunity. Newer variants of adenoviruses, however, have been developed and evaluated in preclinical studies and clinical trials and appear to be less immunogenic (Morse, Chaudhry, et al., 2013; Osada et al., 2009).

Bacteria and yeast have shown some promise as vaccine vectors in preclinical studies and are also being evaluated in clinical trials. Heat-killed recombinant *Saccharomyces cerevisiae* is inherently nonpathogenic, can be easily propagated and purified, and is very stable. Recombinant yeasts have been shown to activate maturation of human DCs and to present both class I and class II epitopes of transgenes (Remondo et al., 2009; Wansley et al., 2008). Surprisingly, it appears that these vectors can be administered multiple times without eliciting host-neutralizing activity (Wansley et al., 2008). Attenuated recombinant *Listeria monocytogenes* (Lm) bacteria have also been shown to target DCs, and, like viral and yeast vectors, they stimulate both innate and adaptive immune responses (Singh & Paterson, 2006). Although DNA vaccine platforms have shown promise in preclinical studies (Kaplan et al., 2006; Xiang et al., 2008), early clinical trials have been disappointing. Their exact mode of action is not known at this time. However, new constructs and methods of administration may enhance their utility.

The use of whole-tumor cell vaccines has the advantage of presenting the patient's immune system with a range of both known and undefined TAAs as immunogens. However, this same property also potentially diminishes the relative level of expression of a particular TAA or group of TAAs and their presentation and processing by APCs. The use of a killed whole-tumor cell vaccine is usually accompanied by an immune stimulant such as GM-CSF, or Bacillus Calmette-Guerin (BCG) adjuvant. Autologous tumor cell vaccines have the advantage of presenting to the immune system the unique set of TAAs, such as particular point mutations or fusion gene products, from a given patient's own tumor (Hoover et al., 1993). Because this technology depends on the availability of tumor biopsies, it is feasible for few tumor types and stages. In one promising variation of this technique, however, DCs and autologous tumor cells are fused together before immunization of the patient (Avigan et al., 2004). DC–tumor cell fusions combine the unique properties of whole-tumor cell vaccines with the enhanced antigen-presenting power of DCs. Alternatively, allogeneic whole-tumor cell vaccines, which typically contain two or three established and characterized human tumor cell lines of a given tumor type, are being used to overcome many logistical limitations of autologous tumor cell vaccines. The GVAX vaccine platforms (Emens et al.,

2009; Laheru et al., 2008; Lutz et al., 2011), which contain allogeneic pancreatic, prostate, or breast tumor cells, are a testament to the ability to provide such a vaccine for multicenter evaluation.

Anti-idiotype vaccines are directed against specific antibodies found on the surface of B-lymphoma cells (Belardelli, Ferrantini, Parmiani, Schlom, & Garaci, 2004; Inoges et al., 2006; Schuster et al., 2011) and have the advantage of targeting a unique tumor-specific antigen. A disadvantage is that their generation and production are quite labor intensive in that, to date, each anti-idiotype vaccine has been patient specific. However, some researchers have shown that these patient-specific vaccines can be produced in less than a month (reviewed in Bendandi, 2009).

3.1. An example of optimizing vaccine potency

The spectrum of therapeutic cancer vaccines is as wide as the spectrum of chemotherapeutics and small molecule targeted therapies. One approach to optimizing vaccine potency, for example, is to endow the vaccine with as much immunostimulatory potential as possible. The large genome of poxviruses allows one to add transgenes for one or more TAAs and three T-cell costimulatory molecules. The generation of a robust host response to a weak "self-antigen" such as a TAA requires at least two signals. B7.1 (CD80) is one of the most studied costimulatory molecules in its interaction with its CD28 ligand on T cells. Initial studies demonstrated that the admixing of recombinant vaccinia (rV)-TAA with rV-B7.1 resulted in enhanced TAA-specific T-cell responses and antitumor immunity compared to either vector alone (Hodge et al., 1995; Kalus et al., 1999). Additional studies were conducted with recombinant vaccinia viruses containing other T-cell costimulatory molecules including LFA-3, CD70, ICAM-1, 4-1BBL, and OX-40L (Kudo-Saito et al., 2006; Lorenz, Kantor, Schlom, & Hodge, 1999a; Lorenz, Kantor, Schlom, & Hodge, 1999b; Uzendoski, Kantor, Abrams, Schlom, & Hodge, 1997). Each was shown to enhance antigen-specific T-cell responses, but the combined use of three specific costimulatory molecules (B7.1, ICAM-1, and LFA-3) acted synergistically to further enhance antigen-specific T-cell responses (Fig. 2.1). Each molecule binds to a different ligand on T cells and is known to signal through different pathways. This TRIad of COstimulatory Molecules has been designated TRICOM (Table 2.4). Attempts to add a fourth costimulatory molecule resulted in either a minimal enhancement or reduced immunogenicity to the TAA transgene.

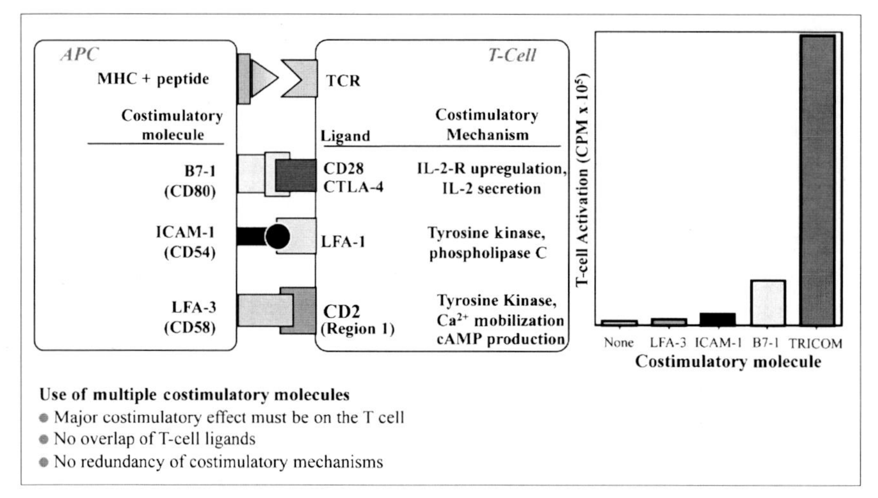

Figure 2.1 The three costimulatory molecules in TRICOM (B7.1, ICAM-1, and LFA-3) act synergistically in enhancing antigen-specific T-cell responses. Each molecule has a distinct ligand on T cells. *Adapted from Schlom, Hodge, et al. (2013). Elsevier Ltd.*

Table 2.4 TRICOM: TRIad of COstimulatory Molecules

Costimulatory molecule	Ligand on T cell
B7-1 (CD80)	CD28/CTLA-4
ICAM-1 (CD54)	LFA-1
LFA-3 (CD58)	CD2
TRICOM = B7-1/ICAM-1/LFA-3	
CEA/TRICOM = CEA/B7-1/ICAM-1/LFA-3	
CEA/MUC-1/TRICOM = CEA/MUC-1/B7-1/ICAM-1/LFA-3 (PANVAC)	
PSA/TRICOM = PSA/B7-1/ICAM-1/LFA-3 (PROSTVAC)	
All vaccines contain:	rV- as a prime vaccine
	avipox (fowlpox, rF-) as multiple booster vaccines
	CEA, MUC-1, and PSA transgenes all contain enhancer agonists

One of the most widely expressed TAAs is CEA (Schlom, Tsang, Hodge, & Greiner, 2001). One of the issues addressed early on was to optimize induction of CEA-specific $CD4^+$ and $CD8^+$ T-cell responses in a host tolerant to CEA and in which CEA is a self-antigen. Since mice do not express human CEA, CEA transgenic (CEA.Tg) mice are employed (Clarke, Mann, Simpson, Rickard-Dickson, & Primus, 1998). These mice express CEA, as do humans, in fetal tissue and in some adult gut tissues. They also express CEA in sera at levels similar to patients with CEA positive tumors. The challenge was to define the best delivery system to break tolerance to CEA in these mice, and to go on to kill tumors engineered to express human CEA. A recombinant vaccinia expressing the CEA transgene (designated rV-CEA) was constructed and demonstrated superiority to other forms of CEA-targeted therapy (Irvine, Kantor, & Schlom, 1993; Kass et al., 1999). Subsequent studies (Kass et al., 1999) also demonstrated the superiority of rV-CEA versus CEA protein in inducing antitumor responses. This study and numerous others have dispelled the belief of "antigenic competition" in using poxvirus vectors, that is, that the poxviral epitopes would interfere with the TAA transgene epitopes for T-cell activation (Larocca & Schlom, 2011).

Evaluation of different vaccine strategies (Aarts, Schlom, & Hodge, 2002; Grosenbach, Barrientos, Schlom, & Hodge, 2001) in the stringent CEA.Tg animal model also demonstrated that (a) a diversified vaccination protocol consisting of primary vaccination with rV-CEA–TRICOM followed by boosting with recombinant fowlpox (rF)-CEA–TRICOM is more efficacious than homogeneous vaccination with either vector and more efficacious than the use of these vectors with one or no costimulatory molecules (Fig. 2.2); (b) continued boosting with vaccine was required to maintain CEA-specific T-cell responses. These strategies were combined to optimally treat CEA-expressing carcinoma liver metastases in CEA.Tg mice (Aarts et al., 2002; Grosenbach et al., 2001).

3.2. Vaccine/vaccine combinations

Different vaccine platforms directed against the same tumor antigen have been shown to enhance host immunity in different ways and thus may have an additive or synergistic antitumor effect. Using a recombinant poxviral TRICOM vaccine and a yeast (*S. cerevisiae*) vaccine, studies evaluated T-cell populations induced by these two diverse platforms in terms of serum cytokine response, T-cell gene expression, T-cell receptor phenotype, and

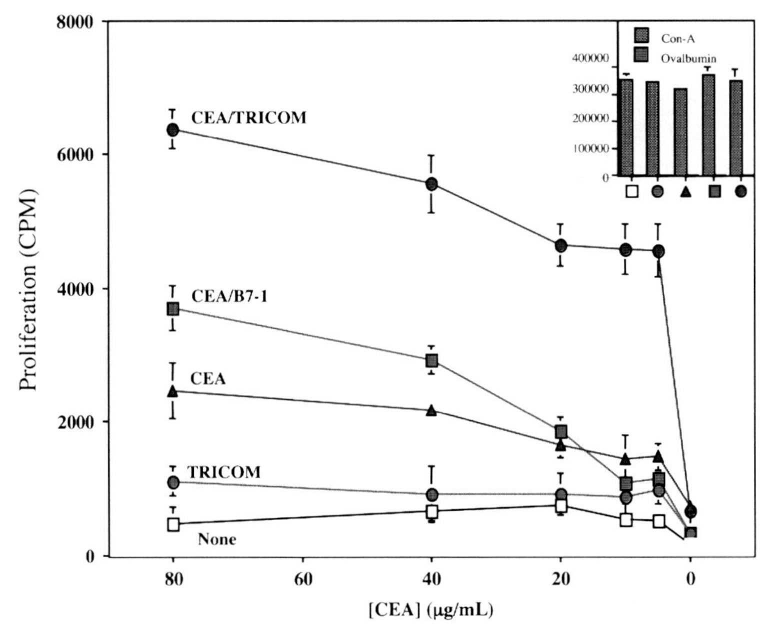

Figure 2.2 CEA-specific lymphoproliferation of T cells from CEA transgenic mice vaccinated with TRICOM vector (without the CEA transgene); rV-, rF-CEA; rV-, rF-CEA-B7.1; or rV-, rF-CEA–TRICOM vectors. *Adapted from Aarts et al. (2002).* (See the color plate.)

antigen-specific cytokine expression (Boehm, Higgins, Franzusoff, Schlom, & Hodge, 2010). T-cell avidity and T-cell-mediated antigen-specific tumor cell lysis demonstrated that vaccination with rV-, rF-CEA–TRICOM or heat-killed yeast-CEA elicits T-cell populations with both shared and unique phenotypic and functional characteristics. Furthermore, both the antigen and the vector played a role in the induction of distinct T-cell populations. These studies (Boehm et al., 2010) thus provide the rationale for future clinical studies investigating concurrent or sequential administration of different vaccine platforms targeting a single antigen. Moreover, studies involving the concurrent or sequential use of vaccines targeting different antigens expressed on a heterogeneous tumor mass are clearly warranted.

4. ANIMAL MODELS TO EVALUATE CANCER VACCINES: PROS AND CONS

The use of animal models to evaluate different forms of immunotherapy, including vaccine therapy, can be extremely helpful in certain cases and "less than useful" in others. For example, murine models can be extremely useful in evaluating the potency of one vaccine platform versus another and in determining the mode of action of a given vaccine platform. They are also useful in the evaluation of therapies employing vaccines in combination with other forms of immunotherapy, or other modes of therapeutic intervention, where scheduling of each intervention can potentially be important and can be monitored. There are, however, several major drawbacks in the use of animal models; in particular, murine transplantable tumors grow at an extremely fast rate, in which the time from transplantation to death is usually a matter of weeks. There is a mindset by some that one must be able to "cure large tumor masses" for any immunotherapy to be effective. The clinical correlate of this is chemotherapy or the adoptive transfer of hundreds of millions to billions of T cells to melanoma patients with large tumor masses. The appropriate utilization of therapeutic cancer vaccines, however, is in the adjuvant or neoadjuvant setting or in patients with minimal residual metastatic disease. Many of these patients have no measurable disease by conventional scanning technologies, but have micrometastatic disease and/or a high risk of recurrence. The evaluation of vaccines and vaccine strategies in animal models thus does not require the use of mice with a large tumor burden, where the tumor mass is a high percentage of the mass of the host. Indeed, the FDA-approved checkpoint inhibitor ipilimumab is ineffective in some models as a monotherapy in mice bearing nonimmunogenic tumors (Hodge et al., 2005). Moreover, both the Provenge vaccine and ipilimumab were FDA approved on the basis of increased survival and not on the basis of improved time to progression (TTP) or reduction of tumor burden employing RECIST criteria.

Animal models are also inappropriate in the analysis of immunogenicity of human TAAs. It is well known that the immunogenicity of a given antigen may differ greatly between different species and even among different strains of mice. Even the use of HLA-A2 Tg mice may not provide an appropriate answer as to immunogenicity in humans due to the fact that the T-cell receptor repertoire of mice (including A2 Tg) and humans differs appreciably. It is suggested that the appropriate model to determine the potential immunogenicity of a given tumor antigen in humans, short of a clinical trial, is the ability of a given peptide/protein or vaccine to activate human T cells *in vitro*, which in turn have the ability to lyse human tumors expressing that antigen. Alternatively, analysis of PBMCs from patients who have received vaccine (not containing the tumor antigen in question), or checkpoint inhibitor monoclonal antibody, can be performed posttreatment to evaluate the ability to generate T cells to the tumor antigen in question, that is, for evidence of epitope spreading/antigen cascade.

5. TYPES OF IMMUNOTHERAPY

5.1. Vaccine therapy and adoptive T-cell transfer: A study in contrasts

Different modes of immunotherapy that target the same tumor antigen can lead to quite distinct results. This is exemplified in a series of studies, both preclinical and clinical involving vaccine therapy and adoptively transferred T cells, both of which target the TAA CEA. Preclinical studies of both modalities used human CEA.Tg mice where CEA is a self-antigen. In one study, it was indicated that vector (devoid of costimulatory molecules)- or DNA-based vaccine directed against CEA was incapable of rejecting CEA positive tumors (Bos et al., 2008). The study went on to report that "effective tumor targeting is only achieved by adoptive transfer of T cells." However, such treatment resulted in severe intestinal autoimmune pathology associated with weight loss and mortality. In a clinical study (Parkhurst et al., 2011), autologous T cells were genetically engineered to express a T-cell receptor directed against a specific human CEA epitope and T cells were adoptively transferred to three patients with metastatic colorectal cancer. All patients experienced profound decreases in serum CEA levels (74–99%) and one patient had an objective regression of cancer metastatic to the lung and liver. However, a severe transient inflammatory colitis represented a dose-limiting toxicity that was induced in all three patients. The authors concluded that this study demonstrated the limitations of using CEA as a target for cancer immunotherapy.

These preclinical and clinical studies are contrasted with the use of vector-based vaccines that contain transgenes for CEA and three costimulatory molecules (designated TRICOM). Using the same CEA. Tg mice described above, tumor-bearing mice received a prime vaccination with rV-CEA–TRICOM and multiple booster vaccinations with rF-CEA–TRICOM. Antitumor immunity leading to cure in approximately 60% of mice was achieved in the absence of any evidence of autoimmunity (Hodge, Grosenbach, et al., 2003). Indeed, no evidence of toxicity was seen employing an array of clinical serum and urine chemistry assays and a comprehensive histopathologic evaluation of all tissues after 1 year. Similar results were also obtained employing CEA–TRICOM vaccination in CEA.Tg mice crossed with mice bearing a mutation in the APC gene. These mice developed spontaneous intestinal tumors. Vaccination resulted in improved overall survival (OS) in the absence of autoimmunity (Greiner, Zeytin, Anver, & Schlom, 2002). A third study (Zeytin et al., 2004) demonstrated that CEA–TRICOM vaccination in combination with Celecoxib elicits antitumor immunity and long-term survival in CEA.Tg/MIN mice in the absence of autoimmunity. Several clinical studies have now shown evidence of antitumor immunity in metastatic cancer patients employing CEA–TRICOM- and CEA–MUC1–TRICOM-based vaccines in the absence of autoimmunity (Marshall et al., 2005; Morse, Niedzwiecki, et al., 2013). In a Phase II trial, patients with metastatic colorectal cancer to the liver and/or lung were vaccinated with PANVAC vaccine (rV-, rF-CEA–MUC1–TRICOM) following metastasectomy (Morse, Niedzwiecki, et al., 2013). At approximately 40 months' follow-up, 90% of the vaccinated patients survived versus approximately 47% in the contemporary control group. OS of colorectal cancer patients after metastasectomy in five other trials ranged from 28% to 58% (Andres et al., 2008; Arru et al., 2008; Choti et al., 2002; House et al., 2010; Pawlik et al., 2005; Sasaki et al., 2005). Moreover, no evidence of autoimmunity was reported in the patients receiving PANVAC. This trial will be discussed in detail below. These studies demonstrated the balance that can indeed be achieved between the induction of an antitumor immune response to a self-antigen and the absence of autoimmunity, and illustrated the distinctions between different forms of immunotherapy targeting the same antigen.

6. THE IMPORTANCE OF ANTIGEN CASCADE IN VACCINE-MEDIATED THERAPEUTIC RESPONSES

Studies have been undertaken (Kudo-Saito, Garnett, Wansley, Schlom, & Hodge, 2007; Kudo-Saito, Schlom, Camphausen, Coleman, & Hodge, 2005) to determine the range of specific immune responses associated with vaccination-mediated tumor regression. In one model, CEA.Tg mice bearing CEA^{+} tumors were vaccinated with the CEA–TRICOM subcutaneous/intratumoral (s.c./i.t.) regimen, and antitumor (Fig. 2.3, top panel) and T-cell immune responses were assessed. These studies showed that CEA needed to be present in both the vaccine and the tumor for therapeutic effects. T-cell responses could be detected not only to CEA (the antigen encoded in the vaccine) but also to other antigens expressed on the tumor: wild-type p53 and an

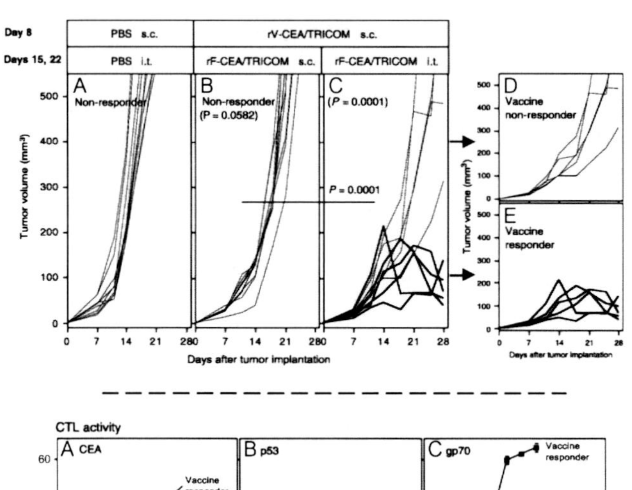
Day 8
Days 15, 22
PBS s.c.
rV-CEA/TRICOM s.c.
PBS i.t.
rF-CEA/TRICOM s.c.
rF-CEA/TRICOM i.t.
A Non-responder
B Non-responder (P = 0.0582)
C (P = 0.0001)
P = 0.0001
D Vaccine non-responder
E Vaccine responder
Tumor volume (mm³)
Days after tumor implantation

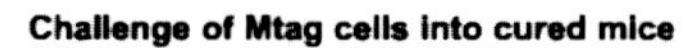

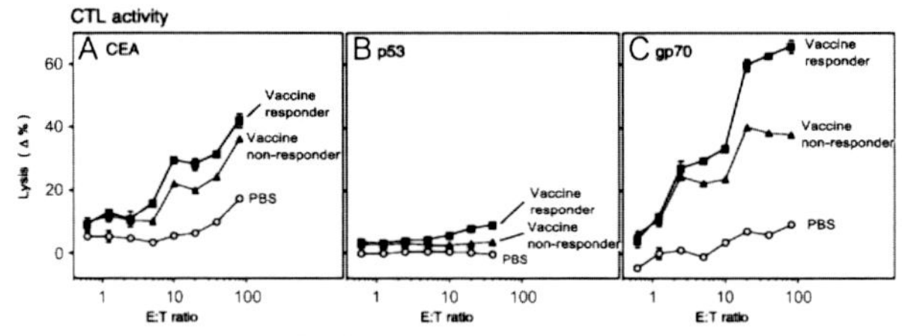
CTL activity
A CEA
B p53
C gp70
Vaccine responder
Vaccine non-responder
PBS
Lysis (Δ %)
E:T ratio

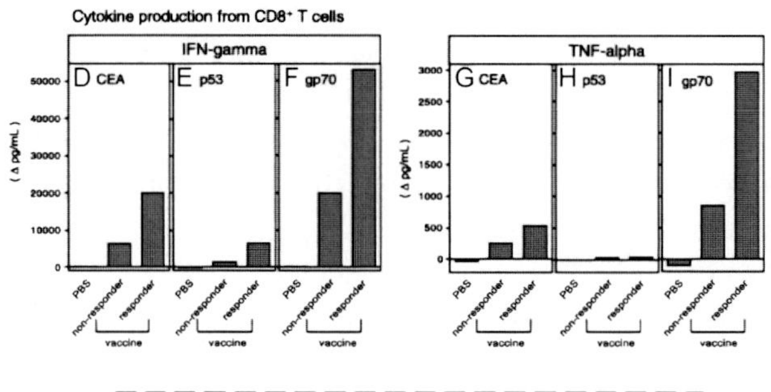
Cytokine production from CD8⁺ T cells
IFN-gamma
TNF-alpha
D CEA
E p53
F gp70
G CEA
H p53
I gp70
(Δ pg/mL)
PBS
non-responder
responder
vaccine

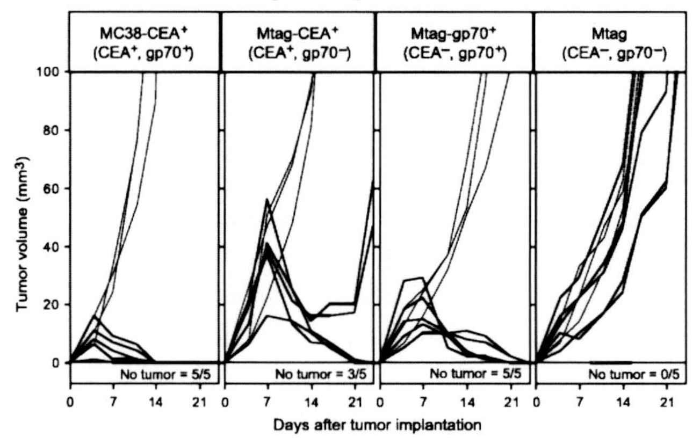
Challenge of Mtag cells into cured mice
MC38-CEA⁺ (CEA⁺, gp70⁺)
Mtag-CEA⁺ (CEA⁺, gp70⁻)
Mtag-gp70⁺ (CEA⁻, gp70⁺)
Mtag (CEA⁻, gp70⁻)
Tumor volume (mm³)
No tumor = 5/5
No tumor = 3/5
No tumor = 5/5
No tumor = 0/5
Days after tumor implantation

endogenous retroviral epitope of gp70 (Fig. 2.3, middle panel A, B, C). Moreover, the magnitude of CD8$^+$ T-cell immune responses to gp70 was far greater than that induced to CEA or p53. Finally, the predominant T-cell population infiltrating the regressing CEA$^+$ tumor after therapy was specific for gp70 (Fig. 2.3, middle panel D–I). Challenge of cured mice with tumors expressing only CEA, only gp70, both antigens, or none showed that the predominant antitumor effect was due to gp70, that is, the cascade antigen (Fig. 2.3, lower panel). Clinical studies in breast cancer patients also showed that there was a correlation with clinical benefit for those patients who demonstrated the phenomenon of antigen cascade/epitope spreading in PBMCs postvaccination (Disis, 2009, 2011; Hardwick & Chain, 2011; Walter et al., 2012). These studies showed that the breadth and magnitude of antitumor immune cascades to multiple antigens can be critical in the therapy of established tumors. These studies also indicate the potential utility of vaccines in addressing the issue of tumor cell phenotypic heterogeneity.

Figure 2.3 The importance of vaccine-induced antigen cascade in antitumor immunity. Top panel: CEA transgenic (Tg) mice were transplanted subcutaneously (s.c.) with MC38-CEA$^+$ tumors on day 0. (A) Control mice were vaccinated with PBS vehicle s.c. on day 8 and intratumorally (i.t.) on days 15 and 22. (B) Mice were vaccinated s.c. with rV-CEA/TRICOM on day 8 and then boosted s.c. with rF-CEA–TRICOM on days 15 and 22. (C) Mice were vaccinated s.c. with rV-CEA–TRICOM on day 8 and then boosted i.t. with rF-CEA–TRICOM on days 15 and 22. *P* values on day 28 compared with the PBS control group. Mice in (C) were separated into two groups (D and E) based on the tumor volume and were used for subsequent immunologic analyses after tumor transplantation. Middle panel: Induction of CD8$^+$ T-cell responses to CEA, p53, and gp70 after the CEA–TRICOM vaccination. Splenic lymphocytes from CEA.Tg mice were used 29 days after tumor transplantation. (A) CEA-specific CTL activity. (B) p53-specific CTL activity. (C) gp70-specific CTL activity. Control mice treated with PBS (○), nonresponders to CEA/TRICOM vaccine therapy (▲), and responders to CEA/TRICOM vaccine therapy (▪). (D–F) Antigen-specific IFN-γ production from CD8$^+$ T cells. (G–I) Antigen-specific tumor necrosis factor-α production from CD8$^+$ T cells. Bottom panel: CEA.Tg mice were vaccinated with CEA–TRICOM as described. Cured mice (see panel (E)) were challenged with tumor cells that were CEA$^+$gp70$^+$, CEA$^+$gp70^{neg}, CEA^{neg}gp70$^+$, or CEA^{neg}gp70^{neg}. The results demonstrate that some of the antitumor effects can be attributed to CEA in the original vaccination, but the most potent antitumor effects are those directed against the tumor-associated cascade antigen gp70 not in the vaccine. As a control, age/sex-matched CEA.Tg mice were implanted with the same tumors (thin lines). *Adapted from Kudo-Saito, Schlom, and Hodge (2005).*

7. TRICOM-BASED VACCINES: CLINICAL STUDIES

Three of the most widely studied human TAAs are CEA, MUC-1, and PSA. CEA is overexpressed in a wide range of human carcinomas, including gastrointestinal, breast, lung, pancreatic, medullary thyroid, ovarian, and prostate. MUC-1 is a tumor-associated mucin, which is overexpressed and hypoglycosylated in all human carcinomas as well as in acute myeloid leukemia and multiple myeloma. The elegant studies of Kufe and colleagues (Kufe, 2009; Raina et al., 2011), as well as others, have demonstrated that the C-terminus of MUC-1 functions as an oncogene. Numerous preclinical studies and recent clinical studies have demonstrated the importance of the induction of $CD8^+$ T-cell responses in vaccine-mediated antitumor immunity. Both the number and avidity of T cells can contribute to tumor cell lysis. Indeed, it has been shown that high avidity T cells can lyse targets with up to 1000-fold lower peptide–MHC complexes than low avidity T cells (Derby, Alexander-Miller, Tse, & Berzofsky, 2001; Hodge et al., 2005; Oh et al., 2003). Since the majority of tumor antigens are "self-antigens," they will by nature induce lower avidity T cells. Even some gene products of somatic mutations, such as point-mutated ras, will generate T cells of much lower avidity when compared to T cells induced by microbial antigens such as influenza. For this reason, strategies have been undertaken to enhance both the number and avidity of T cells to TAAs. One such strategy is the design of enhancer agonist epitopes. The PROSTVAC vaccine (rV-, rF-PSA–TRICOM) contains an enhancer epitope for PSA, and the PANVAC vaccine (rV-, rF-CEA–MUC1–TRICOM) contains enhancer epitopes for both CEA and MUC1 (Table 2.4). Both PROSTVAC and PANVAC are "off-the-shelf" vaccines that can be easily distributed for multicenter clinical trials (Fig. 2.4).

We conducted the first TRICOM trial in humans (Marshall et al., 2005) with rV-, rF-CEA–TRICOM vaccines. Twenty-three out of 58 patients (40%) had stable disease for at least 4 months, with 14 (24%) of these patients having prolonged stable disease (>6 months). Eleven patients had decreasing or stable serum CEA, and one patient had a pathologic complete response. Enhanced CEA-specific T-cell responses were observed in the majority of patients tested.

An early clinical study (Gulley et al., 2008) treated 25 patients with multiple types of cancer with the poxviral vaccine regimen consisting of the genes for CEA and MUC-1, along with TRICOM, engineered into

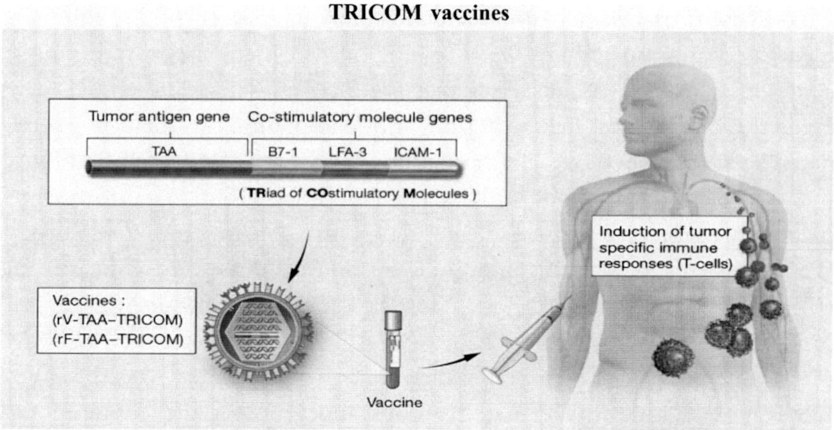

TRICOM vaccines

Figure 2.4 The "off-the-shelf" nature of TRICOM vaccines containing transgenes for one or more tumor-associated antigens (TAAs) and three T-cell costimulatory molecule transgenes. Prime and booster vaccinations are given subcutaneously. *Adapted from Schlom, Hodge, et al. (2013). Elsevier Ltd.* (See the color plate.)

vaccinia (PANVAC-V) as a prime vaccination and into fowlpox (PANVAC-F) as a booster vaccination. The vaccine regimen was well tolerated. Immune responses to MUC-1 and/or CEA were seen following vaccination in nine of 16 patients tested. A breast cancer patient had a confirmed decrease of >20% in the size of large liver metastases, and a patient with clear cell ovarian cancer and symptomatic ascites had a durable (18 months) clinical response radiographically and biochemically.

Another study (Mohebtash et al., 2011) was conducted to obtain preliminary evidence of clinical response in metastatic breast and ovarian cancer patients with PANVAC. Twenty-six patients were enrolled and given monthly vaccinations. These patients were heavily pretreated, with 21 of 26 patients having had three or more prior chemotherapy regimens. Side effects were largely limited to mild injection-site reactions. For the 12 breast cancer patients enrolled, median OS was 13.7 months. Four patients had stable disease beyond their first restaging. One patient had a complete response by RECIST and remained on study for over 37 months. Another patient with metastatic disease confined to the mediastinum had a reduction in a mediastinal mass and was on study for 10 months. Patients with stable or responding disease had fewer prior therapies and lower tumor marker levels than patients with no evidence of response. Further studies to confirm these results are warranted.

A Phase II trial was also conducted in colorectal cancer patients following metastasectomy (surgical removal of lung or liver metastases). In this multicenter trial (Morse, Niedzwiecki, et al., 2013), 74 patients who had no evidence of disease after resection and completion of their physician-determined perioperative chemotherapy were vaccinated with PANVAC (i.e., with vaccine alone or with vaccine-modified DCs). Data from a prospectively registered, comparable contemporary control group of colorectal cancer patients who had undergone metastasectomy were also available (Morse, Niedzwiecki, et al., 2013). The 2-year relapse-free survival was similar in all groups: 47% for the DC-PANVAC group, 55% for the PANVAC group, and 55% for the contemporary control group. However, the 2-year OS was 95% for the vaccinated groups and 75% for the contemporary control group; after approximately 40 months of follow-up, 67 of 74 (90%) of the vaccinated patients survived versus approximately 47% in the contemporary control group; the data for 3- to 5-year survival of colorectal cancer patients after metastasectomy in five other trials range from 28% to 58% (Andres et al., 2008; Arru et al., 2008; Choti et al., 2002; House et al., 2010; Pawlik et al., 2005; Sasaki et al., 2005). A randomized trial is necessary to confirm these results. It is of interest, however, that this is yet another example of a vaccine trial that shows little or no evidence of an improvement in relapse-free survival, yet has an apparent benefit in OS (Hoos et al., 2010).

8. PROSTATE CANCER CLINICAL TRIALS

While the vast majority of prior and ongoing vaccine trials have been conducted in patients with metastatic melanoma, several characteristics render carcinoma patients with minimal disease burden or in the adjuvant and/or neoadjuvant setting amenable to vaccine therapy. Prostate cancer can be considered as a prototype disease for the evaluation of therapeutic cancer vaccines (Madan, Gulley, Fojo, & Dahut, 2010): (a) prostate cancer is generally an indolent disease that may not lead to metastatic disease or death for over a decade or more; consequently, time is often required for a vaccine to generate a sufficient immune response capable of curtailing disease growth; (b) prostate cancer cells express a variety of well-characterized TAAs; (c) the serum marker PSA can be used to identify patients with minimal tumor burden and those responding to therapy; and (d) a well-defined nomogram, the Halabi nomogram (Halabi et al., 2003), can be used at presentation of metastatic disease to predict probable response to standard-of-care chemotherapy and/or hormone therapy.

8.1. PSA–TRICOM (PROSTVAC) studies

A series of clinical trials were conducted to determine the safety and efficacy of PSA-based vaccines. A Phase I study was first conducted to demonstrate the safety of rV-, rF-PSA–TRICOM (PROSTVAC) in patients with prostate cancer (Arlen et al., 2007). PROSTVAC-VF was then evaluated for prolongation of progression-free survival (PFS) and OS in a 43-center randomized, controlled, and blinded Phase II study (Kantoff, Schuetz, et al., 2010). In total, 125 patients with minimally symptomatic mCRPC were randomly assigned. Patients were allocated (2:1) to PROSTVAC-VF plus GM-CSF or to control empty vectors plus saline injections. Eighty-two patients received PROSTVAC-VF and 40 received control vectors. Patient characteristics were similar in both groups. The primary endpoint was PFS, which was similar in the two groups ($P=0.6$). However, at 3 years poststudy, PROSTVAC-VF patients had a better OS with 25 (30%) of 82 alive versus 7 (17%) of 40 controls, a longer median survival by 8.5 months (25.1 months for vaccinated patients vs. 16.6 months for controls), an estimated hazard ratio of 0.56 (95% CI, 0.37–0.85), and a stratified log-rank $P=0.0061$ (Fig. 2.5A). PROSTVAC-VF immunotherapy was well tolerated and associated with a 44% reduction in the death rate and an 8.5-month improvement in median OS in men with mCRPC. These results compare quite favorably with those of the Phase III trials results of Sipuleucel-T vaccine in a similar patient population (Fig. 2.5B). The provocative data with the PROSTVAC vaccine provided evidence of clinically meaningful benefit; a 1200-patient global Phase III study is ongoing.

In a concurrent Phase II study (Gulley, Arlen, Madan, et al., 2010), 32 patients were vaccinated once with rV-PSA–TRICOM and received boosters with rF-PSA–TRICOM. Twelve of the patients showed declines in serum PSA postvaccination, and 2 of 12 showed decreases in index lesions. Median OS was 26.6 months (predicted median OS by the Halabi nomogram was 17.4 months) (Fig. 2.6A). Patients with greater PSA-specific T-cell responses showed a trend ($P=0.055$) toward enhanced survival (Fig. 2.6B). There was no difference in T-cell responses or survival in cohorts of patients receiving GM-CSF versus no GM-CSF. In a concurrent Phase II trial conducted at the same institution (NCI), the Halabi nomogram

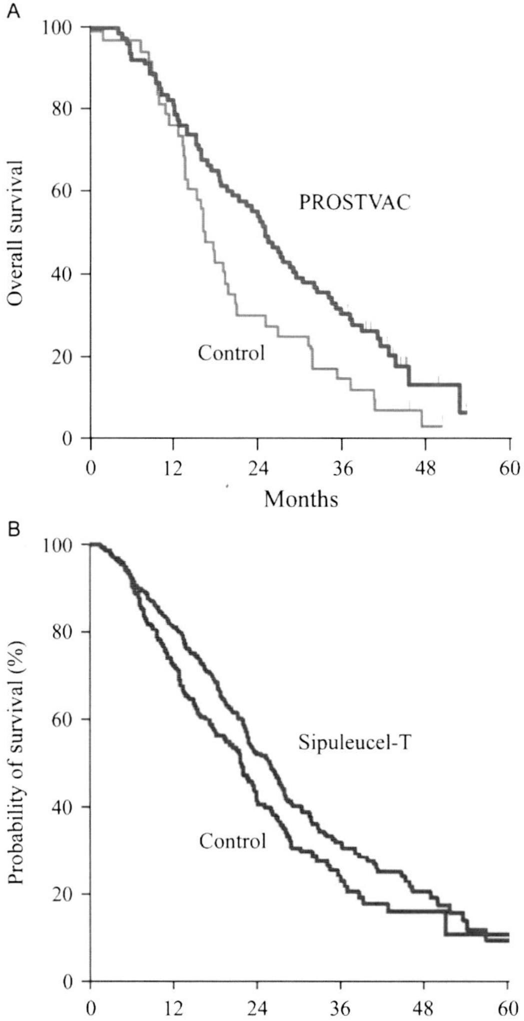

Figure 2.5 (A) Overall survival (OS) of a 43-center placebo-controlled randomized Phase II study of PROSTVAC vaccination. Kaplan–Meier estimator for PROSTVAC (rV-, rF-PSA-TRICOM) arm is shown as a red line and estimator for the control arm is a blue line. The small vertical tic marks show the censoring times. The estimated median OS is 25.1 months for the PROSTVAC arm and 16.6 months for the control arm ($P=0.006$). (B) Overall survival in patients with metastatic castrate-resistant prostate cancer using the Spiluleucel-T vaccine versus control. Sipuleucel-T improved patients' OS (hazard ratio for death $=0.78$; $P=0.03$). The placebo control consisted of cultured antigen-presenting cells (APCs) from leukapheresis, without prostatic acid phosphatase–granulocyte macrophage colony-stimulating factor (PAP–GM-CSF) antigen. Per the trial protocol, the control group could receive cryopreserved APCs with antigen upon disease progression. *Panel (A): Adapted from Kantoff, Schuetz, et al. (2010). Panel (B): Adapted from Kantoff, Higano, et al. (2010).* (See the color plate.)

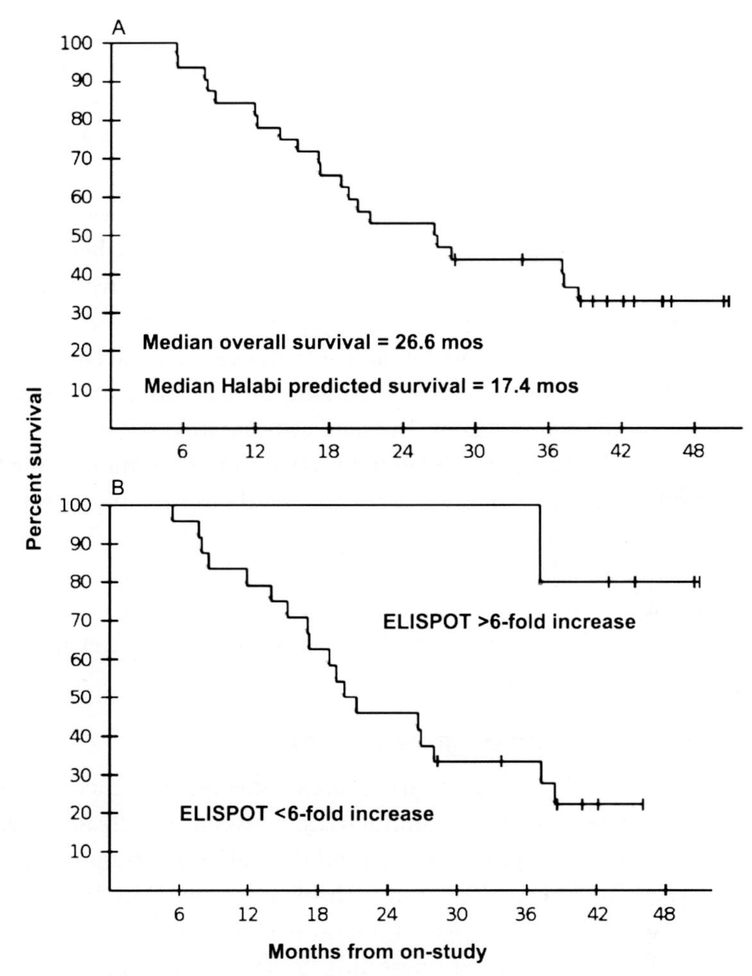

Figure 2.6 (A) The Kaplan–Meier curve for the patients ($n=32$) with metastatic prostatic cancer vaccinated with rV-, rF-PSA–TRICOM (PROSTVAC) demonstrates (A) a median overall survival (OS) of 26.6 months. (B) There was a strong trend in the ability to mount a sixfold increase in PSA-T cells postvaccine and an increase in OS. *Adapted from Gulley, Arlen, Madan, et al. (2010).*

accurately predicted survival in a similar patient population receiving the standard-of-care drug docetaxel. In the vaccine trial, patients with a Halabi-predicted survival (HPS) (Halabi et al., 2003) of <18 months (median predicted 12.3 months) had an actual median OS of 14.6 months, while those with an HPS of ≥ 18 months (median-predicted survival 20.9 months) will meet or exceed 37.3 months, with 12 of 15 patients living longer than predicted ($P=0.035$). Regulatory T-cell (Treg)-suppressive function was also shown to decrease following vaccine in patients surviving longer than predicted and increase in patients surviving less than predicted (Gulley, Arlen, Madan, et al., 2010). Trends were also observed in increased effector:Treg ($CD4^+CD25^+CD127^{neg}FoxP3^+CTLA4^+$) ratios post- versus prevaccination with OS versus HPS. This hypothesis-generating study provided evidence that patients with more indolent mCRPC (HPS $\geq$ 18 months) may best benefit from vaccine therapy.

It has also been hypothesized that vaccine therapy will have significant benefit in patients with minimal disease recurrence. To test this, hormone naive patients ($n=50$) with biochemical recurrence were vaccinated with PROSTVAC (DiPaola et al., 2009). Among 29 patients with follow-up >6 months, the PSA progression-free rate at 6 months (the primary end-point) was 66%. Pretreatment PSA slope was 0.17 log PSA/month (median PSA doubling time 4.4 months), in contrast to the on-treatment slope of 0.12 log PSA/month (median PSA doubling time 7.7 months), $P=0.002$. PROSTVAC vaccine can thus be administered safely with preliminary evidence of patient benefit in the multi-institutional cooperative group setting to patients with minimal disease volume (DiPaola et al., 2009).

8.2. Vaccines and tumor growth rates

One of the confounding phenomena involving therapeutic cancer vaccines is the disconnect, seen in several clinical trials, between an increase in OS in vaccinated patients (vs. control cohorts) and the lack of a corresponding statistical increase in TTP and/or tumor shrinkage using RECIST criteria. Indeed, in the two FDA-approved immunotherapeutics (Provenge vaccine and ipilimumab), a statistical increase in OS was not accompanied by an increase in TTP, as mentioned above. This was also seen in the PROSTVAC trial in metastatic prostate cancer patients. This phenomenon and the disconnect with results observed with cytotoxic drugs have now been studied (Stein et al., 2011).

In patients who are treated with traditional cytotoxic agents, improved TTP is believed to be a prerequisite for improved OS. A recent study (Stein et al., 2011) evaluated tumor regression and growth rates in four chemotherapy trials and one vaccine trial in patients with mCRPC. Figure 2.7A illustrates the growth rate constants observed in that study. Cytotoxic agents affect the tumor only during the period of administration; soon after the drug is discontinued, due to drug resistance or toxicity, tumor shrinkage ceases and the growth rate of the tumor increases (Fig. 2.7A, line b). With vaccine

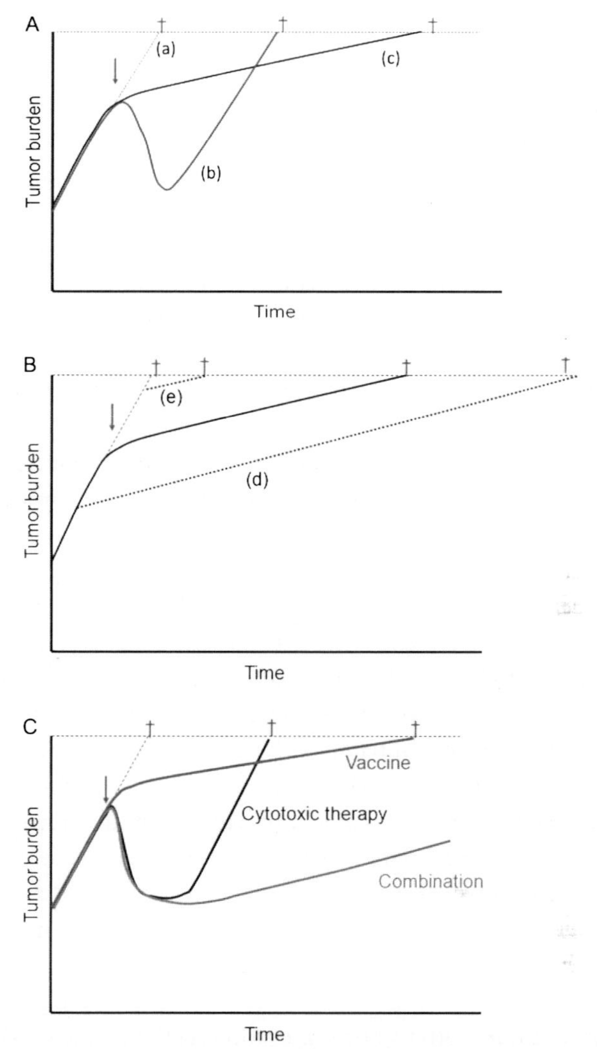

Figure 2.7 Tumor growth rates following chemotherapy versus vaccine therapy. (A) Average tumor growth rates and time to death in patients with metastatic prostate cancer, from five clinical trials (four with chemotherapy and one with PROSTVAC vaccine, also known as PSA–TRICOM). Growth rate of tumor if no therapy is initiated (line a). An examination of five clinical trials (four with chemotherapy and one with PSA–TRICOM (PROSTVAC) vaccine) in patients with metastatic prostate cancer demonstrated that with the use of chemotherapy, there was an initial tumor reduction, but that the growth rate of tumors at relapse (line b) was similar to the initial tumor growth rate prior to therapy; this is contrasted with the reduction in tumor growth rate following vaccine therapy (line c). Thus, for patients with little or no tumor reduction (and thus virtually no increase in time to progression), an increase in survival was observed. †Time to death. (B) This phenomenon could potentially be enhanced if vaccine therapy is initiated

(Continued)

therapy, the mechanism of action and kinetics of clinical response appear to be quite different (Stein et al., 2011). Therapeutic vaccines do not directly target the tumor; rather, they target the immune system. Immune responses often take time to develop and can potentially be enhanced by continued booster vaccinations. Any resulting tumor cell lysis as a consequence of vaccination can lead to cross-priming of additional TAAs; this can take place via the uptake of lysed tumor cells by host APCs, which activate more T cells to other antigens in the tumor, thus broadening the immune repertoire (a phenomenon known as "antigen cascade" or "epitope spreading"). This broader, and perhaps more relevant, immune response may also take some time to develop. This phenomenon is demonstrated in the preclinical model depicted in Fig. 2.3 (Kudo-Saito, Schlom, Hodge, 2005).

Although a vaccine may not induce any substantial reduction in tumor burden, vaccines as monotherapy have the potential to apply antitumor activity over a long period, resulting in a slower tumor growth rate (Fig. 2.7A, line c). This deceleration in growth rate may continue for months or years and, more importantly, through subsequent therapies. This process can thus lead to clinically significant improved OS, often with little or no difference in TTP and a low rate of, or lack of, objective response (Fig. 2.7A, line c). Thus, treating patients with a vaccine when they have a lower tumor burden, as compared with a greater tumor burden, may result in far better outcomes (Fig. 2.7B, line d vs. line e). Early clinical trials with vaccine may have been terminated prematurely with the observance of tumor progression before sufficient vaccine boosts could be administered. The realization of this phenomenon has now led to modifications in how vaccine clinical trials are designed and to new "immune response criteria" for immunotherapy (Hoos et al., 2010). It is also believed that the combined use of vaccine and cytotoxic therapy, or with a small molecule targeted therapeutic, may result in both tumor regression (via the cytotoxic therapy) and reduced tumor growth rate (via vaccine therapy) (Fig. 2.7C) (Gulley, Arlen, Hodge, et al., 2010; Gulley et al., 2011; Madan, Mohebtash, et al., 2010; Stein et al., 2011). These concepts will be discussed below.

Figure 2.7—Cont'd earlier in disease progression or in patients with low tumor burden metastatic disease (line d), but would have minimal effect in patients with large tumor burden (line e). (C) Additional therapies received with vaccine may take advantage of both modalities. *Adapted from data in Stein et al. (2011), Gulley et al. (2011), and Madan, Gulley, et al. (2010). Panel (C): Adapted from Schlom (2012). Oxford University Press.* (See the color plate.)

8.3. A case report: Analysis of a prostate cancer patient over an 18-year period

This patient was 60 years old when initially diagnosed with prostate cancer on the basis of a PSA level of 11.9 ng/mL (Rojan, Funches, Regan, Gulley, & Bubley, 2013) (Fig. 2.8). A prostate biopsy specimen revealed Gleason 7 adenocarcinoma. He was treated with radical prostatectomy and was found to have disease involving both lobes of his prostate extending outside the capsule to the inferior margin (stage T3a). After radical prostatectomy, his PSA level became undetectable (<0.2 ng/mL) and remained so for 2 years. His PSA level then began to rise over the next 4 years. A scan demonstrated abnormal activity only in the prostate bed, and he was treated with external beam radiation therapy. However, his PSA level continued to rise without any apparent effect of the radiation on his PSA doubling time.

Seven and a half years after prostatectomy, his PSA level continued to rise. He enrolled in an Eastern Cooperative Oncology Group Phase II clinical trial and received four doses of a PSA–TRICOM vaccine. Five months after receiving his last vaccination, his PSA level had risen to 12.9 ng/mL before gradually declining and reaching a nadir of 4.3 ng/mL (a 67%

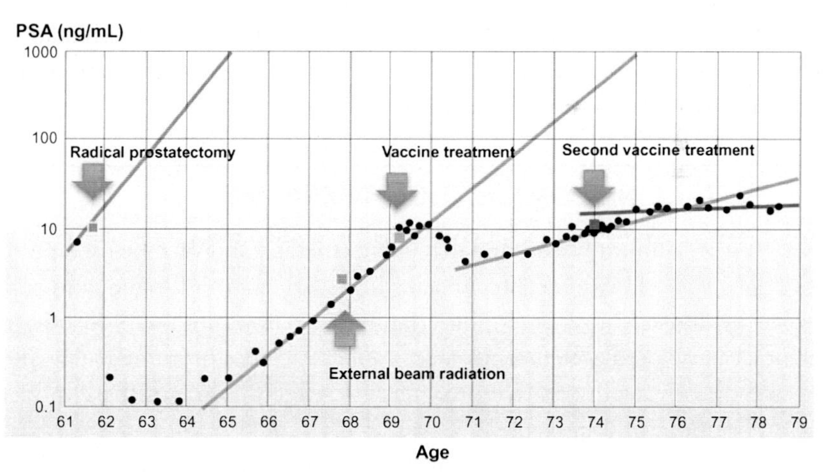

Figure 2.8 Tumor growth rate of a patient with metastatic prostate cancer receiving PSA–TRICOM vaccinations. Prostate-specific antigen (PSA) values (*y*-axis) are in ng/mL and plotted on the natural log scale; PSA was measured at two different institutions as denoted by the two colors. Time (*x*-axis) is in years relative to prostatectomy in 1993. XRT, radiation therapy. *Adapted from Rojan et al. (2013).* (See the color plate.)

decrease 9.2 years after prostatectomy). This nadir was short lived and his PSA level began to rise within 1 year. However, in contrast to his prevaccine PSA doubling time, his doubling time slowed to 2.6 years over the next 5 and 6 years. Approximately 11 years after prostatectomy and 4 years after initial vaccination, a compassionate-use single-patient institutional review board application was approved and he was vaccinated with PROSTVAC vaccine. He subsequently experienced a slowing in his PSA velocity. Since then, his PSA level has declined spontaneously without intervention. His most recent PSA level was 3.5 ng/mL (19.3 years after prostatectomy). This patient originally had a PSA doubling time of 10 months and as such would have been expected to have a median actuarial time to metastases of 8 years (from the time of PSA elevation) and a median actuarial time to death after development of metastases of 5 additional years (Pound et al., 1999). However, he has not required hormonal treatment or chemotherapy in almost 18 years since his biochemical recurrence, and therefore his course has greatly exceeded expectations. Because spontaneous declines in PSA levels are rare in the absence of therapy, it is believed that the ongoing PSA decline in this patient was caused by the combined effect of both sets of vaccinations. Of note, the patient's testosterone levels have been in the low-normal range and much greater than castration levels, arguing against waning testosterone levels causing PSA decline. Anti-PSA antibodies that could decrease PSA levels by pulling PSA out of circulation were not detected on multiple occasions. It is of interest that the decline after the second series of vaccinations did not occur for approximately 3.5 years, consistent with observations that responses to vaccines can be quite delayed (Gulley, 2013).

9. VACCINE COMBINATION THERAPIES

It was commonly assumed that chemotherapy and vaccine therapy are not compatible. However, preclinical studies have demonstrated, and early clinical evidence is now emerging, that multiple nonimmune modalities of therapy can be used concurrently with cancer vaccines or immediately after vaccine therapy, with additive or synergistic effects. Some of these modalities are summarized in Table 2.5.

Certain chemotherapeutic agents can enhance vaccine-mediated T-cell killing by several different mechanisms. Oxaliplatin and anthracyclines, such as doxorubicin, will induce what has been termed "immunogenic tumor cell death," which results in enhanced cross-priming of TAAs by DCs and subsequent activation of T cells (Kepp et al., 2011; Locher et al., 2010; Tesniere

Table 2.5 Vaccines in combination with other therapeutic modalities

Modality	Mechanism of action to enhance vaccine efficacy
Hormonal therapy	Thymic regeneration and induction of naive T cells
Chemotherapy	"Immunogenic" tumor cell death
	Alterations in tumor cell phenotype
	Enriched effector:regulator cell ratios
Radiation	Alterations in tumor cell phenotype
Small molecule targeted therapeutics	Alterations in tumor cell phenotype
	Enriched effector:regulator cell ratios
Monoclonal antibodies	Enhanced ADCC
Imids[a] (lenalidomide)	Stimulate T-cell proliferation

ADCC, antibody-dependent cell-mediated cytotoxicity.
[a]Imids are a novel class of immunomodulators.

et al., 2010; Zitvogel et al., 2010). Other chemotherapies have been shown to suppress certain types of immune regulatory cells and/or enhance the ratio of effector T cells to suppressor T cells (as will be discussed below). Certain chemotherapeutic agents and small molecule targeted therapies have also been shown to induce "immunogenic modulation" of tumor cells. This results in an increased expression of TAAs, peptide–MHC complexes, adhesion molecules, and death receptors such as Fas on the surface of tumor cells and thus renders the tumor cells more susceptible to vaccine-induced T-cell killing (Garnett et al., 2004; Kodumudi et al., 2010); this same phenomenon has been observed when tumor cells have been exposed to external beam radiation, radiolabeled monoclonal antibodies, and chelated bone-seeking radionuclides (Chakraborty et al., 2004; Chakraborty, Gelbard, et al., 2008; Gelbard et al., 2006).

Hormonal therapy that is used in the treatment of several different stages of prostate cancer has been shown to induce thymic regeneration and the induction of naive T cells (Lee, Hakim, & Gress, 2010; Sportes et al., 2010; Williams & Gress, 2008); it is at this interval that vaccine therapy may be most effective (Arredouani et al., 2010; Sanda et al., 1999).

A clinical study has demonstrated increased levels of infiltrating T cells in prostate cancer biopsies post- (vs. pre-) hormonal therapy.

10. COMBINATION THERAPIES—PRECLINICAL STUDIES

There have been numerous studies combining different forms of radiation with vaccine therapy. Below are some examples.

10.1. Vaccine and radiation synergy

A synergy has been demonstrated between local radiation of tumor and vaccine therapy (Chakraborty et al., 2004; Kudo-Saito, Schlom, Camphausen, et al., 2005). CEA.Tg mice with developing MC38 murine carcinoma cells transfected with CEA were treated with rV-, rF-CEA–TRICOM. One dose of 8-Gy radiation to tumor induced upregulation of the death receptor Fas (CD95) *in situ* for up to 11 days. When vaccine therapy and local radiation of tumor were used in combination, as opposed to individually, dramatic and significant cures were achieved (Chakraborty et al., 2004; Kudo-Saito, Schlom, Camphausen, et al., 2005). This was shown to be mediated by the engagement of the Fas/Fas ligand pathway.

One of the issues with radiation therapy is the nonlethal dose of radiation that some tumor cells will receive due to issues of normal tissue toxicity or their location outside the field being irradiated. Twenty-three human carcinoma cell lines (12 colon, 7 lung, and 4 prostate) were examined for their response to non-lytic doses of radiation (Garnett et al., 2004). Seventy-two hours postirradiation, changes in the expression of surface molecules involved in T-cell-mediated immune attack such as specific TAAs and MHC class I, and Fas were examined. Twenty-one of the 23 (91%) cell lines upregulated one or more of these surface molecules postirradiation. Overall, the results of this study suggested that nonlethal doses of radiation can be used to make human tumors more amenable to T-cell attack. Another study (Chakraborty, Wansley, et al., 2008) explored the possibility that exposure to palliative doses of a radiopharmaceutical agent could also alter the phenotype of tumor cells to render them more susceptible to T-cell-mediated killing. LNCaP tumor cells exposed to ^{153}Sm-EDTMP, which is used to treat pain due to bone metastasis, also upregulated the surface molecules such as Fas, CEA, MUC-1, MHC class I, and ICAM-1, and rendered LNCaP cells more susceptible to killing by CTLs specific for PSA, CEA, and MUC-1.

10.2. Vaccines in combination with chemotherapy

Taxanes are commonly used to treat breast, prostate, and lung cancers, among others. One murine *in vivo* study (Garnett, Schlom, & Hodge, 2008), for example, showed that docetaxel modulates $CD4^+$, $CD8^+$, $CD19^+$, NK (natural killer), and Treg populations in nontumor-bearing mice, and docetaxel combined with CEA–TRICOM vaccination is superior to either agent alone at reducing tumor burden.

10.3. Vaccine in combination with small molecule targeted therapies

Studies (Farsaci, Higgins, & Hodge, 2012; Finke et al., 2008) have investigated the immunomodulatory effects of sunitinib to rationally design a potential combinational platform with vaccine therapy. In one study, the effect of differently timed combinations of sunitinib and CEA–TRICOM vaccine in CEA.Tg mice was evaluated. *In vivo*, one cycle of sunitinib caused bimodal immune effects: (a) decreased regulatory cells during the 4 weeks of treatment and (b) an immunosuppression rebound during the 2 weeks of treatment interruption. Continuous sunitinib followed by vaccine, however, increased intratumoral infiltration of antigen-specific T lymphocytes, decreased immunosuppressant Tregs and myeloid-derived suppressor cells, reduced tumor volumes, and increased survival. These studies showed that the immunomodulatory activity of continuous sunitinib can create a more immune-permissive environment for combination therapy with vaccine. Small molecule BCL-2 inhibitors are being examined as monotherapy in Phase I/II clinical trials for several types of tumors. Activated mature $CD8^+$ T lymphocytes were shown (Farsaci et al., 2010) to be more resistant to a BCL-2 inhibitor as compared to early-activated cells. *In vivo*, optimal antitumor activity was obtained when the BCL-2 inhibitor was given after vaccination so as to not negatively impact the induction of vaccine-mediated immunity; this resulted in an increase in the $CD8^+$:Treg ratio and significant reduction of pulmonary tumor nodules (Fig. 2.9). These studies (Farsaci et al., 2010; Kim et al., 2014) and others also undermine the importance of scheduling of a given "nonimmune" therapeutic with vaccine therapy.

10.4. The effect of nonimmune therapeutic interventions on immune cells

As discussed below, the interplay of host immunity with many standard-of-care therapies such as chemotherapy, radiation, hormonal therapy, and small

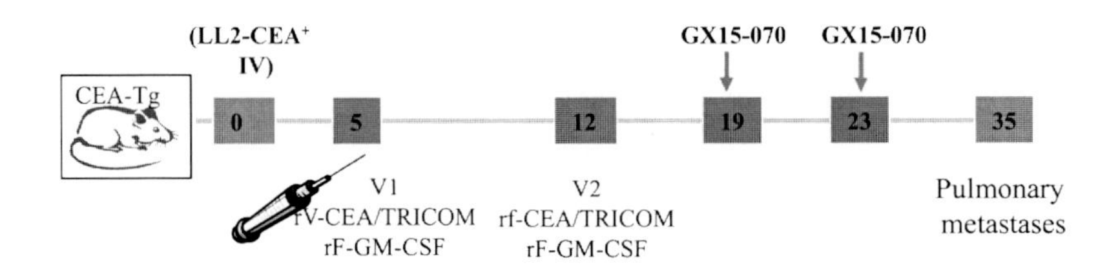

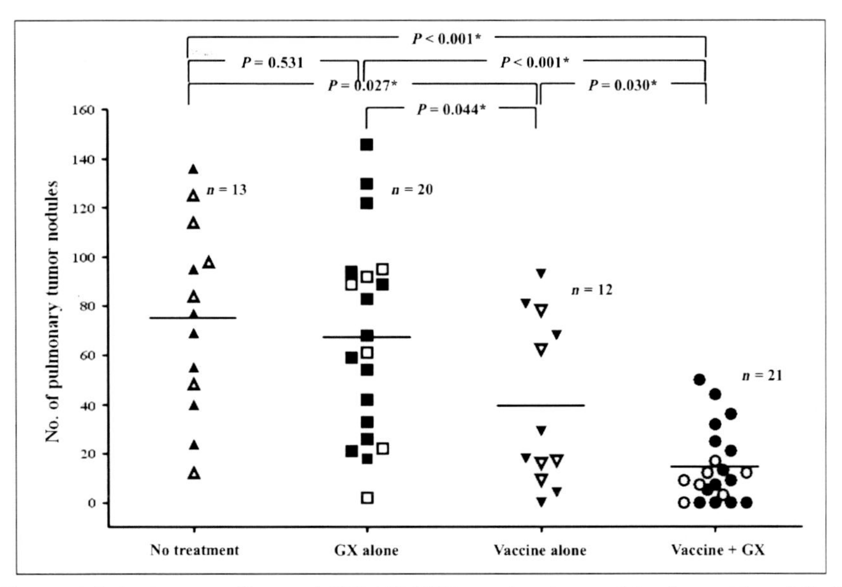

Figure 2.9 Antitumor effects of CEA–TRICOM vaccine in combination with a BCL-2 inhibitor (GX15-070) exploiting the differential effect of the pan BCL-2 inhibitor on Tregs versus effector cells. *Adapted from Schlom, Hodge, et al. (2013). Elsevier Ltd.* (See the color plate.)

molecule targeted therapeutics is now becoming apparent through several lines of investigation. For many cancer types, the specific immune infiltrate in the primary tumor is a strong and independent predictor of response to subsequent therapies and is thus a strong prognostic indicator (see Fridman, Pages, Sautes-Fridman, & Galon, 2012; Galluzzi, Senovilla, Zitvogel, & Kroemer, 2012; Galon et al., 2012; Jochems & Schlom, 2011 for reviews). Preclinical studies and some clinical studies have shown that certain chemotherapeutic agents and small molecule targeted therapeutics can have differential effects on specific components of the immune system that can lead to enhanced or reduced antitumor effects (Adotevi et al., 2010; Emens et al., 2009; Finke et al., 2008; Ko et al., 2010, 2009; Vanneman & Dranoff, 2012). These phenomena have potentially important implications in designing clinical trials of the combined use of "nonimmune" therapies with immunotherapeutic agents such as cancer vaccines.

Many preclinical studies and some clinical studies have provided evidence that Tregs play an important role in inhibiting immune responses to active immunotherapy protocols employing agents such as therapeutic cancer vaccines or checkpoint inhibitors. Studies have also demonstrated that both the number and suppressive function of Tregs on effector T cells need to be investigated. To provide insight toward the possibility of how active immune therapies can be employed in combination with non-immune standard-of-care therapies of carcinoma patients, both number and function of Tregs obtained from peripheral blood of cancer patients was investigated both prior to and during therapy with two chemotherapy regimens and two targeted therapy regimens. These studies showed that tamoxifen plus GnRH treatment had minimal effects on Tregs in breast cancer patients, and the effect of sunitinib had differential effects on Tregs among patients with metastatic renal carcinomas (Roselli et al., 2013). However, the use of the two chemotherapy regimens, docetaxel in patients with both metastatic prostate cancer and metastatic breast cancer (Fig. 2.10), and cisplatin plus vinorelbine in patients with non-small cell lung cancer (Fig. 2.11), each resulted in statistically significant increases in $CD4^{+}$:Treg

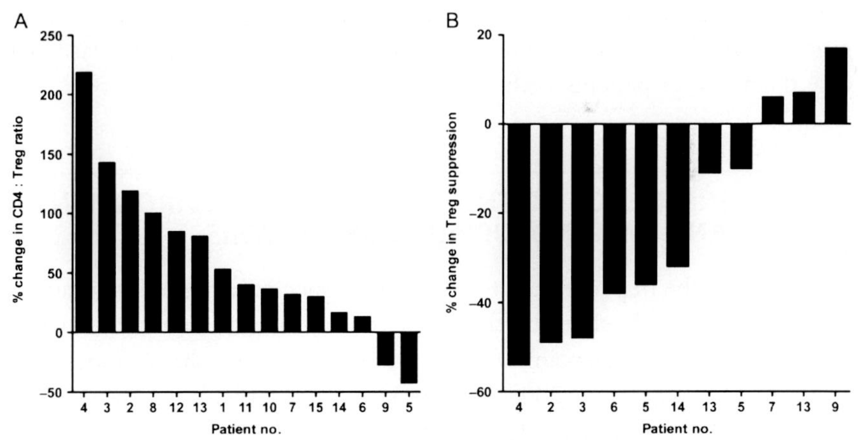

Figure 2.10 Changes in Teff:Tregs ratios and suppressive activity of Tregs during therapy with docetaxel in patients with hormone refractory prostate cancer. (A) Waterfall plot of the change in the ratio of Teff:Tregs during therapy with docetaxel in patients with hormone refractory prostate cancer. Peripheral blood samples were collected prior to therapy and before starting cycle II. (B) Waterfall plot of the change in suppressive activity of Tregs during therapy with docetaxel. *Adapted from Roselli et al. (2013).*

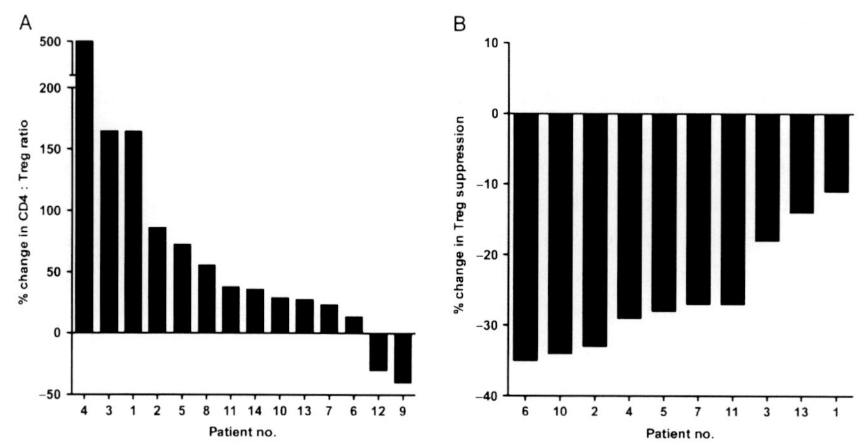

Figure 2.11 Changes in Teff:Tregs ratios and suppressive activity of Tregs in non-small cell lung cancer (NSCLC) patients before and during therapy with cisplatin plus vinorelbine. (A) Waterfall plot of the change in the ratio of Teff:Tregs in NSCLC patients before and during therapy with cisplatin plus vinorelbine. Patients with NSCLC were treated in the adjuvant setting, postsurgery. PBMCs were collected from peripheral blood at baseline and postcycle III. (B) Waterfall plot of the change in suppressive activity of Tregs from NSCLC patients before and during therapy with cisplatin plus vinorelbine. *Adapted from Roselli et al. (2013).*

ratios and reduced functional suppressive activity of Tregs posttherapy in the majority of patients (Roselli et al., 2013). These studies provide further rationale for the use of vaccines in combination with certain standard-of-care therapies for patients with carcinomas.

11. INFLUENCE OF THE TUMOR MICROENVIRONMENT AND IMMUNOSUPPRESSIVE FACTORS

One of the major reasons for the limited success of therapeutic cancer vaccines to date is likely to be the negative influence of the tumor microenvironment and other immunosuppressive factors (Cham, Driessens, O'Keefe, & Gajewski, 2008; Gajewski, Meng, Blank, et al., 2006; Gajewski, Meng, & Harlin, 2006). Preclinical studies have shown that the interstitial pressure within a large tumor mass diminishes diffusion of macromolecules, such as antibodies, and effector cells, such as T cells (Carmeliet & Jain, 2011; Fukumura, Duda, Munn, & Jain, 2010). Most solid tumors also lack T-cell costimulatory molecules. Thus, when activated T cells, especially those of relatively low avidity directed against self-antigens, bind to tumors lacking costimulatory molecules, they are anergized and lose lytic capacity. Similarly, it has been shown in preclinical models of chronic viral infection that T cells chronically exposed to viral antigen can become exhausted (Kim & Ahmed, 2010; Mueller & Ahmed, 2009). The inhibitory co-receptor programmed death 1 (PD-1) has been shown to be present on such exhausted T cells (Vezys et al., 2011), and PDL1 on the surface of tumor cells can anergize tumor-infiltrating T cells. The use of cancer vaccines with checkpoint inhibitors such as anti-CTLA4 (ipilimumab), anti-PD1, and anti-PDL1 is thus a quite fertile area of investigation. One such clinical trial combining PROSTVAC vaccine with ipilimumab will be discussed below.

12. VACCINE COMBINATION THERAPIES—CLINICAL STUDIES

Several hypothesis-generating clinical trials were first conducted in prostate cancer patients with vaccine in combination with hormonal therapy, radiation, or chemotherapy (Arlen et al., 2006, 2005; Gulley et al., 2005; Madan et al., 2008). In one trial, 42 nonmetastatic prostate cancer patients were randomized to receive vaccine (rV-PSA + rV-B7.1 followed by rF-PSA boosts) versus second-line antiandrogen therapy with nilutamide (Arlen et al., 2005). A survival analysis at 6.5 years from the initiation of therapy on this trial has been reported (Madan et al., 2008). Median survival exhibited a trend toward improvement for patients initially randomized to the vaccine arm (median 5.1 vs. 3.4 years). These data suggested that patients with more indolent disease may derive clinical benefit from vaccine alone or vaccine before second-line hormone therapy compared with hormone therapy alone or hormone therapy followed by vaccine. A study is currently enrolling patients with nonmetastatic CRPC on testosterone suppression therapy who have a rising PSA. The first 26 patients enrolled were evaluated. Median TTP was 223 days for flutamide + PSA–TRICOM versus 85 days for flutamide alone (Bilusic et al., 2011). Two Phase II trials are also ongoing in both early and metastatic prostate cancer where patients are randomized to receive the newly FDA-approved androgen receptor antagonist enzalutamide ± PROSTVAC vaccine (NCT01867333; NCT01875250).

A combination therapy randomized Phase II study has recently been completed in metastatic prostate cancer patients with bone metastases. Patients received the bone-seeking radionuclide conjugate ^{153}Sm ± PROSTVAC vaccine. While this was a small trial, there was a clear trend in improved TTP in the combination arm (Heery et al., 2013).

Preclinical studies in mice have shown that CTLA4 blockade can, among other reported activities, increase T-cell avidity, leading to enhanced T-cell-mediated immune responses to the vaccine (Allison et al., 1998; Egen, Kuhns, & Allison, 2002; Hodge et al., 2005).

Ipilimumab (Yervoy) is an antagonistic anti-CTLA4 monoclonal antibody that blocks the activity of CTLA4 (Beer et al., 2008; Small et al., 2007; Wolchok et al., 2010). A Phase III randomized trial of ipilimumab in patients with metastatic melanoma showed a significant improvement in OS, but no significant improvement in TTP, relative to an active control group (Hodi et al., 2010). The PSA–TRICOM vaccine is designed to enhance T-cell costimulation through enhanced expression of the transgenes of three T-cell costimulatory molecules (CD58, CD80, and ICAM1) on APCs engaging their respective ligands on T cells. CD80 is known to react with CD28 on T cells for positive costimulation and with CTLA4 for immune inhibition. The antagonist monoclonal antibody anti-CTLA4 was designed to interrupt this negative signal and enhance immunity. It was thus unclear how a vaccine, such as PSA–TRICOM, with its positive costimulation, would interact in terms of safety and efficacy with an anti-CTLA4 monoclonal antibody designed to block negative costimulatory signals, especially in view of the severe immune-related adverse events noted in some patients receiving anti-CTLA4 alone. A study was conducted to assess fixed doses of PSA–TRICOM with escalating doses of ipilimumab, with the aim of establishing the safety and tolerability of these combined treatments. The results showed that the combination of a vaccine that enhances immune costimulation with an immune checkpoint inhibitor does not seem to be associated with increased immune-related adverse events compared with ipilimumab alone. There also appeared to be a survival benefit in patients receiving PROSTVAC + the 10 mg/kg dose of ipilimumab versus lower doses of ipilimumab (Madan et al., 2012) (Fig. 2.12). The results compare quite favorably with the results of a Phase II study employing PROSTVAC alone in a similar population. There also appeared to be a greater serum PSA response in the chemotherapy naïve patients in the combination study (Madan et al., 2012) (Fig. 2.13). A Phase III trial of ipilimumab with radiation in advanced metastatic prostate cancer did not show a statistical survival benefit, that is, only a 1.2-month difference in OS versus the placebo arm (Gerritsen, 2013). The results of this trial using ipilimumab alone, which is in contrast with the results of two trials using vaccine plus ipilimumab (Madan et al., 2012; van den Eertwegh et al., 2012), provide evidence that

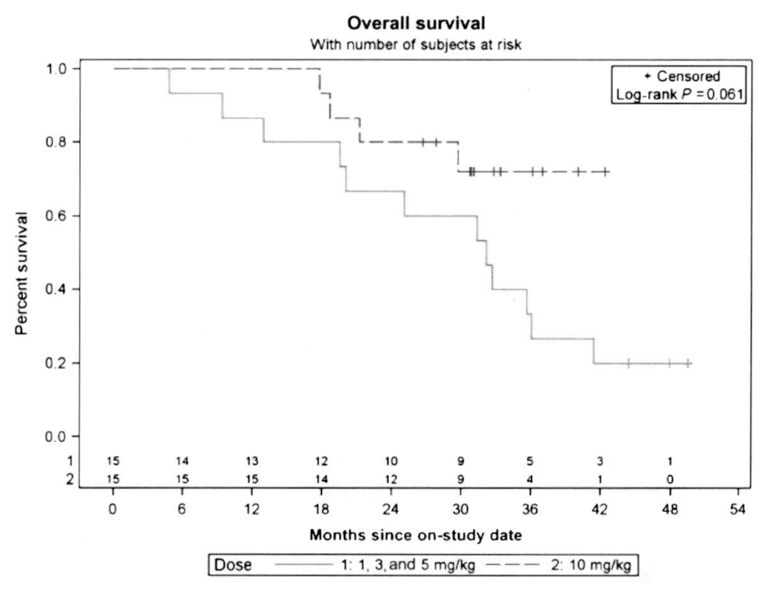

Figure 2.12 There was a trend toward improved overall survival in chemotherapy naïve metastatic prostate cancer patients treated with PROSTVAC vaccine and the 10 mg/kg dose of ipilimumab relative to lower dose levels of ipilimumab. *Adapted from Madan et al. (2012).* (See the color plate.)

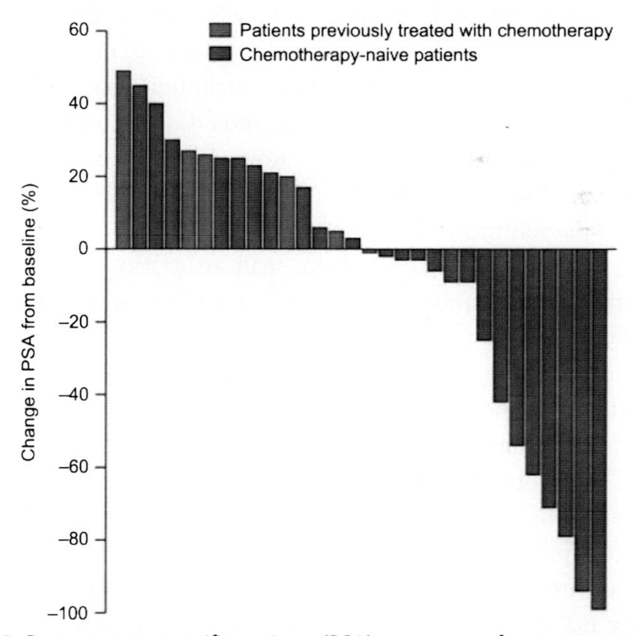

Figure 2.13 Best prostate-specific antigen (PSA) responses after treatment with PROSTVAC vaccine plus ipilimumab. 25% of patients had PSA declines >50% posttreatment. *Adapted from Madan et al. (2012).*

the combination of ipilimumab + vaccine may be more efficacious than the use of either agent alone. While the various combination therapy trials discussed above all provided preliminary evidence of improved patient outcome when vaccine is added to another therapy, these results must be considered as only hypothesis generating. Larger randomized trials are required to substantiate such findings.

13. BIOMARKERS

The most common biomarker used in vaccine therapy trials has been the immune response of patients to TAAs post- versus prevaccination. Most trials have analyzed antibody responses to TAAs and/or analyzed PBMCs for $CD8^+$ and/or $CD4^+$ responses to the TAA in the vaccine using enzyme-linked immunosorbent spot- or fluorescence-activated cell sorting-based assays for cytokine production, or by binding of a peptide tetramer complex to the surface of T cells. Preclinical studies have shown, however, that the level of cytokine production by a T cell is not always associated with its lytic capacity. Because of the limited availability of samples, however, few studies have actually been able to measure the lytic capacity of T cells. Moreover, with the exception of melanoma, tumor biopsy specimens—the more appropriate site to obtain TAA-specific T cells—are usually unavailable in many trials. Recently, a more comprehensive analysis of immune cell subsets from PBMCs has occurred in some studies including, in addition to T-cell responses, analyses of Tregs, MDSCs, NK, and DCs (Butterfield et al., 2011; Disis, 2011). Ratios of effector to regulatory cells have also been analyzed (Gulley, Arlen, Madan, et al., 2010). Numerous studies have also used analyses of multiple serum cytokines and chemokines.

Bioassays from several studies have seen associations between clinical outcome and a given immune assay; however, these results are far from having identified any one assay as a "surrogate" for clinical benefit. Potential reasons for this may be that PBMC analyses may not reflect which immune cells are actually at the tumor site. Few studies have actually analyzed the "antigen cascade" phenomenon, where the true correlate of clinical benefit may be a T-cell population directed against a TAA not in the vaccine, but generated via cross-priming. Only recently has survival benefit emerged as a prominent endpoint in many vaccine studies for comprehensive correlative analyses with survival. The diversity of the immune responses among individuals may not allow any one marker or set of markers as a surrogate for clinical response. Even analyses of patients' immune response to influenza have been confounding. While antibodies to flu hemagglutinin are used as a "surrogate" for protection in population studies, they do not predict protection from flu for an individual vaccine (Nakaya et al., 2011).

14. VACCINE TARGETS INVOLVED IN TUMOR PROGRESSION AND DRUG RESISTANCE

Much attention continues to be paid to the development of small molecule targeted therapeutics and monoclonal antibodies that target gene products involved in tumor initiation, that is, oncogenes, and to tumor suppressor genes. Equally important, however, are those genes and gene products involved in tumor progression, that is, those gene products associated with tumor invasion, metastasis, and drug resistance. The phenomena of solid tumor cell "stemness" and EMT are such processes. Evidence is now emerging that "cancer stem cells" are more "plastic" than originally believed, and the distinction between cancer cell "stemness" and EMT is becoming more blurred. EMT is a reversible process during which cells switch from a polarized, epithelial phenotype into a highly motile, mesenchymal phenotype (Kalluri & Weinberg, 2009; Thiery & Sleeman, 2006). While EMT is a normal process during embryogenesis and organogenesis, numerous observations now support the concept that EMT also plays an essential role in the progression of carcinomas (Thiery, 2002). During the metastasis of carcinomas, tumor cells must undertake a series of sequential steps that will allow them to detach from the primary tumor mass and to finally reach the distant sites of metastasis. By undergoing EMT, tumor cells can acquire the ability to move and to invade the surrounding tissues, two fundamental properties for tumor dissemination. In addition, several reports are now indicating that tumor cells undergoing EMT also acquire stem cell-like features and mechanisms of resistance to cell death (Arumugam et al., 2009; Vega et al., 2004). The induction of EMT in various cancer cell lines, for example, has been shown to positively correlate with resistance to radiation (Kurrey et al., 2009), chemotherapy (Yang et al., 2006), and epidermal growth factor receptor kinase inhibitors (Thomson et al., 2005). Many of the molecules known to be mediators of EMT or stemness are transcription factors and/or intracellular molecules. Thus for these to be targets of vaccine-mediated therapy, one must demonstrate that they are processed intracellularly and transported to the cell surface in the context of peptide–MHC complexes in both APCs and targeted tumor cells. Two examples of such potential targets are Brachyury and the C-terminus of MUC1.

Studies have demonstrated that Brachyury functions as a master regulator of EMT in human carcinoma cells. The upregulation of Brachyury in human epithelial cancer cells results in morphological changes representative of EMT, including the acquisition of a fibroblast-like morphology, the loss of the epithelial markers E-cadherin and Plakoglobin, and enhanced levels of the mesenchymal proteins Fibronectin, N-cadherin, and Vimentin (Fernando et al., 2010) (Fig. 2.14A). As a consequence of this phenotypic

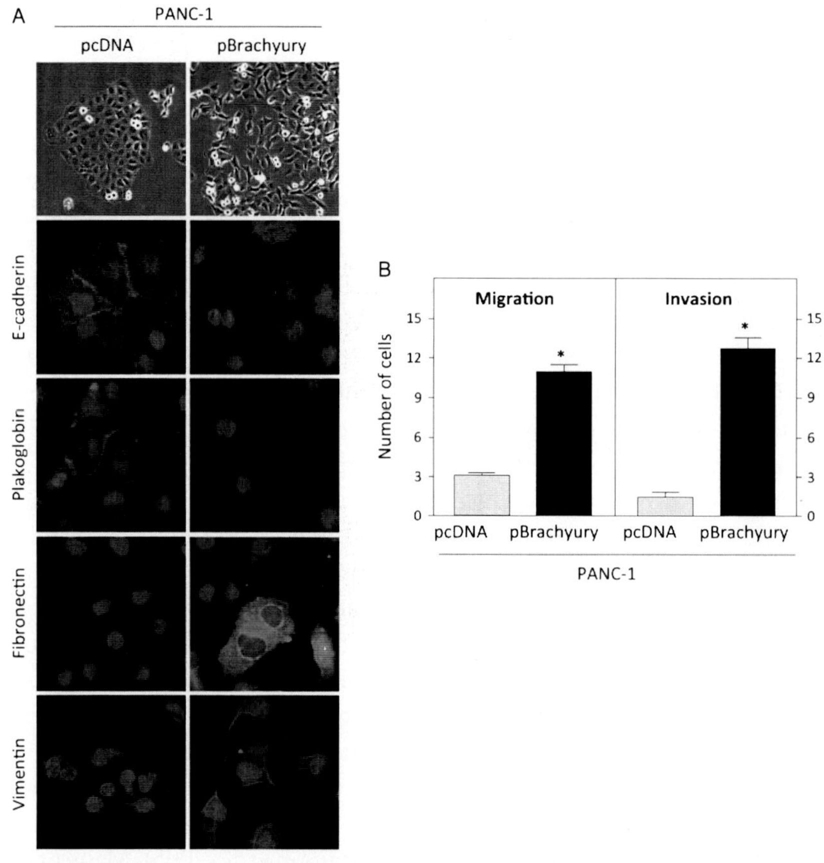

Figure 2.14 Brachyury induces epithelial-to-mesenchymal transition (EMT) in human carcinoma cells. (A) Pancreatic carcinoma PANC-1 cells were stably transfected with a control pcDNA or a vector encoding for full-length Brachyury protein (pBrachyury). Top panels: bright field images of cells grown on plastic surface. Bottom panels: immunofluorescence analysis of EMT markers in cells grown on cover glasses. The green signal represents the staining of the corresponding protein, and the blue signal represents the DAPI-stained nuclei. (B) *In vitro* cell migration and ECM invasion assays. [$^*P < 0.05$]. *Adapted from Fernando et al. (2010).* (See the color plate.)

switch, human carcinoma cells undergoing Brachyury-mediated EMT acquire enhanced motility and the ability to invade (Fernando et al., 2010) (Fig. 2.14B). Xenograft experiments conducted in immunocompromised mice demonstrated that Brachyury expression does not affect the growth of the primary tumor. Inhibition of Brachyury, however, resulted in a significant impairment on the ability of tumor cells to disseminate from the subcutaneous tumor to the site of metastasis, and a reduced ability to establish experimental lung metastasis after intravenous injection (Fig. 2.15).

Overexpression of Brachyury in human carcinoma cells has been shown to enhance resistance of tumor cells to chemotherapy and radiation, while silencing of Brachyury in tumor cells that naturally express high levels of this transcription factor has resulted in enhanced susceptibility to both types of therapeutics, *in vitro* (Fernando et al., 2010; Palena & Schlom, 2013; Roselli et al., 2012).

It has been demonstrated that human $CD8^+$ Brachyury-specific T cells could be expanded *in vitro* from PBMCs of cancer patients and normal donors by using a 9-mer peptide of Brachyury that specifically binds to the HLA-A2 molecule (Palena et al., 2007). Brachyury-specific T-cell lines

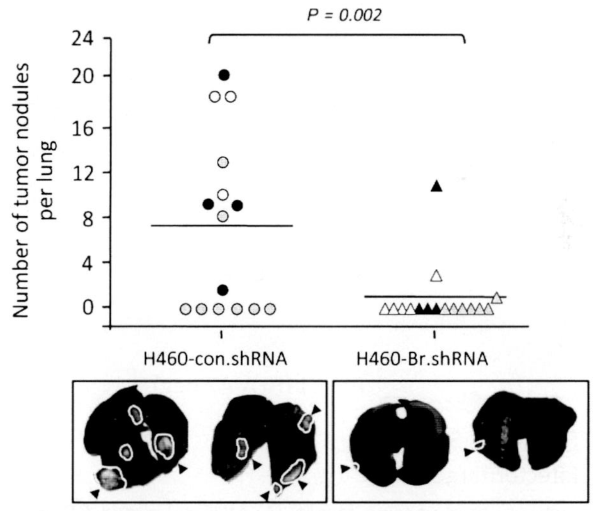

Figure 2.15 Brachyury controls tumor dissemination and metastasis. Athymic mice were inoculated with human H460 lung carcinoma cells transfected as indicated via tail vein. Forty-five days after tumor implantation, animals were euthanized and lungs were evaluated for tumor nodules. Circles denote con.shRNA and triangles denote Br.shRNA. Two representative lungs from each group are shown for comparison. White outlines and black arrowheads point to tumor masses. *Adapted from Fernando et al. (2010).*

have been shown to lyse carcinoma cells (Fernando et al., 2010; Roselli et al., 2012) (Fig. 2.16). Brachyury-specific T cells have also been expanded from the blood of cancer patients (Palena et al., 2007) and those T cells can lyse Brachyury-positive carcinoma cells in an MHC-restricted manner.

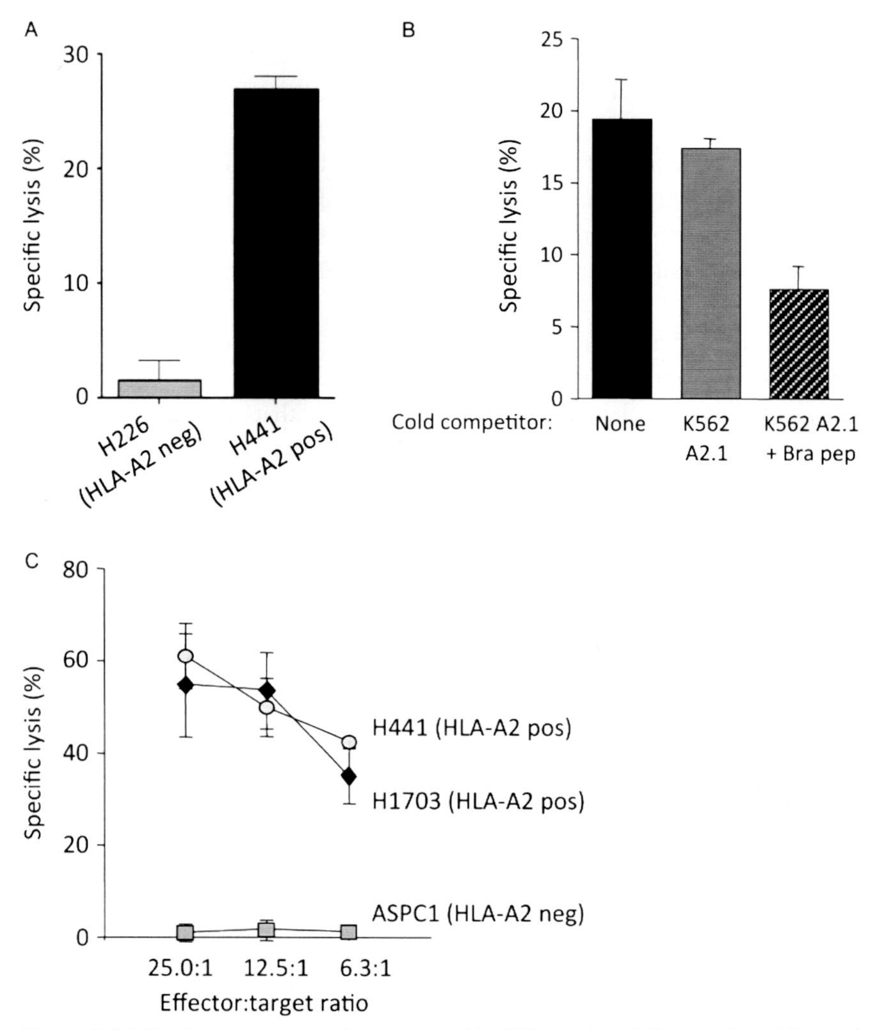

Figure 2.16 Brachyury as a vaccine target. (A) MHC-restricted CTL-mediated lysis of H226 (HLA-A2 negative) and H441 (HLA-A2 positive) lung carcinoma cells with a Brachyury-specific T-cell line. (B) Lysis of H441 tumor cells with a Brachyury-specific T-cell line derived from a prostate cancer patient in the presence of cold, competitor K562 A2.1 cells unpulsed or pulsed with the specific Brachyury peptide (Bra pep). (C) Lysis of H441 and H1703 lung carcinoma and control ASPC1 cells by tetramer-isolated, $CD8^+$ Brachyury-specific T-cell line derived from a different prostate cancer patient. *Adapted from Roselli et al. (2012).*

A Phase I clinical study with a Brachyury vaccine is ongoing in patients with carcinomas, and future Phase II clinical studies employing Brachyury-based vaccines are anticipated.

The C-terminus of MUC1 has been shown by several groups to be extremely important in the initiation and progression of a range of human neoplasms. Overexpression of MUC1-C makes it possible for malignant cells of epithelial or hematopoietic origin to exploit this physiologic stress response and thus stimulate their expansion and survival (Uchida, Raina, Kharbanda, & Kufe, 2013). The MUC1-C oncoprotein has also been shown to induce tamoxifen and herceptin resistance in human breast tumor cells (Fessler, Wotkowicz, Mahanta, & Bamdad, 2009; Kharbanda, Rajabi, Jin, Raina, & Kufe, 2013). MUC1-C-induced transcriptional programs have also been shown to be associated with tumorigenesis and predict a poor outcome in breast and lung cancer patients (Lacunza et al., 2010; MacDermed et al., 2010; Pitroda, Khodarev, Beckett, Kufe, & Weichselbaum, 2009). The MUC1-C oncoprotein has also been shown to confer androgen-independent growth in human prostate cancer cells, regulate survival of pancreatic cancer cells (Banerjee et al., 2012), and enhance invasiveness of pancreatic cancer cells by inducing EMT (Roy et al., 2011).

Seven novel CTL epitopes in the MUC1-C region of MUC1 have been identified recently along with enhancer agonists for each of these epitopes (Jochems et al., 2014). This was demonstrated by the greater ability of the agonist peptide, compared to its corresponding native peptide, to generate MUC1-C-specific T-cell lines, enhance IFN-γ production by T cells, and lyse human tumor cell targets endogenously expressing the native epitope in an MHC-restricted manner. T-cell lines were able to be generated from PBMCs of numerous cancer patients, employing the MUC1 agonist peptides. The MUC1-C agonist epitopes span class I MHC HLA-A2, -A3, and -A24, which encompass the majority of the population. The studies provide the rationale for immunotherapy clinical studies employing a range of vaccines that target Brachyury and/or the agonist epitopes of the C-terminus of MUC1 and thus target the biologically relevant processes of cancer cell progression and drug resistance.

15. CONCLUDING REMARKS

This chapter was designed to provide an overview of the progress in numerous different aspects of cancer vaccine design, development, and clinical application. It is anticipated that therapeutic cancer vaccines will eventually be employed in the management of numerous cancer types and stages, principally in the neoadjuvant and/or adjuvant settings, and in patients with evidence of, or the potential of, nominal residual metastatic disease. The low level of toxicity renders cancer vaccines ideal for combination therapies. Evidence is now mounting that, when used in appropriate scheduling regimens, cancer vaccines can be used in combination with certain chemotherapeutic agents, radiation, hormone therapy, and certain small molecule targeted therapeutics. Indeed, it is being shown that these "nonimmune"-based therapies can have an immune-enhancing component, by either altering the tumor cell to render it more susceptible to T-cell-mediated attack, lysing tumor cells in a manner to further enhance immunity, or altering the balance of immune effector cells over immune regulatory cells. As other nonvaccine immunotherapies, such as checkpoint inhibitors, are developed, they will undoubtedly be employed to enhance vaccine efficacy.

REFERENCES

Aarts, W. M., Schlom, J., & Hodge, J. W. (2002). Vector-based vaccine/cytokine combination therapy to enhance induction of immune responses to a self-antigen and antitumor activity. *Cancer Research*, *62*, 5770–5777.

Adotevi, O., Pere, H., Ravel, P., Haicheur, N., Badoual, C., Merillon, N., et al. (2010). A decrease of regulatory T cells correlates with overall survival after sunitinib-based antiangiogenic therapy in metastatic renal cancer patients. *Journal of Immunotherapy*, *33*, 991–998.

Allison, J. P., Chambers, C., Hurwitz, A., Sullivan, T., Boitel, B., Fournier, S., et al. (1998). A role for CTLA-4-mediated inhibitory signals in peripheral T cell tolerance? *Novartis Foundation Symposium*, *215*, 92–98, discussion 98–102, 186–190.

Andres, A., Majno, P. E., Morel, P., Rubbia-Brandt, L., Giostra, E., Gervaz, P., et al. (2008). Improved long-term outcome of surgery for advanced colorectal liver metastases: Reasons and implications for management on the basis of a severity score. *Annals of Surgical Oncology*, *15*, 134–143.

Arlen, P. M., Gulley, J. L., Parker, C., Skarupa, L., Pazdur, M., Panicali, D., et al. (2006). A randomized phase II study of concurrent docetaxel plus vaccine versus vaccine alone in metastatic androgen-independent prostate cancer. *Clinical Cancer Research*, *12*, 1260–1269.

Arlen, P. M., Gulley, J. L., Todd, N., Lieberman, R., Steinberg, S. M., Morin, S., et al. (2005). Antiandrogen, vaccine and combination therapy in patients with nonmetastatic hormone refractory prostate cancer. *The Journal of Urology*, *174*, 539–546.

Arlen, P. M., Skarupa, L., Pazdur, M., Seetharam, M., Tsang, K. Y., Grosenbach, D. W., et al. (2007). Clinical safety of a viral vector based prostate cancer vaccine strategy. *The Journal of Urology*, *178*, 1515–1520.

Arredouani, M. S., Tseng-Rogenski, S. S., Hollenbeck, B. K., Escara-Wilke, J., Leander, K. R., Defeo-Jones, D., et al. (2010). Androgen ablation augments human HLA2.1-restricted T cell responses to PSA self-antigen in transgenic mice. *Prostate*, *70*, 1002–1011.

Arru, M., Aldrighetti, L., Castoldi, R., Di Palo, S., Orsenigo, E., Stella, M., et al. (2008). Analysis of prognostic factors influencing long-term survival after hepatic resection for metastatic colorectal cancer. *World Journal of Surgery*, *32*, 93–103.

Arumugam, T., Ramachandran, V., Fournier, K. F., Wang, H., Marquis, L., Abbruzzese, J. L., et al. (2009). Epithelial to mesenchymal transition contributes to drug resistance in pancreatic cancer. *Cancer Research*, *69*, 5820–5828.

Avigan, D., Vasir, B., Gong, J., Borges, V., Wu, Z., Uhl, L., et al. (2004). Fusion cell vaccination of patients with metastatic breast and renal cancer induces immunological and clinical responses. *Clinical Cancer Research*, *10*, 4699–4708.

Banchereau, J., Palucka, A. K., Dhodapkar, M., Burkeholder, S., Taquet, N., Rolland, A., et al. (2001). Immune and clinical responses in patients with metastatic melanoma to CD34(+) progenitor-derived dendritic cell vaccine. *Cancer Research*, *61*, 6451–6458.

Banchereau, J., & Steinman, R. M. (1998). Dendritic cells and the control of immunity. *Nature*, *392*, 245–252.

Banerjee, S., Mujumdar, N., Dudeja, V., Mackenzie, T., Krosch, T. K., Sangwan, V., et al. (2012). MUC1c regulates cell survival in pancreatic cancer by preventing lysosomal permeabilization. *PLoS One*, 7, e43020.

Beer, T., Slovin, S. F., Higano, C. S., Tejwani, S., Dorff, T. B., Stankevich, E., et al. (2008). Phase I trial of ipilimumab (IPI) alone or in combination with radiotherapy (XRT) in patients with metastatic castration resistant prostate cancer (mCRPC). *Journal of Clinical Oncology*, *26*, 15s, abstr 5004.

Belardelli, F., Ferrantini, M., Parmiani, G., Schlom, J., & Garaci, E. (2004). International meeting on cancer vaccines: How can we enhance efficacy of therapeutic vaccines? *Cancer Research*, *64*, 6827–6830.

Bendandi, M. (2009). Idiotype vaccines for lymphoma: Proof-of-principles and clinical trial failures. *Nature Reviews. Cancer*, *9*, 675–681.

Bilusic, M., Gulley, J., Heery, C., Apolo, A. B., Arlen, P. M., Rauckhorst, M., et al. (2011). A randomized phase II study of flutamide with or without PSA-TRICOM in nonmetastatic castration-resistant prostate cancer. American Society of Clinical Oncology 2011 Genitourinary Symposium. *Journal of Clinical Oncology*, *29*(Suppl. 7), abstr 163.

Boehm, A. L., Higgins, J., Franzusoff, A., Schlom, J., & Hodge, J. W. (2010). Concurrent vaccination with two distinct vaccine platforms targeting the same antigen generates phenotypically and functionally distinct T-cell populations. *Cancer Immunology, Immunotherapy*, *59*, 397–408.

Bos, R., van Duikeren, S., Morreau, H., Franken, K., Schumacher, T. N., Haanen, J. B., et al. (2008). Balancing between antitumor efficacy and autoimmune pathology in T-cell-mediated targeting of carcinoembryonic antigen. *Cancer Research*, *68*, 8446–8455.

Brichard, V. G., & Lejeune, D. (2008). Cancer immunotherapy targeting tumour-specific antigens: Towards a new therapy for minimal residual disease. *Expert Opinion on Biological Therapy*, *8*, 951–968.

Butterfield, L. H., Palucka, A. K., Britten, C. M., Dhodapkar, M. V., Hakansson, L., Janetzki, S., et al. (2011). Recommendations from the iSBTc-SITC/FDA/NCI Workshop on Immunotherapy Biomarkers. *Clinical Cancer Research*, *17*, 3064–3076.

Butts, C., Murray, N., Maksymiuk, A., Goss, G., Marshall, E., Soulieres, D., et al. (2005). Randomized phase IIB trial of BLP25 liposome vaccine in stage IIIB and IV non-small-cell lung cancer. *Journal of Clinical Oncology*, *23*, 6674–6681.

Carmeliet, P., & Jain, R. K. (2011). Principles and mechanisms of vessel normalization for cancer and other angiogenic diseases. *Nature Reviews. Drug Discovery*, *10*, 417–427.

Chakraborty, M., Abrams, S. I., Coleman, C. N., Camphausen, K., Schlom, J., & Hodge, J. W. (2004). External beam radiation of tumors alters phenotype of tumor cells to render them susceptible to vaccine-mediated T-cell killing. *Cancer Research, 64*, 4328–4337.

Chakraborty, M., Gelbard, A., Carrasquillo, J. A., Yu, S., Mamede, M., Paik, C. H., et al. (2008). Use of radiolabeled monoclonal antibody to enhance vaccine-mediated antitumor effects. *Cancer Immunology, Immunotherapy, 57*, 1173–1183.

Chakraborty, M., Wansley, E. K., Carrasquillo, J. A., Yu, S., Paik, C. H., Camphausen, K., et al. (2008). The use of chelated radionuclide (samarium-153-ethylenediaminetetramethylenephosphonate) to modulate phenotype of tumor cells and enhance T cell-mediated killing. *Clinical Cancer Research, 14*, 4241–4249.

Cham, C. M., Driessens, G., O'Keefe, J. P., & Gajewski, T. F. (2008). Glucose deprivation inhibits multiple key gene expression events and effector functions in CD8+ T cells. *European Journal of Immunology, 38*, 2438–2450.

Cheever, M. A., Allison, J. P., Ferris, A. S., Finn, O. J., Hastings, B. M., Hecht, T. T., et al. (2009). The prioritization of cancer antigens: A national cancer institute pilot project for the acceleration of translational research. *Clinical Cancer Research, 15*, 5323–5337.

Choti, M. A., Sitzmann, J. V., Tiburi, M. F., Sumetchotimetha, W., Rangsin, R., Schulick, R. D., et al. (2002). Trends in long-term survival following liver resection for hepatic colorectal metastases. *Annals of Surgery, 235*, 759–766.

Clarke, P., Mann, J., Simpson, J. F., Rickard-Dickson, K., & Primus, F. J. (1998). Mice transgenic for human carcinoembryonic antigen as a model for immunotherapy. *Cancer Research, 58*, 1469–1477.

Derby, M., Alexander-Miller, M., Tse, R., & Berzofsky, J. (2001). High-avidity CTL exploit two complementary mechanisms to provide better protection against viral infection than low-avidity CTL. *The Journal of Immunology, 166*, 1690–1697.

Dhodapkar, M. V., & Dhodapkar, K. M. (2011). Spontaneous and therapy-induced immunity to pluripotency genes in humans: Clinical implications, opportunities and challenges. *Cancer Immunology, Immunotherapy, 60*, 413–418.

Dhodapkar, K. M., Feldman, D., Matthews, P., Radfar, S., Pickering, R., Turkula, S., et al. (2010). Natural immunity to pluripotency antigen OCT4 in humans. *Proceedings of the National Academy of Sciences of the United States of America, 107*, 8718–8723.

DiPaola, R. S., Chen, Y., Bubley, G. J., Hahn, N. M., Stein, M., Schlom, J., et al. (2009). A phase II study of PROSTVAC-V (vaccinia)/TRICOM and PROSTVAC-F (fowlpox)/TRICOM with GM-CSF in patients with PSA progression after local therapy for prostate cancer: Results of ECOG 9802. In *American Society of Clinical Oncology 2009 genitourinary cancers symposium*. abstr 108.

Disis, M. L. (2009). Enhancing cancer vaccine efficacy via modulation of the tumor microenvironment. *Clinical Cancer Research, 15*, 6476–6478.

Disis, M. L. (2011). Immunologic biomarkers as correlates of clinical response to cancer immunotherapy. *Cancer Immunology, Immunotherapy, 60*, 433–442.

Dzutsev, A. H., Belyakov, I. M., Isakov, D. V., Margulies, D. H., & Berzofsky, J. A. (2007). Avidity of CD8 T cells sharpens immunodominance. *International Immunology, 19*, 497–507.

Egen, J. G., Kuhns, M. S., & Allison, J. P. (2002). CTLA-4: New insights into its biological function and use in tumor immunotherapy. *Nature Immunology, 3*, 611–618.

Emens, L. A., Asquith, J. M., Leatherman, J. M., Kobrin, B. J., Petrik, S., Laiko, M., et al. (2009). Timed sequential treatment with cyclophosphamide, doxorubicin, and an allogeneic granulocyte-macrophage colony-stimulating factor-secreting breast tumor vaccine: A chemotherapy dose-ranging factorial study of safety and immune activation. *Journal of Clinical Oncology, 27*, 5911–5918.

Farsaci, B., Higgins, J. P., & Hodge, J. W. (2012). Consequence of dose scheduling of sunitinib on host immune response elements and vaccine combination therapy. *International Journal of Cancer, 130*, 1948–1959.

Farsaci, B., Sabzevari, H., Higgins, J. P., Di Bari, M. G., Takai, S., Schlom, J., et al. (2010). Effect of a small molecule BCL-2 inhibitor on immune function and use with a recombinant vaccine. *International Journal of Cancer, 127*, 1603–1613.

Fernando, R. I., Litzinger, M., Trono, P., Hamilton, D. H., Schlom, J., & Palena, C. (2010). The T-box transcription factor Brachyury promotes epithelial-mesenchymal transition in human tumor cells. *The Journal of Clinical Investigation, 120*, 533–544.

Fessler, S. P., Wotkowicz, M. T., Mahanta, S. K., & Bamdad, C. (2009). MUC1* is a determinant of trastuzumab (Herceptin) resistance in breast cancer cells. *Breast Cancer Research and Treatment, 118*, 113–124.

Finke, J. H., Rini, B., Ireland, J., Rayman, P., Richmond, A., Golshayan, A., et al. (2008). Sunitinib reverses type-1 immune suppression and decreases T-regulatory cells in renal cell carcinoma patients. *Clinical Cancer Research, 14*, 6674–6682.

Finn, O. J., Gantt, K. R., Lepisto, A. J., Pejawar-Gaddy, S., Xue, J., & Beatty, P. L. (2011). Importance of MUC1 and spontaneous mouse tumor models for understanding the immunobiology of human adenocarcinomas. *Immunologic Research, 50*, 261–268.

Frazer, I. H., Lowy, D. R., & Schiller, J. T. (2007). Prevention of cancer through immunization: Prospects and challenges for the 21st century. *European Journal of Immunology, 37*(Suppl. 1), S148–S155.

Freedman, A., Neelapu, S. S., Nichols, C., Robertson, M. J., Djulbegovic, B., Winter, J. N., et al. (2009). Placebo-controlled phase III trial of patient-specific immunotherapy with mitumprotimut-T and granulocyte-macrophage colony-stimulating factor after rituximab in patients with follicular lymphoma. *Journal of Clinical Oncology, 27*, 3036–3043.

Fridman, W. H., Pages, F., Sautes-Fridman, C., & Galon, J. (2012). The immune contexture in human tumours: Impact on clinical outcome. *Nature Reviews. Cancer, 12*, 298–306.

Fukumura, D., Duda, D. G., Munn, L. L., & Jain, R. K. (2010). Tumor microvasculature and microenvironment: Novel insights through intravital imaging in pre-clinical models. *Microcirculation, 17*, 206–225.

Gajewski, T. F., Meng, Y., Blank, C., Brown, I., Kacha, A., Kline, J., et al. (2006). Immune resistance orchestrated by the tumor microenvironment. *Immunological Reviews, 213*, 131–145.

Gajewski, T. F., Meng, Y., & Harlin, H. (2006). Immune suppression in the tumor microenvironment. *Journal of Immunotherapy, 29*, 233–240.

Galluzzi, L., Senovilla, L., Zitvogel, L., & Kroemer, G. (2012). The secret ally: Immunostimulation by anticancer drugs. *Nature Reviews. Drug Discovery, 11*, 215–233.

Galon, J., Pages, F., Marincola, F. M., Thurin, M., Trinchieri, G., Fox, B. A., et al. (2012). The immune score as a new possible approach for the classification of cancer. *Journal of Translational Medicine, 10*, 1.

Garnett, C. T., Palena, C., Chakraborty, M., Tsang, K. Y., Schlom, J., & Hodge, J. W. (2004). Sublethal irradiation of human tumor cells modulates phenotype resulting in enhanced killing by cytotoxic T lymphocytes. *Cancer Research, 64*, 7985–7994.

Garnett, C. T., Schlom, J., & Hodge, J. W. (2008). Combination of docetaxel and recombinant vaccine enhances T-cell responses and antitumor activity: Effects of docetaxel on immune enhancement. *Clinical Cancer Research, 14*, 3536–3544.

Gelbard, A., Garnett, C. T., Abrams, S. I., Patel, V., Gutkind, J. S., Palena, C., et al. (2006). Combination chemotherapy and radiation of human squamous cell carcinoma of the head and neck augments CTL-mediated lysis. *Clinical Cancer Research, 12*, 1897–1905.

Gerritsen, W. R. (2013). CA184-043: A randomized, multicenter, double-blind phase 3 trial comparing overall survival (OS) in patients (pts) with post-docetaxel castration-resistant prostate cancer (CRPC) and bone metastases treated with ipilimumab (ipi) vs placebo

(pbo), each following single-dose radiotherapy (RT). In *The European cancer congress, September 27–October 1*. abstr 2850.
Gilewski, T. A., Ragupathi, G., Dickler, M., Powell, S., Bhuta, S., Panageas, K., et al. (2007). Immunization of high-risk breast cancer patients with clustered sTn-KLH conjugate plus the immunologic adjuvant QS-21. *Clinical Cancer Research, 13*, 2977–2985.
Gnjatic, S., Old, L. J., & Chen, Y. T. (2009). Autoantibodies against cancer antigens. *Methods in Molecular Biology, 520*, 11–19.
Gnjatic, S., Ritter, E., Buchler, M. W., Giese, N. A., Brors, B., Frei, C., et al. (2010). Seromic profiling of ovarian and pancreatic cancer. *Proceedings of the National Academy of Sciences of the United States of America, 107*, 5088–5093.
Gnjatic, S., Wheeler, C., Ebner, M., Ritter, E., Murray, A., Altorki, N. K., et al. (2009). Seromic analysis of antibody responses in non-small cell lung cancer patients and healthy donors using conformational protein arrays. *Journal of Immunological Methods, 341*, 50–58.
Greiner, J. W., Zeytin, H., Anver, M. R., & Schlom, J. (2002). Vaccine-based therapy directed against carcinoembryonic antigen demonstrates antitumor activity on spontaneous intestinal tumors in the absence of autoimmunity. *Cancer Research, 62*, 6944–6951.
Grosenbach, D. W., Barrientos, J. C., Schlom, J., & Hodge, J. W. (2001). Synergy of vaccine strategies to amplify antigen-specific immune responses and antitumor effects. *Cancer Research, 61*, 4497–4505.
Gulley, J. L. (2013). Therapeutic vaccines: The ultimate personalized therapy? *Hum Vaccines & Immunotherapeutics, 9*, 219–221.
Gulley, J. L., Arlen, P. M., Bastian, A., Morin, S., Marte, J., Beetham, P., et al. (2005). Combining a recombinant cancer vaccine with standard definitive radiotherapy in patients with localized prostate cancer. *Clinical Cancer Research, 11*, 3353–3362.
Gulley, J. L., Arlen, P. M., Hodge, J. W., & Schlom, J. (2010). Vaccines and immunostimulants. In D. Kufe (Ed.), *Holland-Frei cancer medicine* (8th ed., pp. 725–736). Shelton, CT: People's Medical Publishing House—USA (Chapter 57).
Gulley, J. L., Arlen, P. M., Madan, R. A., Tsang, K. Y., Pazdur, M. P., Skarupa, L., et al. (2010). Immunologic and prognostic factors associated with overall survival employing a poxviral-based PSA vaccine in metastatic castrate-resistant prostate cancer. *Cancer Immunology, Immunotherapy, 59*, 663–674.
Gulley, J. L., Arlen, P. M., Tsang, K. Y., Yokokawa, J., Palena, C., Poole, D. J., et al. (2008). Pilot study of vaccination with recombinant CEA-MUC-1-TRICOM poxviral-based vaccines in patients with metastatic carcinoma. *Clinical Cancer Research, 14*, 3060–3069.
Gulley, J. L., Madan, R. A., & Schlom, J. (2011). The impact of tumour volume on potential efficacy of therapeutic vaccines. *Current Oncology, 18*, e150–e157.
Halabi, S., Small, E. J., Kantoff, P. W., Kattan, M. W., Kaplan, E. B., Dawson, N. A., et al. (2003). Prognostic model for predicting survival in men with hormone-refractory metastatic prostate cancer. *Journal of Clinical Oncology, 21*, 1232–1237.
Hardwick, N., & Chain, B. (2011). Epitope spreading contributes to effective immunotherapy in metastatic melanoma patients. *Immunotherapy, 3*, 731–733.
Heery, C. R., Madan, R. A., Bilusic, M., Kim, J. W., Singh, N. K., Rauckhorst, M., et al. (2013). A phase II randomized clinical trial of samarium-153 EDTMP (Sm-153) with or without PSA-tricom vaccine in metastatic castration-resistant prostate cancer (mCRPC) after docetaxel. In *ASCO Genitourinary cancers symposium, February 14–16, Orlando, FL. Journal of Clinical Oncology*, Vol. 31, Suppl. 6; abstr 102.
Higano, C. S., Schellhammer, P. F., Small, E. J., Burch, P. A., Nemunaitis, J., Yuh, L., et al. (2009). Integrated data from 2 randomized, double-blind, placebo-controlled, phase 3 trials of active cellular immunotherapy with sipuleucel-T in advanced prostate cancer. *Cancer, 115*, 3670–3679.

Hodge, J. W., Chakraborty, M., Kudo-Saito, C., Garnett, C. T., & Schlom, J. (2005). Multiple costimulatory modalities enhance CTL avidity. *The Journal of Immunology*, *174*, 5994–6004.
Hodge, J. W., Grosenbach, D. W., Aarts, W. M., Poole, D. J., & Schlom, J. (2003). Vaccine therapy of established tumors in the absence of autoimmunity. *Clinical Cancer Research*, *9*, 1837–1849.
Hodge, J. W., Higgins, J., & Schlom, J. (2009). Harnessing the unique local immunostimulatory properties of modified vaccinia Ankara (MVA) virus to generate superior tumor-specific immune responses and antitumor activity in a diversified prime and boost vaccine regimen. *Vaccine*, *27*, 4475–4482.
Hodge, J. W., McLaughlin, J. P., Abrams, S. I., Shupert, W. L., Schlom, J., & Kantor, J. A. (1995). Admixture of a recombinant vaccinia virus containing the gene for the costimulatory molecule B7 and a recombinant vaccinia virus containing a tumor-associated antigen gene results in enhanced specific T-cell responses and antitumor immunity. *Cancer Research*, *55*, 3598–3603.
Hodge, J. W., Poole, D. J., Aarts, W. M., Gomez Yafal, A., Gritz, L., & Schlom, J. (2003). Modified vaccinia virus ankara recombinants are as potent as vaccinia recombinants in diversified prime and boost vaccine regimens to elicit therapeutic antitumor responses. *Cancer Research*, *63*, 7942–7949.
Hodi, F. S., O'Day, S. J., McDermott, D. F., Weber, R. W., Sosman, J. A., Haanen, J. B., et al. (2010). Improved survival with ipilimumab in patients with metastatic melanoma. *The New England Journal of Medicine*, *363*, 711–723.
Hoos, A., Eggermont, A. M., Janetzki, S., Hodi, F. S., Ibrahim, R., Anderson, A., et al. (2010). Improved endpoints for cancer immunotherapy trials. *Journal of the National Cancer Institute*, *102*, 1388–1397.
Hoover, H. C., Jr., Brandhorst, J. S., Peters, L. C., Surdyke, M. G., Takeshita, Y., Madariaga, J., et al. (1993). Adjuvant active specific immunotherapy for human colorectal cancer: 6.5-year median follow-up of a phase III prospectively randomized trial. *Journal of Clinical Oncology*, *11*, 390–399.
House, M. G., Ito, H., Gonen, M., Fong, Y., Allen, P. J., DeMatteo, R. P., et al. (2010). Survival after hepatic resection for metastatic colorectal cancer: Trends in outcomes for 1,600 patients during two decades at a single institution. *Journal of the American College of Surgeons*, *210*, 744–752, 752–755.
Hua, W., Yao, Y., Chu, Y., Zhong, P., Sheng, X., Xiao, B., et al. (2011). The CD133+ tumor stem-like cell-associated antigen may elicit highly intense immune responses against human malignant glioma. *Journal of Neuro-Oncology*, *105*, 149–157.
Inoges, S., Rodriguez-Calvillo, M., Zabalegui, N., Lopez-Diaz de Cerio, A., Villanueva, H., Soria, E., et al. (2006). Clinical benefit associated with idiotypic vaccination in patients with follicular lymphoma. *Journal of the National Cancer Institute*, *98*, 1292–1301.
Irvine, K., Kantor, J., & Schlom, J. (1993). Comparison of a CEA-recombinant vaccinia virus, purified CEA, and an anti-idiotypic antibody bearing the image of a CEA epitope in the treatment and prevention of CEA-expressing tumors. *Vaccine Research*, *2*, 79–94.
Jochems, C., & Schlom, J. (2011). Tumor-infiltrating immune cells and prognosis: The potential link between conventional cancer therapy and immunity. *Experimental Biology and Medicine (Maywood, NJ)*, *236*, 567–579.
Jochems, C., Tucker, J., Vergati, M., Boyerinas, B., Gulley, J. L., Schlom, J., et al. (2014). Identification and characterization of agonist epitopes of the MUC1-C oncoprotein. *Cancer Immunology, Immunotherapy*, *63*, 161–174.
Kalluri, R., & Weinberg, R. A. (2009). The basics of epithelial-mesenchymal transition. *The Journal of Clinical Investigation*, *119*, 1420–1428.
Kalus, R. M., Kantor, J. A., Gritz, L., Gomez Yafal, A., Mazzara, G. P., Schlom, J., et al. (1999). The use of combination vaccinia vaccines and dual-gene vaccinia vaccines to

enhance antigen-specific T-cell immunity via T-cell costimulation. *Vaccine, 17*, 893–903.

Kantoff, P. W., Higano, C. S., Shore, N. D., Berger, E. R., Small, E. J., Penson, D. F., et al. (2010). Sipuleucel-T immunotherapy for castration-resistant prostate cancer. *The New England Journal of Medicine, 363*, 411–422.

Kantoff, P. W., Schuetz, T. J., Blumenstein, B. A., Glode, L. M., Bilhartz, D. L., Wyand, M., et al. (2010). Overall survival analysis of a phase II randomized controlled trial of a Poxviral-based PSA-targeted immunotherapy in metastatic castration-resistant prostate cancer. *Journal of Clinical Oncology, 28*, 1099–1105.

Kaplan, C. D., Kruger, J. A., Zhou, H., Luo, Y., Xiang, R., & Reisfeld, R. A. (2006). A novel DNA vaccine encoding PDGFRbeta suppresses growth and dissemination of murine colon, lung and breast carcinoma. *Vaccine, 24*, 6994–7002.

Karbach, J., Neumann, A., Atmaca, A., Wahle, C., Brand, K., von Boehmer, L., et al. (2011). Efficient in vivo priming by vaccination with recombinant NY-ESO-1 protein and CpG in antigen naive prostate cancer patients. *Clinical Cancer Research, 17*, 861–870.

Kass, E., Schlom, J., Thompson, J., Guadagni, F., Graziano, P., & Greiner, J. W. (1999). Induction of protective host immunity to carcinoembryonic antigen (CEA), a self-antigen in CEA transgenic mice, by immunizing with a recombinant vaccinia-CEA virus. *Cancer Research, 59*, 676–683.

Kaufman, H. L., Wang, W., Manola, J., DiPaola, R. S., Ko, Y. J., Sweeney, C., et al. (2004). Phase II randomized study of vaccine treatment of advanced prostate cancer (E7897): A trial of the Eastern Cooperative Oncology Group. *Journal of Clinical Oncology, 22*, 2122–2132.

Kemp, T. J., Garcia-Pineres, A., Falk, R. T., Poncelet, S., Dessy, F., Giannini, S. L., et al. (2008). Evaluation of systemic and mucosal anti-HPV16 and anti-HPV18 antibody responses from vaccinated women. *Vaccine, 26*, 3608–3616.

Kemp, T. J., Hildesheim, A., Safaeian, M., Dauner, J. G., Pan, Y., Porras, C., et al. (2011). HPV16/18 L1 VLP vaccine induces cross-neutralizing antibodies that may mediate cross-protection. *Vaccine, 29*, 2011–2014.

Kepp, O., Galluzzi, L., Martins, I., Schlemmer, F., Adjemian, S., Michaud, M., et al. (2011). Molecular determinants of immunogenic cell death elicited by anticancer chemotherapy. *Cancer Metastasis Reviews, 30*, 61–69.

Kharbanda, A., Rajabi, H., Jin, C., Raina, D., & Kufe, D. (2013). Oncogenic MUC1-C promotes tamoxifen resistance in human breast cancer cells. *Molecular Cancer Research, 11*, 714–723.

Kim, P. S., & Ahmed, R. (2010). Features of responding T cells in cancer and chronic infection. *Current Opinion in Immunology, 22*, 223–230.

Kim, P. S., Jochems, C., Grenga, I., Donahue, R. N., Tsang, K. Y., Gulley, J. L., et al. (2014). Pan-Bcl-2 inhibitor, GX15–070 (Obatoclax), decreases human T regulatory lymphocytes while preserving effector T lymphocytes: a rationale for its use in combination immunotherapy. *The Journal of Immunology, 192*, 2622–2633.

Ko, J. S., Rayman, P., Ireland, J., Swaidani, S., Li, G., Bunting, K. D., et al. (2010). Direct and differential suppression of myeloid-derived suppressor cell subsets by sunitinib is compartmentally constrained. *Cancer Research, 70*, 3526–3536.

Ko, J. S., Zea, A. H., Rini, B. I., Ireland, J. L., Elson, P., Cohen, P., et al. (2009). Sunitinib mediates reversal of myeloid-derived suppressor cell accumulation in renal cell carcinoma patients. *Clinical Cancer Research, 15*, 2148–2157.

Kodumudi, K. N., Woan, K., Gilvary, D. L., Sahakian, E., Wei, S., & Djeu, J. Y. (2010). A novel chemoimmunomodulating property of docetaxel: Suppression of myeloid-derived suppressor cells in tumor bearers. *Clinical Cancer Research, 16*, 4583–4594.

Kudo-Saito, C., Garnett, C. T., Wansley, E. K., Schlom, J., & Hodge, J. W. (2007). Intratumoral delivery of vector mediated IL-2 in combination with vaccine results in enhanced T cell avidity and anti-tumor activity. *Cancer Immunology, Immunotherapy, 56*, 1897–1910.

Kudo-Saito, C., Hodge, J. W., Kwak, H., Kim-Schulze, S., Schlom, J., & Kaufman, H. L. (2006). 4-1BB ligand enhances tumor-specific immunity of poxvirus vaccines. *Vaccine, 24*, 4975–4986.

Kudo-Saito, C., Schlom, J., Camphausen, K., Coleman, C. N., & Hodge, J. W. (2005). The requirement of multimodal therapy (vaccine, local tumor radiation, and reduction of suppressor cells) to eliminate established tumors. *Clinical Cancer Research, 11*, 4533–4544.

Kudo-Saito, C., Schlom, J., & Hodge, J. W. (2005). Induction of an antigen cascade by diversified subcutaneous/intratumoral vaccination is associated with antitumor responses. *Clinical Cancer Research, 11*, 2416–2426.

Kufe, D. W. (2009). Mucins in cancer: Function, prognosis and therapy. *Nature Reviews. Cancer, 9*, 874–885.

Kurrey, N. K., Jalgaonkar, S. P., Joglekar, A. V., Ghanate, A. D., Chaskar, P. D., Doiphode, R. Y., et al. (2009). Snail and slug mediate radioresistance and chemoresistance by antagonizing p53-mediated apoptosis and acquiring a stem-like phenotype in ovarian cancer cells. *Stem Cells, 27*, 2059–2068.

Lacunza, E., Baudis, M., Colussi, A. G., Segal-Eiras, A., Croce, M. V., & Abba, M. C. (2010). MUC1 oncogene amplification correlates with protein overexpression in invasive breast carcinoma cells. *Cancer Genetics and Cytogenetics, 201*, 102–110.

Laheru, D., Lutz, E., Burke, J., Biedrzycki, B., Solt, S., Onners, B., et al. (2008). Allogeneic granulocyte macrophage colony-stimulating factor-secreting tumor immunotherapy alone or in sequence with cyclophosphamide for metastatic pancreatic cancer: A pilot study of safety, feasibility, and immune activation. *Clinical Cancer Research, 14*, 1455–1463.

Larocca, C., & Schlom, J. (2011). Viral vector-based therapeutic cancer vaccines. *Cancer Journal, 17*, 359–371.

Lee, D. K., Hakim, F. T., & Gress, R. E. (2010). The thymus and the immune system: Layered levels of control. *Journal of Thoracic Oncology, 5*, S273–S276.

Locher, C., Conforti, R., Aymeric, L., Ma, Y., Yamazaki, T., Rusakiewicz, S., et al. (2010). Desirable cell death during anticancer chemotherapy. *Annals of the New York Academy of Sciences, 1209*, 99–108.

Lorenz, M. G., Kantor, J. A., Schlom, J., & Hodge, J. W. (1999a). Anti-tumor immunity elicited by a recombinant vaccinia virus expressing CD70 (CD27L). *Human Gene Therapy, 10*, 1095–1103.

Lorenz, M. G., Kantor, J. A., Schlom, J., & Hodge, J. W. (1999b). Induction of anti-tumor immunity elicited by tumor cells expressing a murine LFA-3 analog via a recombinant vaccinia virus. *Human Gene Therapy, 10*, 623–631.

Luiten, R. M., Kueter, E. W., Mooi, W., Gallee, M. P., Rankin, E. M., Gerritsen, W. R., et al. (2005). Immunogenicity, including vitiligo, and feasibility of vaccination with autologous GM-CSF-transduced tumor cells in metastatic melanoma patients. *Journal of Clinical Oncology, 23*, 8978–8991.

Lutz, E., Yeo, C. J., Lillemoe, K. D., Biedrzycki, B., Kobrin, B., Herman, J., et al. (2011). A lethally irradiated allogeneic granulocyte-macrophage colony stimulating factor-secreting tumor vaccine for pancreatic adenocarcinoma. A Phase II trial of safety, efficacy, and immune activation. *Annals of Surgery, 253*, 328–335.

Lyerly, H.K., Hobeika, A., Niedzwiecki, D., Osada, T., Marshall, J., Garrett, C. R., et al. (2011). A dendritic cell-based vaccine effects on T-cell responses compared with a viral vector vaccine when administered to patients following resection of colorectal metastases in a randomized Phase II study. American Society of Clinical Oncology 2011 Annual Meeting. *Journal of Clinical Oncology, 29*(Suppl.) (abstract).

MacDermed, D. M., Khodarev, N. N., Pitroda, S. P., Edwards, D. C., Pelizzari, C. A., Huang, L., et al. (2010). MUC1-associated proliferation signature predicts outcomes in lung adenocarcinoma patients. *BMC Medical Genomics*, *3*, 16.

MacDonald, G. H., & Johnston, R. E. (2000). Role of dendritic cell targeting in Venezuelan equine encephalitis virus pathogenesis. *Journal of Virology*, *74*, 914–922.

Madan, R. A., Gulley, J. L., Fojo, T., & Dahut, W. L. (2010). Therapeutic cancer vaccines in prostate cancer: The paradox of improved survival without changes in time to progression. *The Oncologist*, *15*, 969–975.

Madan, R. A., Gulley, J. L., Schlom, J., Steinberg, S. M., Liewehr, D. J., Dahut, W. L., et al. (2008). Analysis of overall survival in patients with nonmetastatic castration-resistant prostate cancer treated with vaccine, nilutamide, and combination therapy. *Clinical Cancer Research*, *14*, 4526–4531.

Madan, R. A., Mohebtash, M., Arlen, P. M., Vergati, M., Rauckhorst, M., Steinberg, S. M., et al. (2012). Ipilimumab and a poxviral vaccine targeting prostate-specific antigen in metastatic castration-resistant prostate cancer: A phase 1 dose-escalation trial. *The Lancet Oncology*, *13*, 501–508.

Madan, R. A., Mohebtash, M., Schlom, J., & Gulley, J. L. (2010). Therapeutic vaccines in metastatic castration-resistant prostate cancer: Principles in clinical trial design. *Expert Opinion on Biological Therapy*, *10*, 19–28.

Marshall, J. L., Gulley, J. L., Arlen, P. M., Beetham, P. K., Tsang, K. Y., Slack, R., et al. (2005). Phase I study of sequential vaccinations with fowlpox-CEA(6D)-TRICOM alone and sequentially with vaccinia-CEA(6D)-TRICOM, with and without granulocyte-macrophage colony-stimulating factor, in patients with carcinoembryonic antigen-expressing carcinomas. *Journal of Clinical Oncology*, *23*, 720–731.

Marshall, J. L., Hoyer, R. J., Toomey, M. A., Faraguna, K., Chang, P., Richmond, E., et al. (2000). Phase I study in advanced cancer patients of a diversified prime-and-boost vaccination protocol using recombinant vaccinia virus and recombinant nonreplicating avipox virus to elicit anti-carcinoembryonic antigen immune responses. *Journal of Clinical Oncology*, *18*, 3964–3973.

Mine, T., Matsueda, S., Li, Y., Tokumitsu, H., Gao, H., Danes, C., et al. (2009). Breast cancer cells expressing stem cell markers CD44+ CD24 lo are eliminated by Numb-1 peptide-activated T cells. *Cancer Immunology, Immunotherapy*, *58*, 1185–1194.

Mohebtash, M., Tsang, K. Y., Madan, R. A., Huen, N. Y., Poole, D. J., Jochems, C., et al. (2011). A pilot study of MUC-1/CEA/TRICOM poxviral-based vaccine in patients with metastatic breast and ovarian cancer. *Clinical Cancer Research*, *17*, 7164–7173.

Morse, M. A., Chaudhry, A., Gabitzsch, E. S., Hobeika, A. C., Osada, T., Clay, T. M., et al. (2013). Novel adenoviral vector induces T-cell responses despite anti-adenoviral neutralizing antibodies in colorectal cancer patients. *Cancer Immunology, Immunotherapy*, *62*, 1293–1301.

Morse, M. A., Niedzwiecki, D., Marshall, J. L., Garrett, C., Chang, D. Z., Aklilu, M., et al. (2013). A randomized phase II study of immunization with dendritic cells modified with poxvectors encoding CEA and MUC1 compared with the same poxvectors plus GM-CSF for resected metastatic colorectal cancer. *Annals of Surgery*, *258*, 879–886.

Morse, M., Niedzwiecki, D., Marshall, J., Garrett, C. R., Chang, D. Z., Aklilu, M., et al. (2011). Survival rates among patients vaccinated following resection of colorectal cancer metastases in a Phase II randomized study compared with contemporary controls. American Society of Clinical Oncology 2011 Annual Meeting. *Journal of Clinical Oncology, 29* (Suppl.) (abstract).

Moss, B. (1996). Genetically engineered poxviruses for recombinant gene expression, vaccination, and safety. *Proceedings of the National Academy of Sciences of the United States of America*, *93*, 11341–11348.

Mueller, S. N., & Ahmed, R. (2009). High antigen levels are the cause of T cell exhaustion during chronic viral infection. *Proceedings of the National Academy of Sciences of the United States of America*, *106*, 8623–8628.

Nakaya, H. I., Wrammert, J., Lee, E. K., Racioppi, L., Marie-Kunze, S., Haining, W. N., et al. (2011). Systems biology of vaccination for seasonal influenza in humans. *Nature Immunology*, *12*, 786–795.

NCT01867333. Enzalutamide with or without vaccine therapy for advanced prostate cancer. http://www.clinicaltrials.gov/ct2/show/NCT01867333?term=NCT01867333&rank=1.

NCT01875250. Enzalutamide in combination with PSA-TRICOM in patients with non-metastatic castration sensitive prostate cancer. http://www.clinicaltrials.gov/ct2/show/NCT01875250?term=NCT01875250&rank=1.

Oh, S., Hodge, J. W., Ahlers, J. D., Burke, D. S., Schlom, J., & Berzofsky, J. A. (2003). Selective induction of high avidity CTL by altering the balance of signals from APC. *The Journal of Immunology*, *170*, 2523–2530.

Okada, H., Kalinski, P., Ueda, R., Hoji, A., Kohanbash, G., Donegan, T. E., et al. (2011). Induction of CD8+ T-cell responses against novel glioma-associated antigen peptides and clinical activity by vaccinations with {alpha}-type 1 polarized dendritic cells and polyinosinic-polycytidylic acid stabilized by lysine and carboxymethylcellulose in patients with recurrent malignant glioma. *Journal of Clinical Oncology*, *29*, 330–336.

Okur, F. V., Yvon, E., Biagi, E., Dotti, G., Carrum, G., Heslop, H., et al. (2011). Comparison of two CD40-ligand/interleukin-2 vaccines in patients with chronic lymphocytic leukemia. *Cytotherapy*, *13*, 1128–1139.

Osada, T., Yang, X. Y., Hartman, Z. C., Glass, O., Hodges, B. L., Niedzwiecki, D., et al. (2009). Optimization of vaccine responses with an E1, E2b and E3-deleted Ad5 vector circumvents pre-existing anti-vector immunity. *Cancer Gene Therapy*, *16*, 673–682.

Palena, C., Polev, D. E., Tsang, K. Y., Fernando, R. I., Litzinger, M., Krukovskaya, L. L., et al. (2007). The human T-box mesodermal transcription factor Brachyury is a candidate target for T-cell-mediated cancer immunotherapy. *Clinical Cancer Research*, *13*, 2471–2478.

Palena, C., & Schlom, J. (2013). Target: Brachyury, a master driver of epithelial-to-mesenchymal transition (EMT). In: J. Marshall (Ed.), *Encyclopedia of Cancer Therapeutic Targets*. Berlin Heidelberg: Springer-Verlag.

Parkhurst, M. R., Yang, J. C., Langan, R. C., Dudley, M. E., Nathan, D. A., Feldman, S. A., et al. (2011). T cells targeting carcinoembryonic antigen can mediate regression of metastatic colorectal cancer but induce severe transient colitis. *Molecular Therapy*, *19*, 620–626.

Pawlik, T. M., Scoggins, C. R., Zorzi, D., Abdalla, E. K., Andres, A., Eng, C., et al. (2005). Effect of surgical margin status on survival and site of recurrence after hepatic resection for colorectal metastases. *Annals of Surgery*, *241*, 722–724, discussion 722–714.

Pejawar-Gaddy, S., Rajawat, Y., Hilioti, Z., Xue, J., Gaddy, D. F., Finn, O. J., et al. (2010). Generation of a tumor vaccine candidate based on conjugation of a MUC1 peptide to polyionic papillomavirus virus-like particles. *Cancer Immunology, Immunotherapy*, *59*, 1685–1696.

Pitroda, S. P., Khodarev, N. N., Beckett, M. A., Kufe, D. W., & Weichselbaum, R. R. (2009). MUC1-induced alterations in a lipid metabolic gene network predict response of human breast cancers to tamoxifen treatment. *Proceedings of the National Academy of Sciences of the United States of America*, *106*, 5837–5841.

Polyak, K., & Weinberg, R. A. (2009). Transitions between epithelial and mesenchymal states: Acquisition of malignant and stem cell traits. *Nature Reviews. Cancer*, *9*, 265–273.

Pound, C. R., Partin, A. W., Eisenberger, M. A., Chan, D. W., Pearson, J. D., & Walsh, P. C. (1999). Natural history of progression after PSA elevation following radical prostatectomy. *JAMA, 281*, 1591–1597.

Ragupathi, G., Damani, P., Srivastava, G., Srivastava, O., Sucheck, S. J., Ichikawa, Y., et al. (2009). Synthesis of sialyl Lewis(a) (sLe (a), CA19-9) and construction of an immunogenic sLe(a) vaccine. *Cancer Immunology, Immunotherapy, 58*, 1397–1405.

Raina, D., Kosugi, M., Ahmad, R., Panchamoorthy, G., Rajabi, H., Alam, M., et al. (2011). Dependence on the MUC1-C oncoprotein in non-small cell lung cancer cells. *Molecular Cancer Therapeutics, 10*, 806–816.

Remondo, C., Cereda, V., Mostbock, S., Sabzevari, H., Franzusoff, A., Schlom, J., et al. (2009). Human dendritic cell maturation and activation by a heat-killed recombinant yeast (Saccharomyces cerevisiae) vector encoding carcinoembryonic antigen. *Vaccine, 27*, 987–994.

Rojan, A., Funches, R., Regan, M. M., Gulley, J. L., & Bubley, G. J. (2013). Dramatic and prolonged PSA response after retreatment with a PSA vaccine. *Clinical Genitourinary Cancer, 11*, 362–364.

Roselli, M., Cereda, V., di Bari, M., Formica, V., Spila, A., Jochems, C., et al. (2013). Effects of conventional therapeutic interventions on the number and function of regulatory T cells. *OncoImmunology, 2*(10), e27025.

Roselli, M., Fernando, R. I., Guadagni, F., Spila, A., Alessandroni, J., Palmirotta, R., et al. (2012). Brachyury, a driver of the epithelial-mesenchymal transition, is overexpressed in human lung tumors: An opportunity for novel interventions against lung cancer. *Clinical Cancer Research, 18*, 3868–3879.

Roy, L. D., Sahraei, M., Subramani, D. B., Besmer, D., Nath, S., Tinder, T. L., et al. (2011). MUC1 enhances invasiveness of pancreatic cancer cells by inducing epithelial to mesenchymal transition. *Oncogene, 30*, 1449–1459.

Salazar, L. G., Wallace, D., Mukherjee, P., Higgins, D., Childs, J., Bates, N., et al. (2009). HER-2/neu (HER2) specific T-cell immunity in patients with HER2+ inflammatory breast cancer (IBC) and prognosis. American Society of Clinical Oncology 2009 Annual Meeting. *Journal of Clinical Oncology, 27*(15s), abstr 3057.

Sanda, M. G., Smith, D. C., Charles, L. G., Hwang, C., Pienta, K. J., Schlom, J., et al. (1999). Recombinant vaccinia-PSA (PROSTVAC) can induce a prostate-specific immune response in androgen-modulated human prostate cancer. *Urology, 53*, 260–266.

Sasaki, A., Iwashita, Y., Shibata, K., Matsumoto, T., Ohta, M., & Kitano, S. (2005). Analysis of preoperative prognostic factors for long-term survival after hepatic resection of liver metastasis of colorectal carcinoma. *Journal of Gastrointestinal Surgery, 9*, 374–380.

Schlom, J. (2012). Therapeutic cancer vaccines: Current status and moving forward. *Journal of the National Cancer Institute, 104*, 599–613.

Schlom, J., Hodge, J., Palena, C., Greiner, J., Tsang, K.-Y., Farsaci, B., et al. (2013). Recombinant TRICOM-based therapeutic cancer vaccines: Lessons learned. In G. Prendergast & E. Jaffee (Eds.), *Cancer immunotherapy: Immune suppression and tumor growth* (2nd ed., pp. 309–331). Elsevier Inc/Academic Press (Chapter 20).

Schlom, J., Palena, C., Gulley, J., Greiner, J., Tsang, K.-Y., Madan, R., et al. (2013). The use of T-cell costimulation to enhance the immunogenicity of tumors. In S. Gerson & E. Lattime (Eds.), *Gene therapy of cancer* (3rd ed., pp. 315–334). Elsevier (Chapter 22).

Schlom, J., Tsang, K. Y., Hodge, J. W., & Greiner, J. W. (2001). Carcinoembryonic antigen as a vaccine target. In R. C. Rees & A. Robins (Eds.), *Cancer immunology: Immunology in medicine series* (pp. 73–100). Norwell, MA: Kluwer Academic Publishers.

Schuster, S. J., Neelapu, S. S., Gause, B. L., Janik, J. E., Muggia, F. M., Gockerman, J. P., et al. (2011). Vaccination with patient-specific tumor-derived antigen in first remission improves disease-free survival in follicular lymphoma. *Journal of Clinical Oncology, 29*, 2787–2794.

Schwartzentruber, D. J., Lawson, D. H., Richards, J. M., Conry, R. M., Miller, D. M., Treisman, J., et al. (2011). gp100 peptide vaccine and interleukin-2 in patients with advanced melanoma. *The New England Journal of Medicine, 364*, 2119–2127.

Singh, R., & Paterson, Y. (2006). Listeria monocytogenes as a vector for tumor-associated antigens for cancer immunotherapy. *Expert Review of Vaccines, 5*, 541–552.

Small, E. J., Tchekmedyian, N. S., Rini, B. I., Fong, L., Lowy, I., & Allison, J. P. (2007). A pilot trial of CTLA-4 blockade with human anti-CTLA-4 in patients with hormone-refractory prostate cancer. *Clinical Cancer Research, 13*, 1810–1815.

Sosman, J. A., Carrillo, C., Urba, W. J., Flaherty, L., Atkins, M. B., Clark, J. I., et al. (2008). Three phase II cytokine working group trials of gp100 (210M) peptide plus high-dose interleukin-2 in patients with HLA-A2-positive advanced melanoma. *Journal of Clinical Oncology, 26*, 2292–2298.

Spisek, R., Kukreja, A., Chen, L. C., Matthews, P., Mazumder, A., Vesole, D., et al. (2007). Frequent and specific immunity to the embryonal stem cell-associated antigen SOX2 in patients with monoclonal gammopathy. *The Journal of Experimental Medicine, 204*, 831–840.

Sportes, C., Babb, R. R., Krumlauf, M. C., Hakim, F. T., Steinberg, S. M., Chow, C. K., et al. (2010). Phase I study of recombinant human interleukin-7 administration in subjects with refractory malignancy. *Clinical Cancer Research, 16*, 727–735.

Stein, W. D., Gulley, J. L., Schlom, J., Madan, R. A., Dahut, W., Figg, W. D., et al. (2011). Tumor regression and growth rates determined in five intramural NCI prostate cancer trials: The growth rate constant as an indicator of therapeutic efficacy. *Clinical Cancer Research, 17*, 907–917.

Tesniere, A., Schlemmer, F., Boige, V., Kepp, O., Martins, I., Ghiringhelli, F., et al. (2010). Immunogenic death of colon cancer cells treated with oxaliplatin. *Oncogene, 29*, 482–491.

Thiery, J. P. (2002). Epithelial-mesenchymal transitions in tumour progression. *Nature Reviews. Cancer, 2*, 442–454.

Thiery, J. P., & Sleeman, J. P. (2006). Complex networks orchestrate epithelial-mesenchymal transitions. *Nature Reviews. Molecular Cell Biology, 7*, 131–142.

Thomson, S., Buck, E., Petti, F., Griffin, G., Brown, E., Ramnarine, N., et al. (2005). Epithelial to mesenchymal transition is a determinant of sensitivity of non-small-cell lung carcinoma cell lines and xenografts to epidermal growth factor receptor inhibition. *Cancer Research, 65*, 9455–9462.

Uchida, Y., Raina, D., Kharbanda, S., & Kufe, D. (2013). Inhibition of the MUC1-C oncoprotein is synergistic with cytotoxic agents in the treatment of breast cancer cells. *Cancer Biology & Therapy, 14*, 127–134.

Uzendoski, K., Kantor, J. A., Abrams, S. I., Schlom, J., & Hodge, J. W. (1997). Construction and characterization of a recombinant vaccinia virus expressing murine intercellular adhesion molecule-1: Induction and potentiation of antitumor responses. *Human Gene Therapy, 8*, 851–860.

van den Eertwegh, A. J., Versluis, J., van den Berg, H. P., Santegoets, S. J., van Moorselaar, R. J., van der Sluis, T. M., et al. (2012). Combined immunotherapy with granulocyte-macrophage colony-stimulating factor-transduced allogeneic prostate cancer cells and ipilimumab in patients with metastatic castration-resistant prostate cancer: A phase 1 dose-escalation trial. *The Lancet Oncology, 13*, 509–517.

Vanneman, M., & Dranoff, G. (2012). Combining immunotherapy and targeted therapies in cancer treatment. *Nature Reviews. Cancer, 12*, 237–251.

Vega, S., Morales, A. V., Ocana, O. H., Valdes, F., Fabregat, I., & Nieto, M. A. (2004). Snail blocks the cell cycle and confers resistance to cell death. *Genes & Development, 18*, 1131–1143.

Vezys, V., Penaloza-Macmaster, P., Barber, D. L., Ha, S. J., Konieczny, B., Freeman, G. J., et al. (2011). 4-1BB signaling synergizes with programmed death ligand 1 blockade to

augment CD8 T cell responses during chronic viral infection. *The Journal of Immunology*, *187*, 1634–1642.
von Mehren, M., Arlen, P., Gulley, J., Rogatko, A., Cooper, H. S., Meropol, N. J., et al. (2001). The influence of granulocyte macrophage colony-stimulating factor and prior chemotherapy on the immunological response to a vaccine (ALVAC-CEA B7.1) in patients with metastatic carcinoma. *Clinical Cancer Research*, 7, 1181–1191.
von Mehren, M., Arlen, P., Tsang, K. Y., Rogatko, A., Meropol, N., Cooper, H. S., et al. (2000). Pilot study of a dual gene recombinant avipox vaccine containing both carcinoembryonic antigen (CEA) and B7.1 transgenes in patients with recurrent CEA-expressing adenocarcinomas. *Clinical Cancer Research*, *6*, 2219–2228.
Walter, S., Weinschenk, T., Stenzl, A., Zdrojowy, R., Pluzanska, A., Szczylik, C., et al. (2012). Multipeptide immune response to cancer vaccine IMA901 after single-dose cyclophosphamide associates with longer patient survival. *Nature Medicine*, *18*, 1254–1261.
Wansley, E. K., Chakraborty, M., Hance, K. W., Bernstein, M. B., Boehm, A. L., Guo, Z., et al. (2008). Vaccination with a recombinant Saccharomyces cerevisiae expressing a tumor antigen breaks immune tolerance and elicits therapeutic antitumor responses. *Clinical Cancer Research*, *14*, 4316–4325.
Wheeler, C. J., & Black, K. L. (2011). Vaccines for glioblastoma and high-grade glioma. *Expert Review of Vaccines*, *10*, 875–886.
Williams, K. M., & Gress, R. E. (2008). Immune reconstitution and implications for immunotherapy following haematopoietic stem cell transplantation. *Best Practice & Research. Clinical Haematology*, *21*, 579–596.
Wolchok, J. D., Neyns, B., Linette, G., Negrier, S., Lutzky, J., Thomas, L., et al. (2010). Ipilimumab monotherapy in patients with pretreated advanced melanoma: A randomised, double-blind, multicentre, phase 2, dose-ranging study. *The Lancet Oncology*, *11*, 155–164.
Xiang, R., Luo, Y., Niethammer, A. G., & Reisfeld, R. A. (2008). Oral DNA vaccines target the tumor vasculature and microenvironment and suppress tumor growth and metastasis. *Immunological Reviews*, *222*, 117–128.
Yang, A. D., Fan, F., Camp, E. R., van Buren, G., Liu, W., Somcio, R., et al. (2006). Chronic oxaliplatin resistance induces epithelial-to-mesenchymal transition in colorectal cancer cell lines. *Clinical Cancer Research*, *12*, 4147–4153.
Zeytin, H. E., Patel, A. C., Rogers, C. J., Canter, D., Hursting, S. D., Schlom, J., et al. (2004). Combination of a poxvirus-based vaccine with a cyclooxygenase-2 inhibitor (celecoxib) elicits antitumor immunity and long-term survival in CEA.Tg/MIN mice. *Cancer Research*, *64*, 3668–3678.
Zitvogel, L., Kepp, O., Senovilla, L., Menger, L., Chaput, N., & Kroemer, G. (2010). Immunogenic tumor cell death for optimal anticancer therapy: The calreticulin exposure pathway. *Clinical Cancer Research*, *16*, 3100–3104.

CHAPTER THREE

IKK/Nuclear Factor-kappaB and Oncogenesis: Roles in Tumor-Initiating Cells and in the Tumor Microenvironment

Jennifer W. Bradford, Albert S. Baldwin[1]

Lineberger Comprehensive Cancer Center, University of North Carolina School of Medicine, Chapel Hill, North Carolina, USA

[1]Corresponding author: e-mail address: abaldwin@med.unc.edu

Contents

Abstract

The IKK/nuclear factor-kappaB pathway (NF-κB) is critical in proper immune function, cell survival, apoptosis, cellular proliferation, synaptic plasticity, and even memory. While NF-κB is crucial for both innate and adaptive immunity, defective regulation of this master transcriptional regulator is seen in a variety of diseases including autoimmune disease, neurodegenerative disease, and, important to this review, cancer. While NF-κB functions in cancer to promote a number of critical oncogenic functions, here we discuss the importance of the NF-κB signaling pathway in contributing to cancer through promotion of the tumor microenvironment and through maintenance/expansion of tumor-initiating cells, processes that appear to be functionally interrelated.

Advances in Cancer Research, Volume 121
ISSN 0065-230X
http://dx.doi.org/10.1016/B978-0-12-800249-0.00003-2

1. INTRODUCTION

1.1. NF-κB family members

First described almost 30 years ago as a DNA-binding activity involved in the regulation of immunoglobulin κ light-chain gene expression (Sen & Baltimore, 1986), mammalian nuclear factor-kappaB pathway (NF-κB) is a family of five highly conserved transcription factors (RelA/p65, RelB, c-Rel, p50 [NF-κB1/p105 precursor], and p52 [NF-κB2/p100 precursor]) that form different homo- and heterodimers to regulate target gene expression (Baldwin, 2012; Hayden & Ghosh, 2012; Fig. 3.1). These proteins

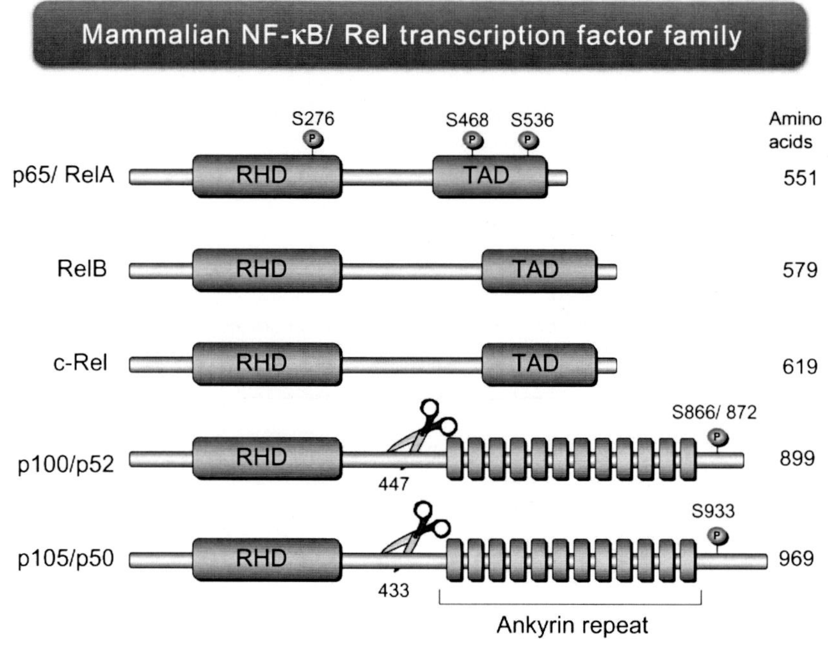

Figure 3.1 NF-κB family members consist of five highly conserved transcription factors including p65/RelA, RelB, c-Rel, p100/p52, and p105/p50. NF-κB members bind as hetero- or homodimers to activate transcription of downstream targets. The rel homology domain (RHD) in the N-terminus mediates DNA binding and binding to other NF-κB family members, while the transcriptional activation domain (TAD) in the C-terminus is critical for transcriptional activity. Phosphorylation occurs where indicated, which are marks of activity. Cleavage of p100 and p105 occur via the ubiquitin–proteasome system and produces the active p52 and p50 proteins, respectively. *Figure used with permission from Wilson (2009).* (See the color plate.)

contain an approximate 300 amino acid conserved Rel homology domain that contains sequences for nuclear localization, DNA binding, dimerization, and interaction with the inhibitor of kappaB (IκB) proteins (Baldwin, 2012; Hayden & Ghosh, 2012). RelA, RelB, and c-Rel have C-terminal domains that contain transcriptional activation domains, while full-length p100 and p105 contain IκB-like ankyrin repeats (DiDonato, Mercurio, & Karin, 2012). NF-κB proteins, p50 and p52, are produced by proteolytic cleavage of precursors p105 and p100, respectfully. c-Rel is the cellular homologue of v-Rel, the transforming gene of avian reticuloendotheliosis virus. In *Drosophila*, there are three NF-κB family members (Dorsal, Dif, and Relish) that promote dorsoventral patterning in early development and innate immune signaling.

1.2. NF-κB regulation

Two distinct regulatory pathways are known to control NF-κB activation: the canonical and noncanonical pathways (Fig. 3.2). The canonical pathway is controlled through an IKK complex which consists of the catalytic subunits, IKKβ and IKKα, and the regulatory and scaffold subunit, NEMO (IKKγ) (Ghosh, May, & Kopp, 1998; Israel, 2000; Karin & Ben-Neriah, 2000). Under resting conditions, the RelA/p50 heterodimer is held inactive by the IκB proteins. IκB physically blocks the nuclear localization sequence of RelA/p65 which leads to inactivation of the heterodimer. Various stimuli including lipopolysaccharide (LPS) and cytokines such as IL-1β and TNF trigger a signaling cascade through receptor-induced signaling to activate the IKK complex, with IKKβ functioning as the dominant kinase in this cascade (DiDonato et al., 2012). This leads to phosphorylation of IκBα on Ser32/Ser36, resulting in rapid IκBα ubiquitination and proteasome-dependent degradation. Once RelA/p50 is free from IκBα inhibition, it is able to accumulate in the nucleus and bind to kappaB sites within promoters and regulatory regions of genes that regulate apoptosis, the inflammatory response, and cell proliferation (DiDonato et al., 2012; Fig. 3.3). Along with phosphorylation and degradation of IκBα, posttranslational modifications of RelA/p65, including acetylation and methylation, can modulate NF-κB activity (Yang, Tajkhorshid, & Chen, 2010). Additionally, NF-κB activates transcription of the gene encoding its own inhibitor, IκBα, thus providing a negative feedback loop for additional control.

The noncanonical pathway, however, is controlled through an IKKα complex and leads to activation of the p52–RelB heterodimer

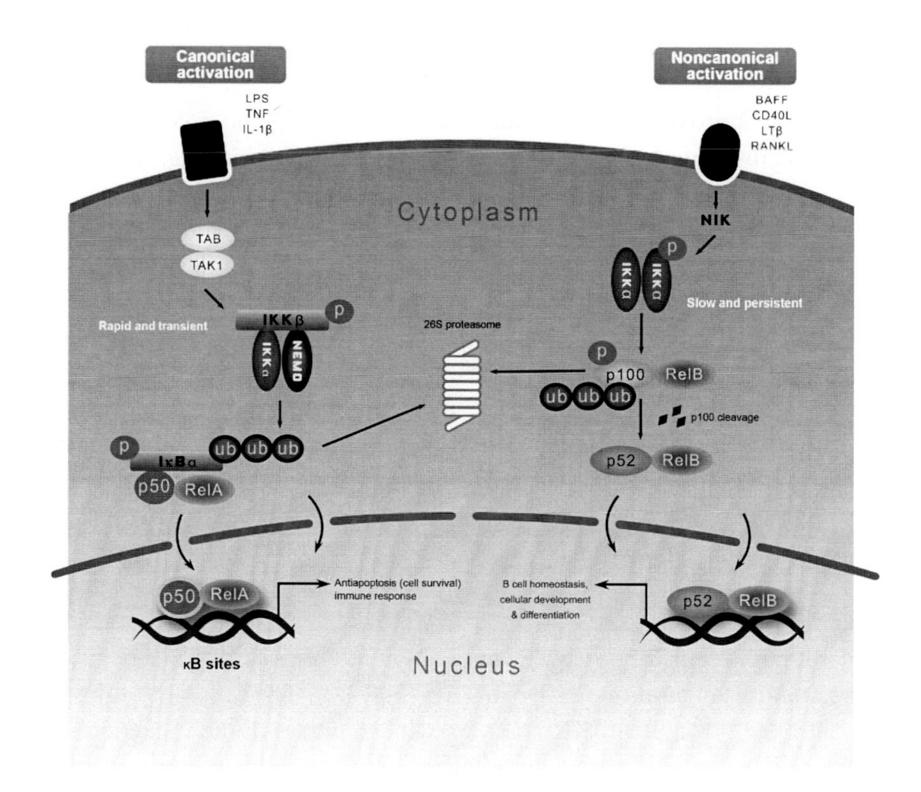

Figure 3.2 The canonical and noncanonical NF-κB pathways. Upon activation of the canonical pathway IKKβ phosphorylates IκBα, which leads to its degradation via the ubiquitin–proteasome system. The RelA/p50 heterodimer is then free to enter the nucleus and promote target gene expression. The noncanonical pathway is activated by LTβ, BAFF, and RANK ligands which causes NIK to phosphorylate the IKKα homodimer. p100 of the p100/RelB heterodimer is then phosphorylated and cleaved to p52. The subsequently formed p52/RelB heterodimer is then free to enter the nucleus to promote target gene expression. (See the color plate.)

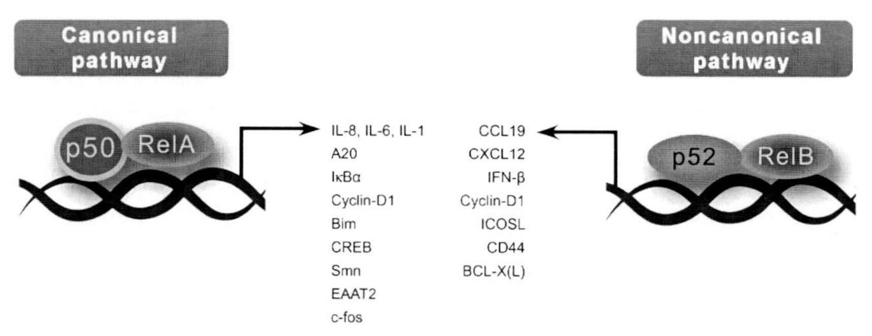

Figure 3.3 Downstream targets activated by NF-κB. Activation of the canonical NF-κB pathway leads to transcriptional activation of genes involved in anti-apoptosis and the immune response, while activation of the noncanonical pathway is involved in B cell homeostasis, cellular development, and differentiation. Note that some factors can be activated by both the canonical and noncanonical pathways such as Cyclin-D1. This is just a shortlist of the downstream targets of NF-κB. (See the color plate.)

(Hayden & Ghosh, 2004, 2012). Ligands of BAFF, CD40, and LTβ-R ligands activate the noncanonical pathway by activating NIK which phosphorylates and activates IKKα homodimers, leading to phosphorylation and subsequent proteolytic processing of p100 to p52 (Senftleben et al., 2001). The transcriptionally active RelB–p52 heterodimer can then enter the nucleus to promote transcription of a variety of genes involved in B cell homeostasis and cellular development and differentiation (see Fig. 3.3).

1.3. NF-κB-independent functions for IKK

While IKK is critical for activation of NF-κB complexes downstream of cytokine signaling and through oncoprotein expression, evidence has been presented that IKK can phosphorylate and regulate critical regulatory proteins involved in distinct signaling pathways. For example, it was reported that IKKβ can phosphorylate the tumor suppressor p53 leading to its destabilization (Xia et al., 2009). This finding is consistent with a prosurvival function for IKK signaling, blocking the potential apoptotic functions of p53. IKKβ was shown to phosphorylate the forkhead transcription factor Foxo3a, promoting its ubiquitination and proteolysis (Hu et al., 2004). IKKα phosphorylates the p27/Kip1 cdk inhibitor to promote tumor-initiating cells (TICs) in ErbB2-driven breast cancer (Zhang et al., 2013). Downstream of activated Akt, in a variety of cancers and in response to insulin exposure, IKKα interacts with mTORC1 to drive its activity (Dan, Adli, & Baldwin, 2007). Interestingly, this interaction reciprocally promotes IKK activity to activate NF-κB (Dan et al., 2008). These functions of IKK are clearly relevant to oncogenesis as blocking p53 function, suppressing p27 activity, and driving mTORC1 activity should promote tumorigenic potential of most cancers.

1.4. NF-κB and cancer

NF-κB is activated in a variety of cancers as detected by phosphorylation of IκBα and RelA, elevated levels of nuclear canonical and/or noncanonical forms of NF-κB, phosphorylation of IKK, and elevated expression of NF-κB target genes such as IL-6 and a variety of chemokines/cytokines. Most oncoproteins, such as oncogenic Ras alleles, and growth factor receptors (such as EGFR and ErbB2/Her2) are known to lead to NF-κB signaling (Finco et al., 1997; Merkhofer, Cogswell, & Baldwin, 2010). In fact, IKK/NF-κB is required for efficient cell transformation and tumorigenesis induced by oncogenic Ras (Basseres, Ebbs, Levantini, & Baldwin, 2010; Ling et al.,

2012; Mayo et al., 1997; Meylan et al., 2009; Xia et al., 2012). Loss of tumor suppressors such as PTEN and p53 have been shown to lead to elevated NF-κB activity. Interestingly, loss of p53 was shown to lead to a glucose-driven modification of IKK by O-GlcNac to increase its activity (Kawauchi, Araki, Tobiume, & Tanaka, 2008, 2009). Thus, many oncogenic signaling pathways lead to activation of IKK and NF-κB (see Table 3.1).

In most solid tumors, members of the IKK/NF-κB pathway are rarely mutated, presumably because this pathway functions downstream of oncogenic signaling pathways that are found mutated. These upstream activating mutations typically activate IKK/NF-κB along with additional signaling cascades such as Akt or MEK/Erk. In this regard, Akt signaling can promote IKK/NF-κB signaling. Importantly, both canonical and noncanonical NF-κB have been shown to be important for K-Ras-driven pancreatic cancer cell growth and survival. Downstream of oncogenic K-Ras, GSK-3α coordinates the stability of the TAK1/TAB complex to drive canonical IKK activation and promotes p100 to p52 processing to drive noncanonical NF-κB activation (Bang, Wilson, Ryan, Yeh, & Baldwin, 2013).

Genes encoding NF-κB pathway members have been found to be amplified in a variety of cancers and include *TRAF6*, *IKBKB*, *IKBKG*, *IRAK1*, and *RIPK1* (Beroukhim et al., 2010). IKKβ and IKKγ, members of the important

Table 3.1 Oncogenic signaling pathways that lead to activation of IKK/NF-κB

Oncogenic pathway	NF-κB pathway driven	Cancer type	References
Ras	NF-κB	–	Mayo et al. (1997)
p62, downstream of Ras	IKK/NF-κB	–	Duran et al. (2008)
p53 (loss)	IKK	–	Kawauchi et al. (2009)
Her2/Erb2	NIK/IKK	Breast cancer	Merkhofer et al. (2010)
Her2/Erb2	NIK/IKKα	Breast cancer TICs	Zhang et al. (2013)
Genetic lesions	Noncanonical NF-κB	Mantle cell lymphoma	Rahal et al. (2014)
CARD11	IKK/NF-κB	Diffuse large B lymphoma	Lenz et al. (2008)
Akt, loss of PTEN	IKK/NF-κB	Prostate	Dan et al. (2008)

While members of the NF-κB transcription factor family are rarely mutated in cancer, there are many upstream pathways and other genomic alterations that can lead to downstream activation of IKK and NF-κB in cancer. A few of these oncogenic pathways are listed.

IKK complex, are also amplified in certain cancers as is the IKK-related kinase IKKε (Beroukhim et al., 2010; Boehm et al., 2007). In hematologic malignancies such as multiple myeloma and diffuse large B cell lymphoma, both canonical and noncanonical signaling pathways exhibit activating mutations (Annunziata et al., 2007; Keats et al., 2007; Lenz et al., 2008).

In cancer, NF-κB is proposed to contribute to oncogenesis through the induction of genes encoding proteins involved in suppressing apoptosis, promoting invasion and angiogenesis, and enhancing epithelial–mesenchymal transition (EMT) (Baldwin, 2001; Basseres & Baldwin, 2006). As described earlier, IKK can promote oncogenic phenotypes separate from its ability to promote NF-κB activation. NF-κB can have both tumor suppressive (Perkins & Gilmore, 2006) and oncogenic (Basseres & Baldwin, 2006) properties and is at the forefront of studies on inflammation and cancer (Greten et al., 2004; Pikarsky et al., 2004). As the topic of NF-κB in cancer is very expansive, we will focus on two areas of research in this review that have not been extensively reviewed: NF-κB involvement in the tumor microenvironment and in TICs.

2. TUMOR MICROENVIRONMENT

2.1. Background on the tumor microenvironment

Tumors are heterogenic and are composed of tumor cells, which are heterogenous themselves, and noncancerous cells that are recruited into the tumor. A wide array of noncancerous cells associated with tumors make up the tumor stroma/microenvironment including macrophages, fibroblasts, T cells, B cells, neutrophils, etc. Research on the tumor microenvironment has seen an explosion of interest in recent years as it has become apparent that an environment rich in cells of the immune system such as macrophages and T cells, and nonimmune cells like fibroblasts promote tumor growth, aggressive properties like metastasis, resistance to chemotherapy, and relapse (Sun et al., 2012). The tumor microenvironment is further complicated by the presence of various cytokines, hormones, and growth factors produced by both tumor and stromal cells, many of which work to promote tumor survival, growth, and metastasis. Work on the tumor microenvironment has debunked the classical cell-autonomous view of cancer, and new therapies are aimed at targeting stromal cells in addition to cancer cells. The NF-κB pathway has been implicated in a variety of tumor-promoting roles within the stroma and may be a viable therapeutic target for these cells (Fig. 3.4).

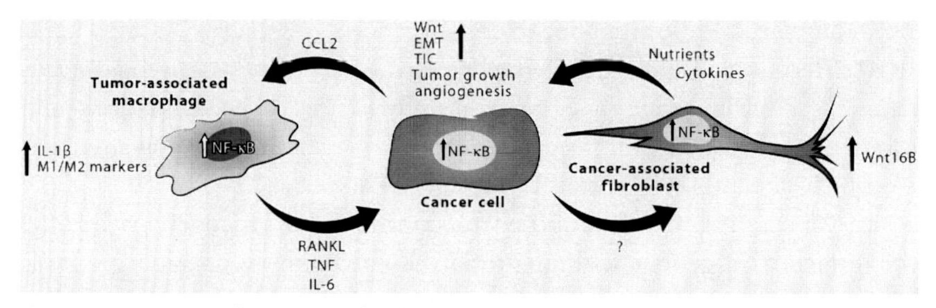

Figure 3.4 NF-κB pathway is involved in stromal communication with cancer cells. Stromal cells like macrophages and fibroblasts are influenced by cancer cells, partially by NF-κB, which promotes altered stromal cell phenotype and function. Through the NF-κB pathway, tumor stroma can drive aggressive properties in cancer cells like EMT, invasion, and TIC expansion. (See the color plate.)

2.2. NF-κB and tumor-associated macrophages

2.2.1 Macrophage phenotypes

It is becoming recognized that the canonical NF-κB signaling pathway is a central player not only in tumor progression but also in tumor-associated macrophage (TAM) regulation. Macrophages are a heterogeneous population of phagocytic leukocytes and are required for proper immune function, tissue remodeling, and repair. Classical activation of macrophages by bacterial products like LPS and IFNγ results in Toll-like receptor signaling and subsequent NF-κB activation. Classically activated macrophages adopt a proinflammatory, M1 phenotype that is characterized by expression of iNOS, MHC class II molecules, antigen presentation, and inflammatory cytokines like IL-12^{high}, IL-6, and IL-1β. Macrophages should normally target and destroy cancer cells, but macrophages that are recruited to tumors by the release of chemoattractants including CSF-1, CC chemokines, and VEGF (Biswas & Lewis, 2010; Mantovani, Allavena, Sica, & Balkwill, 2008) are often activated to an immunosuppressive (M2) phenotype (Pollard, 2008). Protumorigenic M2 macrophages are characterized by the production of matrix metalloproteinases, anti-inflammatory factors such as IL-10^{high} and IL-12^{low}, are able to promote angiogenesis and tissue remodeling activity, and have poor antigen presenting ability (Gordon, 2003; Lewis & Pollard, 2006). As tumor grade/stage increase, macrophages tend to become more M2 in phenotype, but tumors often contain TAMs with mixed M1/M2 markers (Biswas et al., 2006; Movahedi et al., 2010; Van Ginderachter et al., 2006; Fig. 3.5). TAMs are often present in large numbers in a variety of tumors (composing upwards of half of the total invasive breast cancer tumor mass), which is associated with poor prognosis in both breast

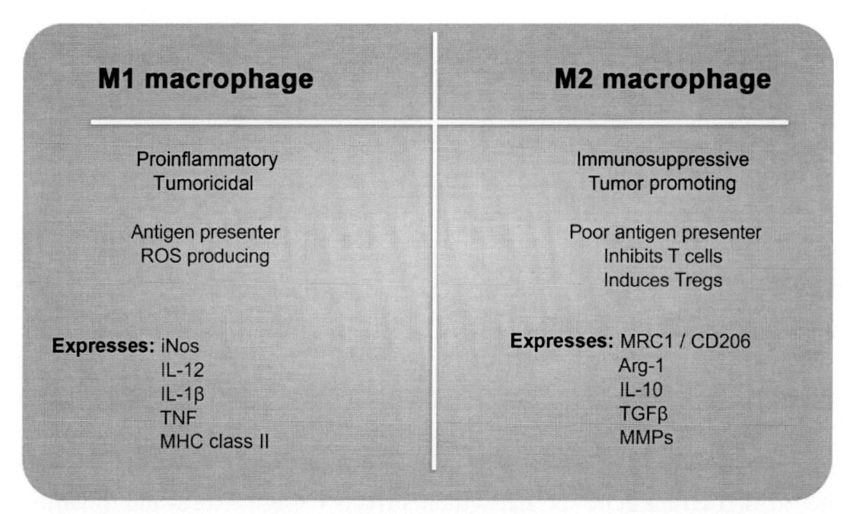

Figure 3.5 M1/M2 macrophage activation. Macrophages are recruited to tumors and are often activated to an either classical (M1) or an alternative (M2) phenotype. M1 macrophages are proinflammatory and tumoricidal in nature, while M2 macrophages are immunosuppressive and promote the tumor.

cancer (Anton & Glod, 2009; Laoui et al., 2011; Mahmoud et al., 2012) and pancreatic cancer (Kurahara et al., 2011).

2.2.2 NF-κB and TAMs

TAMs have the ability to promote cancer cell migration, tumor growth, and angiogenesis, and the NF-κB pathway may play a role in all of these functions. The NF-κB pathway can inhibit normal M1 activation when overexpressed in macrophages and therefore may decrease tumor cell phagocytosis. This is important as TAMs are known to have high levels of the NF-κB member p50 (Saccani et al., 2006). It has been demonstrated for some time now that depletion of macrophages from the tumor site can decrease progression of the tumor. Macrophages can impart invasive properties on breast cancer cells like invasion by secreting TNFα, which activates NF-κB and JNK in breast cancer cells. This crosstalk can be inhibited by the use of a TNFα neutralizing antibody (Hagemann et al., 2005). Alternatively, ovarian cancer cells are capable of activating macrophages *in vitro* to a phenotype similar to what is as seen in ovarian cancer TAMs, which is dependent on TNFα and NF-κB (Hagemann et al., 2006). Studies like these demonstrate the bidirectional impact that NF-κB signaling has on both cancer cells and TAMs. To study TAMs more specifically, a LysMCre/floxed IKKβ mouse model was created where IKKβ is deleted in the myeloid lineage (Greten et al., 2004). It was found that IKKβ deletion in the myeloid lineage reduced colon tumor incidence and tumor size in a mouse model of colitis-associated cancer (Greten et al., 2004). Francis Balkwill and colleague's important paper on the reeducation of TAMs demonstrated that NF-κB signaling can maintain an immunosuppressive (M2) TAM phenotype. It was found that the NF-κB signaling in TAMs was being driven by malignant epithelial cells and that inhibition of IKKβ in macrophages could increase tumoricidal activity of the macrophages and decrease tumor growth in xenograft models (Hagemann et al., 2008).

Metastasis-associated macrophages (MAMs) are a subpopulation of stromal macrophage that has the ability to promote cancer cell extravasation and metastatic cell growth (Wynn, Chawla, & Pollard, 2013). It was found that pulmonary metastatic tumor cells and the surrounding stroma release the NF-κB target gene CCL2, which attracts CCR2 expressing inflammatory monocytes (Qian et al., 2011). These inflammatory monocytes develop into MAMs and promote metastasis. Inhibiting the CCL2/CCR2 signaling pathway by siRNA or neutralizing antibodies against CCL2 reduced lung colonization in experimental metastasis assays and reduced transendothelial migration of 4173 cells *in vitro*. Importantly, CCL2 blockade after intravenous injection of MDA-MB231 cells reduced lung tumor burden and increased mouse survival (Qian et al., 2011). Another independent study confirmed that lymphoma-derived mesenchymal stromal cells (L-MSCs) were effective at recruiting macrophages via CCR2 and that bone marrow mesenchymal cells treated with TNFα (a potent activator on NF-κB) were able to acquire the same macrophage recruiting properties as L-MSCs (Ren et al., 2012). These studies indicate the importance of the NF-κB target gene CCL2 (and NF-κB itself) in recruitment of macrophages and in inflammatory macrophage seeding and promotion of lung metastases.

It has been shown that Wnt signaling and promotion of colon cancer cells can also be promoted by macrophages through the NF-κB pathway. More specifically, TAMs and the NF-κB-dependent target gene IL-1β can activate NF-κB-dependent PDK1/AKT signaling in colon cancer cells, thus inactivating GSK3β, which enhances Wnt signaling and tumor promotion (Kaler, Godasi, Augenlicht, & Klampfer, 2009).

2.2.3 Noncanonical NF-κB signaling in TAM promotion of cancer

While noncanonical NF-κB signaling in TAM regulation has not received the same attention as the canonical pathway, several studies do reveal a role for this pathway in the ability of macrophages to drive cancer. The first indication that the noncanonical pathway may been involved is that macrophages express many noncanonical NF-κB activation ligands, such as LTβ, LTα, and RANKL, and also many of the receptors for these ligands (Biswas & Lewis, 2010; Bonizzi & Karin, 2004). In a model of prostate cancer, an inactivating IKKα mutation slows prostate cancer progression and inhibits metastasis in TRAMP mice, which express a prostate specific SV40 T antigen (Luo et al., 2007). Gene analysis revealed that IKKα was promoting metastasis by inhibiting the *Maspin* gene. IKKα and the noncanonical pathway is proposed to be driven in these prostate cancer cells by RANKL-expressing inflammatory cells (Luo et al., 2007).

2.2.4 Tumor inhibiting role for NF-κB signaling in the tumor microenvironment?

All of these studies mentioned earlier demonstrate how macrophages promote cancer through the NF-κB pathway. However, it has also been found that activation of NF-κB may be beneficial in a cancer setting, which leads to the question of whether NF-κB expression during a specific window of tumor progression is beneficial in inhibiting tumorigenesis. In a chemically induced (diethylnitrosamine) mouse model of liver cancer, mice that lacked IKKβ in hepatocytes exhibited increased hepatocarcinogenesis which correlated with increased ROS production and JNK activation (Maeda, Kamata, Luo, Leffert, & Karin, 2005). However, in mice that lacked IKKβ in both hepatocytes and Kupffer cells (liver macrophages), hepatocarcinogenesis was fourfold lower than in control mice (Maeda et al., 2005). These cellular differences could be due to the important roll Kupffer cells play in providing essential NF-κB dependent mitogens to the hepatocellular carcinoma (Qian & Pollard, 2010), which is blocked by NF-κB signaling ablation.

In a more recent study, Tet-inducible mice were produced to express either IKKβ or dominant negative IκBα in macrophages to either activate or inactivate NF-κB signaling, respectively (Connelly et al., 2011). Using these mice, it was found that activation of NF-κB in macrophages during the seeding period reduced mammary tumor metastasis to the lung (Connelly et al., 2011). Whether these observations are specific to certain mouse models of cancer remains to be seen but should be further investigated as NF-κB inhibition as part of cancer therapy could have aversive outcomes during certain stages of disease.

2.3. NF-κB and tumor-associated T lymphocytes

T cells are vital members of the adaptive immune system and are responsible for cell-mediated actions including activating macrophages via Th1 and Th2 cytokines, helping B cells produce antibodies, and killing cells infected with viruses and other pathogens. The presence of T cells within a tumor setting is associated with improved patient survival (Fridman et al., 2011; Zhang et al., 2003), and as a result, immunotherapy to increase T cell function in cancer patients is currently showing promise. A recent article by Amer Beg's group demonstrates an important tumor suppressive role for NF-κB by enhancing T cell recruitment, T cell-mediated immune surveillance, and antitumor responses in lung cancer (Hopewell et al., 2013). This was demonstrated in a Lewis lung carcinoma mouse model that constitutively expressed IKKβ and had an induced OVA-specific CD8 T cell response. In this model, lung tumors initially grew but were later rejected, while control tumors (LLC model with CD8 T cell response but no constitutively active IKKβ) grew without restriction (Hopewell et al., 2013). NF-κB activity was also found to be strongly associated with T cell infiltration in human lung tumors. Their results indicate that NF-κB is required for a full antitumor CD8 T cell response in murine and human lung tumors and may be partially due to NF-κB enhanced expression of the T cell chemokine CCL2 (Hopewell et al., 2013). This result contrasts with the previously mentioned tumor-promoting role CCL2 plays in recruiting TAMs and MAMs, which promotes metastasis.

2.4. NF-κB promotes recruitment of regulatory T lymphocytes in lung cancer

Regulatory T lymphocytes (Tregs) are a subpopulation of T cell ($FoxP3^+CD4^+CD25^+$) that help control immune responses and prevent autoimmunity by regulating immune system homeostasis. The positive functions of Tregs in preventing autoimmunity can quickly be negated, however, if they continue to suppress normal T cell function. Constitutive NF-κB expression in lung epithelial cells dramatically increased Treg numbers in IKTA mice (express the constitutively active form of IKKβ) (Zaynagetdinov et al., 2012). This study found that NF-κB overexpression was sufficient to cause lung tumors in a urethane mouse model and that NF-κB overexpression during the promotion of tumorigenesis is the critical timeframe for cancer promotion (Zaynagetdinov et al., 2012). Depletion of

Tregs in the IKTA mouse model resulted in improved tumor burden, implicating the importance of functional T cells for tumor cell clearance.

2.5. NF-κB and cancer-associated fibroblasts

Like immune cells, fibroblasts are also recruited to tumors and can promote aggressive properties in the cancer cells. In fact, cancer-associated fibroblasts (CAFs) are metabolically (Chaudhri et al., 2013) and phenotypically (Bhowmick, Neilson, & Moses, 2004) different from normal fibroblasts and provide epithelial cancer cells with vital nutrients. This is especially important in supporting a growing tumor before a blood supply is established (Martinez-Outschoorn et al., 2013). The NF-κB pathway has also been implicated in CAF promotion of a variety of cancers including pancreatic, mammary, skin, and prostate cancers. Doug Hanahan's group found that NF-κB was driving an inflammatory signature in CAFs from mouse and human skin, mammary, and pancreatic tumors. CAFs from these tumors mediated tumor-promoting inflammation, macrophage recruitment, neovascularization, and tumor growth, all of which was abolished when NF-κB was inhibited (Erez, Truitt, Olson, Arron, & Hanahan, 2010). Another more recent study found that NF-κB activates Wnt16B in prostate stromal fibroblasts, which promoted EMT in neoplastic prostate cells (Sun et al., 2012). Even more interesting is the finding of WNT16B promoted cancer cell survival following cytotoxic therapy in a cell nonautonomous manner (Sun et al., 2012).

3. TICs/CANCER STEM CELLS

Solid tumors are typically heterogeneous, being comprised of cells with different phenotypic and signaling properties. One subset of cells, known as tumor-initiating cells (TICs) or cancer stem cells, exhibit properties of self-renewal and tumorigenic potential when delivered to immunodeficient mice recipients (Charafe-Jauffret, Ginestier, & Birnbaum, 2009; D'Angelo & Wicha, 2010; Ginestier et al., 2010; Fig. 3.6). Additionally, these cells exhibit an EMT phenotype, are chemo/radioresistant, and are invasive. In certain characteristics, TICs share similarities with embryonic and adult stem cells. While it is proposed that TICs give rise to the more differentiated cancer cells within a tumor, evidence has been presented that these differentiated cancer cells can also give rise to TICs. Given the importance of IKK and NF-κB in cancer, it is not surprising that research from several different groups has converged on a concept that these signaling modules could regulate the so-called TIC phenotype.

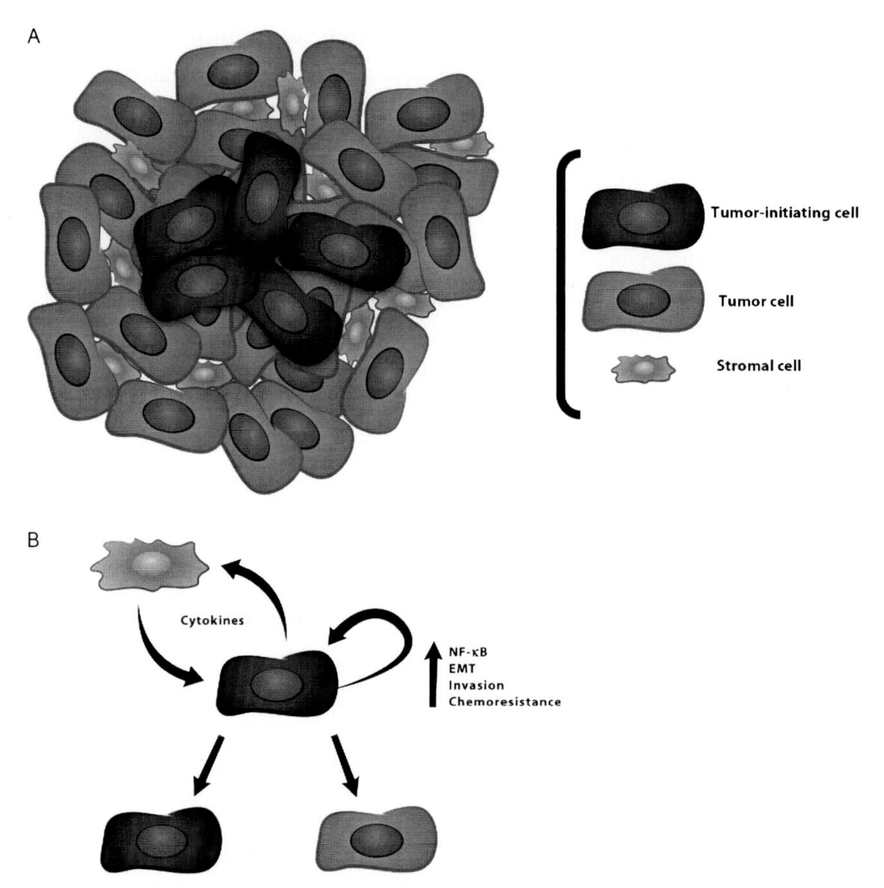

Figure 3.6 Tumor-initiating cells are a subpopulation of chemoresistant cancer cell. Tumor-initiating cells (TICs) are a subpopulation of cancer cell that are in part driven by canonical and noncanonical NF-κB signaling, and are known to be resistant to chemotherapies. (A) TICs are able to self-renew, can also produce non-TIC cancer cells, and are influenced by tumor stroma (B). (See the color plate.)

3.1. TICs and NF-κB

Early evidence for the involvement of IKK/NF-κB in TICs was shown by Karin and colleagues (Cao, Luo, & Karin, 2007), who showed that IKKα is required for the self-renewal of TICs from carcinogen-induced and Her2-driven breast cancers. Subsequently, Pestell and colleagues (Liu et al., 2010) found that inducible suppression of NF-κB (via expression of the super-repressor form of IκBα) in the adult mammary epithelium delayed the onset and number of Her2-driven breast tumors. Gene expression analysis showed that inhibition of NF-κB blocked expression of genes associated with stem cells, including Nanog and Sox2, and blocked TIC expansion *in vitro*.

Recently, Zhang et al. (2013) showed that IKKα promotes ErbB2/Her2-driven TICs via phosphorylation and induction of nuclear export of the cdk inhibitor p27 (Zhang et al., 2013). Our group showed that both IKKα and IKKβ promote TICs from basal-like and claudin-low breast cancer cells (Kendellen, Bradford, Lawrence, Clark, & Baldwin, 2013). In this regard, canonical and noncanonical NF-κB subunits drive TIC maintenance, and TICs (characterized by the CD44+ group) exhibited elevated NF-κB and IKK activity as compared to the CD44− subset. Additionally, we showed that NF-κB promotes EMT in breast cancer cells, a process known to be involved in TIC maintenance/expansion. Evidence was presented that inflammatory cytokines regulated by NF-κB and working in autocrine/paracrine pathways promotes TIC properties.

Rajasekhar, Studer, Gerald, Socci, and Scher (2011) found that NF-κB activation is elevated in a subset of prostate tumor cells with tumor-initiating properties (Rajasekhar et al., 2011). These cells exhibit low PSA and AR levels. The use of certain inhibitors that can affect NF-κB activation blocked secondary-sphere formation of these cells. Schwitalla et al. (2013) showed that NF-κB modulates Wnt signaling and that ablation of RelA/p65 in intestinal epithelium retards crypt stem cell expansion (Schwitalla et al., 2013). Elevated NF-κB signaling enhances Wnt activation and induces dedifferentiation of non-stem cells that acquire tumor-initiating capacity. We showed that RelA is important in maintaining hematopoietic stem cells (HSCs), as knockout of RelA in this compartment blocks levels of early progenitors (Stein & Baldwin, 2013). Evidence was presented that NF-κB controls the expression of genes associated with HSCs. For example, loss of RelA leads to downregulation of expression of Mpl which is the receptor for thrombopoietin, a factor involved in promoting HSCs. Consistent with the involvement of RelA in driving early stemness, knockout of RelA led to expression of monocyte differentiation markers.

3.2. TICs, cytokines, and tumor microenvironment

It is well established that hematopoietic stem cells are influenced by their microenvironment; thus, it is important to consider that tumor microenvironment could have effects on TICs. As with normal stem cells, TICs are regulated by both intrinsic and extrinsic signals. Evidence indicates that tumor development and cellular hierarchy coevolve with the tumor microenvironment during the process of tumorigenesis (Polyak, Haviv, & Campbell, 2009). Other evidence supports the concept that paracrine signaling coordinates tumorigenic progression, including maintenance of TICs.

As described earlier, tumor cells produce and secrete factors that attract and regulate many of the cells that constitute the tumor microenvironment. In this regard, immunomodulatory cells provide both stimulatory and inhibitory effects on tumor development and progression (Mantovani, 2009; Yu, Pardoll, & Jove, 2009). Understanding mechanisms derived from the microenvironment that promote TICs could have significant potential relative to new therapeutic approaches for cancer.

Both IL-6 and IL-1 were shown to regulate breast TIC self-renewal, and the genes encoding these cytokines are known to be regulated by NF-κB (Ginestier et al., 2010; Iliopoulos, Hirsch, Wang, & Struhl, 2011). IL-6 is known to link inflammation and malignant transformation in a pathway involving NF-κB, Lin28, and the Let-7 miRNA (Iliopoulos, Hirsch, & Struhl, 2009). Korkaya, Liu, and Wicha (2011) showed that IL-6 controls resistance to trastuzumab in Her2+ breast cancer by expanding the TIC population. Our data showed that inflammatory cytokines controlled by NF-κB promote/expand the TIC phenotype of basal-like and claudin-low breast cancer cells (Kendellen et al., 2013). In that study, IL-1 and IL-6, but not IL-8, promoted TIC expansion. Iliopoulos et al. (2011) showed that inducible formation of breast cancer TICs occurred by IL-6 secretion (Iliopoulos et al., 2011). Additionally, evidence has been presented that STAT3 (e.g., activated downstream of IL-6) functions in inflammation-driven cancer (Iliopoulos, Jaeger, Hirsch, Bulyk, & Struhl, 2010), and it is known that STAT3 and NF-κB can function to promote expression of certain pro-oncogenic genes (Lee et al., 2009; Yu et al., 2009).

4. CONCLUSIONS

The NF-κB pathway plays a dynamic role in various cancers and is becoming more understood as a key player in stromal interactions and in TIC function within the tumor. The influence of NF-κB can be seen in the form of tumor progression and other aggressive properties like TIC expansion and invasion. This highlights the fact that NF-κB is a viable target for therapeutic options for both cancer and stromal cells.

ACKNOWLEDGMENTS

We thank members of the Baldwin lab for insightful discussion pertaining to this review, and we thank Scott Bradford for assisting with the figures. The authors' work is funded by NCI Grants CA73756 and CA75080, NIH Grants AI35098 and F32 CA162628-01, and the Waxman Cancer Research Foundation. The authors declare no conflict of interest.

REFERENCES

Annunziata, C. M., Davis, R. E., Demchenko, Y., Bellamy, W., Gabrea, A., Zhan, F., et al. (2007). Frequent engagement of the classical and alternative NF-kappaB pathways by diverse genetic abnormalities in multiple myeloma. *Cancer Cell, 12*, 115–130.

Anton, K., & Glod, J. (2009). Targeting the tumor stroma in cancer therapy. *Current Pharmaceutical Biotechnology, 10*, 185–191.

Baldwin, A. S. (2001). Control of oncogenesis and cancer therapy resistance by the transcription factor NF-kappaB. *The Journal of Clinical Investigation, 107*, 241–246.

Baldwin, A. S. (2012). Regulation of cell death and autophagy by IKK and NF-kappaB: Critical mechanisms in immune function and cancer. *Immunological Reviews, 246*, 327–345.

Bang, D., Wilson, W., Ryan, M., Yeh, J. J., & Baldwin, A. S. (2013). GSK-3alpha promotes oncogenic KRAS function in pancreatic cancer via TAK1-TAB stabilization and regulation of noncanonical NF-kappaB. *Cancer Discovery, 3*, 690–703.

Basseres, D. S., & Baldwin, A. S. (2006). Nuclear factor-kappaB and inhibitor of kappaB kinase pathways in oncogenic initiation and progression. *Oncogene, 25*, 6817–6830.

Basseres, D. S., Ebbs, A., Levantini, E., & Baldwin, A. S. (2010). Requirement of the NF-kappaB subunit p65/RelA for K-Ras-induced lung tumorigenesis. *Cancer Research, 70*, 3537–3546.

Beroukhim, R., Mermel, C. H., Porter, D., Wei, G., Raychaudhuri, S., Donovan, J., et al. (2010). The landscape of somatic copy-number alteration across human cancers. *Nature, 463*, 899–905.

Bhowmick, N. A., Neilson, E. G., & Moses, H. L. (2004). Stromal fibroblasts in cancer initiation and progression. *Nature, 432*, 332–337.

Biswas, S. K., Gangi, L., Paul, S., Schioppa, T., Saccani, A., Sironi, M., et al. (2006). A distinct and unique transcriptional program expressed by tumor-associated macrophages (defective NF-kappaB and enhanced IRF-3/STAT1 activation). *Blood, 107*, 2112–2122.

Biswas, S. K., & Lewis, C. E. (2010). NF-kappaB as a central regulator of macrophage function in tumors. *Journal of Leukocyte Biology, 88*, 877–884.

Boehm, J. S., Zhao, J. J., Yao, J., Kim, S. Y., Firestein, R., Dunn, I. F., et al. (2007). Integrative genomic approaches identify IKBKE as a breast cancer oncogene. *Cell, 129*, 1065–1079.

Bonizzi, G., & Karin, M. (2004). The two NF-kappaB activation pathways and their role in innate and adaptive immunity. *Trends in Immunology, 25*, 280–288.

Cao, Y., Luo, J. L., & Karin, M. (2007). IkappaB kinase alpha kinase activity is required for self-renewal of ErbB2/Her2-transformed mammary tumor-initiating cells. *Proceedings of the National Academy of Sciences of the United States of America, 104*, 15852–15857.

Charafe-Jauffret, E., Ginestier, C., & Birnbaum, D. (2009). Breast cancer stem cells: Tools and models to rely on. *BMC Cancer, 9*, 202.

Chaudhri, V. K., Salzler, G. G., Dick, S. A., Buckman, M. S., Sordella, R., Karoly, E. D., et al. (2013). Metabolic alterations in lung cancer-associated fibroblasts correlated with increased glycolytic metabolism of the tumor. *Molecular Cancer Research, 11*, 579–592.

Connelly, L., Barham, W., Onishko, H. M., Chen, L., Sherrill, T. P., Zabuawala, T., et al. (2011). NF-kappaB activation within macrophages leads to an anti-tumor phenotype in a mammary tumor lung metastasis model. *Breast Cancer Research, 13*, R83.

Dan, H. C., Adli, M., & Baldwin, A. S. (2007). Regulation of mammalian target of rapamycin activity in PTEN-inactive prostate cancer cells by I kappa B kinase alpha. *Cancer Research, 67*, 6263–6269.

Dan, H. C., Cooper, M. J., Cogswell, P. C., Duncan, J. A., Ting, J. P., & Baldwin, A. S. (2008). Akt-dependent regulation of NF-{kappa}B is controlled by mTOR and Raptor in association with IKK. *Genes & Development, 22*, 1490–1500.

D'Angelo, R. C., & Wicha, M. S. (2010). Stem cells in normal development and cancer. *Progress in Molecular Biology and Translational Science, 95*, 113–158.

DiDonato, J. A., Mercurio, F., & Karin, M. (2012). NF-kappaB and the link between inflammation and cancer. *Immunological Reviews, 246*, 379–400.

Duran, A., Linares, J. F., Galvez, A. S., Wikenheiser, K., Flores, J. M., Diaz-Meco, M. T., et al. (2008). The signaling adaptor p62 is an important NF-kappaB mediator in tumorigenesis. *Cancer Cell, 13*, 343–354.

Erez, N., Truitt, M., Olson, P., Arron, S. T., & Hanahan, D. (2010). Cancer-associated fibroblasts are activated in incipient neoplasia to orchestrate tumor-promoting inflammation in an NF-kappaB-dependent manner. *Cancer Cell, 17*, 135–147.

Finco, T. S., Westwick, J. K., Norris, J. L., Beg, A. A., Der, C. J., & Baldwin, A. S., Jr. (1997). Oncogenic Ha-Ras-induced signaling activates NF-kappaB transcriptional activity, which is required for cellular transformation. *The Journal of Biological Chemistry, 272*, 24113–24116.

Fridman, W. H., Galon, J., Pages, F., Tartour, E., Sautes-Fridman, C., & Kroemer, G. (2011). Prognostic and predictive impact of intra- and peritumoral immune infiltrates. *Cancer Research, 71*, 5601–5605.

Ghosh, S., May, M. J., & Kopp, E. B. (1998). NF-kappa B and Rel proteins: Evolutionarily conserved mediators of immune responses. *Annual Review of Immunology, 16*, 225–260.

Ginestier, C., Liu, S., Diebel, M. E., Korkaya, H., Luo, M., Brown, M., et al. (2010). CXCR1 blockade selectively targets human breast cancer stem cells in vitro and in xenografts. *The Journal of Clinical Investigation, 120*, 485–497.

Gordon, S. (2003). Alternative activation of macrophages. *Nature Reviews. Immunology, 3*, 23–35.

Greten, F. R., Eckmann, L., Greten, T. F., Park, J. M., Li, Z. W., Egan, L. J., et al. (2004). IKKbeta links inflammation and tumorigenesis in a mouse model of colitis-associated cancer. *Cell, 118*, 285–296.

Hagemann, T., Lawrence, T., McNeish, I., Charles, K. A., Kulbe, H., Thompson, R. G., et al. (2008). "Re-educating" tumor-associated macrophages by targeting NF-kappaB. *The Journal of Experimental Medicine, 205*, 1261–1268.

Hagemann, T., Wilson, J., Burke, F., Kulbe, H., Li, N. F., Pluddemann, A., et al. (2006). Ovarian cancer cells polarize macrophages toward a tumor-associated phenotype. *Journal of Immunology, 176*, 5023–5032.

Hagemann, T., Wilson, J., Kulbe, H., Li, N. F., Leinster, D. A., Charles, K., et al. (2005). Macrophages induce invasiveness of epithelial cancer cells via NF-kappa B and JNK. *Journal of Immunology, 175*, 1197–1205.

Hayden, M. S., & Ghosh, S. (2004). Signaling to NF-kappaB. *Genes & Development, 18*, 2195–2224.

Hayden, M. S., & Ghosh, S. (2012). NF-kappaB, the first quarter-century: Remarkable progress and outstanding questions. *Genes & Development, 26*, 203–234.

Hopewell, E. L., Zhao, W., Fulp, W. J., Bronk, C. C., Lopez, A. S., Massengill, M., et al. (2013). Lung tumor NF-kappaB signaling promotes T cell-mediated immune surveillance. *The Journal of Clinical Investigation, 123*, 2509–2522.

Hu, M. C., Lee, D. F., Xia, W., Golfman, L. S., Ou-Yang, F., Yang, J. Y., et al. (2004). IkappaB kinase promotes tumorigenesis through inhibition of forkhead FOXO3a. *Cell, 117*, 225–237.

Iliopoulos, D., Hirsch, H. A., & Struhl, K. (2009). An epigenetic switch involving NF-kappaB, Lin28, Let-7 MicroRNA, and IL6 links inflammation to cell transformation. *Cell, 139*, 693–706.

Iliopoulos, D., Hirsch, H. A., Wang, G., & Struhl, K. (2011). Inducible formation of breast cancer stem cells and their dynamic equilibrium with non-stem cancer cells via

IL6 secretion. *Proceedings of the National Academy of Sciences of the United States of America, 108*, 1397–1402.

Iliopoulos, D., Jaeger, S. A., Hirsch, H. A., Bulyk, M. L., & Struhl, K. (2010). STAT3 activation of miR-21 and miR-181b-1 via PTEN and CYLD are part of the epigenetic switch linking inflammation to cancer. *Molecular Cell, 39*, 493–506.

Israel, A. (2000). The IKK complex: An integrator of all signals that activate NF-kappaB? *Trends in Cell Biology, 10*, 129–133.

Kaler, P., Godasi, B. N., Augenlicht, L., & Klampfer, L. (2009). The NF-kappaB/AKT-dependent Induction of Wnt signaling in colon cancer cells by macrophages and IL-1beta. *Cancer Microenvironment*, Epub ahead of print.

Karin, M., & Ben-Neriah, Y. (2000). Phosphorylation meets ubiquitination: The control of NF-[kappa]B activity. *Annual Review of Immunology, 18*, 621–663.

Kawauchi, K., Araki, K., Tobiume, K., & Tanaka, N. (2008). p53 regulates glucose metabolism through an IKK-NF-kappaB pathway and inhibits cell transformation. *Nature Cell Biology, 10*, 611–618.

Kawauchi, K., Araki, K., Tobiume, K., & Tanaka, N. (2009). Loss of p53 enhances catalytic activity of IKKbeta through O-linked beta-N-acetyl glucosamine modification. *Proceedings of the National Academy of Sciences of the United States of America, 106*, 3431–3436.

Keats, J. J., Fonseca, R., Chesi, M., Schop, R., Baker, A., Chng, W. J., et al. (2007). Promiscuous mutations activate the noncanonical NF-kappaB pathway in multiple myeloma. *Cancer Cell, 12*, 131–144.

Kendellen, M. F., Bradford, J. W., Lawrence, C. L., Clark, K. S., & Baldwin, A. S. (2013). Canonical and non-canonical NF-kappaB signaling promotes breast cancer tumor-initiating cells. *Oncogene, 33*, 1297–1305.

Korkaya, H., Liu, S., & Wicha, M. S. (2011). Breast cancer stem cells, cytokine networks, and the tumor microenvironment. *The Journal of Clinical Investigation, 121*, 3804–3809.

Kurahara, H., Shinchi, H., Mataki, Y., Maemura, K., Noma, H., Kubo, F., et al. (2011). Significance of M2-polarized tumor-associated macrophage in pancreatic cancer. *The Journal of Surgical Research, 167*, e211–e219.

Laoui, D., Movahedi, K., Van Overmeire, E., Van den Bossche, J., Schouppe, E., Mommer, C., et al. (2011). Tumor-associated macrophages in breast cancer: Distinct subsets, distinct functions. *The International Journal of Developmental Biology, 55*, 861–867.

Lee, H., Herrmann, A., Deng, J. H., Kujawski, M., Niu, G., Li, Z., et al. (2009). Persistently activated Stat3 maintains constitutive NF-kappaB activity in tumors. *Cancer Cell, 15*, 283–293.

Lenz, G., Davis, R. E., Ngo, V. N., Lam, L., George, T. C., Wright, G. W., et al. (2008). Oncogenic CARD11 mutations in human diffuse large B cell lymphoma. *Science, 319*, 1676–1679.

Lewis, C. E., & Pollard, J. W. (2006). Distinct role of macrophages in different tumor microenvironments. *Cancer Research, 66*, 605–612.

Ling, J., Kang, Y., Zhao, R., Xia, Q., Lee, D. F., Chang, Z., et al. (2012). KrasG12D-induced IKK2/beta/NF-kappaB activation by IL-1alpha and p62 feedforward loops is required for development of pancreatic ductal adenocarcinoma. *Cancer Cell, 21*, 105–120.

Liu, M., Sakamaki, T., Casimiro, M. C., Willmarth, N. E., Quong, A. A., Ju, X., et al. (2010). The canonical NF-kappaB pathway governs mammary tumorigenesis in transgenic mice and tumor stem cell expansion. *Cancer Research, 70*, 10464–10473.

Luo, J. L., Tan, W., Ricono, J. M., Korchynskyi, O., Zhang, M., Gonias, S. L., et al. (2007). Nuclear cytokine-activated IKKalpha controls prostate cancer metastasis by repressing Maspin. *Nature, 446*, 690–694.

Maeda, S., Kamata, H., Luo, J. L., Leffert, H., & Karin, M. (2005). IKKbeta couples hepatocyte death to cytokine-driven compensatory proliferation that promotes chemical hepatocarcinogenesis. *Cell, 121*, 977–990.

Mahmoud, S. M., Lee, A. H., Paish, E. C., Macmillan, R. D., Ellis, I. O., & Green, A. R. (2012). Tumour-infiltrating macrophages and clinical outcome in breast cancer. *Journal of Clinical Pathology, 65*, 159–163.

Mantovani, A. (2009). Cancer: Inflaming metastasis. *Nature, 457*, 36–37.

Mantovani, A., Allavena, P., Sica, A., & Balkwill, F. (2008). Cancer-related inflammation. *Nature, 454*, 436–444.

Martinez-Outschoorn, U. E., Curry, J. M., Ko, Y. H., Lin, Z., Tuluc, M., Cognetti, D., et al. (2013). Oncogenes and inflammation rewire host energy metabolism in the tumor microenvironment: RAS and NFkappaB target stromal MCT4. *Cell Cycle, 12*, 2580–2597.

Mayo, M. W., Wang, C. Y., Cogswell, P. C., Rogers-Graham, K. S., Lowe, S. W., Der, C. J., et al. (1997). Requirement of NF-kappaB activation to suppress p53-independent apoptosis induced by oncogenic Ras. *Science, 278*, 1812–1815.

Merkhofer, E. C., Cogswell, P., & Baldwin, A. S. (2010). Her2 activates NF-kappaB and induces invasion through the canonical pathway involving IKKalpha. *Oncogene, 29*, 1238–1248.

Meylan, E., Dooley, A. L., Feldser, D. M., Shen, L., Turk, E., Ouyang, C., et al. (2009). Requirement for NF-kappaB signalling in a mouse model of lung adenocarcinoma. *Nature, 462*, 104–107.

Movahedi, K., Laoui, D., Gysemans, C., Baeten, M., Stange, G., Van den Bossche, J., et al. (2010). Different tumor microenvironments contain functionally distinct subsets of macrophages derived from Ly6C(high) monocytes. *Cancer Research, 70*, 5728–5739.

Perkins, N. D., & Gilmore, T. D. (2006). Good cop, bad cop: The different faces of NF-kappaB. *Cell Death and Differentiation, 13*, 759–772.

Pikarsky, E., Porat, R. M., Stein, I., Abramovitch, R., Amit, S., Kasem, S., et al. (2004). NF-kappaB functions as a tumour promoter in inflammation-associated cancer. *Nature, 431*, 461–466.

Pollard, J. W. (2008). Macrophages define the invasive microenvironment in breast cancer. *Journal of Leukocyte Biology, 84*, 623–630.

Polyak, K., Haviv, I., & Campbell, I. G. (2009). Co-evolution of tumor cells and their microenvironment. *Trends in Genetics, 25*, 30–38.

Qian, B. Z., Li, J., Zhang, H., Kitamura, T., Zhang, J., Campion, L. R., et al. (2011). CCL2 recruits inflammatory monocytes to facilitate breast-tumour metastasis. *Nature, 475*, 222–225.

Qian, B. Z., & Pollard, J. W. (2010). Macrophage diversity enhances tumor progression and metastasis. *Cell, 141*, 39–51.

Rahal, R., Frick, M., Romero, R., Korn, J. M., Kridel, R., Chun Chan, F., et al. (2014). Pharmacological and genomic profiling identifies NF-kappaB-targeted treatment strategies for mantle cell lymphoma. *Nature Medicine, 20*, 87–92.

Rajasekhar, V. K., Studer, L., Gerald, W., Socci, N. D., & Scher, H. I. (2011). Tumour-initiating stem-like cells in human prostate cancer exhibit increased NF-kappaB signalling. *Nature Communications, 2*, 162.

Ren, G., Zhao, X., Wang, Y., Zhang, X., Chen, X., Xu, C., et al. (2012). CCR2-dependent recruitment of macrophages by tumor-educated mesenchymal stromal cells promotes tumor development and is mimicked by TNFalpha. *Cell Stem Cell, 11*, 812–824.

Saccani, A., Schioppa, T., Porta, C., Biswas, S. K., Nebuloni, M., Vago, L., et al. (2006). p50 nuclear factor-kappaB overexpression in tumor-associated macrophages inhibits M1 inflammatory responses and antitumor resistance. *Cancer Research, 66*, 11432–11440.

Schwitalla, S., Fingerle, A. A., Cammareri, P., Nebelsiek, T., Goktuna, S. I., Ziegler, P. K., et al. (2013). Intestinal tumorigenesis initiated by dedifferentiation and acquisition of stem-cell-like properties. *Cell, 152*, 25–38.

Sen, R., & Baltimore, D. (1986). Multiple nuclear factors interact with the immunoglobulin enhancer sequences. *Cell, 46*, 705–716.

Senftleben, U., Cao, Y., Xiao, G., Greten, F. R., Krahn, G., Bonizzi, G., et al. (2001). Activation by IKKalpha of a second, evolutionary conserved, NF-kappa B signaling pathway. *Science, 293*, 1495–1499.

Stein, S. J., & Baldwin, A. S. (2013). Deletion of the NF-kappaB subunit p65/RelA in the hematopoietic compartment leads to defects in hematopoietic stem cell function. *Blood, 121*, 5015–5024.

Sun, Y., Campisi, J., Higano, C., Beer, T. M., Porter, P., Coleman, I., et al. (2012). Treatment-induced damage to the tumor microenvironment promotes prostate cancer therapy resistance through WNT16B. *Nature Medicine, 18*, 1359–1368.

Van Ginderachter, J. A., Movahedi, K., Hassanzadeh Ghassabeh, G., Meerschaut, S., Beschin, A., Raes, G., et al. (2006). Classical and alternative activation of mononuclear phagocytes: Picking the best of both worlds for tumor promotion. *Immunobiology, 211*, 487–501.

Wilson, W., III. (2009). The regulation of constitutive NF-κB activity by glycogen synthase kinase-3 in pancreatic cancer. In *ProQuest*. UNC-Chapel Hill Dissertation.

Wynn, T. A., Chawla, A., & Pollard, J. W. (2013). Macrophage biology in development, homeostasis and disease. *Nature, 496*, 445–455.

Xia, Y., Padre, R. C., De Mendoza, T. H., Bottero, V., Tergaonkar, V. B., & Verma, I. M. (2009). Phosphorylation of p53 by IkappaB kinase 2 promotes its degradation by beta-TrCP. *Proceedings of the National Academy of Sciences of the United States of America, 106*, 2629–2634.

Xia, Y., Yeddula, N., Leblanc, M., Ke, E., Zhang, Y., Oldfield, E., et al. (2012). Reduced cell proliferation by IKK2 depletion in a mouse lung-cancer model. *Nature Cell Biology, 14*, 257–265.

Yang, X. D., Tajkhorshid, E., & Chen, L. F. (2010). Functional interplay between acetylation and methylation of the RelA subunit of NF-kappaB. *Molecular and Cellular Biology, 30*, 2170–2180.

Yu, H., Pardoll, D., & Jove, R. (2009). STATs in cancer inflammation and immunity: A leading role for STAT3. *Nature Reviews. Cancer, 9*, 798–809.

Zaynagetdinov, R., Stathopoulos, G. T., Sherrill, T. P., Cheng, D. S., McLoed, A. G., Ausborn, J. A., et al. (2012). Epithelial nuclear factor-kappaB signaling promotes lung carcinogenesis via recruitment of regulatory T lymphocytes. *Oncogene, 31*, 3164–3176.

Zhang, L., Conejo-Garcia, J. R., Katsaros, D., Gimotty, P. A., Massobrio, M., Regnani, G., et al. (2003). Intratumoral T cells, recurrence, and survival in epithelial ovarian cancer. *The New England Journal of Medicine, 348*, 203–213.

Zhang, W., Tan, W., Wu, X., Poustovoitov, M., Strasner, A., Li, W., et al. (2013). A NIK-IKKalpha module expands ErbB2-induced tumor-initiating cells by stimulating nuclear export of p27/Kip1. *Cancer Cell, 23*, 647–659.

CHAPTER FOUR

The Rb–E2F Transcriptional Regulatory Pathway in Tumor Angiogenesis and Metastasis

Courtney Schaal[1], Smitha Pillai[1], Srikumar P. Chellappan[2]
Department of Tumor Biology, H. Lee Moffitt Cancer Center and Research Institute, Tampa, Florida, USA
[1]These authors contributed equally to this work.
[2]Corresponding author: e-mail address: srikumar.chellappan@moffitt.org

Contents

Abstract

The retinoblastoma tumor suppressor protein Rb plays a major role in regulating G1/S transition and is a critical regulator of cell proliferation. Rb protein exerts its growth regulatory properties mainly by physically interacting with the transcriptionally active members of the E2F transcription factor family, especially E2Fs 1, 2, and 3. Given its critical role in regulating cell proliferation, it is not surprising that Rb is inactivated in almost all tumors, either through the mutation of Rb gene itself or through the mutations of its upstream regulators including K-Ras and INK4. Recent studies have revealed a significant role for Rb and its downstream effectors, especially E2Fs, in regulating various aspects of

Advances in Cancer Research, Volume 121
ISSN 0065-230X
http://dx.doi.org/10.1016/B978-0-12-800249-0.00004-4

tumor progression, angiogenesis, and metastasis. Thus, components of the Rb–E2F pathway have been shown to regulate the expression of genes involved in angiogenesis, including VEGF and VEGFR, genes involved in epithelial–mesenchymal transition including E-cadherin and ZEB proteins, and genes involved in invasion and migration like matrix metalloproteinases. Rb has also been shown to play a major role in the functioning of normal and cancer stem cells; further, Rb and E2F appear to play a regulatory role in the energy metabolism of cancer cells. These findings raise the possibility that mutational events that initiate tumorigenesis by inducing uncontrolled cell proliferation might also contribute to the progression and metastasis of cancers through the mediation of the Rb–E2F transcriptional regulatory pathway. This review highlights these recent studies on tumor promoting functions of the Rb–E2F pathway.

1. INTRODUCTION

E2F transcription factors comprise a family of extensively studied transcription factors which bind to multiple target gene promoters regulating their expression and are perhaps most well known for their function in cell cycle progression and transition into S phase, DNA synthesis, and cellular proliferation (Chen, Tsai, & Leone, 2009). There are nine known members of the E2F family, including E2F1–E2F3a which are strong transcriptional activators, E2F3b–E2F5 which are passive repressors, and E2Fs 6–8 which have been established as active repressors of transcription (DeGregori & Johnson, 2006; Lammens, Li, Leone, & De Veylder, 2009). E2Fs 1–3 are regulated by the retinoblastoma (Rb) tumor suppressor protein, which physically interacts with their transcriptional activation domain, while its related family members p107 and p130 physically interact with E2Fs 4–5 proteins to modulate transcriptional activity of target gene promoters (Chen et al., 2009). Typically, when a cell is in a quiescent state, hypophosphorylated Rb associates with E2F acting to repress transcriptional activity (Chellappan, Hiebert, Mudryj, Horowitz, & Nevins, 1991; Hiebert, Chellappan, Horowitz, & Nevins, 1992). Upon mitogenic signaling, cyclin D complexes with CDK4/6 and initiates a series of sequential phosphorylations of the Rb protein, resulting in its inactivation and disassociation from E2Fs allowing E2F to activate gene transcription (Nevins, 2001; Sherr & McCormick, 2002). The Rb–E2F pathway has been linked to cell cycle regulation, DNA synthesis, and proliferation through the findings that E2F regulates and activates a majority of genes involved in these processes including nucleotide biosynthesis enzymes such as thymidylate synthase (TS) and thymidine kinase (TK), DNA replication factors such as DNA polymerase alpha, and proteins critical to progression through cell cycle such as cyclin E and CDK2, as well as a large number of other genes (Frolov & Dyson, 2004; Morris & Dyson, 2001; Nevins, 2001). The Rb–E2F pathway is known to be disrupted in nearly all human cancers, resulting in prooncogenic effects (Di Fiore, D'Anneo, Tesoriere, & Vento, 2013; Indovina, Marcelli, Casini, Rizzo, & Giordano, 2013; Nevins, 2001; Sherr & McCormick, 2002). Specifically, Rb is frequently inactivated due to mutation or loss, allowing for increased E2F transcriptional activity ultimately resulting in increased proliferation, which is a critical feature of cancer (Hanahan & Weinberg, 2000, 2011). Indeed, inactivation of the Rb protein by mutation in the RB gene itself, or mutations of upstream signaling molecules, or through the interaction of viral oncoproteins occurs in virtually all tumors (Di Fiore et al., 2013). While E2Fs may be classically known for promoting proliferation, they have also been shown to have roles in other biological processes such as apoptosis, and more recently, the Rb–E2F pathway has been implicated in angiogenesis, epithelial-to-mesenchymal transition (EMT), migration, invasion, and metastasis (Indovina et al., 2013); all features of more advanced, aggressive cancers. This pathway has also been implicated in maintenance of stemness in subpopulations of cells with classical stem cell properties (Takahashi, Sasaki, & Kitajima, 2012), which may also be linked to cancer promotion. While it has become clear that Rb can regulate multiple cellular functions independent of E2F1 (Dick & Rubin, 2013), and while E2F transcription factors can be regulated by molecules other than Rb (Schlisio, Halperin, Vidal, & Nevins, 2002; Wang, Nath, Adlam, & Chellappan, 1999; Wang, Nath, Fusaro, & Chellappan, 1999; Zhou, Srinivasan, Nawaz, & Slingerland, 2013), the repression of E2Fs 1–3 by Rb protein regulates the expression of genes involved in various cellular processes, affecting the normal biology of cells as well as their oncogenic transformation. We discuss the role of the Rb–E2F pathway in angiogenesis, metastasis, and stemness in this review.

2. Rb–E2F PATHWAY IN ANGIOGENESIS

Angiogenesis, the development of new blood vessels from preexisting vessels is a critical step in tumor progression and metastasis (Bergers & Benjamin, 2003). The formation of a functional vascular system ensures adequate supply of oxygen and nutrients to the tumor, enabling the tumor mass to grow beyond a critical size. This complex multistep process is controlled by the balance between factors that promote angiogenesis (proangiogenic factors) and factors that suppress angiogenesis (antiangiogenic factors). During tumor development, induction of angiogenic switch is a key step determined by proangiogenic factors overcoming antiangiogenic factors in the tumor microenvironment. It is well established that many oncogenes and tumor suppressor genes regulate angiogenesis during tumor progression by modulating the expression of these factors. Recent studies have revealed that the Rb–E2F pathway plays critical roles in different aspects of angiogenesis.

2.1. E2F-mediated regulation of vascular endothelial growth factor

The role of E2F1 has been examined in E2F1$^{-/-}$ mice. *In vivo* studies using E2F1$^{-/-}$ mice displayed enhanced angiogenesis, endothelial cell proliferation, and reperfusion in a hind limb ischemia model, resulting from elevated VEGFA expression (Qin et al., 2006). The proangiogenic phenotype brought about by E2F1 deficiency is thought to be the result of overproduction of vascular endothelial growth factor (VEGF) and could be prevented by VEGF blockade. Under hypoxic conditions, E2F1 downregulates the expression of VEGF promoter activity by associating with p53 and specifically downregulating expression of VEGF but not other hypoxia-inducible genes, suggesting a promoter structure context-dependent regulatory mechanism. However, regulation of VEGF promoter by p53 is more complex than was originally proposed and the underlying mechanisms remain unclear and controversial (Berger et al., 2010; Nag, Qin, Srivenugopal, Wang, & Zhang, 2013; Teodoro, Evans, & Green, 2007). A more recent study demonstrated rapid induction of VEGF transcription by p53 upon exposure to hypoxia by binding to a highly conserved and functional p53-binding site within VEGF promoter in a HIF-1α-dependent manner (Farhang Ghahremani et al., 2013). However, during sustained hypoxia, p53 indirectly downregulated VEGF expression through Rb pathway in a p21-dependent fashion (Farhang Ghahremani et al., 2013). Recent studies have also revealed E2F1-mediated repression of VEGFA transcription in a p53-independent manner by transcriptional activation of the serine/arginine-rich splicing protein (SR protein) SC35, a splicing factor of VEGFA. E2F1 altered the ratio of proangiogenic versus antiangiogenic isoforms of VEGFA by inducing the expression of SC35 in *in vitro* as well as in xenograft models in nude mice (Merdzhanova et al., 2010). Thus, it appears that E2F1 can affect the expression of VEGF through the involvement of other transcription factors and splicing factors.

2.2. Rb–E2F and Id proteins in angiogenesis

Growing evidence in recent years links Id (inhibitor of differentiation/DNA binding) proteins to Rb/E2F1 pathway in modulating tumor neoangiogenesis. The Id family consists of four proteins (Id1–4) belonging to helix–loop–helix transcription factors that lack a DNA-binding domain (Perk, Iavarone, & Benezra, 2005; Ruzinova & Benezra, 2003). Id proteins associate with basic HLH transcription factors and prevent them from binding to DNA or forming active heterodimers thereby regulating a number of crucial cellular processes such as differentiation, proliferation, and cell death (Ruzinova & Benezra, 2003). Among Id family members, only Id2 has been reported to interact with and inhibit Rb during mouse embryogenesis, establishing its role as a target of Rb during development of the nervous system and hematopoiesis (Iavarone, Garg, Lasorella, Hsu, & Israel, 1994; Toma, El-Bizri, Barnabe-Heider, Aloyz, & Miller, 2000). In Rb mutant mice ($Rb^{-/+}$), pituitary carcinogenesis is dependent on the expression of Id2, which was necessary and sufficient for the expression of VEGF, thus promoting angiogenesis by functioning as a master regulator of VEGF. In addition to impaired VEGF expression, Rb mutant, Id2 null ($Rb^{+/-}$ and $Id2^{-/-}$) pituitary tumors showed inhibition of HIF-1α expression, contributing to the angiogenic switch of Rb-null pituitary tumors (Lasorella, Rothschild, Yokota, Russell, & Iavarone, 2005).

Induction of the Rb-related protein p130 was previously described to inhibit angiogenesis *in vivo*, correlating with the downregulation of VEGF mRNA levels and protein expression (Claudio et al., 2001). Tumors from Rb2 overexpressing mice showed decreased blood vessel density and regression of established tumors. Since this study did not identify any Rb responsive site in the promoter, Rb2 is speculated to regulate VEGF expression by an indirect mechanism. It is tempting to imagine that its association to Id proteins might be one of the mechanisms contributing to angiogenesis regulation. Further, a recent study identified Id4 as a transcriptional target of gain of function p53 mutants R175H, R273H, and R280K which form a complex with E2F1; this complex could assemble on Id4 promoter and induce Id4 expression. The Id4 protein could bind to and stabilize mRNAs encoding proangiogenic cytokines IL8 and GRO-α leading to enhanced angiogenic potential of breast cancer cells expressing mutant p53 (Fontemaggi et al., 2009). These studies suggest a potential antiangiogenic role for Rb in multiple systems.

2.3. Rb and HIF-1α

In contrast to the antiangiogenic role of Rb, it is shown to interact with and activate HIF-1α, a major transcriptional activator of genes that coordinate physiological and pathological responses toward hypoxia (Budde, Schneiderhan-Marra, Petersen, & Brune, 2005). Under normoxic conditions, HIF-1α is rapidly ubiquitinated and subjected to proteosomal degradation (Ahluwalia & Tarnawski, 2012; Baldewijns et al., 2010; Semenza, 2008). Cells respond to hypoxia by inducing protective mechanisms such as stabilization of HIF-1α. In response to hypoxia, HIF-1 heterodimer containing HIF-1α and constitutively expressed HIF-1β (ARNT) binds to hypoxia responsive elements present in the enhancer or promoter regions of hypoxia-inducible genes including VEGF (Labrecque, Prefontaine, & Beischlag, 2013). A series of *in vitro* and *in vivo* experiments demonstrated the physical interaction between Rb and HIF-1α and the overexpression of Rb under normoxia increased transcriptional activation by HIF-1α. Interestingly, HIF-1α was found to prevent the transcriptional repressor function of Rb, demonstrating a reciprocal relationship in their transcriptional regulatory functions. Thus, it appears that the binding of Rb to HIF-1α enhances the transcriptional activity of HIF-1α but also suggests that HIF-1α is one of the non-E2F targets of Rb. These findings raise the possibility that Rb can affect the induction of angiogenesis directly and indirectly, through its interaction with multiple factors.

2.4. E2F-mediated transcriptional regulation of angiogenic factors and their receptors

Initial attempts to identify E2F1-inducible gene targets using cDNA microarrays revealed several genes associated with the process of angiogenesis such as FGF2, MMP16, and VEGFB, which showed significantly elevated expression (Stanelle, Stiewe, Theseling, Peter, & Putzer, 2002). Further, a recent study identified E2F1 as an enhancer of angiogenesis through the regulation of VEGFC/VEGFR3 signaling axis in tumors, which cooperatively activated platelet-derived growth factor B (PDGF-B) expression (Engelmann et al., 2013). Activation or forced expression of E2F1 in a number of cell lines led to the upregulation of VEGFR3 and its ligand VEGFC which could be abolished by E2F1 depletion (Engelmann et al., 2013). E2F1-dependent VEGFR3 expression was crucial for tumor cells to enhance angiogenic tubule formation *in vitro* and also neovascularization in mice. Under hypoxic conditions, VEGFC/VEGFR3 levels were upregulated in an E2F1-dependent manner and resulted in enhanced levels of the proangiogenic cytokine PDGF-B (Engelmann et al., 2013).

A number of growth factor receptors involved in proliferation, angiogenesis, embryonic development, and tumorigenesis have been demonstrated as a new category of E2F target genes (Minato et al., 2007; Tashiro, Minato, Maruki, Asagiri, & Imoto, 2003). Expression of FGF receptor 2 protein in mid-to-late G1 phase is induced by E2F1, 2, and 3, and this transcriptional activation could be inhibited by Rb (Tashiro et al., 2003). In addition, when cyclin D1 overexpressing rodent fibroblasts were treated with bFGF (basic fibroblast growth factor), there was increase in cell cycle progression, which was at least in part facilitated by elevated FGFR1 expression through Rb–E2F pathway (Kanai, Tashiro, Maruki, Minato, & Imoto, 2009; Tashiro et al., 2003). Similarly, the expression of PDGFR-α was found to be transcriptionally regulated by E2F1 (Minato et al., 2007).

Studies from our lab demonstrate that E2Fs not only regulate VEGFA transcription in cells that secrete angiogenic factors but also regulate expression of VEGF receptors in endothelial cells (Pillai, Kovacs, & Chellappan, 2010). Human vascular endothelial cells when stimulated with VEGFA led to the inactivation or hyperphosphorylation of Rb, resulting in E2F1-mediated transcription of VEGF receptors, FLT1 (VEGFR1) and KDR (VEGFR2). There was enhanced recruitment of E2F1 in the promoters of FLT1 and KDR with concomitant dissociation of Rb in multiple endothelial cells after VEGF stimulation; depletion of E2F1 in endothelial cells prevented tubulogenesis in matrigel, but depletion of E2F4 had no effect. VEGF stimulation also resulted in the acetylation of E2F1 contributing to its enhanced transcriptional activity indicating the existence of a positive feedback loop regulating E2F1 acetylation and VEGF receptor expression. The acetylation of E2F1 associated with VEGF stimulation was predominantly mediated by p300/CBP-associated factor, PCAF. Interestingly, depletion of PCAF, like depletion of E2F1, disrupted angiogenic tubule formation in matrigel (Pillai et al., 2010).

Endothelial cells when stimulated with proangiogenic cytokines such as VEGF can activate pathways involving Rb and E2Fs. VEGF stimulation of aortic endothelial cells resulted in E2F-mediated transcriptional activation of metallothionein 1G (hMT1G) gene, a protein involved in metal metabolism and detoxification and recently demonstrated to be involved in regulation of angiogenesis (Joshi, Ordonez-Ercan, Dasgupta, & Chellappan, 2005; Miyashita & Sato, 2005; Zbinden et al., 2010). Further, binding of

E2F1–3 in the promoter region of metallothionein gene was enhanced upon VEGF stimulation with a concomitant dissociation of Rb from the promoter region. Given the proangiogenic role of metallothioneins, it is tempting to speculate that VEGF-mediated induction of MT1 by Rb–E2F pathway might be contributing to the process of angiogenesis and tumor progression (Joshi et al., 2005). Collectively, these studies indicate that transcriptionally active E2Fs, especially E2F1, can induce genes that are involved in promoting angiogenesis.

The positive regulation of proangiogenic factors by E2F1 is in contrast with earlier studies where E2F1$^{-/-}$ mice displayed enhanced angiogenesis of tumor xenografts and increased endothelial cell proliferation through induction of VEGFA (Qin et al., 2006). Interestingly, E2F1 has been shown to directly regulate the transcription of thrombospondin 1, an inhibitor of angiogenesis, which could be repressed by Rb implying that the Rb–E2F1 pathway might contribute to inhibition of angiogenesis by modulating TSP1 expression (Ji, Zhang, & Xiao, 2010). It is therefore apparent that the role of E2F1 in vascular biology is more complex than expected, and further studies are needed to understand the molecular events leading to these conflicting observations.

2.5. Raf-1-mediated inactivation of Rb contributes to VEGF-induced angiogenesis

Our earlier studies had shown that the kinase Raf-1 (C-Raf) physically interacts with Rb very early in the cell cycle, facilitating the phosphorylation and inactivation of Rb (Wang, Ghosh, & Chellappan, 1998). The binding of Raf-1 preceded the association of cyclin D with Rb protein, and this binding was necessary for the subsequent inactivation of Rb by cyclins (Dasgupta et al., 2004). These studies also indicated a direct link between the Raf-1 kinase and the Rb protein, independent of the MEK–ERK pathways. An eight-amino acid peptide could disrupt the binding of Raf-1 to Rb and inhibit cell cycle progression (Dasgupta et al., 2004); similarly, a small molecule inhibitor of the Rb–Raf-1 interaction, called RRD-251, had similar antitumor activities (Kinkade et al., 2008). These studies have gained relevance in the context of angiogenesis as well; elegant studies published from David Cheresh's lab established the role of Raf-1 kinase as a crucial regulator of endothelial cell survival during angiogenesis (Alavi, Hood, Frausto, Stupack, & Cheresh, 2003; Hood & Cheresh, 2002). These studies demonstrated that differential activation of Raf-1 in response to bFGF and VEGF, resulting in protection from distinct pathways of apoptosis in human endothelial cells and chick embryo vasculature. bFGF activated Raf-1 through p21-activated protein kinase-1 phosphorylation of serines 338 and 339, resulting in Raf-1 mitochondrial translocation and endothelial cell protection from the intrinsic pathway of apoptosis, independent of the mitogen-activated protein kinase kinase-1 (MEK1). In contrast, VEGF activated Raf-1 via Src kinase, leading to phosphorylation of tyrosines 340 and 341 and MEK1-dependent protection from extrinsic-mediated apoptosis. In addition, inhibition of Raf-1 by delivering a mutant Raf using nanoparticles into the angiogenic vessels in tumor bearing mice ultimately led to tumor cell apoptosis and sustained regression of established primary and metastatic tumors (Hood & Cheresh, 2002).

Our earlier studies had shown that Rb is inactivated by Raf-1 kinase upon VEGF stimulation and disruption of the binding of Raf-1 to Rb using an eight-amino acid peptide could inhibit VEGF-dependent angiogenesis *in vitro* (Dasgupta et al., 2004). In addition, this peptide could inhibit tumor growth when delivered intratumorally in nude mouse models. The tumors from mice treated with the peptide showed a significant reduction in neovasculature as seen by CD31 staining. Further, RRD-251, which is a small molecule inhibitor of Rb–Raf-1 interaction, could efficiently inhibit angiogenesis *in vitro* and in tumors (Davis & Chellappan, 2008; Kinkade et al., 2008; Singh, Johnson, & Chellappan, 2010). The fact that cell cycle regulators like Rb and E2F can regulate angiogenic tubule formation is not surprising since endothelial cell proliferation, a crucial step in this process might be regulated by this pathway (Dasgupta et al., 2004); further, it is likely that the regulation of VEGFR expression by E2F1 might also be affected by the inhibition of E2F1 function.

2.6. Atypical E2Fs and angiogenesis

The E2F family of transcription factors consists of nine members, and these members are classically grouped as either activators (E2F1–3a) or repressors (E2F3b–8) (Chen et al., 2009). However, recent studies suggest that E2Fs can function as either activators or repressors of transcription depending on the cellular context, target genes, and cofactors involved (Chong et al., 2009; Lee, Bhinge, & Iyer, 2011). The classical E2Fs (E2F1–6) contain one DNA-binding domain and form heterodimers with DP proteins. The more recently identified members, E2F7 and E2F8, referred to as atypical E2Fs as they possess two DNA-binding domains, form homo- or heterodimers, and regulate transcription in a DP-independent manner

(Lammens et al., 2009). Although deletion of E2F7/8 in mice resulted in embryonic death by day E11.5, they did not show any defects in proliferation (Li et al., 2008); instead, these mice displayed massive apoptosis and vascular defects at E10.5. Interestingly, apoptosis, but not vascular defects, were rescued by additional deletion of E2F1 or p53 in these double knock-out mice, indicating that E2F7/8 regulate vascular integrity through an alternate mechanism (Li et al., 2008). Further studies revealed that E2F7 and E2F8 could directly bind to the promoter of VEGFA and activate transcription, independent of canonical E2F-binding elements (Bakker, Weijts, Westendorp, & de Bruin, 2013; Weijts et al., 2012). Instead, E2F7/8 form a transcriptional complex with HIF-1 to stimulate VEGFA promoter induction. These results uncover an unexpected link between atypical E2Fs and HIF1–VEGFA pathway, adding to the complexity of the molecular mechanisms by which Rb–E2F pathway regulates angiogenesis.

2.7. E2Fs in miRNA-mediated regulation of angiogenesis

Exposure of cells to hypoxia leads to massive changes in gene expression, including that of microRNAs (Shen, Li, Jia, Piazza, & Xi, 2013). MicroRNAs are short noncoding RNAs that modulate the stability and/or translation potential of their targets (Hao et al., 2014). Recent evidence suggests that E2Fs can act as both targets of miRNAs and regulators of their biogenesis during hypoxic/ischemic angiogenesis (Biyashev & Qin, 2011). MicroRNA profiling of ovarian cancer cells identified miR-210 as one of the most prominent microRNA consistently induced under hypoxic conditions. Further, biocomputational analysis and *in vitro* assays demonstrated that E2F3 is regulated by miR-210; induction of miR-210 effectively downregulated E2F3 expression at protein level (Giannakakis et al., 2008). However, further functional analysis is needed to establish a direct link between miR-210 regulation of E2F3 in angiogenesis. Another study revealed miR-23b as a regulator of endothelial cell growth during pulsatile shear flow-induced growth arrest by modulating Rb–E2F pathway (Wang et al., 2010). Inhibition of miR-23b could reverse pulsatile shear-induced E2F1 expression and Rb hypophosphorylation and attenuated G0/G1 arrest. Even though this mechanism is more relevant in understanding cardiovascular homeostasis, similar regulatory mechanisms might regulate endothelial cell growth during tumor angiogenesis.

The above studies clearly indicate that the Rb–E2F transcriptional regulatory pathway modulates angiogenesis at multiple levels, by affecting the expression of a variety of growth factors, their receptors as well as other regulators of angiogenesis. It is also clear that Rb and E2F family members can regulate these processes independently of each other, or as a canonical pathway. It appears that the regulation of angiogenesis by this pathway is fine-tuned by various extracellular stimuli including hypoxia and might execute the regulation positively or negatively depending on the microenvironment and the physiological needs of the cells in question. These observations also raise the possibility that targeting the Rb–E2F pathway might be a viable option to combat aberrant angiogenesis associated with tumor growth.

3. Rb–E2F PATHWAY AND TUMOR METASTASIS

Metastasis is the end result of an evolutionary process of malignant transformation and adaptation of cancer cells and is the major cause of cancer-related deaths (Siegel, Naishadham, & Jemal, 2013). While a significant amount of information is available on the events that regulate metastasis in specific cancer types (Aladowicz et al., 2013; Canel, Serrels, Frame, & Brunton, 2013; Horm & Schroeder, 2013; Perlikos, Harrington, & Syrigos, 2013; Proulx & Detmar, 2013), and various mouse models have been developed to study metastasis (Saxena & Christofori, 2013), the underlying molecular mechanisms of metastasis remain poorly understood (Chiang & Massague, 2008). Initially, the progression of tumors is dependent on the acquisition of properties that allow cells to survive in less than ideal environments, which they normally would not tolerate well. Once a tumor is initiated, genetic reprogramming of cells both within the tumor itself as well as in the microenvironment, results in phenotypic changes that give rise to a more aggressive form of cancer. Such changes include EMT (Creighton, Gibbons, & Kurie, 2013; Kalluri & Weinberg, 2009; Tsai & Yang, 2013), increased vasculature recruitment and angiogenesis, and increased migratory and invasive capacity; these features create ideal parameters for cells from the primary tumor site to escape into circulation and home to distal metastatic sites (Chiang & Massague, 2008; De Craene & Berx, 2013). It is thought that cancer stem cells, also known as tumor-initiating cells, may have a major role in metastasis. This is because of their ability to survive in harsh environments, in addition to their ability to self-renew and also give rise to heterogeneous tumor populations (Scheel & Weinberg, 2012; Visvader & Lindeman, 2008, 2012). Recent studies have attempted to elucidate the cellular alterations involved in the acquisition of a cancer stem cell phenotype, in order to target these alterations for cancer therapy.

Upon oncogenic transformation, genetic mutations and aberrant signaling give rise to cells with unlimited proliferative and growth capacity, as well as the ability to evade apoptosis. Typically, tumor cells are encapsulated within a basement membrane and recruit new vasculature to that site to supply the oxygen and nutrients that are needed. As the tumor progresses, certain cell populations within the tumor acquire the ability to break down this membrane and leave this initial "primary" tumor site, intravasate through the new vasculature into the blood stream and into systemic circulation (Scheel & Weinberg, 2012). After surviving in the bloodstream, some of these cells are able to extravasate from blood vessels into distant tissues and organs where they then act to initiate secondary "metastatic" tumor sites (Chaffer & Weinberg, 2011). What causes these cells to home to specific distal organs, and how they are capable of surviving the entire metastatic process is not yet fully elucidated. It is thought that in order to leave the primary tumor and intravasate into the bloodstream that cells undergo EMT gene reprogramming giving them a survival advantage until they reach their destination and extravasate, after which mesenchymal-to-epithelial reprogramming is thought to occur, allowing the cell to establish itself in its new environment (Valastyan & Weinberg, 2011). The reversible process of EMT is thought to be a critical feature for migration, invasion, and ultimately tumor metastasis. The Rb–E2F pathway appears to facilitate multiple steps of the metastatic process.

3.1. Expression of Rb and E2F in metastatic cancers

A number of studies have investigated the gene expression profiles of metastatic tumors in comparison to corresponding primary tumors in an attempt to identify genes which act as biomarkers of metastatic cells, and more intriguingly, genes which are causally linked to metastasis with functional implications (Arpino et al., 2013; Bidard, Pierga, Soria, & Thiery, 2013; Gonzalez-Gonzalez et al., 2013; Hudson, Kulp, & Loots, 2013). In a study looking at breast tumors, eight publicly available gene expression datasets were analyzed to assess genes which were significantly differently regulated in metastasizing tumors compared to nonmetastasizing tumors from over 1200 breast cancer patients (Thomassen, Tan, & Kruse, 2008). This analysis revealed upregulation of cell cycle pathways, which indicated increased proliferative potential of these cells, and E2F transcription factors were identified as being involved in metastasis. In fact, several of the gene sets found to be upregulated in metastatic samples contained multiple E2F recognition sites and have been previously associated with E2Fs (Thomassen et al., 2008). It should be noted that a previous study of E2F expression levels in metastatic breast cancer patient samples produced contradictory results; this study reported reduced levels of E2F1 and E2F4 in samples of metastatic breast cancer (Ho, Calvano, Bisogna, & Van Zee, 2001). However, this study used only 10 matched patient samples from one location compared to the 1200 patient samples from multiple locations assessed in the more recent study (Thomassen et al., 2008). In addition, another study reported E2F4 levels to be increased in more advanced and metastatic breast cancer patient samples (Rakha, Pinder, Paish, Robertson, & Ellis, 2004).

In a series of studies conducted on colorectal cancer, E2F expression levels were also found to be increased in metastatic tumors as compared to primary tumors (Banerjee et al., 2000, 1998; Blanchet, Annicotte, & Fajas, 2009; Enders, 2004; Gorlick et al., 1998). In an initial study, it was found that metastatic colorectal tumors of the lung had increased levels of TS, an enzyme whose expression is regulated by E2F transcription factors (Gorlick et al., 1998). Another group also found that overexpression of E2F1 in human fibrosarcoma cells resulted in a corresponding increase in TS (Banerjee et al., 1998). In this context, it was then investigated whether E2F1 expression levels were also altered in metastatic colorectal tumor samples from patients, and indeed, E2F1 expression was found to be increased. This was more pronounced in metastatic lesions to the lung than to the liver (Banerjee et al., 2000). Further, additional studies from the same group revealed that colorectal cancers display amplification of chromosome 20q region that encodes the E2F1 gene and that E2F1 had increased copy number and that the greatest gains in E2F1 copy number were seen in metastatic colorectal cancers as revealed by comparative genomic hybridization and DNA PCR analysis of patient samples (Iwamoto et al., 2004). In support of these findings, translocation of chromosome 20q11-encoding E2F1 has also been associated with malignant melanoma, and increased copy number of E2F1 is seen in lymph node metastasis of melanoma compared to primary tumors (Nelson et al., 2006). Results similar to that seen in colorectal cancer were also obtained in studies conducted on human lung neuroendocrine tumors. Here, the expression of E2F1 and its target Skp2 were found to be correlated and increased in high-grade carcinomas, and overexpression of E2F1 and Skp2 was associated with advanced stage disease and nodal metastasis (Salon et al., 2007). In another study, E2F1 was found to be expressed at the same level in both primary tumors and lymph node metastasis of human gastric tumor samples (Yasui et al., 1999). This suggests that the increased expression of E2F in metastasis may be cancer-type specific.

A number of studies have additionally looked at Rb expression patterns in primary versus metastatic tumors in multiple cancer types. In a study done on hepatocellular carcinoma patient samples, it was observed that Rb expression was not only altered in poorly differentiated metastatic carcinomas but also absent in nearly half of all of the metastatic lesions, and the altered expression/absence was significantly greater in metastatic compared to primary hepatocellular carcinoma samples (Hui, Li, Makuuchi, Takayama, & Kubota, 1999). In another study, loss of Rb was found to associate with an increase in Cox2 expression as well as increased chance of recurring invasive basal-like breast cancer (Gauthier et al., 2007). Further, additional reports showed that Cox2 overexpression resulted in increased motility and invasion in human breast cancer cell lines *in vitro*, with a concomitant increase in pro-UPA expression, which is an enzyme capable of degrading basement membrane facilitating migration and invasion (Singh, Berry, Shoher, Ramakrishnan, & Lucci, 2005). Similar results were demonstrated in human colon cancer cells, where cells overexpressing Cox2 had increased invasive capacity corresponding to increased expression of matrix-degrading MMPs 1 and 2 (Tsujii, Kawano, & DuBois, 1997). Association of increased Cox2 expression with Rb loss may indicate a potential mechanism by which Rb suppresses metastatic potential. It may be concluded that while expression studies have demonstrated that Rb and E2F expression is altered in metastatic tumors compared to primary tumors, more evidence is needed to establish that alterations in the Rb/E2F pathway is functionally relevant to metastasis rather than just as an associated biomarker. To this extent, a number of studies have been conducted *in vitro* and *in vivo* to further elucidate their role in this process.

3.2. Rb and E2F in metastasis—Mouse models

In attempts to understand the role of the Rb–E2F pathway in cancer, Rb- and E2F-deficient mouse models have been established. In one such model, heterozygous Rb mice were shown to be prone to develop tumors. Of the tumors that form, medullary thyroid carcinomas that arise tend not to develop secondary cancers or metastasize and are therefore less aggressive tumors arising from this genetic background (Schreiber, Vormbrock, & Ziebold, 2012; Ziebold, Lee, Bronson, & Lees, 2003). Interestingly, when

E2F3 is mutated in the heterozygous Rb mouse background, more aggressive and metastatic medullary thyroid tumors developed (Schreiber et al., 2012; Ziebold et al., 2003). Opposing results were seen in pituitary carcinomas arising from Rb heterozygous E2F3-deficient mice, as there was a marked reduction in pituitary tumor size and increased survival in mice lacking E2F3 (Ziebold et al., 2003). Gene ChIP microarray methods were used to identify gene sets enriched in metastatic tumors from this mouse model followed by experiments to verify the candidates identified; the results suggested that E2Fs directly regulate putative metastasis markers (Schreiber et al., 2012). In an additional study using a Myc-inducible model of breast cancer in mice, knockout of E2F2 resulted in decreased formation of tumors with EMT characteristics as well as reduced Ras activation which has been shown to regulate E2F activity (Fujiwara, Yuwanita, Hollern, & Andrechek, 2011; Yoon, Shin, & Mercurio, 2006). Correspondingly, low expression of E2F2 in human breast cancer samples was found to correlate with increased relapse-free survival (Fujiwara et al., 2011). Rb family members, in addition to Rb itself, have been shown to be involved in potentiating metastasis in a mouse model of small cell lung cancer (SCLC) (Schaffer et al., 2010). In this model, it was shown that triple-conditional knockout of Rb, its family member p130, as well as p53 resulted in high incidence of metastasis to the liver, pulmonary lymph nodes, and kidney; all of which are common sites of metastasis for human SCLC, and this incidence of metastasis was not observed in double Rb/p53 knock-out mice (Schaffer et al., 2010). This suggests that inactivation of multiple Rb family members may be required for metastasis, potentially due to redundant or compensatory roles of these proteins.

3.3. Rb- and E2F-mediated regulation of metastatic properties of cancer cells

A significant amount of work has been done to elucidate the role of Rb and E2F in modulating EMT, invasion and migration of cancer cells in culture. In such experiments, overexpression of the activating E2Fs has been shown to increase invasive capacity of cells. In a head and neck cancer model, cell lines determined to be invasive *in vivo* displayed higher levels of E2F1 compared to noninvasive cell lines (Zhang, Liu, Johnson, & Klein-Szanto, 2000). To assess whether E2F1 expression enhanced invasiveness, noninvasive cell lines with low levels of E2F1 were transfected with E2F1 expression vectors resulting in significantly increased invasion of these cells (Zhang et al., 2000). In a different study investigating the invasion of breast carcinoma, E2Fs 1–3 were shown to mediate H-Ras-dependent invasion of these cells *in vitro* (Yoon et al., 2006). It is known that active Ras proteins play a critical role in cell invasion and metastasis and that a number of cancers have aberrantly activated Ras, including invasive breast carcinomas. In this study, Ras was shown to mediate increased expression and activity of the activating E2Fs 1–3, which in turn stimulated invasion; however, the inhibitory E2Fs 4 and 5 were not found to impact invasion (Yoon et al., 2006). Further, the mechanism through which E2Fs 1–3 could induce invasion appeared to occur via their ability to increase expression of the β4 integrin subunit, a component of the α6, β4 integrin involved in cell-to-matrix adhesion, known to enhance carcinoma invasion (Yoon et al., 2006).

In addition to E2Fs playing a role in Ras-mediated migration, invasion, and metastasis, Rb has also been shown to be involved (Liu et al., 2013). In mouse embryonic fibroblasts (MEFs), cells with mutant Rb family members (Rb1, p107, and p130) lead to tumor initiation in part through induction of the EMT transcription factor, ZEB1. Activation of ZEB1 is thought to facilitate carcinoma invasion and metastasis by allowing for a more migratory *mesenchymal* phenotype and is typically repressed by Rb. Similarly, MEFs with aberrantly activated Ras signaling lead to tumor initiation also in part through inducing expression of ZEB1 (Liu et al., 2013). In this study, loss of Rb was found to be required for Ras induction of ZEB1 and tumor formation and that concurrent loss of Rb and activation of Ras resulted in a more aggressive tumor phenotype with increased invasiveness (Liu et al., 2013). In an invasive breast carcinoma model, a group which had previously demonstrated that depletion of Rb results in downregulation of the epithelial marker E-cadherin initiating EMT, went on to further demonstrate that the inactivation of Rb in human breast cancer cell lines results in a mesenchymal morphology and highly invasive capacity in part due to upregulation of ZEB proteins (Arima et al., 2012, 2008). ZEB proteins ZEB1 and ZEB2, which are transcriptional repressors of E-cadherin, were found to be markedly upregulated in Rb-inactive cells and knockdown of ZEB resulted in decreased invasiveness with corresponding increase in epithelial markers (Arima et al., 2012). Aside from ZEB and E-cadherin, the Rb–E2F pathway has been shown to mediate expression of other EMT markers. In our studies on non-SCLC (NSCLC) cell lines, E2F1 was found to bind to and activate the promoters of mesenchymal genes fibronectin and vimentin, increasing their expression and this further contributed to an EMT-like phenotype necessary for subsequent metastasis (Pillai, Trevino, et al., in preparation).

Interestingly, while Rb status in tumor cells themselves has an impact on cell behavior, it has also been shown that the inactivation of Rb in stromal fibroblasts promote the invasion of associated epithelial cells (Pickard et al., 2012). A dynamic interplay exists between a tumor and its surrounding stroma, and the stroma contributes significantly to the development and progression of tumors through secretion of signaling molecules such as growth factors and cytokines, as well as through cell-to-cell interactions. In this study, it was demonstrated that Rb and Rb-dependent signaling pathways are frequently inactivated in oropharyngeal cancer-associated fibroblasts of the stroma, and this observation is further supported by the finding that when immortalized epithelial cells were cocultured with Rb-inactivated fibroblasts in three-dimensional *in vitro* cultures, the epithelial cells became invasive (Pickard et al., 2012). In an effort to elucidate the mechanism by which Rb inactivation results in invasion, cytokine arrays were used to identify differences in secreted factors between Rb-depleted fibroblasts and fibroblasts with functional Rb. This revealed an increase in secretion of the keratinocyte growth factor 7 (KGF) by Rb-inactivated cells, which was further demonstrated to increase invasiveness when added to cell culture systems (Pickard et al., 2012). It was then additionally shown that KGF induces the protease MMP1 in an AKT–Ets2-dependent manner, contributing to basement membrane degradation. This is a critical step in the metastatic cascade, allowing tumor cells to leave the primary site to intravasate into circulation. Thus, the status and activity of Rb in the tumor cells as well as the stromal fibroblasts might determine the invasive and metastatic properties of the tumors.

More recently, the histone demethylase Rb-binding protein 2 (RBP2) has been implicated not only in lung cancer progression but also in metastasis (Teng et al., 2013). RBP2 typically acts to repress its target gene promoters by removing methyl groups from di- and tri-methylated and active promoters; but occasionally, RBP2 has been shown to activate transcription as well. When RBP2 was depleted in human lung cancer cell lines *in vitro*, there was a marked reduction in cell motility, migration, and invasion (Teng et al., 2013). Further, there was reduced metastasis when RBP2-depleted cells were injected into the tail vein of mice (Teng et al., 2013). To assess which genes may be playing a role in the ability of RBP2 to impact cell motility, migration, invasion, and metastasis, a microarray analysis was conducted comparing cell lines expressing RBP2 to those with RBP2 knocked down. This analysis revealed integrin β1 (ITGB1) to be greatly decreased in RBP2 knock-down cells (Teng et al., 2013). ITGB1, which is known to regulate cell-to-matrix interactions and play a role in lung cancer metastasis, was shown to be a direct target activated by RBP2 (Teng et al., 2013). Collectively, these results demonstrate that RBP2 plays an important role in lung cancer metastasis, in part through the activation of ITGB1. In an additional study conducted in mammary epithelial cells, another Rb-associated protein, RbAp46, was shown to be involved in metastatic processes (Li & Wang, 2006). RbAp46 is known to have multiple roles as part of a histone deacetylase complex which acts to repress transcription, as a histone acetyltransferase that binds to H2A- and H4-stimulating HAT activity, and as a subunit of the chromosome remodeling NuRD complex. This study further demonstrated that overexpression of RbAp46 results in upregulation of mesenchymal markers and downregulation of epithelial markers consistent with the EMT phenotype thought to be essential for metastasis and was further shown to increase migration and invasion *in vitro* (Li & Wang, 2006).

While multiple studies have implicated the Rb/E2F pathway in tumor cell migration, invasion, and metastasis, the mechanism underlying their involvement remains poorly understood. In a study conducted using osteosarcoma cell lines, cDNA microarray screening was used to identify E2F1 target genes (Stanelle et al., 2002). Among the genes identified, those involved in angiogenesis, invasion, and metastasis were revealed; specifically, the E2F1 induction of MMP16 was notably apparent. MMP16 is known to activate MMP2, which has an established role in regulation of invasion and metastasis (Stanelle et al., 2002). In the initial steps leading to metastasis, tumors rely on matrix-degrading proteins such as matrix metalloproteases (MMPs) to break down the basement membrane barrier, allowing tumor cells to escape into the surrounding microenvironment and vasculature for dissemination. These cells also acquire mesenchymal phenotypes characteristic of EMT, leading to decreased cell-to-cell and cell-to-matrix adhesion permitting increased motility. In a NSCLC model, it was shown that E2F transcription factors could regulate MMP genes involved in invasion and metastasis (Johnson et al., 2012). In this study, predicted E2F1-binding sites were found on 23 human MMP gene promoters; further, E2F1 and Rb associated with MMP9, MMP14, and MMP15 promoters in chromatin immunoprecipitation assays. E2F1 was shown to induce the expression of MMP9, MMP14, and MMP15, and the induction of MMP9 resulted in enhanced gelatinase activity (Johnson et al., 2012). It was further demonstrated that E2F1 and E2F3 are required for MMP9 and MMP14 gene expression resulting in collagen degradation and that the depletion of E2F1 and E2F3 resulted in decreased migratory and invasive capacity of

NSCLC cells *in vitro* (Johnson et al., 2012). In the same study, it was shown that disruption of Rb–Raf interactions, which typically contribute to the inactivation of Rb allowing for E2F transcriptional activity, resulted in repression of MMP transcription as well as inhibition of invasion and migration *in vitro* (Johnson et al., 2012). To further support these findings, it was then shown that the disruption of Rb–Raf-1 interaction using small molecule inhibitor RRD-251 resulted in decreased metastatic lung colonization in NSCLC mouse models (Johnson et al., 2012). These results further implicate a direct role for the Rb–E2F pathway in metastatic processes, by promoting the expression of genes involved in invasion and migration as depicted in Fig. 4.1.

3.4. Rb–E2F pathway in cancer stem cells

Stem cells are defined by their ability to self-renew to maintain stem cell populations over time, ability to give rise to multiple differentiated cell lineages, and ability to maintain a balance between stem cell number, proliferation, differentiation, and death (Galderisi, Cipollaro, & Giordano, 2006). Recently, cancer cells with such properties have been identified and termed "cancer stem cells" and are thought to be involved in tumor metastasis as well as dormancy, which can lead to metastasis (Visvader & Lindeman, 2008). A number of reports have linked regulators of cell cycle, including Rb, with maintenance of stem cells and their functions such as self-renewal, lineage specification, and maintenance of a senescent state (Calo et al., 2010; Chen et al., 2007; Conklin, Baker, & Sage, 2012; Galderisi et al., 2006; Sage, 2012; Wenzel et al., 2007). This suggests that the regulation of stem cell or stem-like cell populations within tumors by Rb may be an additional mechanism by which the Rb/E2F pathway is involved in metastatic processes.

Embryonic stem cells (ESCs) have unlimited self-renewal potential while maintaining pluripotency as do adult stem cells, but unlike adult stem cells ESCs have relatively short G1 phase of cell cycle and arrest of G1 phase is thought to be permissive for differentiation (Burdon, Chambers, Stracey, Niwa, & Smith, 1999; Burdon, Smith, & Savatier, 2002). The stringent regulation of the duration of G1 phase in these cells is thought to be critical in determining whether they will differentiate, proliferate, senesce, undergo apoptosis, or enter quiescence (Blomen & Boonstra, 2007). Rb restricts G1/S transition of cell cycle and has been shown to be involved in maintenance of self-renewal in ESCs (Conklin et al., 2012; Conklin & Sage, 2009).

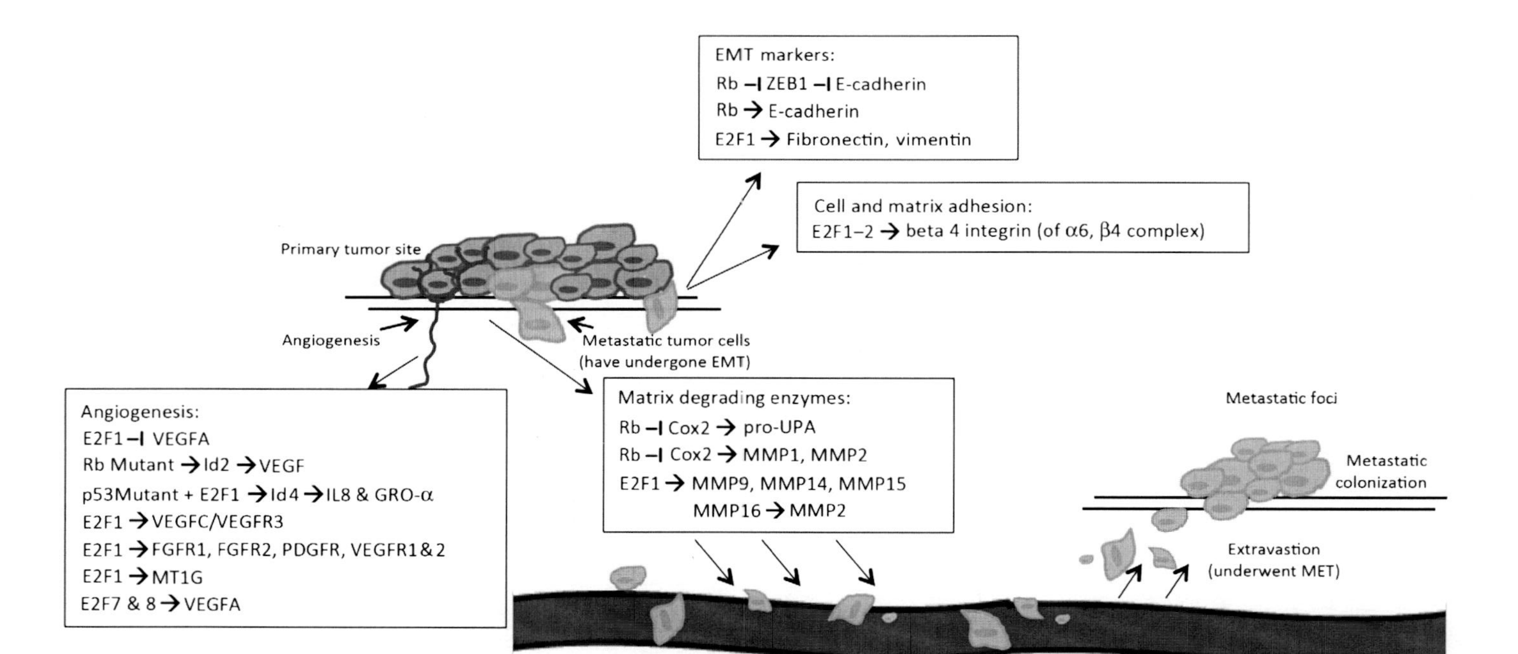

Figure 4.1 Representation of angiogenesis and metastasis by the Rb–E2F pathway. Although E2F1 inhibits angiogenesis in mouse models, growing evidence suggests a possible proangiogenic role for E2F1, enhances angiogenesis by transcriptionally upregulating growth factors and cytokines like PDGF-B, IL8, and Groα, as well as receptors like FGFR, FLT1, KDR, and PDGFRα. In Rb mutant mice, Id2 activated VEGF expression resulting in enhanced angiogenesis. Rb suppresses EMT by inducing epithelial E-cadherin, in part through repression of ZEB1; Rb-inactive cancer show increased ZEB1 and decreased E-cadherin. E2F1 can induce the mesenchymal fibronectin and vimentin promoters as well as induce β4 integrin and matrix-degrading enzymes MMPs 9,14,15, and 16. Cox2, which induces matrix-degrading enzymes including pro-UPA, MMP1, and MMP2, is repressed by Rb; conversely, Rb inactivation leads to Cox2 expression resulting in increased expression of these enzymes. → Indicates activation of expression, whereas ⊣ Indicates inhibition of expression. (See the color plate.)

In human ESCs (hESCs), overexpression of unphosphorylated thus activated Rb resulted in decreased cell proliferation, diminished capacity to remain undifferentiated, and increased cell death (Conklin et al., 2012). Further, activation of Rb in hESCs was found to be associated with decreased expression of E2F target genes and increased expression of embryonic development and cellular differentiation markers, ultimately resulting in decreased self-renewal (Conklin et al., 2012). Interestingly, inactivation of all three Rb family members in hESCs also resulted in growth inhibition, expansion, and apoptosis resulting in decreased self-renewal. This indicates too much or too little Rb activity can abrogate self-renewal of hESCs, suggesting the need for homeostatic Rb activity in order to maintain pluripotency (Conklin et al., 2012). Further, other molecules in the Rb pathway have been found to associate with the vital ES cell transcription factors, Oct4, Sox2, Klf4, Nanog, and c-myc, known to be necessary to maintain pluripotency (Yamanaka, 2009). Specifically, E2F has been shown to regulate Oct4 (Chavez et al., 2009), and Sox2 and Nanog have been shown to bind to the E2F3 gene promoter, although this has not been functionally validated (Boyer et al., 2005). Cdk6 and cdc25a have also been shown to decrease in ESCs when Nanog is knocked down (Conklin & Sage, 2009), and Nanog binds to Cdk6 and cdc25a promoters, although this too has yet to be functionally validated (Zhang et al., 2009). In mouse ESCs (mESCs), an Oct4-Nipp1/Ccnf-PP1-pRb axis, controlling self-renewal and differentiation, has recently been identified (Schoeftner, Scarola, Comisso, Schneider, & Benetti, 2013). This study revealed that Oct4 activates expression of Nipp1 and Ccnf which act to inhibit the PP1 complex, which results in the hyperphosphorylation and thus inactivation of Rb. This sequence of events promotes mESC self-renewal and is further under the control of miR-335 which acts to regulate Oct4 and Rb (Schoeftner et al., 2013). It was shown that during differentiation of these cells, miR-335 is upregulated and acts to suppress Oct4 leading to the repression of the Oct4-Nipp1/Ccnf-PP1-pRb axis, dephosphorylation of Rb, and exit from self-renewal (Schoeftner et al., 2013). In an additional study, it was shown that E2F and Myc drive miR-17–92 clusters regulating G1/S phase transition which is required for ESC self-renewal and proliferation (Gunaratne, 2009). The authors of this study further postulate that any perturbation in key ESC miRNA networks could be hallmarks of cancer stem cells as well, acting to maintain their self-renewal potential (Gunaratne, 2009). Further, E2F1 has been shown to bind to the gene promoter of BMI1, which encodes for polycomb group transcription factors involved in neuroblastoma tumor formation; BMI1 is also involved in ESC development and self-renewal (Nowak et al., 2006). This links E2F1 to BMI1, which has dual roles in tumorigenicity as well as stem cells properties, raising the possibility that BMI1 could also be involved in maintenance of cancer stem cells, at least in neuroblastomas (Nowak et al., 2006). Collectively, this data demonstrate a role for the Rb in ESC self-renewal and raise the possibility of a role in cancer stem cell self-renewal.

In addition to self-renewal, Rb has been implicated in the differentiation of stem cells. Initial studies done in Rb and Rb family knock-out mice demonstrated severely impaired development in early embryonic stages, including impaired neurogenesis, hematopoiesis, and myogenesis (Clarke et al., 1992; Jacks et al., 1992; Lee et al., 1992, 1994). In a conditional knock-out study which simultaneously inactivated all three Rb family members, Rb, p107, and p130, it was found that these triple knock-out mice develop cell-intrinsic myeloproliferation stemming from hyperproliferation of early hematopoietic progenitors. There was a concomitant increase in lymphoid progenitor apoptosis and loss of quiescence of hematopoietic stem cells, inferring that the Rb family members are needed to maintain hematopoietic stem cell quiescence and the balance between lymphoid and myeloid cell fates (Viatour et al., 2008). Rb has also been shown to interact with lineage-specific transcription factors during differentiation of progenitor stem cells, suggesting that Rb plays a role in the regulation of genes involved in the differentiation of multiple cell lineages (Sage, 2012). Rb has been shown to cooperate with the myogenic transcription factor MyoD to activate muscle-specific gene expression, and lack of Rb results in altered muscle differentiation (Novitch, Mulligan, Jacks, & Lassar, 1996). Further, Rb has been found to play a role in the differentiation of satellite myoblasts, which are myocyte precursors (Li, Coonrod, & Horwitz, 2000). Rb has additionally been found to regulate the activity of cardiogenic transcription factors CMF1 and LEK1, and ESCs derived from Rb-deficient mice have decreased expression of these transcription factors (Papadimou, Menard, Grey, & Puceat, 2005). In hematopoietic cell lineages, Rb has been shown to interact with erythroid transcription factors Id2 and PU1, and Rb can determine the commitment of Cd34+ pluripotent cells into either neutrophilic or monocytic lineages (Bergh, Ehinger, Olsson, Jacobsen, & Gullberg, 1999). Rb acts to repress the adipogenic transcription factors PPAR2γ and C/EBP in mesenchymal stem cells, which have critical roles in lineage specification as well as differentiation (Wang & Scott, 1993). Similarly, Rb has been shown to associate with the osteogenic transcription factor RunX2 in mesenchymal stem cells, which directs differentiation of cells into bone lineages (Calo et al., 2010). Lastly, Rb has been shown to interact with Pdx1 which is associated with pancreatic development (Kim et al., 2011). This data suggest that Rb in involved in regulation of genes involved in the differentiation of multiple different cell lineages from stem cells; the major pathways regulated by Rb are shown in Fig. 4.2.

Additionally, similarities between poorly differentiated cancer cells and ESCs have been identified, and it has been shown that inactivation of Rb in mouse fibroblasts allows for the formation of cellular aggregates expressing pluripotent genes which could additionally form tumors in mice (Zbinden et al., 2010). Cancer stem-like cells have certain similar characteristics as normal stem cells such as the capacity for unlimited replication potential, self-renewal, and the ability to give rise to heterogeneous populations of tumor cells. This has led to the belief that some tumor cells are derived from transformed stem cells; hence, these are the cancer stem cells that give rise to tumors (Takahashi et al., 2012). Since most tumors have inactivated Rb, this could potentially be contributing to the aberrant nature of these stem cells as they cannot be efficiently regulated and could be contributing to the generation of heterogeneous tumor populations with unlimited proliferative potential as well as potential to undergo EMT, migrate, invade, and ultimately metastasize to distal organs for recolonization. The plasticity of stem cells to differentiate and perhaps dedifferentiate is a characteristic that would give great advantage to these cells as transformed components of a tumor cell population, especially in the process of metastasis. The role of Rb in stemness has been reviewed in depth elsewhere (Sage, 2012; Takahashi et al., 2012).

3.5. Rb–E2F pathway in energy metabolism

The Rb–E2F pathway has additionally been shown to play a role in metabolic processes (Clem & Chesney, 2012; Nicolay & Dyson, 2013; Nicolay et al., 2013). A growing number of studies have shown evidence of metabolic alterations in cancer development and progression, acquired to meet the needs of aberrant, rapidly proliferating cells (Han et al., 2013; Jang, Kim, & Lee, 2013). These metabolic alterations are now well established as hallmarks of cancer and include changes such as enhanced glycolysis and aerobic glycolysis, lactate production, and increased *de novo* fatty acid, lipid, and macromolecule synthesis (Fritz & Fajas, 2010). Interestingly, it is also established that tumor growth and invasion involve major metabolic alterations of the cell (Fritz & Fajas, 2010; Han et al., 2013). Recent studies

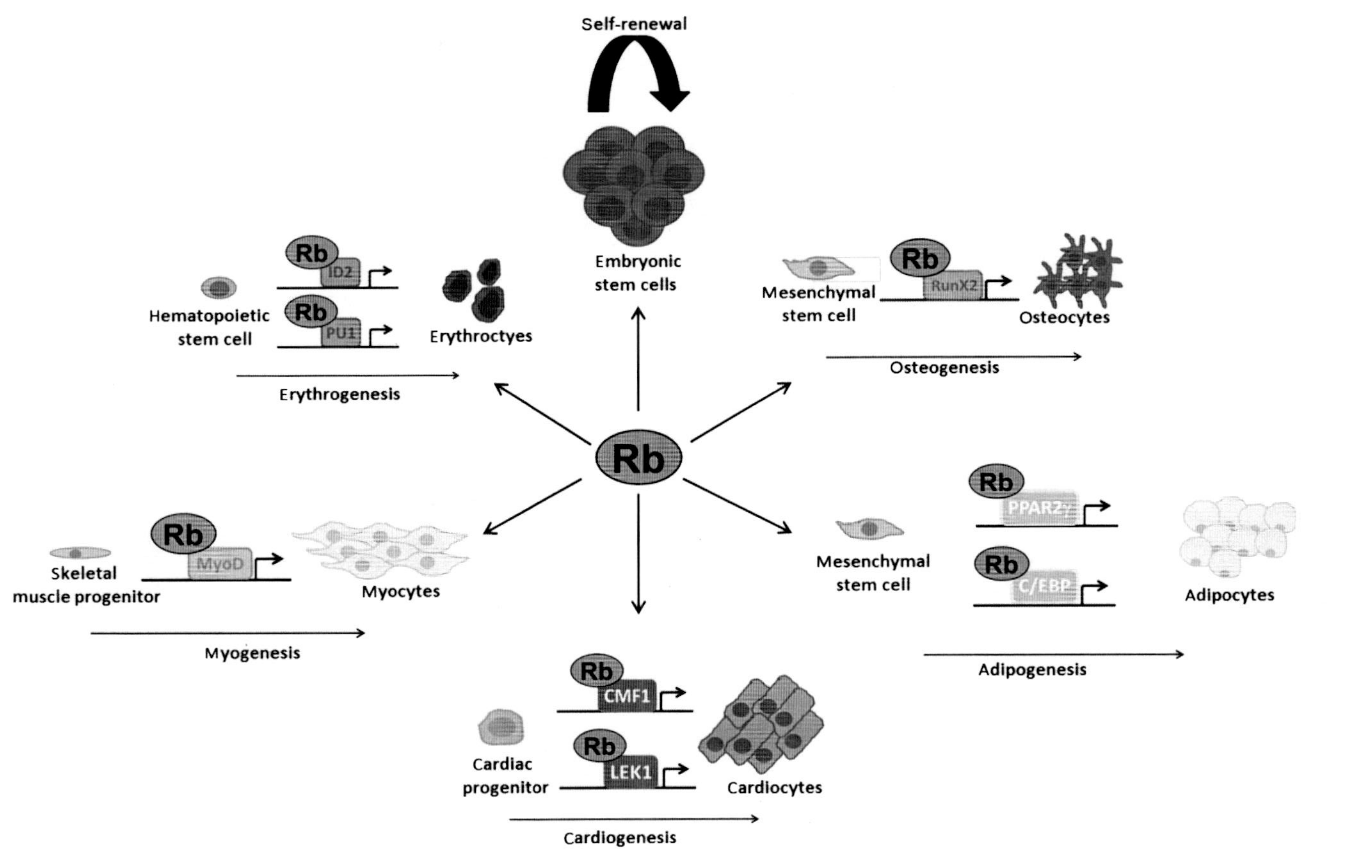

Figure 4.2 Schematic depicting the role of Rb in regulation of genes involved in stemness and lineage specification in multiple cell types. Rb associates with adipogenic factors PPAR2γ and C/EBP, osteogenic factor RunX2, cardiogenic factors CMF1 and LEK1, erythroid factors Id2 and PU1, and the myogenic factor MyoD facilitating differentiation of the indicated progenitors and stem cells to the differentiated cell types. Further, Rb can regulate the self-renewal of embryonic stem cells as well as cancer stem cells, thus playing a critical role in oncogenesis as well as normal development or organisms. (See the color plate.)

demonstrate that the Rb–E2F pathway plays a role in the deregulation of metabolism in cancer, and this may be another mechanism through which Rb–E2F is involved in tumor metastasis.

In a series of studies investigating regulation of lipid metabolism, a role for both E2F and Rb were observed (Dali-Youcef et al., 2007; Fajas, Egler, et al., 2002; Fajas, Landsberg, et al., 2002). E2Fs were shown to regulate adipogenesis via E2F1-mediated induction of the transcription factor PPARγ and E2F4-mediated repression of PPARγ (Fajas, Landsberg, et al., 2002b). PPARγ has been shown to be critical in adipogenesis and end terminal differentiation of adipocytes, and the induction of PPARγ by E2F1 was shown to occur during clonal expansion of adipocytes. At the same time, the repression by E2F4 was shown to occur during the terminal differentiation stages (Fajas, Landsberg, et al., 2002). In contrast, it was simultaneously shown that Rb could repress PPARγ and adipocyte differentiation via recruitment of HDAC3 (Fajas, Egler, et al., 2002). Additional evidence for the role of the Rb/E2F pathway in lipid metabolism was obtained from mouse models, since mice lacking Rb showed increased energy expenditure in adipose tissue, due to enhanced mitochondrial activity (Dali-Youcef et al., 2007). It has also been shown that Rb interacts with SREBPs, which are master regulators of lipid metabolism, repressing their activity; but in the absence of Rb, E2F1 and E2F3 were shown to bind to and activate SREBP promoters (Takahashi et al., 2012).

E2Fs have additionally been shown to play a role in overall energy homeostasis. Initial studies have demonstrated that E2F1-deficient mice are resistant to diet-induced obesity, have impaired glucose homeostasis, and have dysfunctional pancreatic β-islet cells resulting in insufficient insulin secretion in response to glucose resulting in glucose intolerance (Fajas et al., 2004). This suggests a role for E2F1 in glucose-stimulated insulin secretion and overall glucose homeostasis (Fajas et al., 2004). Additionally, Kir6.2, which is known to play a role in insulin secretion, is a direct E2F1 target, and E2F1 modulates insulin secretion of β-islet cells through the transcriptional regulation of Kir6.2 (Blanchet et al., 2009). E2F1 has also been shown to activate the promoter of the F-type PFK2 gene, which expresses an enzyme responsible for catalyzing glycolysis (Darville, Antoine, Mertens-Strijthagen, Dupriez, & Rousseau, 1995). It has further been shown that E2F1 acts to repress genes regulating energy homeostasis and mitochondrial oxidative functions in basal conditions and was responsible for the switch from oxidative to glycolytic metabolism in response to stress (Blanchet et al., 2011). In mice lacking E2F1, there was a marked oxidative phenotype due to the upregulation of genes typically suppressed by E2F1. These include genes involved in the mitochondrial respiratory chain such as Atp5g1, Cox5a, Ndufl c, Sdha, and Uqcr; genes involved in the TCA cycle such as Idk3a; genes involved in respiratory uncoupling such as Ucp1 and Ucp2; and genes involved in fatty acid oxidation such as Acadl, Pdk4, and Cpt-1 (Blanchet et al., 2011). Further, it was found that slow switch oxidative myofiber MyHC-I and MyHC-IIa expression was increased in cells from mice lacking E2F1, while there was a decrease in the expression fast twitch glycolytic myofiber (Blanchet et al., 2011). In additional studies, E21F was shown to regulate glucose homeostasis through inhibition of glucose oxidation via activation of pyruvate dehydrogenase kinase 4 (PDK4), which is a key nutrient sensor (Hsieh, Das, Sambandam, Zhang, & Nahle, 2008). Further, E2F1 negatively regulates oxidative metabolism in skeletal muscle through repression of PGC-1, indicating a role for E2F1 in the metabolic switch from oxidative phosphorylation to aerobic glycolysis common to cancer cells (Fritz & Fajas, 2010). The Rb/E2F pathway has also been shown to be involved in macromolecule synthesis; specifically E2F activates the genes involved in DNA synthesis such as TS, DHFR, and TK (Trimarchi & Lees, 2002). Other review articles cover these aspects in greater depth (Blanchet et al., 2009; Fritz & Fajas, 2010; Nicolay & Dyson, 2013). Thus, it is clear that the Rb–E2F transcriptional regulatory pathway might be affecting the biology of tumor cells by affecting their energy metabolism as well (shown in Fig. 4.3), in addition to many other facets including proliferation, invasion, and stemness.

4. CONCLUSIONS

The studies reviewed in the article demonstrate a clear role for the Rb–E2F transcriptional regulatory pathway in various aspects of tumor progression, angiogenesis, stemness, and metastasis. While the Rb–E2F pathway was classically thought to primarily regulate cell cycle progression, it is now apparent that this pathway can affects multiple facets of the biology of normal and tumor cells. The ability of this pathway to affect angiogenesis, which determines the availability of nutrients to cancer cells, and their ability to modulate the energy metabolism of cancer cells, can be expected to have a major impact on the ability of a tumor to grow and metastasize. Further, the regulation of genes involved in EMT and invasion by the Rb–E2F pathway might also determine the metastatic potential of tumors. The regulation of these varied and divergent pathways by Rb and E2F is depicted in Fig. 4.1; as

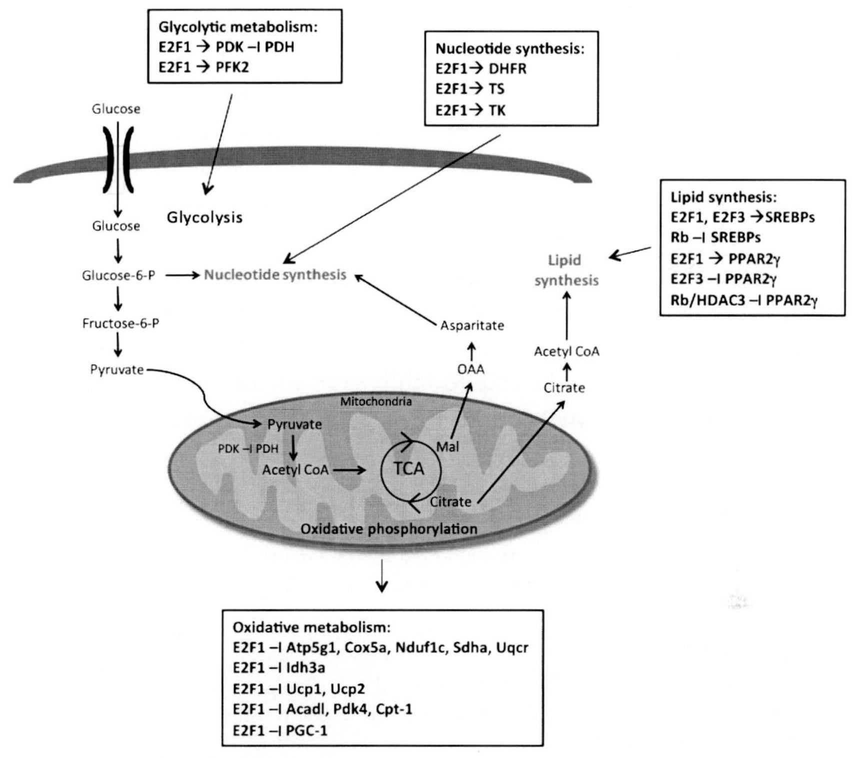

Figure 4.3 Schematic representation of the regulation of metabolic pathways by Rb and E2F1. E2F1 has been implicated in facilitating the metabolic switch from oxidative metabolism to glycolysis in cancer. It represses multiple factors involved in oxidative metabolism including genes involved in the mitochondrial respiratory chain like Atp5g1, Cox5a, Ndufl c, Sdha, and Uqcr, genes involved in the TCA cycle like Idk3a, genes involved in respiratory uncoupling like Ucp1 and Ucp2, as well as genes involved in fatty acid oxidation like Acadl, Pdk4, and Cpt-1. Conversely, E2F1 activates genes involved in glycolytic metabolism including PDK and PFK2. E2F1 can induce the adipogenic factor PPAR2γ, while E2F3 and Rb repress it at different points in the adipogenic process. E2Fs 1 and 3 can also activate SREBPs, which are master regulators of lipid synthesis. E2F1 has further been shown to induce genes involved in DNA synthesis such as dihydrofolate reductase (DHFR), thymidylate synthase (TS), and thymidine kinase (TK). → Indicates activation of expression whereas ⊣ Indicates inhibition of expression.

can be seen, a complicated network of events that promote tumor growth and metastasis are affected by these molecules. Given that the Rb–E2F pathway is altered in various cancers in response to mutations in upstream signaling molecules, it is possible that the same mutations that drive cell cycle progression by inactivating Rb might also be contributing to the progression and metastasis of cancers. These observations highlight the importance of this pathway in tumor biology but also raise the possibility that it can be targeted for the development of novel therapeutic agents. In this context, several molecules that target cyclin-dependent kinases are in various stages of development; newer generations of such inhibitors can be expected to have significant activity as anticancer agents. Similarly, agents like RRD-251, which disrupts the binding of Raf-1 kinase to Rb (Davis & Chellappan, 2008), and inhibitors of E2F transcription factor (Ma et al., 2008) are other promising approaches to target this pathway. It can be imagined that agents that are capable of targeting the Rb–E2F pathway selectively with limited toxicity will be able to inhibit various aspects of oncogenesis, tumor progression, and metastasis and would have immense benefits as therapeutic agents.

ACKNOWLEDGMENTS

The studies in the Chellappan lab are supported by the Grants CA127725 and CA139612 from the NCI. Our apologies to the authors whose work could not be cited directly.

REFERENCES

Ahluwalia, A., & Tarnawski, A. S. (2012). Critical role of hypoxia sensor—HIF-1alpha in VEGF gene activation. Implications for angiogenesis and tissue injury healing. *Current Medicinal Chemistry*, *19*, 90–97.

Aladowicz, E., Ferro, L., Vitali, G. C., Venditti, E., Fornasari, L., & Lanfrancone, L. (2013). Molecular networks in melanoma invasion and metastasis. *Future Oncology*, *9*, 713–726.

Alavi, A., Hood, J. D., Frausto, R., Stupack, D. G., & Cheresh, D. A. (2003). Role of Raf in vascular protection from distinct apoptotic stimuli. *Science*, *301*, 94–96.

Arima, Y., Hayashi, H., Sasaki, M., Hosonaga, M., Goto, T. M., Chiyoda, T., et al. (2012). Induction of ZEB proteins by inactivation of RB protein is key determinant of mesenchymal phenotype of breast cancer. *The Journal of Biological Chemistry*, *287*, 7896–7906.

Arima, Y., Inoue, Y., Shibata, T., Hayashi, H., Nagano, O., Saya, H., et al. (2008). Rb depletion results in deregulation of E-cadherin and induction of cellular phenotypic changes that are characteristic of the epithelial-to-mesenchymal transition. *Cancer Research*, *68*, 5104–5112.

Arpino, G., Generali, D., Sapino, A., Lucia del, M., Frassoldati, A., de Laurentis, M., et al. (2013). Gene expression profiling in breast cancer: A clinical perspective. *Breast*, *22*, 109–120.

Bakker, W. J., Weijts, B. G., Westendorp, B., & de Bruin, A. (2013). HIF proteins connect the RB-E2F factors to angiogenesis. *Transcription*, *4*, 62–66.

Baldewijns, M. M., van Vlodrop, I. J., Vermeulen, P. B., Soetekouw, P. M., van Engeland, M., & de Bruine, A. P. (2010). VHL and HIF signalling in renal cell carcinogenesis. *The Journal of Pathology*, *221*, 125–138.

Banerjee, D., Gorlick, R., Liefshitz, A., Danenberg, K., Danenberg, P. C., Danenberg, P. V., et al. (2000). Levels of E2F-1 expression are higher in lung metastasis of colon cancer as compared with hepatic metastasis and correlate with levels of thymidylate synthase. *Cancer Research*, *60*, 2365–2367.

Banerjee, D., Schnieders, B., Fu, J. Z., Adhikari, D., Zhao, S. C., & Bertino, J. R. (1998). Role of E2F-1 in chemosensitivity. *Cancer Research*, *58*, 4292–4296.

Berger, B., Capper, D., Lemke, D., Pfenning, P. N., Platten, M., Weller, M., et al. (2010). Defective p53 antiangiogenic signaling in glioblastoma. *Neuro-Oncology, 12*, 894–907.
Bergers, G., & Benjamin, L. E. (2003). Tumorigenesis and the angiogenic switch. *Nature Reviews. Cancer, 3*, 401–410.
Bergh, G., Ehinger, M., Olsson, I., Jacobsen, S. E., & Gullberg, U. (1999). Involvement of the retinoblastoma protein in monocytic and neutrophilic lineage commitment of human bone marrow progenitor cells. *Blood, 94*, 1971–1978.
Bidard, F. C., Pierga, J. Y., Soria, J. C., & Thiery, J. P. (2013). Translating metastasis-related biomarkers to the clinic—Progress and pitfalls. *Nature Reviews. Clinical Oncology, 10*, 169–179.
Biyashev, D., & Qin, G. (2011). E2F and microRNA regulation of angiogenesis. *American Journal of Cardiovascular Disease, 1*, 110–118.
Blanchet, E., Annicotte, J. S., & Fajas, L. (2009). Cell cycle regulators in the control of metabolism. *Cell Cycle, 8*, 4029–4031.
Blanchet, E., Annicotte, J. S., Lagarrigue, S., Aguilar, V., Clape, C., Chavey, C., et al. (2011). E2F transcription factor-1 regulates oxidative metabolism. *Nature Cell Biology, 13*, 1146–1152.
Blomen, V. A., & Boonstra, J. (2007). Cell fate determination during G1 phase progression. *Cellular and Molecular Life Sciences, 64*, 3084–3104.
Boyer, L. A., Lee, T. I., Cole, M. F., Johnstone, S. E., Levine, S. S., Zucker, J. P., et al. (2005). Core transcriptional regulatory circuitry in human embryonic stem cells. *Cell, 122*, 947–956.
Budde, A., Schneiderhan-Marra, N., Petersen, G., & Brune, B. (2005). Retinoblastoma susceptibility gene product pRB activates hypoxia-inducible factor-1 (HIF-1). *Oncogene, 24*, 1802–1808.
Burdon, T., Chambers, I., Stracey, C., Niwa, H., & Smith, A. (1999). Signaling mechanisms regulating self-renewal and differentiation of pluripotent embryonic stem cells. *Cells, Tissues, Organs, 165*, 131–143.
Burdon, T., Smith, A., & Savatier, P. (2002). Signalling, cell cycle and pluripotency in embryonic stem cells. *Trends in Cell Biology, 12*, 432–438.
Calo, E., Quintero-Estades, J. A., Danielian, P. S., Nedelcu, S., Berman, S. D., & Lees, J. A. (2010). Rb regulates fate choice and lineage commitment in vivo. *Nature, 466*, 1110–1114.
Canel, M., Serrels, A., Frame, M. C., & Brunton, V. G. (2013). E-cadherin-integrin crosstalk in cancer invasion and metastasis. *Journal of Cell Science, 126*, 393–401.
Chaffer, C. L., & Weinberg, R. A. (2011). A perspective on cancer cell metastasis. *Science, 331*, 1559–1564.
Chavez, L., Bais, A. S., Vingron, M., Lehrach, H., Adjaye, J., & Herwig, R. (2009). In silico identification of a core regulatory network of OCT4 in human embryonic stem cells using an integrated approach. *BMC Genomics, 10*, 314.
Chellappan, S. P., Hiebert, S., Mudryj, M., Horowitz, J. M., & Nevins, J. R. (1991). The E2F transcription factor is a cellular target for the RB protein. *Cell, 65*, 1053–1061.
Chen, D., Opavsky, R., Pacal, M., Tanimoto, N., Wenzel, P., Seeliger, M. W., et al. (2007). Rb-mediated neuronal differentiation through cell-cycle-independent regulation of E2f3a. *PLoS Biology, 5*, e179.
Chen, H. Z., Tsai, S. Y., & Leone, G. (2009). Emerging roles of E2Fs in cancer: An exit from cell cycle control. *Nature Reviews. Cancer, 9*, 785–797.
Chiang, A. C., & Massague, J. (2008). Molecular basis of metastasis. *The New England Journal of Medicine, 359*, 2814–2823.
Chong, J. L., Wenzel, P. L., Saenz-Robles, M. T., Nair, V., Ferrey, A., Hagan, J. P., et al. (2009). E2f1-3 switch from activators in progenitor cells to repressors in differentiating cells. *Nature, 462*, 930–934.

Clarke, A. R., Maandag, E. R., van Roon, M., van der Lugt, N. M., van der Valk, M., Hooper, M. L., et al. (1992). Requirement for a functional Rb-1 gene in murine development. *Nature*, *359*, 328–330.

Claudio, P. P., Stiegler, P., Howard, C. M., Bellan, C., Minimo, C., Tosi, G. M., et al. (2001). RB2/p130 gene-enhanced expression down-regulates vascular endothelial growth factor expression and inhibits angiogenesis in vivo. *Cancer Research*, *61*, 462–468.

Clem, B. F., & Chesney, J. (2012). Molecular pathways: Regulation of metabolism by RB. *Clinical Cancer Research: An Official Journal of the American Association for Cancer Research*, *18*, 6096–6100.

Conklin, J. F., Baker, J., & Sage, J. (2012). The RB family is required for the self-renewal and survival of human embryonic stem cells. *Nature Communications*, *3*, 1244.

Conklin, J. F., & Sage, J. (2009). Keeping an eye on retinoblastoma control of human embryonic stem cells. *Journal of Cellular Biochemistry*, *108*, 1023–1030.

Creighton, C. J., Gibbons, D. L., & Kurie, J. M. (2013). The role of epithelial-mesenchymal transition programming in invasion and metastasis: A clinical perspective. *Cancer Management and Research*, *5*, 187–195.

Dali-Youcef, N., Mataki, C., Coste, A., Messaddeq, N., Giroud, S., Blanc, S., et al. (2007). Adipose tissue-specific inactivation of the retinoblastoma protein protects against diabesity because of increased energy expenditure. *Proceedings of the National Academy of Sciences of the United States of America*, *104*, 10703–10708.

Darville, M. I., Antoine, I. V., Mertens-Strijthagen, J. R., Dupriez, V. J., & Rousseau, G. G. (1995). An E2F-dependent late-serum-response promoter in a gene that controls glycolysis. *Oncogene*, *11*, 1509–1517.

Dasgupta, P., Sun, J., Wang, S., Fusaro, G., Betts, V., Padmanabhan, J., et al. (2004). Disruption of the Rb—Raf-1 interaction inhibits tumor growth and angiogenesis. *Molecular and Cellular Biology*, *24*, 9527–9541.

Davis, R. K., & Chellappan, S. (2008). Disrupting the Rb-Raf-1 interaction: A potential therapeutic target for cancer. *Drug News & Perspectives*, *21*, 331–335.

De Craene, B., & Berx, G. (2013). Regulatory networks defining EMT during cancer initiation and progression. *Nature Reviews. Cancer*, *13*, 97–110.

DeGregori, J., & Johnson, D. G. (2006). Distinct and overlapping roles for E2F family members in transcription, proliferation and apoptosis. *Current Molecular Medicine*, *6*, 739–748.

Dick, F. A., & Rubin, S. M. (2013). Molecular mechanisms underlying RB protein function. *Nature Reviews. Molecular Cell Biology*, *14*, 297–306.

Di Fiore, R., D'Anneo, A., Tesoriere, G., & Vento, R. (2013). RB1 in cancer: Different mechanisms of RB1 inactivation and alterations of pRb pathway in tumorigenesis. *Journal of Cellular Physiology*, *228*, 1676–1687.

Enders, G. H. (2004). Colon cancer metastasis: Is E2F-1 a driving force? *Cancer Biology & Therapy*, *3*, 400–401.

Engelmann, D., Mayoli-Nussle, D., Mayrhofer, C., Furst, K., Alla, V., Stoll, A., et al. (2013). E2F1 promotes angiogenesis through the VEGF-C/VEGFR-3 axis in a feedback loop for cooperative induction of PDGF-B. *Journal of Molecular Cell Biology*, *5*, 391–403.

Fajas, L., Annicotte, J. S., Miard, S., Sarruf, D., Watanabe, M., & Auwerx, J. (2004). Impaired pancreatic growth, beta cell mass, and beta cell function in E2F1 (-/-) mice. *The Journal of Clinical Investigation*, *113*, 1288–1295.

Fajas, L., Egler, V., Reiter, R., Hansen, J., Kristiansen, K., Debril, M. B., et al. (2002). The retinoblastoma-histone deacetylase 3 complex inhibits PPARgamma and adipocyte differentiation. *Developmental Cell*, *3*, 903–910.

Fajas, L., Landsberg, R. L., Huss-Garcia, Y., Sardet, C., Lees, J. A., & Auwerx, J. (2002). E2Fs regulate adipocyte differentiation. *Developmental Cell*, *3*, 39–49.

Farhang Ghahremani, M., Goossens, S., Nittner, D., Bisteau, X., Bartunkova, S., Zwolinska, A., et al. (2013). p53 promotes VEGF expression and angiogenesis in the absence of an intact p21-Rb pathway. *Cell Death and Differentiation, 20*, 888–897.

Fontemaggi, G., Dell'Orso, S., Trisciuoglio, D., Shay, T., Melucci, E., Fazi, F., et al. (2009). The execution of the transcriptional axis mutant p53, E2F1 and ID4 promotes tumor neo-angiogenesis. *Nature Structural & Molecular Biology, 16*, 1086–1093.

Fritz, V., & Fajas, L. (2010). Metabolism and proliferation share common regulatory pathways in cancer cells. *Oncogene, 29*, 4369–4377.

Frolov, M. V., & Dyson, N. J. (2004). Molecular mechanisms of E2F-dependent activation and pRB-mediated repression. *Journal of Cell Science, 117*, 2173–2181.

Fujiwara, K., Yuwanita, I., Hollern, D. P., & Andrechek, E. R. (2011). Prediction and genetic demonstration of a role for activator E2Fs in Myc-induced tumors. *Cancer Research, 71*, 1924–1932.

Galderisi, U., Cipollaro, M., & Giordano, A. (2006). The retinoblastoma gene is involved in multiple aspects of stem cell biology. *Oncogene, 25*, 5250–5256.

Gauthier, M. L., Berman, H. K., Miller, C., Kozakeiwicz, K., Chew, K., Moore, D., et al. (2007). Abrogated response to cellular stress identifies DCIS associated with subsequent tumor events and defines basal-like breast tumors. *Cancer Cell, 12*, 479–491.

Giannakakis, A., Sandaltzopoulos, R., Greshock, J., Liang, S., Huang, J., Hasegawa, K., et al. (2008). miR-210 links hypoxia with cell cycle regulation and is deleted in human epithelial ovarian cancer. *Cancer Biology & Therapy, 7*, 255–264.

Gonzalez-Gonzalez, M., Garcia, J. G., Montero, J. A., Fernandez, L. M., Bengoechea, O., Munez, O. B., et al. (2013). Genomics and proteomics approaches for biomarker discovery in sporadic colorectal cancer with metastasis. *Cancer Genomics & Proteomics, 10*, 19–25.

Gorlick, R., Metzger, R., Danenberg, K. D., Salonga, D., Miles, J. S., Longo, G. S., et al. (1998). Higher levels of thymidylate synthase gene expression are observed in pulmonary as compared with hepatic metastases of colorectal adenocarcinoma. *Journal of Clinical Oncology, 16*, 1465–1469.

Gunaratne, P. H. (2009). Embryonic stem cell microRNAs: Defining factors in induced pluripotent (iPS) and cancer (CSC) stem cells? *Current Stem Cell Research & Therapy, 4*, 168–177.

Han, T., Kang, D., Ji, D., Wang, X., Zhan, W., Fu, M., et al. (2013). How does cancer cell metabolism affect tumor migration and invasion? *Cell Adhesion & Migration, 7*, 395–403.

Hanahan, D., & Weinberg, R. A. (2000). The hallmarks of cancer. *Cell, 100*, 57–70.

Hanahan, D., & Weinberg, R. A. (2011). Hallmarks of cancer: The next generation. *Cell, 144*, 646–674.

Hao, J., Zhao, S., Zhang, Y., Zhao, Z., Ye, R., Wen, J., et al. (2014). Emerging role of microRNAs in cancer and cancer stem cells. *Journal of Cellular Biochemistry, 115*, 605–610.

Hiebert, S. W., Chellappan, S. P., Horowitz, J. M., & Nevins, J. R. (1992). The interaction of RB with E2F coincides with an inhibition of the transcriptional activity of E2F. *Genes & Development, 6*, 177–185.

Ho, G. H., Calvano, J. E., Bisogna, M., & Van Zee, K. J. (2001). Expression of E2F-1 and E2F-4 is reduced in primary and metastatic breast carcinomas. *Breast Cancer Research and Treatment, 69*, 115–122.

Hood, J. D., & Cheresh, D. A. (2002). Targeted delivery of mutant Raf kinase to neovessels causes tumor regression. *Cold Spring Harbor Symposia on Quantitative Biology, 67*, 285–291.

Horm, T. M., & Schroeder, J. A. (2013). MUC1 and metastatic cancer: Expression, function and therapeutic targeting. *Cell Adhesion & Migration, 7*, 187–198.

Hsieh, M. C., Das, D., Sambandam, N., Zhang, M. Q., & Nahle, Z. (2008). Regulation of the PDK4 isozyme by the Rb-E2F1 complex. *The Journal of Biological Chemistry*, *283*, 27410–27417.

Hudson, B. D., Kulp, K. S., & Loots, G. G. (2013). Prostate cancer invasion and metastasis: Insights from mining genomic data. *Briefings in Functional Genomics*, *12*, 397–410.

Hui, A. M., Li, X., Makuuchi, M., Takayama, T., & Kubota, K. (1999). Over-expression and lack of retinoblastoma protein are associated with tumor progression and metastasis in hepatocellular carcinoma. *International Journal of Cancer*, *84*, 604–608.

Iavarone, A., Garg, P., Lasorella, A., Hsu, J., & Israel, M. A. (1994). The helix-loop-helix protein Id-2 enhances cell proliferation and binds to the retinoblastoma protein. *Genes & Development*, *8*, 1270–1284.

Indovina, P., Marcelli, E., Casini, N., Rizzo, V., & Giordano, A. (2013). Emerging roles of RB family: New defense mechanisms against tumor progression. *Journal of Cellular Physiology*, *228*, 525–535.

Iwamoto, M., Banerjee, D., Menon, L. G., Jurkiewicz, A., Rao, P. H., Kemeny, N. E., et al. (2004). Overexpression of E2F-1 in lung and liver metastases of human colon cancer is associated with gene amplification. *Cancer Biology & Therapy*, *3*, 395–399.

Jacks, T., Fazeli, A., Schmitt, E. M., Bronson, R. T., Goodell, M. A., & Weinberg, R. A. (1992). Effects of an Rb mutation in the mouse. *Nature*, *359*, 295–300.

Jang, M., Kim, S. S., & Lee, J. (2013). Cancer cell metabolism: Implications for therapeutic targets. *Experimental & Molecular Medicine*, *45*, e45.

Ji, W., Zhang, W., & Xiao, W. (2010). E2F-1 directly regulates thrombospondin 1 expression. *PLoS One*, *5*, e13442.

Johnson, J. L., Pillai, S., Pernazza, D., Sebti, S. M., Lawrence, N. J., & Chellappan, S. P. (2012). Regulation of matrix metalloproteinase genes by E2F transcription factors: Rb-Raf-1 interaction as a novel target for metastatic disease. *Cancer Research*, *72*, 516–526.

Joshi, B., Ordonez-Ercan, D., Dasgupta, P., & Chellappan, S. (2005). Induction of human metallothionein 1G promoter by VEGF and heavy metals: Differential involvement of E2F and metal transcription factors. *Oncogene*, *24*, 2204–2217.

Kalluri, R., & Weinberg, R. A. (2009). The basics of epithelial-mesenchymal transition. *The Journal of Clinical Investigation*, *119*, 1420–1428.

Kanai, M., Tashiro, E., Maruki, H., Minato, Y., & Imoto, M. (2009). Transcriptional regulation of human fibroblast growth factor receptor 1 by E2F-1. *Gene*, *438*, 49–56.

Kim, Y. C., Kim, S. Y., Mellado-Gil, J. M., Yadav, H., Neidermyer, W., Kamaraju, A. K., et al. (2011). RB regulates pancreas development by stabilizing Pdx1. *The EMBO Journal*, *30*, 1563–1576.

Kinkade, R., Dasgupta, P., Carie, A., Pernazza, D., Carless, M., Pillai, S., et al. (2008). A small molecule disruptor of Rb/Raf-1 interaction inhibits cell proliferation, angiogenesis, and growth of human tumor xenografts in nude mice. *Cancer Research*, *68*, 3810–3818.

Labrecque, M. P., Prefontaine, G. G., & Beischlag, T. V. (2013). The aryl hydrocarbon receptor nuclear translocator (ARNT) family of proteins: Transcriptional modifiers with multi-functional protein interfaces. *Current Molecular Medicine*, *13*, 1047–1065.

Lammens, T., Li, J., Leone, G., & De Veylder, L. (2009). Atypical E2Fs: New players in the E2F transcription factor family. *Trends in Cell Biology*, *19*, 111–118.

Lasorella, A., Rothschild, G., Yokota, Y., Russell, R. G., & Iavarone, A. (2005). Id2 mediates tumor initiation, proliferation, and angiogenesis in Rb mutant mice. *Molecular and Cellular Biology*, *25*, 3563–3574.

Lee, B. K., Bhinge, A. A., & Iyer, V. R. (2011). Wide-ranging functions of E2F4 in transcriptional activation and repression revealed by genome-wide analysis. *Nucleic Acids Research*, *39*, 3558–3573.

Lee, E. Y., Chang, C. Y., Hu, N., Wang, Y. C., Lai, C. C., Herrup, K., et al. (1992). Mice deficient for Rb are nonviable and show defects in neurogenesis and haematopoiesis. *Nature, 359*, 288–294.

Lee, E. Y., Hu, N., Yuan, S. S., Cox, L. A., Bradley, A., Lee, W. H., et al. (1994). Dual roles of the retinoblastoma protein in cell cycle regulation and neuron differentiation. *Genes & Development, 8*, 2008–2021.

Li, F. Q., Coonrod, A., & Horwitz, M. (2000). Selection of a dominant negative retinoblastoma protein (RB) inhibiting satellite myoblast differentiation implies an indirect interaction between MyoD and RB. *Molecular and Cellular Biology, 20*, 5129–5139.

Li, J., Ran, C., Li, E., Gordon, F., Comstock, G., Siddiqui, H., et al. (2008). Synergistic function of E2F7 and E2F8 is essential for cell survival and embryonic development. *Developmental Cell, 14*, 62–75.

Li, G. C., & Wang, Z. Y. (2006). Constitutive expression of RbAp46 induces epithelial-mesenchymal transition in mammary epithelial cells. *Anticancer Research, 26*, 3555–3560.

Liu, Y., Sanchez-Tillo, E., Lu, X., Huang, L., Clem, B., Telang, S., et al. (2013). Sequential inductions of the ZEB1 transcription factor caused by mutation of Rb and then Ras proteins are required for tumor initiation and progression. *The Journal of Biological Chemistry, 288*, 11572–11580.

Ma, Y., Kurtyka, C. A., Boyapalle, S., Sung, S. S., Lawrence, H., Guida, W., et al. (2008). A small-molecule E2F inhibitor blocks growth in a melanoma culture model. *Cancer Research, 68*, 6292–6299.

Merdzhanova, G., Gout, S., Keramidas, M., Edmond, V., Coll, J. L., Brambilla, C., et al. (2010). The transcription factor E2F1 and the SR protein SC35 control the ratio of pro-angiogenic versus antiangiogenic isoforms of vascular endothelial growth factor-A to inhibit neovascularization in vivo. *Oncogene, 29*, 5392–5403.

Minato, Y., Tashiro, E., Kanai, M., Nihei, Y., Kodama, Y., & Imoto, M. (2007). Transcriptional regulation of a new variant of human platelet-derived growth factor receptor alpha transcript by E2F-1. *Gene, 403*, 89–97.

Miyashita, H., & Sato, Y. (2005). Metallothionein 1 is a downstream target of vascular endothelial zinc finger 1 (VEZF1) in endothelial cells and participates in the regulation of angiogenesis. *Endothelium, 12*, 163–170.

Morris, E. J., & Dyson, N. J. (2001). Retinoblastoma protein partners. *Advances in Cancer Research, 82*, 1–54.

Nag, S., Qin, J., Srivenugopal, K. S., Wang, M., & Zhang, R. (2013). The MDM2-p53 pathway revisited. *Journal of Biomedical Research, 27*, 254–271.

Nelson, M. A., Reynolds, S. H., Rao, U. N., Goulet, A. C., Feng, Y., Beas, A., et al. (2006). Increased gene copy number of the transcription factor E2F1 in malignant melanoma. *Cancer Biology & Therapy, 5*, 407–412.

Nevins, J. R. (2001). The Rb/E2F pathway and cancer. *Human Molecular Genetics, 10*, 699–703.

Nicolay, B. N., & Dyson, N. J. (2013). The multiple connections between pRB and cell metabolism. *Current Opinion in Cell Biology, 25*, 735–740.

Nicolay, B. N., Gameiro, P. A., Tschop, K., Korenjak, M., Heilmann, A. M., Asara, J. M., et al. (2013). Loss of RBF1 changes glutamine catabolism. *Genes & Development, 27*, 182–196.

Novitch, B. G., Mulligan, G. J., Jacks, T., & Lassar, A. B. (1996). Skeletal muscle cells lacking the retinoblastoma protein display defects in muscle gene expression and accumulate in S and G2 phases of the cell cycle. *The Journal of Cell Biology, 135*, 441–456.

Nowak, K., Kerl, K., Fehr, D., Kramps, C., Gessner, C., Killmer, K., et al. (2006). BMI1 is a target gene of E2F-1 and is strongly expressed in primary neuroblastomas. *Nucleic acids research, 34*, 1745–1754.

Papadimou, E., Menard, C., Grey, C., & Puceat, M. (2005). Interplay between the retinoblastoma protein and LEK1 specifies stem cells toward the cardiac lineage. *The EMBO Journal, 24*, 1750–1761.

Perk, J., Iavarone, A., & Benezra, R. (2005). Id family of helix-loop-helix proteins in cancer. *Nature Reviews Cancer, 5*, 603–614.

Perlikos, F., Harrington, K. J., & Syrigos, K. N. (2013). Key molecular mechanisms in lung cancer invasion and metastasis: A comprehensive review. *Critical Reviews in Oncology/Hematology, 87*, 1–11.

Pickard, A., Cichon, A. C., Barry, A., Kieran, D., Patel, D., Hamilton, P., et al. (2012). Inactivation of Rb in stromal fibroblasts promotes epithelial cell invasion. *The EMBO Journal, 31*, 3092–3103.

Pillai, S., Kovacs, M., & Chellappan, S. (2010). Regulation of vascular endothelial growth factor receptors by Rb and E2F1: Role of acetylation. *Cancer Research, 70*, 4931–4940.

Proulx, S. T., & Detmar, M. (2013). Molecular mechanisms and imaging of lymphatic metastasis. *Experimental Cell Research, 319*, 1611–1617.

Qin, G., Kishore, R., Dolan, C. M., Silver, M., Wecker, A., Luedemann, C. N., et al. (2006). Cell cycle regulator E2F1 modulates angiogenesis via p53-dependent transcriptional control of VEGF. *Proceedings of the National Academy of Sciences of the United States of America, 103*, 11015–11020.

Rakha, E. A., Pinder, S. E., Paish, E. C., Robertson, J. F., & Ellis, I. O. (2004). Expression of E2F-4 in invasive breast carcinomas is associated with poor prognosis. *The Journal of Pathology, 203*, 754–761.

Ruzinova, M. B., & Benezra, R. (2003). Id proteins in development, cell cycle and cancer. *Trends in Cell Biology, 13*, 410–418.

Sage, J. (2012). The retinoblastoma tumor suppressor and stem cell biology. *Genes & Development, 26*, 1409–1420.

Salon, C., Merdzhanova, G., Brambilla, C., Brambilla, E., Gazzeri, S., & Eymin, B. (2007). E2F-1, Skp2 and cyclin E oncoproteins are upregulated and directly correlated in high-grade neuroendocrine lung tumors. *Oncogene, 26*, 6927–6936.

Saxena, M., & Christofori, G. (2013). Rebuilding cancer metastasis in the mouse. *Molecular Oncology, 7*, 283–296.

Schaffer, B. E., Park, K. S., Yiu, G., Conklin, J. F., Lin, C., Burkhart, D. L., et al. (2010). Loss of p130 accelerates tumor development in a mouse model for human small-cell lung carcinoma. *Cancer Research, 70*, 3877–3883.

Scheel, C., & Weinberg, R. A. (2012). Cancer stem cells and epithelial-mesenchymal transition: Concepts and molecular links. *Seminars in Cancer Biology, 22*, 396–403.

Schlisio, S., Halperin, T., Vidal, M., & Nevins, J. R. (2002). Interaction of YY1 with E2Fs, mediated by RYBP, provides a mechanism for specificity of E2F function. *The EMBO Journal, 21*, 5775–5786.

Schoeftner, S., Scarola, M., Comisso, E., Schneider, C., & Benetti, R. (2013). An Oct4-pRb axis, controlled by MiR-335, integrates stem cell self-renewal and cell cycle control. *Stem Cells, 31*, 717–728.

Schreiber, C., Vormbrock, K., & Ziebold, U. (2012). Genes involved in the metastatic cascade of medullary thyroid tumours. *Methods in Molecular Biology, 878*, 217–228.

Semenza, G. L. (2008). Hypoxia-inducible factor 1 and cancer pathogenesis. *IUBMB Life, 60*, 591–597.

Shen, G., Li, X., Jia, Y. F., Piazza, G. A., & Xi, Y. (2013). Hypoxia-regulated microRNAs in human cancer. *Acta Pharmacologica Sinica, 34*, 336–341.

Sherr, C. J., & McCormick, F. (2002). The RB and p53 pathways in cancer. *Cancer Cell, 2*, 103–112.

Siegel, R., Naishadham, D., & Jemal, A. (2013). Cancer statistics, 2013. *CA: A Cancer Journal for Clinicians, 63*, 11–30.

Singh, B., Berry, J. A., Shoher, A., Ramakrishnan, V., & Lucci, A. (2005). COX-2 overexpression increases motility and invasion of breast cancer cells. *International Journal of Oncology, 26*, 1393–1399.

Singh, S., Johnson, J., & Chellappan, S. (2010). Small molecule regulators of Rb-E2F pathway as modulators of transcription. *Biochimica et Biophysica Acta, 1799*, 788–794.

Stanelle, J., Stiewe, T., Theseling, C. C., Peter, M., & Putzer, B. M. (2002). Gene expression changes in response to E2F1 activation. *Nucleic Acids Research, 30*, 1859–1867.

Takahashi, C., Sasaki, N., & Kitajima, S. (2012). Twists in views on RB functions in cellular signaling, metabolism and stem cells. *Cancer Science, 103*, 1182–1188.

Tashiro, E., Minato, Y., Maruki, H., Asagiri, M., & Imoto, M. (2003). Regulation of FGF receptor-2 expression by transcription factor E2F-1. *Oncogene, 22*, 5630–5635.

Teng, Y. C., Lee, C. F., Li, Y. S., Chen, Y. R., Hsiao, P. W., Chan, M. Y., et al. (2013). Histone demethylase RBP2 promotes lung tumorigenesis and cancer metastasis. *Cancer Research, 73*, 4711–4721.

Teodoro, J. G., Evans, S. K., & Green, M. R. (2007). Inhibition of tumor angiogenesis by p53: A new role for the guardian of the genome. *Journal of Molecular Medicine, 85*, 1175–1186.

Thomassen, M., Tan, Q., & Kruse, T. A. (2008). Gene expression meta-analysis identifies metastatic pathways and transcription factors in breast cancer. *BMC Cancer, 8*, 394.

Toma, J. G., El-Bizri, H., Barnabe-Heider, F., Aloyz, R., & Miller, F. D. (2000). Evidence that helix-loop-helix proteins collaborate with retinoblastoma tumor suppressor protein to regulate cortical neurogenesis. *The Journal of Neuroscience, 20*, 7648–7656.

Trimarchi, J. M., & Lees, J. A. (2002). Sibling rivalry in the E2F family. *Nature Reviews. Molecular Cell Biology, 3*, 11–20.

Tsai, J. H., & Yang, J. (2013). Epithelial-mesenchymal plasticity in carcinoma metastasis. *Genes & Development, 27*, 2192–2206.

Tsujii, M., Kawano, S., & DuBois, R. N. (1997). Cyclooxygenase-2 expression in human colon cancer cells increases metastatic potential. *Proceedings of the National Academy of Sciences of the United States of America, 94*, 3336–3340.

Valastyan, S., & Weinberg, R. A. (2011). Tumor metastasis: Molecular insights and evolving paradigms. *Cell, 147*, 275–292.

Viatour, P., Somervaille, T. C., Venkatasubrahmanyam, S., Kogan, S., McLaughlin, M. E., Weissman, I. L., et al. (2008). Hematopoietic stem cell quiescence is maintained by compound contributions of the retinoblastoma gene family. *Cell Stem Cell, 3*, 416–428.

Visvader, J. E., & Lindeman, G. J. (2008). Cancer stem cells in solid tumours: Accumulating evidence and unresolved questions. *Nature Reviews. Cancer, 8*, 755–768.

Visvader, J. E., & Lindeman, G. J. (2012). Cancer stem cells: Current status and evolving complexities. *Cell Stem Cell, 10*, 717–728.

Wang, K. C., Garmire, L. X., Young, A., Nguyen, P., Trinh, A., Subramaniam, S., et al. (2010). Role of microRNA-23b in flow-regulation of Rb phosphorylation and endothelial cell growth. *Proceedings of the National Academy of Sciences of the United States of America, 107*, 3234–3239.

Wang, S., Ghosh, R. N., & Chellappan, S. P. (1998). Raf-1 physically interacts with Rb and regulates its function: A link between mitogenic signaling and cell cycle regulation. *Molecular and Cellular Biology, 18*, 7487–7498.

Wang, S., Nath, N., Adlam, M., & Chellappan, S. (1999). Prohibitin, a potential tumor suppressor, interacts with RB and regulates E2F function. *Oncogene, 18*, 3501–3510.

Wang, S., Nath, N., Fusaro, G., & Chellappan, S. (1999). Rb and prohibitin target distinct regions of E2F1 for repression and respond to different upstream signals. *Molecular and Cellular Biology, 19*, 7447–7460.

Wang, H., & Scott, R. E. (1993). Inhibition of distinct steps in the adipocyte differentiation pathway in 3 T3 T mesenchymal stem cells by dimethyl sulphoxide (DMSO). *Cell Proliferation, 26*, 55–66.

Weijts, B. G., Bakker, W. J., Cornelissen, P. W., Liang, K. H., Schaftenaar, F. H., Westendorp, B., et al. (2012). E2F7 and E2F8 promote angiogenesis through transcriptional activation of VEGFA in cooperation with HIF1. *The EMBO Journal*, *31*, 3871–3884.

Wenzel, P. L., Wu, L., de Bruin, A., Chong, J. L., Chen, W. Y., Dureska, G., et al. (2007). Rb is critical in a mammalian tissue stem cell population. *Genes & Development*, *21*, 85–97.

Yamanaka, S. (2009). Ekiden to iPS Cells. *Nature Medicine*, *15*, 1145–1148.

Yasui, W., Naka, K., Suzuki, T., Fujimoto, J., Hayashi, K., Matsutani, N., et al. (1999). Expression of p27Kip1, cyclin E and E2F-1 in primary and metastatic tumors of gastric carcinoma. *Oncology Reports*, *6*, 983–987.

Yoon, S. O., Shin, S., & Mercurio, A. M. (2006). Ras stimulation of E2F activity and a consequent E2F regulation of integrin alpha6beta4 promote the invasion of breast carcinoma cells. *Cancer Research*, *66*, 6288–6295.

Zbinden, S., Wang, J., Adenika, R., Schmidt, M., Tilan, J. U., Najafi, A. H., et al. (2010). Metallothionein enhances angiogenesis and arteriogenesis by modulating smooth muscle cell and macrophage function. *Arteriosclerosis, Thrombosis, and Vascular Biology*, *30*, 477–482.

Zhang, S. Y., Liu, S. C., Johnson, D. G., & Klein-Szanto, A. J. (2000). E2F-1 gene transfer enhances invasiveness of human head and neck carcinoma cell lines. *Cancer Research*, *60*, 5972–5976.

Zhang, X., Neganova, I., Przyborski, S., Yang, C., Cooke, M., Atkinson, S. P., et al. (2009). A role for NANOG in G1 to S transition in human embryonic stem cells through direct binding of CDK6 and CDC25A. *The Journal of Cell Biology*, *184*, 67–82.

Zhou, W., Srinivasan, S., Nawaz, Z., & Slingerland, J. M. (2013). ERalpha, SKP2 and E2F-1 form a feed forward loop driving late ERalpha targets and G1 cell cycle progression. *Oncogene*, *192*, 1–13.

Ziebold, U., Lee, E. Y., Bronson, R. T., & Lees, J. A. (2003). E2F3 loss has opposing effects on different pRB-deficient tumors, resulting in suppression of pituitary tumors but metastasis of medullary thyroid carcinomas. *Molecular and Cellular Biology*, *23*, 6542–6552.

CHAPTER FIVE

ATP-Dependent Chromatin Remodeling Complexes as Novel Targets for Cancer Therapy

Kimberly Mayes, Zhijun Qiu, Aiman Alhazmi, Joseph W. Landry[1]
Department of Human and Molecular Genetics, VCU Institute of Molecular Medicine, Massey Cancer Center, Virginia Commonwealth University School of Medicine, Richmond, Virginia, USA
[1]Corresponding author: e-mail address: jwlandry2@vcu.edu

Contents

Abstract

The progression to advanced stage cancer requires changes in many characteristics of a cell. These changes are usually initiated through spontaneous mutation. As a result of these mutations, gene expression is almost invariably altered allowing the cell to acquire tumor-promoting characteristics. These abnormal gene expression patterns are in part enabled by the posttranslational modification and remodeling of nucleosomes in chromatin. These chromatin modifications are established by a functionally diverse family of enzymes including histone and DNA-modifying complexes, histone deposition pathways, and chromatin remodeling complexes. Because the modifications these enzymes deposit are essential for maintaining tumor-promoting gene expression, they have recently attracted much interest as novel therapeutic targets. One class of enzyme that has not generated much interest is the chromatin remodeling complexes. In this review, we will present evidence from the literature that these enzymes have both causal and enabling roles in the transition to advanced stage cancers; as such, they should be seriously considered as high-value therapeutic targets. Previously published strategies for

Advances in Cancer Research, Volume 121
ISSN 0065-230X
http://dx.doi.org/10.1016/B978-0-12-800249-0.00005-6

discovering small molecule regulators to these complexes are described. We close with thoughts on future research, the field should perform to further develop this potentially novel class of therapeutic target.

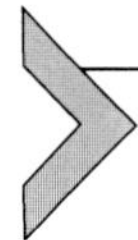

1. INTRODUCTION—THE IMPORTANCE OF GENE EXPRESSION TO CANCER BIOLOGY

Advanced stage cancer occurs when normal cells acquire several tumor-promoting characteristics. These characteristics are broadly organized into eight categories that have been widely accepted as hallmarks of cancer (Hanahan & Weinberg, 2011). At a fundamental level, acquiring the hallmarks of cancer is a consequence of deregulating any one of a number of basic cellular properties including motility, differentiation, proliferation, and viability. In one theory, the acquisition of hallmarks occurs spontaneously as a result of somatic mutation. At a molecular level, these mutations deregulate cellular properties to promote tumor growth (Nowell, 1976). In some cases, these somatic mutations directly cause abnormal gene expression patterns favoring tumor cell growth. A classic example is the t(8;14)(q24; q32) translocation that juxtaposes the immunoglobulin heavy chain locus with the MYC protooncogene resulting in elevated MYC expression (Erikson, ar-Rushdi, Drwinga, Nowell, & Croce, 1983; Taub et al., 1982). In other cases, mutations deregulate circuits resulting in the abnormal regulation of a large number of genes to favor tumor cell growth. Mutations in Ras (RasG12V), which result in tumor-promoting gene regulation, are a classic example (Ayllon & Rebollo, 2000; Wong-Staal, Dalla-Favera, Franchini, Gelmann, & Gallo, 1981). Once established, abnormal gene expression profiles maintain the tumor cell phenotype through a process of oncogene addiction (Weinstein, 2002). Because abnormal gene expression first promotes, and can later become essential for the tumor cell phenotype, it can be assumed that correcting abnormal gene expression would be disadvantageous to the cancer cell, and therefore, a viable therapeutic strategy (Yeh, Toniolo, & Frank, 2013).

Several approaches have been proposed to correct abnormal gene expression in tumors. By changing the DNA sequence, one can attempt to correct abnormal gene expression (Perez-Pinera, Ousterout, & Gersbach, 2012). This can occur by correcting the mutations that promote abnormal gene expression or by changing essential regulatory sequences necessary for tumor-promoting gene expression. Correcting DNA sequences in a patient's tumor cells is certainly plausible but is not yet a practical approach to correcting abnormal gene expression. Alternatively, abnormal gene expression can be corrected by altering any one of several posttranslational modifications found on the chromatin of tumor cells that are important for gene expression (Plass et al., 2013). These diverse posttranslational modifications are commonly, but not accurately, referred to as epigenetic regulatory mechanisms (Ptashne, 2013). In contrast to genetic regulatory mechanisms that are stable, posttranslational modifications on chromatin are dynamic, and in theory, reversible. Posttranslational modifications to chromatin that accompany regulated transcription do not change the DNA sequence but rather are deposited onto chromatin by enzyme catalyzed reactions (Allis, Jenuwein, & Reinberg, 2007). Because posttranslational modifications on chromatin are required for regulating procancer gene expression in tumor cells, do not change the DNA sequence, are in theory reversible, and are established by the actions of enzymes, makes them attractive targets to correct abnormal gene expression as a means of therapy. In this review, we will focus on the therapeutic potential of one class of epigenetic regulator—chromatin remodeling. We will present a growing body of evidence that the enzymes that catalyze ATP-dependent chromatin remodeling not only play an important role as drivers of cancer biology but also operate in concert with oncogenes and tumor suppressors to promote the abnormal gene expression necessary to maintain cancer biology. Current strategies used to develop small molecule regulators to chromatin remodeling complexes are discussed, and we close with a discussion on how the field can further develop this class of enzymes as viable therapeutic targets for cancer treatment.

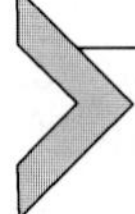

2. AN OVERVIEW OF EPIGENETIC REGULATORY MECHANISMS

In higher eukaryotes, posttranslational modification regulates the function of chromatin. Chromatin is composed of nucleosomes, which in turn are composed of two copies of each of the conical histones H2A, H2B, H3, and H4 wrapped by ~150 bp of DNA (Luger, Mader, Richmond, Sargent, & Richmond, 1997). Nucleosomes are positioned along DNA in a "beads on a string" configuration to create a 10-nm fiber that through internucleosome contacts creates a variety of higher ordered chromatin structures including a molten globule with short stretches of a 30-nm fiber (Kornberg, 1974; Woodcock & Ghosh, 2010). In most cases, these chromatin structures repress DNA accessibility and therefore must be modified to promote essential nuclear processes like transcription, DNA replication, and DNA repair (Laybourn & Kadonaga, 1991). These modifications include histone posttranslational modifications, DNA methylation, incorporation of histone variants, and chromatin remodeling (Allis et al., 2007). Small RNA regulation is another major category of epigenetic regulation that does not operate on chromatin (at least in mammals) and is not discussed in this review. The interplay between chromatin modifications participate as part of a larger "epigenetic code" which is essential for the function of DNA regulatory elements in chromatin. These modifications arise and are subsequently maintained on chromatin through interactions between the enzymes which deposit them and sequence-specific DNA-binding factors (Kouzarides, 2007).

The posttranslational modification of histones occurs by histone modifying enzymes. And just as histones can be modified, the modifications can be removed by de-modifying enzymes. Histones are primarily modified on the N-terminal and C-terminal ends but can be modified throughout their sequence (Kouzarides, 2007). The effect of histone modifications is to alter chromatin structure or recruit additional chromatin-modifying complexes, which in turn regulate nuclear processes like transcription. Common modifications include acetylation, methylation, and phosphorylation, and many of the enzymes which add and remove these modifications have been identified and characterized. Histone modifications are usually found in combinations in chromatin (e.g., H3K9me1 with H3K4me3) so as to provide redundancy in their ability to direct nuclear processes (Wang et al., 2008).

Just like histones, DNA can be modified to provide a stable method of regulating gene expression. In mammals, modifications to DNA usually occur on CpG dinucleotides, but they have also been observed outside of this context (Lister et al., 2009; Ziller et al., 2011). The most common modification to CpG dinucleotides is the methylation of the fifth carbon of cytosine (5mC), catalyzed by the DNMT family of enzymes. 5mC is removed by both passive and active means, the latter including the TET family of enzymes and to a lesser extent APOBEC (Shen & Zhang, 2013). DNA methylation largely represses gene expression by altering nucleosome positioning and stability and by recruiting chromatin-modifying complexes containing subunits that bind methylated DNA (Bartke et al., 2010; Lee & Lee, 2012).

To similar effect, the incorporation of histone variants serves to alter chromatin structure and stability, and it provides unique contact surfaces for the recruitment of chromatin-modifying complexes (Skene & Henikoff, 2013). The incorporation of histone variants can occur through both ATP-dependent and ATP-independent mechanisms. ATP-independent mechanisms include histone chaperones dedicated to the deposition of specific histone variants. Histone chaperones support replication dependent (coupled with DNA replication during S-phase) and independent (outside S-phase) histone deposition pathways. Examples include the ATRX/DAXX pathway for the deposition of histone H3.3 and the HJURP pathway for the deposition of CENP-A (Skene & Henikoff, 2013). The incorporation of histone variants by ATP-dependent mechanisms usually occurs through chromatin remodeling complexes and is the focus of this review.

3. ATP-DEPENDENT CHROMATIN REMODELING

Chromatin remodeling is the process wherein the position, occupancy, or the histone composition of a nucleosome is altered in the chromatin. These activities occur by both ATP-dependent and ATP-independent mechanisms (Aalfs & Kingston, 2000). ATP-independent mechanisms can occur from transcription factors shifting the position of nucleosomes as a consequence of DNA binding, or by the action of histone chaperones, depositing or removing histones from chromatin (De Koning, Corpet, Haber, & Almouzni, 2007; Workman & Kingston, 1992). In contrast to ATP-independent mechanisms, ATP-dependent mechanisms are enzymatic and constitute the majority of the remodeling activity in the cell (Zhang et al., 2011). These activities are usually catalyzed by multisubunit ATP-dependent chromatin remodeling complexes which are organized into SWI/SNF, ISWI, CHD, and INO80 families (Fig. 5.1). ATPases are organized into these families based on the sequence homology of the incorporated ATPase (Becker & Workman, 2013). Each family of ATPase catalyzes distinct remodeling activities: incremental nucleosome sliding on DNA in cis; the creation of DNA loops on the surface of the nucleosome; eviction of histone H2A/H2B dimers; eviction of the histone octamer; or the exchange of histone octamer subunits within the nucleosome to change its composition. Each of these activities alters the accessibility of DNA in the chromatin to DNA-binding factors, which in turn regulates essential nuclear processes like transcription, DNA replication, and DNA repair.

The chromatin remodeling activity of an ATPase usually occurs in the context of a large multisubunit complex, but on rare occasions, they are known to function as single ATPase subunits (Fig. 5.2; Becker & Workman, 2013).

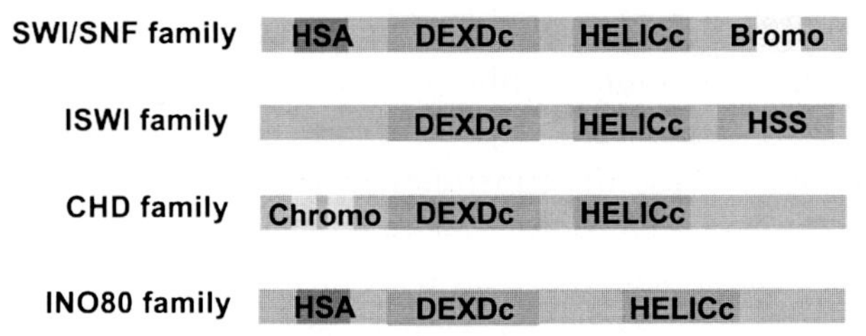

Figure 5.1 *Chromatin remodeling complexes are segregated into distinct families.* The ATPase containing subunits found in chromatin remodeling complexes have common functional domains allowing their segregation into distinct families. All ATPase subunits have distinctive DEXDc ATP-binding domains and a HELICc helicase domain. The combination of these domains convert the chemical energy stored in ATP into mechanical energy used to remodel nucleosomes. The space between these two domains varies among family members and is distinctively long for the INO80 family of ATPases. Other domains which are diagnostic to specific families include the HSA domain (HSA) found in SWI/SNF and INO80 families, bromodomains (bromo) found in SWI/SNF family members, chromodomains (chromo) found in the CHD family, and the HAND–SANT–SLIDE (HSS) cluster found in ISWI family members (Becker & Workman, 2013). (See the color plate.)

Complexes in the SWI/SNF family include the large multisubunit BAF, PBAF, and WINAC complexes which function as coregulators of transcription, and they also aid in the repair of DNA damage (Oya et al., 2009; Wang et al., 1996). Remodeling reactions catalyzed by SWI/SNF family members include simple nucleosome sliding reactions, but they can include more dramatic reactions like creating DNA loops on the surface of nucleosomes or the eviction of H2A/H2B dimers from the nucleosome structure (Bowman, 2010). Like SWI/SNF, the members of the INO80 family of complexes are also large multisubunit complexes which regulate transcription, but they have more prominent functions in DNA damage repair (Morrison & Shen, 2009). These complexes are unique among ATP-dependent remodeling complexes because they can catalyze the exchange of histones from the nucleosome structure. Complexes with the best characterized functions include SRCAP and Tip60/p400, which exchange the H2A/H2B histone dimer found in a canonical nucleosome for a variant H2A.Z/H2B dimer (Gevry, Chan, Laflamme, Livingston, & Gaudreau, 2007; Ruhl et al., 2006; Wong, Cox, & Chrivia, 2007). In contrast to most chromatin remodeling enzymes, these complexes are dedicated histone exchange enzymes, with little nucleosome sliding activities. Like Tip60/p400 and SRCAP, the INO80 complex has been reported to exchange a H2A.Z/H2B histone dimer found in a variant nucleosome for a canonical

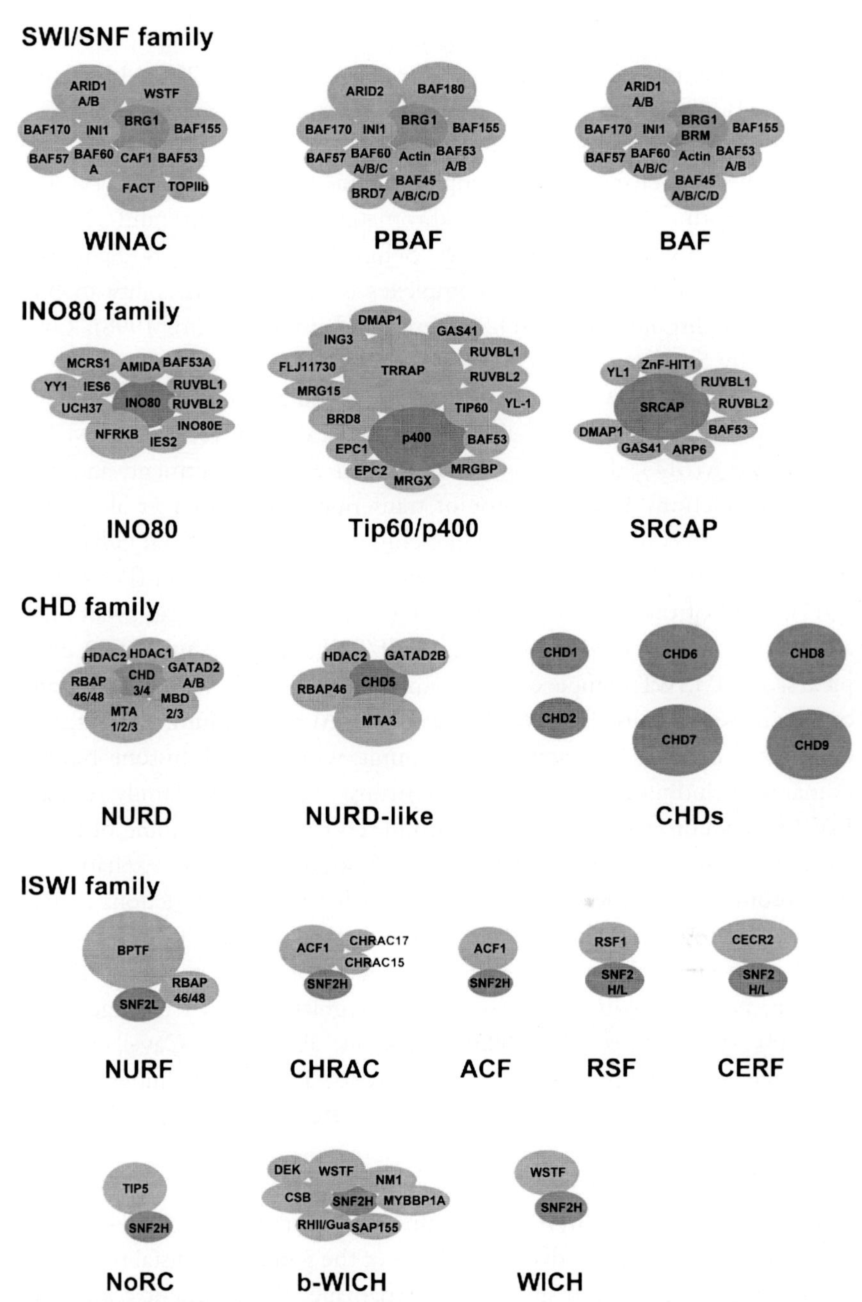

Figure 5.2 *Chromatin remodeling complexes vary subunit composition to increase diversity.* The subunit composition of major characterized chromatin remodeling complexes from the SWI/SNF, INO80, CHD, and ISWI families are shown in cartoon. Complexes are organized into rows based on family. ATPase subunits are shown in red; all other subunits are shown in blue. Subunits which can be differentially incorporated into complexes are separated by a backslash symbol (/). (See the color plate.)

H2A/H2B dimer but also has significant nucleosome sliding activities (Papamichos-Chronakis, Watanabe, Rando, & Peterson, 2011; Udugama, Sabri, & Bartholomew, 2011). The CHD family contains nine different ATPases, the most numerous of the remodeling families (Hall & Georgel, 2007). The best characterized complex in this family is NURD. NURD contains both ATP-dependent chromatin remodeling and histone deacetylase activities (Tong, Hassig, Schnitzler, Kingston, & Schreiber, 1998). A subset of the NURD complexes is unique in its ability to bind 5mC DNA through the MBD2 subunit (Hendrich & Bird, 1998). Once recruited by 5mC enriched DNA, the MBD2–NURD complex promotes the repression of genes through its remodeling and histone deacetylase activities. However, alternative NURD complexes exist that replace the MBD2 subunit for MBD3, decoupling NURD from 5mC recruitment, and promote its functions as an activator of transcription (Gunther et al., 2013). In addition to NURD, a NURD-like complex containing CHD5 has been identified in the brain and acts as a coregulator of transcription (Potts et al., 2011). Most of the complexes containing other CHD ATPases await characterization. As opposed to the large SWI/SNF, INO80, and CHD complexes, most ISWI complexes are comparatively small, consisting of only two or three subunits (Erdel & Rippe, 2011). At a minimum, each of these complexes contains a single large subunit with several histone-binding domains (including PHD and bromodomains) and an ISWI family ATPase. ISWI family chromatin remodeling complexes catalyze the sliding of nucleosomes in short increments without DNA looping, histone exchange, or nucleosome displacement. These complexes have diverse functions including the spacing of nucleosomes after DNA replication (CHRAC, ACF), RNA polymerase elongation (RSF), serving as coregulators of transcription (CERF, NURF, NoRC, b-WICH), and regulating DNA damage repair (WICH) (Banting et al., 2005; Barak et al., 2003; Cavellan, Asp, Percipalle, & Farrants, 2006; LeRoy, Loyola, Lane, & Reinberg, 2000; LeRoy, Orphanides, Lane, & Reinberg, 1998; Poot et al., 2000, 2004; Strohner et al., 2001).

Understanding how these complexes function *in vivo* continues to be an active area of basic research. Key to this endeavor includes understanding how they are recruited to discrete regions of the genome at distal regulatory elements, promoters, and sites of DNA damage. Recruitment commonly occurs through direct interactions with a combination of transcription factors, histone modifications, and DNA sequences. These interactions occur through domains on the surfaces of subunits in the complex, and they likely operate in a combinatorial fashion to confer the necessary affinity and specificity for meaningful recruitment (Lange, Demajo, Jain, & Di Croce, 2011). Domains found in subunits that recognize specific histone modifications include a variety of well-characterized PHD, chromo, and bromodomains, among others. How these domains bind modified histones has been reviewed extensively (Yun, Wu, Workman, & Li, 2011). The incorporation of individual subunits into chromatin remodeling complexes varies based on availability, thus providing a means to generate distinct complexes with unique interaction surfaces. The cell-type-restricted expression of BAF45 and BAF60 subunits from the SWI/SNF family of complexes in embryonic stem cells, neural progenitor cells, and more differentiated neurons presents the best example of this mechanism (Ho & Crabtree, 2010). Like that for histone modifications, the interactions between transcription factors and chromatin remodeling complexes are specific, and in many cases they have been shown to depend on the activation state of the transcription factor (ligand binding as an example) (e.g., see Badenhorst et al., 2005; Fujiki et al., 2005). While quite specific in nature, the interaction surfaces on the chromatin remodeling complexes that participate in these interactions are not as well defined as those that interact with modified histones.

The biological functions of chromatin remodeling complexes during normal development have been extensively studied using mouse models (Ho & Crabtree, 2010). As expected, the subunits of many chromatin remodeling complexes have essential functions during mammalian development. Translating these phenotypes into an understanding of how specific chromatin remodeling complexes function during development has remained a challenge. This challenge occurs largely because subunits of chromatin remodeling complexes are often found in several chromatin-associated complexes. As such, the deletion of a single subunit likely compromises several complexes resulting in pleotropic effects. As an example, RUVBL1 and RUVBL2 are utilized by three chromatin remodeling complexes and at least three other chromatin-associated complexes (Huen et al., 2010). However, studies have been done on several subunits for which all the evidence suggests that they are unique to a single chromatin remodeling complex. Examples include BPTF (ISWI family NURF complex) and INO80 (INO80 family INO80 complex) (Barak et al., 2003; Chen et al., 2011; Jin et al., 2005; Landry et al., 2008; Min et al., 2013). In somewhat of a surprise, several ATPases are not essential for the cell: SWI/SNF family members BRG1 and BRM, CHD family member CHD4, and ISWI family member SNF2L (Bultman et al., 2000; Reyes et al., 1998; Yip et al., 2012;

Yoshida et al., 2008). In other cases including the INO80 and p400 ATPases from the INO80 family and SNF2H ATPase from the ISWI family, the ATPase is important for cell viability (Fujii, Ueda, Nagata, & Fukunaga, 2010; Min et al., 2013; Stopka & Skoultchi, 2003). The importance of chromatin remodeling complexes for development, but not necessarily for cell viability, suggests that their mutation can result in a viable cell with abnormal developmental pathways. Because deregulated developmental pathways are known to contribute to the transition to cancer, it is reasonable to suggest that the mutation of subunits in chromatin remodeling complexes could contribute to cancer-related phenotypes (Hanahan & Weinberg, 2011; Plass et al., 2013).

4. EVIDENCE OF WIDESPREAD ROLES FOR CHROMATIN REMODELING IN HUMAN CANCER

Until recently, chromatin remodeling complexes were not thought to have widespread roles in the establishment and progression of human cancers. The recent availability of comprehensive genome-wide datasets, in the form of exon sequencing and genome-wide expression arrays for many human cancers, has provided an opportunity to discover novel connections between chromatin remodeling complexes and human cancer (Chang et al., 2013).

Mutation is thought to be the driving force behind the progression from normal tissue to advanced stage human cancer. Tumor-promoting mutations can be inherited, but they usually occur spontaneously in the form of somatic mutations that activate oncogenes and repress tumor suppressors in pathways important for tumor development. Frequently mutated genes are identified as driver mutations because of their inferred tumor-promoting potential (Kandoth et al., 2013; Tamborero et al., 2013). Driver mutations frequently occur in upstream components of signaling pathways (receptors), likely because of their potential for widespread effects on cell physiology. With less frequency, mutations in downstream components of signaling pathways (transducers) are identified as driving mutations. Included in this category are mutations in the subunits of chromatin remodeling complexes. As an exercise to quantify the mutation frequency for each chromatin remodeling complex subunit from a wide array of tumors, we mined the COSMIC database of somatic mutations in human cancers (Forbes et al., 2011). Consistent with what was published by others, we observed that subunits of the SWI/SNF family of remodeling complexes are the most prominently mutated of the chromatin remodeling complex families (Fig. 5.3; Table 5.1; Kadoch et al.,

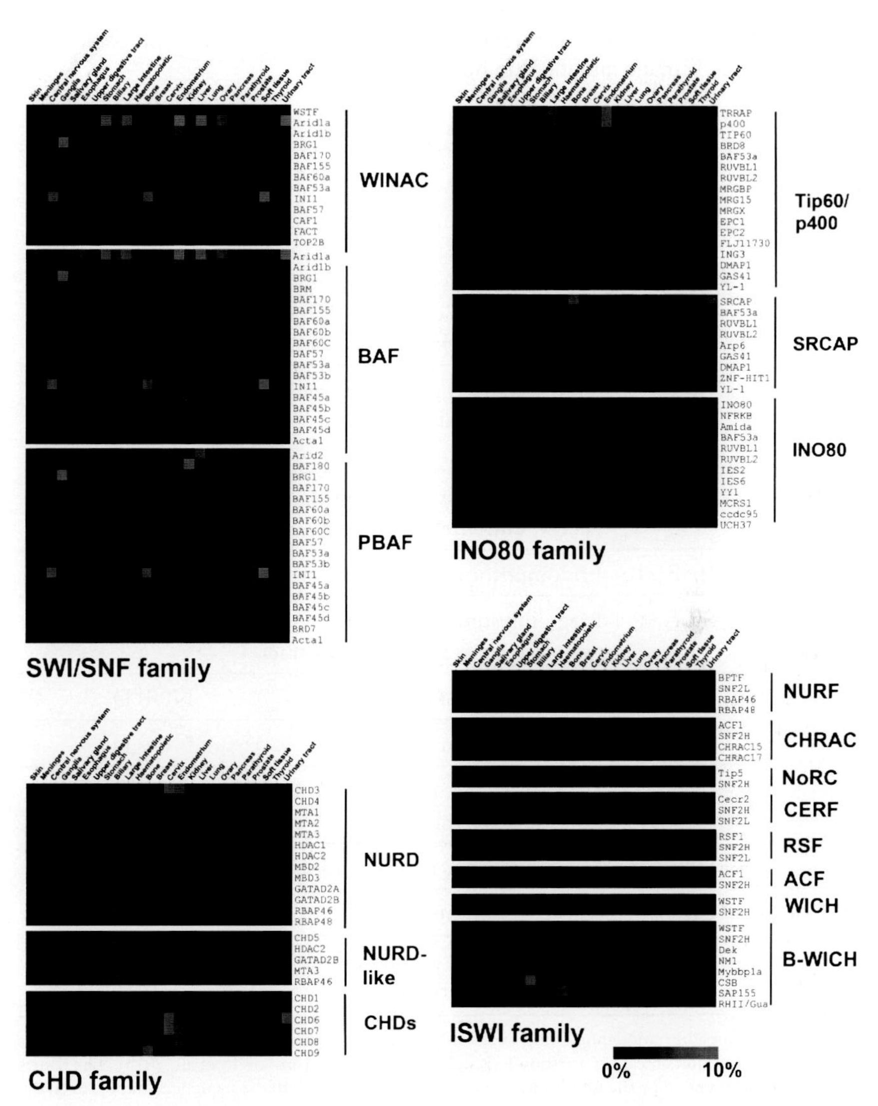

Figure 5.3 *Heat map representation of somatic mutation rate for subunits of chromatin remodeling complexes in human cancers.* The COSMIC database was queried using each of the subunits found in the chromatin remodeling complexes depicted in Fig. 5.2 (Forbes et al., 2011). The frequency of somatic mutation is expressed as a percentage of sequenced tumors from each tissue type. Percentages are not corrected for background mutation rates which vary with individual tumors and tumors from tissue types. The heat map range is shown on the lower right and varies from 0% to 10% of sequenced tumors in the COSMIC database. (See the color plate.)

Table 5.1 Summary of chromatin remodeling subunits identified with probable or potential driver somatic mutations in human tumors

Remodeling family	Subunit	Mutated cancer tissue types	Complexes
SWI/SNF family	WSTF	Large intestine, cervix, endometrium	WINAC, b-WICH, WICH
	ARID1A	Skin, esophagus, stomach, biliary tract, large intestine, cervix, endometrium, liver, ovary, urinary tract, lung	WINAC, BAF
	BRG1	Autonomic ganglia, esophagus, lung, urinary tract, large intestine, endometrium	WINAC, PBAF, BAF
	ARID2	Skin, esophagus, large intestine, cervix, endometrium, liver, lung	PBAF
	BAF180	Endometrium, kidney	PBAF
INO80 family	SRCAP	Large intestine, cervix, bone, endometrium, lung, urinary tract	SRCAP
CHD family	CHD3	Cervix, endometrium	NURD
	CHD4	Stomach, endometrium	NURD
	CHD7	Large intestine, cervix, endometrium, lung, urinary tract	CHD7
	CHD8	Large intestine, cervix, endometrium	CHD8
ISWI family	Tip5	Cervix, endometrium	NoRC
	SAP55	Hematopoietic and lymphoid tissue, endometrium, thyroid	b-WICH

Subunits of chromatin remodeling complexes whose somatic mutation was identified as a probable or potential driver of human cancer are shown (Kandoth et al., 2013; Tamborero et al., 2013). The tissues in which the gene encoding the subunit was mutated in greater than 5% of the sequenced tumors from the COSMIC database are shown. Known chromatin remodeling complexes that each subunit can assemble into are also listed.

2013; Shain & Pollack, 2013). Mutations in the INI1 (also known as SNF5) and ARID1A (also known as BAF250A) components are frequently observed in tumors from diverse tissues including central nervous system, stomach, large intestine, bone, endometrium, liver, soft tissue, ovary, and urinary tract. Additional SWI/SNF subunits include the BRG1 ATPase, which is frequently mutated in tumors from ganglia and to a lesser extent the esophagus, large intestine, lung, and urinary tract. The PBAF-specific BAF180 (also known as PBRM1) and ARID2 subunits are frequently mutated in kidney cancer and to a lesser extent in a variety of tumors from the skin, esophagus, large intestine, cervix, liver, and lung. Interestingly, several of the relatively uncharacterized CHD ATPases were found modestly mutated in a variety of cancers including the stomach, large intestine, cervix, endometrium, lung, and urinary tract.

In addition to somatic mutation rates, the abnormal expression of genes in cancers compared to their normal tissues of origin can indicate roles in tumor development. While changes in gene expression are not considered driving events, they can provide necessary changes to cellular physiology that support cancer phenotypes. Toward this end, we surveyed genome-wide expression datasets from primary tumors using ONCOMINE to determine the frequency of deregulation for each subunit for chromatin remodeling complexes (Fig. 5.4; Table 5.2) (Rhodes et al., 2004). Consistent with their frequent somatic mutation in tumor samples, SWI/SNF family members are frequently deregulated in expression in tumors, thus underscoring their likely importance to cancer biology. The BRG1 ATPase, BAF155, BAF53A, and INI1 are overexpressed in a variety of tumor types including bladder, liver, ovarian, melanoma, leukemia, and myeloma cancers. In addition to the prominent representation of SWI/SNF subunits, subunits from the ISWI, CHD, and INO80 complexes are also deregulated. Subunits of the INO80 family of chromatin remodeling complexes including BAF53A, RUVBL1, and RUVBL2 subunits are overexpressed in bladder, cervix, myeloma, colon, and ovarian cancers. The NURD subunits MTA1 (but not MTA2 or 3), HDAC1 and 2, and pRBAP48 are modestly deregulated in cervix, pancreatic, leukemia, colon, and esophagus cancers. Compared to the other families of remodeling complexes, the ISWI families are only modestly deregulated in cancers including overexpression in head, cervix, and kidney cancers.

The combination of somatic mutation frequency and deregulated expression of the subunits of chromatin remodeling complexes highlights their widespread potential for functional roles in human cancer. Just like characterizing their knockout phenotypes during mouse development, the molecular consequences of mutation or deregulated expression for some of these subunits are complicated by the fact that many are found in multiple protein complexes. In fact, many subunits from the SWI/SNF family, which are frequently mutated or deregulated in expression, are incorporated into several complexes (Tables 5.1 and 5.2). This is less of a problem for the CHD and the INO80 family because many mutated and deregulated subunits are only known to assemble into a single chromatin remodeling complex. However, it must be emphasized that many of these subunits

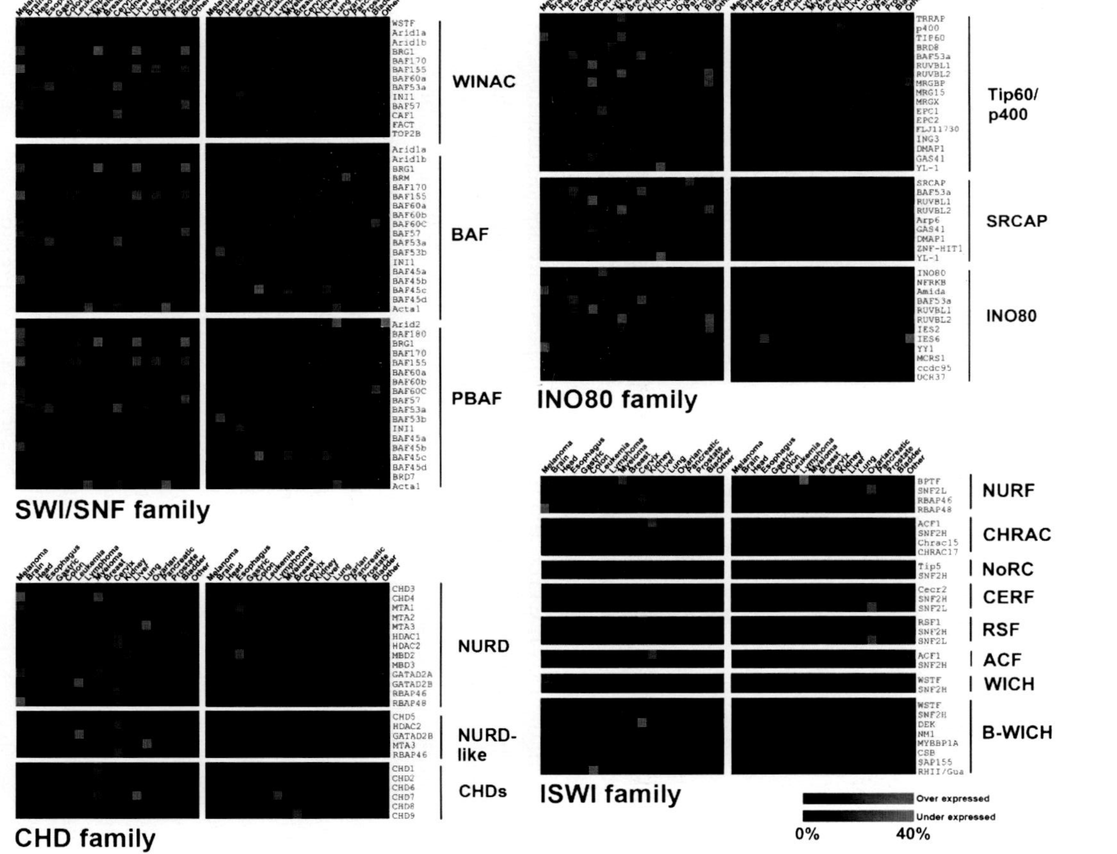

Figure 5.4 *Heat map representation of gene expression changes for subunits of chromatin remodeling complexes in human cancers.* The ONCOMINE database was queried using each of the subunits found in the chromatin remodeling complexes depicted in Fig. 5.2 (Rhodes et al., 2004). Data is expressed as a percentage of unique analyses for which transcript levels of the subunit differ by greater than twofold between tumor and control normal tissue (P value $\leq 1.0 \times 10^{-4}$). Transcripts for which overexpression was observed in tumor tissue are colored in red, and those where an underexpression is observed are shown in green. The heat map range is shown on the lower right and varies from 0% to 40% of unique analyses found in the ONCOMINE database. (See the color plate.)

Table 5.2 Summary of chromatin remodeling subunits identified with frequent deregulated expression in human tumors

Remodeling family	Subunit	Overexpressed	Underexpressed	Complexes
SWI/SNF family	CAF1	Cervix		WINAC
	BAF180	Melanoma		PBAF
	Arid2		Lung	PBAF
	Acta1	Lymphoma, pancreatic		PBAF, BAF
	BAF45b	Melanoma		PBAF, BAF
	BAF53b		Brain	PBAF, BAF
	BAF45c		Colon	PBAF, BAF
	BAF155	Melanoma, liver		WINAC, PBAF, BAF
	BAF57	Bladder		WINAC, PBAF, BAF
	BRG1	Melanoma, myeloma, liver, bladder		WINAC, PBAF, BAF
	BRM		Ovarian	WINAC, PBAF, BAF
	BAF53a	Cervix		WINAC, PBAF, BAF, Tip60/p400, SRCAP INO80
INO80 family	SRCAP	Pancreatic		SRCAP
	Amida	Melanoma		INO80
	IES2	Bladder		INO80
	IES6		Esophagus	INO80
	INO80	Leukemia		INO80
	YY1	Melanoma		INO80
	EPC1	Leukemia		Tip60/p400
	MRGBP	Colon, bladder		Tip60/p400
	TIP60	Melanoma, myeloma		Tip60/p400

Continued

Table 5.2 Summary of chromatin remodeling subunits identified with frequent deregulated expression in human tumors—cont'd

Remodeling family	Subunit	Overexpressed	Underexpressed	Complexes
	TRRAP	Myeloma		Tip60/p400
	YL-1	Liver		Tip60/p400, SRCAP
	RUVBL2	Myeloma, bladder		INO80, Tip60/p400, SRCAP
	RUVBL1	Colon		INO80, Tip60/p400, SRCAP
CHD family	CHD4	Melanoma, myeloma		NURD
	MTA3	Lung		NURD
	MBD2		Esophagus	NURD
	GATAD2B	Leukemia		NURD-Like
	CHD7	Liver	Lymphoma	CHD7
ISWI family	BPTF	Myeloma	Lymphoma	NURF
	DEK	Cervix		b-WICH
	DDX21	Colon		b-WICH
	ACF1	Kidney		CHRAC, ACF
	RBAP48	Melanoma		NURF, NURD
	SNF2L		Ovarian	NURF, RSF, CERF

Subunits of chromatin remodeling complexes whose expression is deregulated in at least 20% of the unique analyses from each tissue type from Fig. 5.4 are listed. Known chromatin remodeling complexes that each subunit can assemble into are also listed.

are found in other nuclear complexes. For example, the Tip60 and TRRAP subunits are found in several histone acetyltransferase complexes (Murr, Vaissiere, Sawan, Shukla, & Herceg, 2007).

In support of these genome-wide analyses, independent reports in the literature have accumulated for several decades and have been used to establish correlative (germ line and somatic mutation, deregulated expression, copy number changes), causative (mouse models), and mechanistic (physical and functional connections to cancer relevant pathways) connections between chromatin remodeling complexes and cancers. Using the significance of genome-wide datasets as a backdrop, we summarize this body of literature, highlighting studies contributing mechanistic insight to the roles for chromatin remodeling complexes in pathways of cancer.

5. A REVIEW OF THE LITERATURE ON CHROMATIN REMODELING AND CANCER

5.1. SWI/SNF family of chromatin remodeling complexes

5.1.1 BRG1/BRM

The catalytic subunits of the SWI/SNF family of chromatin remodeling complexes are the BRG1 and BRM ATPases. Targeted studies reported in the literature from a variety of tumor types observed sporadic mutation of either BRG1 or BRM in human primary tumors (Endo et al., 2013; Gunduz et al., 2005; Medina et al., 2004; Rodriguez-Nieto et al., 2011; Valdman et al., 2003). Because many of the mutations found in primary tumors are missense and heterozygous, their effect on the protein function is unknown. A recent report showed that two BRG1 missense mutations found in human cancers resulted in reduced ATPase activity and increased genome instability, suggesting that at least some of these missense mutations could be detrimental to BRG1 function (Dykhuizen et al., 2013). In contrast to tumors, BRG1 and BRM are frequently mutated in cell lines from several tumor sources including lung, prostate, breast, pancreas, and colon (Decristofaro et al., 2001; Medina et al., 2008; Reisman, Sciarrotta, Wang, Funkhouser, & Weissman, 2003; Wong et al., 2000). Many of these mutations are homozygous truncating or nonsense mutations resulting in an inactive protein. The combination of these results suggest that growth of cancer cells in tissue culture, as opposed to the primary tumor, more frequently selects for the loss of BRG1 and BRM activity.

BRG1 has well-characterized roles in regulating several tumor suppressor and oncogene pathways. BRG1 tumor suppressing roles include directly interacting with and stabilizing p53. These interactions likely recruit BRG1 to the p21 promoter to regulate its expression during DNA damage (Naidu, Love, Imbalzano, Grossman, & Androphy, 2009). In addition to p21, BRG1 is essential for the expression of BRCA1, another known p53 regulated gene (Bochar et al., 2000). BRG1 also directly interacts with BRCA1 but is not necessary for its functions during homologous recombination (Hill, de la Serna, Veal, & Imbalzano, 2004). In addition to its role as a coregulator of p53, BRG1 is known to directly interact with RB to regulate cell cycle arrest (Bartlett, Orvis, Rosson, & Weissman, 2011; Dunaief et al., 1994). Subsequent studies showed that BRG1 regulation of E2F2 requires RB and that the recruitment of BRG1 to the E2F2 promoter requires the RB interacting protein, prohibitin (Wang, Zhang, & Faller, 2002). BRG1 also interacts with other cell cycle regulators including LKB1 (where LKB1 is important for BRG1-dependent growth arrest) and MYC (where it is important for its transactivation activity) (Marignani, Kanai, & Carpenter, 2001; Romero et al., 2012). BRG1 can also regulate Cyclin D1 expression through less well-characterized interactions (Rao et al., 2008). In addition to regulating the cell cycle, BRG1 regulates genes important for metastasis. In these functions, BRG1 interacts with the ZEB1 transcription factor to regulate E-cadherin expression and regulates CD44, a metastasis-associated gene, by an unknown mechanism (Sanchez-Tillo et al., 2010; Strobeck et al., 2001).

Mouse models have demonstrated that BRG1, but not BRM, has tumor suppressing activity. The increased tumor incidences observed in heterozygous BRG1 knockout mice demonstrates that BRG1 haploinsufficiency can promote tumor development (Bultman et al., 2000; Bultman et al., 2008). Because the rate of tumor incidence was not increased by a BRM homozygous knockout background demonstrates that BRG1 is more important of the two SWI/SNF ATPases for tumor development. It is important to note that a functional BRG1 allele is retained in BRG1 heterozygous tumors suggesting that while haploinsufficiency of BRG1 can promote tumor development, there is a selection against loss of heterozygosity (LOH) at BRG1 (Bultman et al., 2008).

5.1.2 INI1

The first clear connection between chromatin remodeling complexes and human cancer was observed in tumors from patients with childhood rhabdoid tumors. These tumors frequently arise in soft tissues of children (brain, kidney, and other soft tissues of the body), are aggressive and poorly responsive to therapies, and as a result are frequently lethal (Haas, Palmer, Weinberg, & Beckwith, 1981). In almost all cases, rhabdoid tumors acquire a somatic mutation in one INI1 allele, followed by a LOH event that inactivates a second functional INI1 allele. There is little other evidence of genome instability or mutation in these tumors (Lee et al., 2012). In more rare circumstances, familial cases can occur as the result of inheriting a defective INI1 allele and acquiring a LOH event at the functional INI1 allele (Biegel et al., 1999, 2000; Sevenet et al., 1999).

The use of mouse models has presented strong evidence that INI1 is a tumor suppressor. As in humans, mice heterozygous for INI1 develop sarcomas closely resembling the human malignant rhabdoid tumors (Klochendler-Yeivin et al., 2000; Roberts, Galusha, McMenamin, Fletcher, & Orkin, 2000). In mice, as in humans, tumors arise by inactivating the second functional INI1 allele through an LOH event (Klochendler-Yeivin et al., 2000). When allele inversion is used to create biallelic inactivation of INI1 in a mosaic of cells, the result is a fully penetrant cancer phenotype inducing rapid onset lymphomas and rhabdoid tumors (Roberts, Leroux, Fleming, & Orkin, 2002). Studies using these mouse models have identified several genes that influence INI1 tumor incidence. Because tumor onset depends on BRG1, aberrant remodeling activity from one or more SWI/SNF family complexes appears to be necessary to promote tumor development. Tumor onset also requires the Cyclin D1 oncogene, a positive regulator of the G1/S-phase transition, suggesting that tumor development promotes a defective G1/S-phase check point (Tsikitis, Zhang, Edelman, Zagzag, & Kalpana, 2005). For less clear reasons, tumor onset requires the polycomb subunit EZH2 (Wilson et al., 2010). The requirement of EZH2 for tumor onset is likely because polycomb promotes protumor gene expression in the absence of INI1. Polycomb and SWI/SNF complexes are known to oppose each other during gene expression in a number of model systems (Francis, Saurin, Shao, & Kingston, 2001; Kennison & Tamkun, 1988; Kia, Gorski, Giannakopoulos, & Verrijzer, 2008). In addition, the cell cycle regulator p53, but not p16Ink4a or RB, inhibits tumor onset suggesting that INI1 loss of function (LOF) triggers p53-dependent cell cycle checkpoints (Isakoff et al., 2005).

In addition to *in vivo* experiments with the mouse, further mechanistic insight into INI1 function has been obtained using tissue culture. The knockout of INI1 in wild-type MEFs results in G2 cell cycle arrest and apoptosis. Correlated with these findings, INI1 knockout regulates the E2F target gene p16Ink4a and upregulates the expression of Cyclin D1, p53, and p21 (Isakoff et al., 2005). Initial work reintroducing INI1 in knockout tumor cells showed that it regulates mitotic checkpoints and chromosome stability via the p16-Cyclin D1/CDK4-pRB-E2F pathway (Vries et al., 2005). Subsequent work reintroducing INI1 into knockout MEFs resulted in increased p21 and p16 expression and G1 arrest. The study showed that INI1 directly interacts with the p21 and p16INK4a promoters, and the study showed that the upregulation of these genes is essential for G1 arrest (Kuwahara, Charboneau, Knudsen, & Weissman, 2010). However, some of these cell cycle defects could be a consequence of INI1 functioning as an essential coregulator of MYC (Cheng et al., 1999). In addition to the cell cycle regulators, INI1 directly regulates the expression of the proapoptotic factor NOXA, possibly contributing to the apoptosis observed with INI1 knockout (Kuwahara, Wei, Durand, & Weissman, 2013).

5.1.3 ARID1A

The SWI/SNF subunit ARID1A (also known as BAF250A) is frequently mutated in several types of human cancer. In some tumor types, AIRID1A is the most frequently mutated gene. For example, AIRID1A mutations approach 50% of the sequenced ovarian clear cell tumors and 30% of sequenced ovarian endometrioid cancers (Jones et al., 2010; Wiegand et al., 2010). In contrast to BRG1 mutations that are frequently missense mutations, many ARID1A mutations may result in LOF gene products as they are nonsense or truncating mutations (Jones et al., 2012). When an ARID1A transgene is used to restore function in ovarian cancer cell lines with ARID1A LOF, the transgene expression suppresses cell line growth both in culture and in mice. In follow-up experiments, this same group showed that ARID1A regulates the expression of the p53 target genes CDKN1A (p21) and SMAD3, and at least part of the suppression of tumor growth by ARID1A involves p21 upregulation. The function of ARID1A in these pathways could be through a direct interaction with p53, an interaction that recruits a SWI/SNF complex to p53 target gene promoters (Guan, Wang, & Shih Ie, 2011). Similar effects were observed in gastric cancer cell lines where expression of ARID1A in cell lines with ARID1A LOF resulted in reduced growth (Zang et al., 2012). In an earlier study, knockdown of ARID1A, but not ARID1B, in immortalized preosteoblast cell line resulted in reduced differentiation and cell cycle arrest. The cell cycle arrest with ARID1A knockdown occurred in the absence of increased p21 expression and downregulation of cyclins (Nagl et al., 2005). These results demonstrate that ARID1A containing complexes are functionally distinct from ARID1B containing complexes with regard to regulation of the cell cycle. This difference could be through differential interactions between ARID1A and ARID1B containing SWI/SNF complexes for E2F transcription factors which are key regulators of the cell cycle (Nagl, Wang, Patsialou, Van Scoy, & Moran, 2007). How mutations in ARID1A can promote the development of tumors *de novo* is largely unknown.

Work using mouse models has not yet shown ARID1A to be involved in cancer. It is known that ARID1A is important for differentiation of ESC

(Gao et al., 2008). An ENU mutagenesis screen has identified a point mutant in ARID1A which prevents the SWI/SNF complex from binding to DNA; this mutation is associated with neural tube closure, heart defects, and extra embryonic vasculature that likely occur through an inability to regulate gene expression (Chandler et al., 2013).

5.1.4 BAF180

At a reduced frequency compared to ARID1A, truncating mutations in BAF180 have been identified in a high percentage of renal cell carcinomas suggesting that it functions as a tumor suppressor in this cancer type (Cancer Genome Atlas Research Network, 2013). Supporting its role as a tumor suppressor, the loss of expression of the BAF180 protein has been confirmed to be a poor prognosis event in renal cell cancer (Pawlowski et al., 2013). LOH events appear to be specific to mutations as promoter hypermethylation of BAF180 does not occur (Ibragimova, Maradeo, Dulaimi, & Cairns, 2013). Like ARID1A, BAF180 is an essential regulator of the p53 target gene p21, likely through direct interactions with p53 and functioning to inhibit the cell cycle and promote senescence (Burrows, Smogorzewska, & Elledge, 2010; Xia et al., 2008). Possibly as a result of its interactions with p53, BAF180 has also been shown to regulate genome stability through functions in regulating DNA damage repair (Niimi, Chambers, Downs, & Lehmann, 2012).

5.1.5 BAF53/BAF57

Both BAF53 and BAF57 are frequently deregulated in several cancer cell types with possible significance to cancer biology. BAF53 is a common subunit to several chromatin remodeling complexes from both the SWI/SNF and INO80 families. In addition to its functions in these complexes, it copurifies with MYC in a complex containing RUVBL1, RUVBL2, and actin. In this same study, the overexpression of BAF53 with deletions in several functional domains has a dominant negative effect on MYC transformation suggesting that these domains are important for this function (Park, Wood, & Cole, 2002). BAF53 is also necessary for the expression and transformation capability of the human papilloma virus E6 and E7 oncogenes. In this function, BAF53 is thought to regulate a unique chromatin structure established at the E6/E7 oncogene integration site and to consequently regulate their expression (Lee, Lee, Kwon, & Kwon, 2011).

Like BAF53, BAF57 is found in each of the SWI/SNF chromatin remodeling complexes. Knockdown of BAF57 in Hela cells results in G2/M arrest, reduced colony formation, and impaired growth in soft agar.

Genome-wide microarray experiments in these cells showed that BAF57 knockdown results in the deregulation of genes (e.g., CCNB1, CENPE, CENPF, and MYC) involved in regulating the G2/M transition (Hah et al., 2010). In prostate cancer cell lines, BAF57 is recruited to androgen receptor (AR)-binding sites in a ligand-dependent manner and regulates many AR gene targets (Link et al., 2005). Through its interactions with AR, BAF57 is thought to regulate key aspects of prostate cancer cell biology. In prostate cancer cell lines, BAF57 overexpression increases cell motility in an alpha 2 integrin-dependent manner and is required for AR-dependent proliferation (Balasubramaniam et al., 2013). The interaction surface on the AR that is important for BAF57 binding was mapped to the DNA-binding domain-hinge region (Link et al., 2008). The function of BAF57 as a coregulator of estrogen receptor alpha (ERα) in breast cancer parallels closely those of the AR in prostate cancer. Like AR, BAF57 directly interacts with the ERα receptor hinge region, and this interaction is stimulated by ligand (Garcia-Pedrero, Kiskinis, Parker, & Belandia, 2006). Also as in AR, BAF57 is required for ERα regulated transcription. Interestingly, the ectopic expression of BAF57 truncation mutants cloned from human breast cancer cells abnormally activates both AR and ERα reporter constructs (Villaronga et al., 2011). These results suggest that BAF57 truncation mutations found in human cancers lead to artificially elevated ER- and AR-regulated gene expression and provide a plausible molecular mechanism for their functions in regulating prostate and breast cancer cell biology.

5.2. INO80 family of chromatin remodeling complexes

5.2.1 TRAAP/p400

TRRAP and p400 are essential subunits of the Tip60/p400 chromatin remodeling complex (Cai et al., 2003). Recently, TRRAP was discovered in an RNAi screen as an essential factor for brain tumor-initiating cellular differentiation (Wurdak et al., 2010). TRRAP knockdown in both tissue culture and mouse tumor models resulted in increased tumor cell differentiation, reduced cell proliferation, and increased sensitivity to apoptotic stimuli. TRRAP has well-documented functions in several pathways relevant to human cancer. TRRAP was originally shown to interact with the MYC and E2F oncogenes, and likely through these direct interactions TRRAP is essential for MYC and E1A transformation (Deleu, Shellard, Alevizopoulos, Amati, & Land, 2001; McMahon, Van Buskirk, Dugan, Copeland, & Cole, 1998). Later, it was discovered that the TRRAP-containing complex essential for E1A transformation contains p400, and that the p400 subunit is also necessary for the E1A transformation (Fuchs et al., 2001). Functional connections between TRRAP and MYC were later reinforced using shRNA screens which again identified TRRAP as an essential factor for MYC transformation activity (Toyoshima et al., 2012). In addition to its functions as a coregulator of MYC and E1A, TRRAP also interacts with acetylated p53 and is essential for p53 to regulate gene expression (Ard et al., 2002; Barlev et al., 2001). In a somewhat surprising role, TRRAP influences β-catenin stability by regulating β-catenin interactions with SKP1/SCF, an E2/E3 ubiquitin ligase complex important for the degradation of β-catenin (Finkbeiner, Sawan, Ouzounova, Murr, & Herceg, 2008). TRRAP regulated stability of β-catenin in turn influences canonical Wnt signaling, a pathway frequently deregulated in human cancers (Clevers, 2006; Polakis, 2000).

In addition to its role as a regulator of transcription, the Tip60/p400 complex has important functions in repairing DNA damage, and through these functions the Tip60/p400 complex could contribute to cancer (Xu & Price, 2011). Defects in DNA damage repair pathways are known to contribute to the progression of cancer by promoting somatic mutations and genome instability (Luijsterburg & van Attikum, 2011). The TRRAP-containing Tip60/p400 complex is recruited to DNA breaks in mammalian cells; once recruited, the Tip60/p400 complex acetylates histones at sites of DNA damage (Ikura et al., 2000). p400 then remodels chromatin by loosening the chromatin structure and promoting access for repair factors like 53BP1, RAD51, and BRCA1 (Murr et al., 2006; Xu et al., 2010).

5.2.2 INO80

While none of the INO80 remodeling complex subunits are significantly mutated in human cancers, they do have functions in telomere stability and DNA damage repair pathways. Through its interactions with the YY1 transcription factor, INO80 has essential functions in homologous recombination repair pathways (Wu et al., 2007). At sites of DNA breaks, INO80 is important for the recruitment of 53BP1 and the 5′-3′ resection of the DNA double-strand break ends (Gospodinov et al., 2011). At least some of these functions are believed to be through the ability of the INO80 complex to regulate gene expression, because the expression of DNA damage repair genes, including RAD54B and XRCC3, are INO80 dependent (Park, Hur, & Kwon, 2010).

Conditional knockout of INO80 in mouse embryonic fibroblasts inhibits cellular proliferation and results in p21-dependent cellular senescence. The decrease in proliferation was not rescued by expressing large

T antigen suggesting that its regulation of growth is not p53 or RB dependent. Knockout cells had increased single-stranded DNA, phosphorylation of CHK1, and defects in telomere structure. INO80 is important for the generation of single-stranded DNA at the telomeres resulting in defects in homology driven DNA repair. INO80 heterozygous knockout mice are not tumor prone, but they do have an increased incidence of soft tissue sarcomas compared to INO80 wild-type mice on a p53 knockout background (Min et al., 2013).

5.2.3 RUVBL1/RUVBL2

The INO80 family members RUVBL1 and RUVBL2 are overexpressed in cancers from many different tissue types (see references in Grigoletto, Lestienne, & Rosenbaum, 2011). Each of the RUVBL proteins directly interacts with the MYC, β-catenin, and E2F1 oncogenes (Bauer et al., 2000; Dugan, Wood, & Cole, 2002; Wood, McMahon, & Cole, 2000). These physical interactions could be significant to the observation that RUVBL1 is necessary for transformation by E1A, β-catenin, and MYC (Dugan et al., 2002; Feng, Lee, & Fearon, 2003; Wood et al., 2000). As a consequence of these interactions, RUVBL proteins regulate the expression of TERT, p21, and KAI-1, which could play a significant role in cancer cell growth. Inducible knockdown of RUVBL2 in tissue culture and xenograft animal tumor models shows that RUVBL2 is essential for growth of several cancer cell types (Menard et al., 2010; Schlabach et al., 2008). The mechanism of these effects could be in part through functions in INO80 family chromatin remodeling complexes; however, the cellular consequences of a RUVBL2 knockdown are likely to be pleotropic due to its widespread use in many signal transduction pathways including mTOR, SMG-1, ATM, ATR, DNA-PKcs to regulate translation, energy metabolism, mRNA decay, and the DNA damage response (Rosenbaum et al., 2013).

5.3. CHD family of chromatin remodeling complexes

5.3.1 CHD3/CHD4

CHD3 and CHD4 are the essential ATPase subunits of the NURD complex. The CHD3 and CHD4 genes themselves are not frequently mutated in human cancers, with the exception that some increased incidence was observed in cancers of the cervix, stomach, and endometrium (Fig. 5.3; Table 5.1). The CHD3/4 remodeling subunits directly interact with NAB2, a corepressor of the EGR family of transcription factors (Srinivasan, Mager, Ward, Mayer, & Svaren, 2006). In prostate cancer, in part by regulating the expression of IGF2, TGFB1, and PDGFA (each of which have roles in tumor progression), EGR1 regulates cell growth, differentiation, and apoptosis (Gitenay & Baron, 2009). In addition to its interactions with NAB2, CHD4 has been shown to interact with ZIP, a transcriptional repressor of genes involved in cell proliferation, survival, and migration (Li, Zhang, et al., 2009).

The CHD4 ATPase, as a component of the NURD remodeling complex, has been shown by several groups to be important for DNA damage repair. Knockdown experiments have shown that CHD4 is essential for the repair of breaks and the survival of the cell after DNA damage (Smeenk et al., 2010). The CHD4 containing NURD complex is recruited to sites of DNA damage by poly ADP-ribose, and once recruited CHD4 is phosphorylated by the ATM kinase. As part of this process, the authors have shown that NURD is important for the deacetylation of p53 to control the G1/S checkpoint through regulated expression of p21, likely through its HDAC1 and HDAC2 subunits (Polo, Kaidi, Baskcomb, Galanty, & Jackson, 2010). At the sites of DNA damage, CHD4 promotes ubiquitination of the BRCA1 complex, possibly through direct interactions with BRCA1, to promote the assembly of the BRCA complex (Larsen et al., 2010; Pan et al., 2012). The inability to assemble functional repair complexes results in elevated ionizing radiation-induced DNA breakage, reduced efficiency of DNA repair, G2/M arrest, and decreased survival.

5.3.2 MTA1

MTA1 is widely reported to be overexpressed in a variety of human cancers and in metastatic tumor cells (especially breast cancer) (see references in Li, Pakala, Nair, Eswaran, & Kumar, 2012). In contrast to its frequent overexpression, there is little evidence that it is significantly mutated in human cancers. Transformation assays have shown that MTA1 is required for MYC-induced transformation (Zhang et al., 2005). Using similar assays, MTA1 overexpression promotes transformation by downregulating Gi12 alpha, a negative regulator of the G protein activation cycle (Ohshiro et al., 2010). The repression of Gi12 alpha results in constitutively active G proteins and as a result, a hyper-stimulation of the mitogenic and procell invasion Ras–Raf pathway. In addition to its roles in promoting MAPK signaling, MTA1 is also a positive regulator of the canonical Wnt signaling pathway (Kumar, Balasenthil, Manavathi, Rayala & Pakala, 2010; Kumar, Balasenthil, Pakala, et al., 2010). MTA1 stimulates the expression of the Wnt1 gene, a ligand of the canonical Wnt signaling pathway, through the repression of the Six3 repressor. The importance of MTA1 for MYC and MAPK-dependent transformation and for Wnt signaling supports its proposed function as a tumor promoter.

Beyond its functions in promoting protumor pathways, MTA1 overexpression also represses tumor suppressor pathways. MTA1 is recruited to the ERα, BRAC1, p21, HIC1, and p19ARF promoters to repress gene expression as a subunit of the NURD complex (Fu et al., 2012; Li et al., 2010, 2011; Marzook et al., 2012; Molli, Singh, Lee, & Kumar, 2008; Van Rechem et al., 2010). Significant mechanistic insight into how MTA1 regulates these promoters was reported recently. MTA1 is methylated by the G9a methyltransferase and demethylated by the LSD1 demethylase (Nair, Li, & Kumar, 2013). The methylation state of MTA1 determines whether it will recruit repressive NURD (methylated MTA1) or activating NURF chromatin remodeling activities (demethylated MTA1). In most cases, MTA1 is recruited to promoters through interactions with transcription factors. For example, MTA1 interacts with the BCL11B repressor in T cell leukemia cells and ERα in breast cancer cells to regulate gene expression (Cismasiu et al., 2005; Mazumdar et al., 2001). Genes important to metastasis are also known targets of MTA1 regulatory activity. MTA1, recruited by c/EBPα to the RNF144A promoter and by TWIST to the E-cadherin promoter, silences the genes and in doing so promotes metastasis (Fu et al., 2011; Marzook et al., 2012). Repression of BRCA1 by MTA1 promotes anchorage-independent invasive growth and metastasis (Molli et al., 2008). The ability of MTA1 to directly interact with p53 provides a mechanism for regulating the p53 target genes p21 and HIC1 (Luo, Su, Chen, Shiloh, & Gu, 2000; Moon, Cheon, & Lee, 2007).

Beyond acting as a coregulator of transcription, there is evidence for MTA1 using posttranscription mechanisms to promote tumor progression. MTA1 overexpression stabilizes p53 independently of the NURD complex by acting as a competing substrate for the COP1 E3 ligase. The stabilization of p53 by MTA1 overexpression promotes the repair of DNA damage in cancer cells (Li, Divijendra Natha Reddy, et al., 2009; Li, Ohshiro, et al., 2009). By recruiting HDAC1, MTA1 stabilizes HIF1α. HIF1α deacetylation by HDAC1 stabilizes the transcription factor and promotes hypoxia-stimulated gene expression, which includes proangiogenesis genes (Moon et al., 2006; Yoo, Kong, & Lee, 2006).

The mouse models used to investigate the functions of MTA1 in cancer have supported much of the work performed in tissue culture.

Overexpression of MTA1 in mouse breast epithelial cells promotes ductal hyperbranching, hyperplasic nodules, and eventually adenocarcinoma development suggesting that it has a causal role in the development of breast cancer (Bagheri-Yarmand, Talukder, Wang, Vadlamudi, & Kumar, 2004). The increase in cell proliferation with MTA1 overexpression in breast epithelial cells results in stimulation of the Wnt pathway, a known positive regulator of the cell cycle (Kumar, Balasenthil, Pakala, et al., 2010). Overexpression of MTA1 using a transgene results in an increase in B-cell lymphomas through the downregulation of p27 and the upregulation of Bcl2 and Cyclin D1 (Bagheri-Yarmand et al., 2007). These results have relevance to humans as MTA1 is frequently overexpressed in human B-cell lymphomas.

5.3.3 HDAC1/HDAC2

Like the MTAs, HDAC1 and HDAC2 are frequently overexpressed but rarely mutated in human cancers. HDAC1 and HDAC2 are known to function as regulators of the cell cycle, angiogenesis, apoptosis, differentiation, and metastasis (Hagelkruys, Sawicka, Rennmayr, & Seiser, 2011). Understanding the context through which HDAC1 and HDAC2 regulate these pathways is complicated by the fact that they are found in many complexes, only a few of which have chromatin remodeling activities (NURD, NURD-like, NoRC). Relevant to its roles in chromatin remodeling complexes, HDAC1 and HDAC2 repress the p21 promoter, likely in context with the NURD complex to regulate the cell cycle (Zupkovitz et al., 2010). Consistent with these observations, knockdown of HDAC1 and HDAC2 in several human cancer cell lines results in upregulation of p21 followed by cell cycle arrest both in tissue culture and in xenograft tumor models (see references in Hagelkruys et al., 2011). In addition to its roles in regulating p21 expression, HDAC1 and HDAC2 play prominent roles in promoting the activity of oncogenic-DNA-binding fusion proteins in hematologic malignancies. In many cases, these fusion proteins bind to promoters of genes to recruit HDAC1 and HDAC2 and repress genes with tumor suppressor activity. NURD directly interacts with the PML–RARα fusion product in human acute promyelocytic leukemias (Morey et al., 2008). PML–RARα recruits NURD with its HDAC activities to the tumor suppressor gene RARB2, promoting its repression.

5.3.4 MBD2

MBD2 is infrequently mutated in human cancers and is not significantly deregulated in expression. It does have important functions in the development of cancers through its ability to recruit the coregulatory activities of the NURD complex to the hypermethylated promoters of tumor suppressor genes. This mechanism is well characterized in several cancer types including colon, lung, prostate, and renal cancers and with a variety of genes including CDKN2A (p21), BubR1, beta-defensin-2, 14-3-3 sigma, and BTG3 tumor suppressors (Magdinier & Wolffe, 2001; Majid et al., 2009; Park et al., 2007; Pulukuri & Rao, 2006; Shukeir, Pakneshan, Chen, Szyf, & Rabbani, 2006; Zhuravel, Shestakova, Glushko, Soldatkina, & Pogrebnoy, 2008). These data suggest the cooperation between DNMTs, MBD2, and HDACs represses tumor suppressor genes.

5.3.5 pRBAP48

pRBAP46/48 proteins are members of a family of WD repeat proteins, proteins which have a highly conserved propeller-like structure. The NURD subunit pRBAP48 is frequently overexpressed in a variety of human cancers (Fig. 5.4). As with several of the other subunits discussed in this review, it is difficult to understand the consequences of its overexpression because pRBAP48 is a subunit of several complexes including histone chaperones, polycomb complexes, the RB family of complexes, and chromatin remodeling complexes like NURD and NURF (Migliori, Mapelli, & Guccione, 2012). When RBAP46 is overexpressed in breast cancer cell lines both colony formation *in vitro* and tumor growth in mice is suppressed (Li, Guan, & Wang, 2003). In addition to these studies, pRBAP48 is a key regulator of the transformation ability of the HPV16 transcription factor in cervical cancer (Kong et al., 2007). The nature of the increased transformation ability is likely through deregulation of the RB and p53 complexes and not through chromatin remodeling.

5.3.6 CHD5

The CHD5 gene is found on chromosome 1q36. This chromosome interval is the most common deletion found in a variety of human cancers, most prominently cancers of the CNS, hematopoietic system, and epithelium (thyroid, colon, cervix, and breast). In an elegant series of chromosome engineering experiments, the CHD5 gene located in this interval was identified as a tumor suppressor (Bagchi et al., 2007). However, in addition to CHD5, several other laboratories have identified additional tumor suppressors in the 1q36 chromosome interval (Henrich, Schwab, & Westermann, 2012). The presence of several tumor suppressors at 1q36 means that the deletion of this interval can result in the loss of several divergent tumor suppressor pathways. Mechanistically, CHD5 tumor suppressor activity was found to be a positive regulator of the p19ARF/p53 pathway (Bagchi et al., 2007). Point mutations in the PHD finger show that the ability to bind unmodified histone H3 is essential for tumor suppressing functions of CHD5 (Paul et al., 2013).

Reduced CHD5 tumor suppressor activities usually occurs by heterozygous mutation of CHD5 or through the repression of its transcription. Rarely is a homozygous deletion of CHD5 found in human cancers. A common method of suppressing CHD5 expression is through promoter CpG hypermethylation (Qu, Dang, & Hou, 2013; Wang et al., 2011; Zhao et al., 2012). Less common methods include the overexpression of microRNA-211 that targets the CHD5 transcript (Cai et al., 2012). CHD5 is predominantly expressed in the brain where it assembles into a NURD-like complex with HDAC2, MTA3, GATAD2B, and RBAP46 and has important functions as a positive and negative regulator of gene expression in neurons (Potts et al., 2011).

5.3.7 CHD1/CHD2/CHD7

Compared with CHD3, CHD4, and CHD5, little is known about the functions of the other CHD ATPases in cancer. CHD1 is frequently mutated in a rare subset of EGR fusion negative prostate cancers (Liu et al., 2012). The selectivity for CHD1 mutation in this subset of prostate cancers could be due to its role as an essential cofactor for AR receptor transcription activity (Burkhardt et al., 2013). AR-stimulated transcription is essential for the expression of EGR, that in turn promotes genome instability and gene fusions to the region. Consequences of CHD1 deletion in prostate cancer cell lines have been investigated and it was shown to promote cell invasiveness (Huang et al., 2012). Mice heterozygous for the hypermorphic allele of CHD2 have defects in hematopoietic stem cell differentiation, accumulation of DNA damage, and increased incidences of lymphomas (Nagarajan et al., 2009). Beyond that, little is known of their molecular mechanisms in cancers. CHD7 was discovered to be significantly mutated in lung tumors from heavy smokers (Pleasance et al., 2010). In these tumors, CHD7 is mutated by either fusions or partial duplications. The significance of these mutations to lung cancers is unknown.

Mouse mutants of CHD7 are embryonic lethal at E10.5 dpc and heterozygous mice have phenotypes similar to humans with CHARGE syndrome (Adams et al., 2007). No cancer phenotypes have been described for these mice.

5.4. ISWI family of chromatin remodeling complexes

5.4.1 SNF2H/SNF2L

The catalytic subunits of the ISWI family of chromatin remodeling complexes are the SNF2H and SNF2L ATPases. Two reports document that SNF2L expression, in contrast to SNF2H, is important for the growth of human cancer cell lines. In these studies, SNF2L was found to be expressed in a wide variety of normal tissues, tumors, and cancer cell lines. Knockdown of SNF2L, but not SNF2H, in several cancer cell lines resulted in reduced cell proliferation, arrest in G2/M, increased DNA damage and activation of DNA damage repair pathways (phosphorylation of CHK1/2, ATM, and p53), and apoptosis (Ye et al., 2009). Follow-up studies identified SNF2LT as a splice variant of SNF2L that lacks its HAND–SANT–SLIDE domains. In these follow-up studies, they observed that knockdown of SNF2LT or the full-length SNF2L gene resulted in similar cancer cell-specific phenotypes (Ye et al., 2012). Similar knockdown studies have shown that SNF2L is an essential regulator of the Wnt signaling pathway (Eckey et al., 2012). In these studies, in contradiction to previous SNF2L studies, SNF2L depletion results in increased proliferation and chemotaxis. This study used genome-wide gene expression profiling to discover that many targets of the Wnt signaling pathway are upregulated with SNF2L knockdown.

Few studies have specifically investigated the role of SNF2H in cancer cell biology. Knockdown studies in several cancer cell lines show that it is not essential for cell proliferation (Ye et al., 2009). SNF2H is a component of several chromatin remodeling complexes with important functions in DNA damage repair, and as such it has been reported to be present at sites of DNA damage (Poot, Bozhenok, van den Berg, Hawkes, & Varga-Weisz, 2005). One study found that the NAD-dependent deacetylase SIRT6 is required for the recruitment of SNF2H to sites of DNA damage (Toiber et al., 2013). Through the use of an SIRT6 knockdown, inefficient DNA damage repair was shown to occur when these factors are not recruited to sites of DNA damage. At sites of DNA damage, SIRT6 deacetylates histone H3K56ac and recruits SNF2H, and the combination of these two events remodels nucleosomes to loosen chromatin structure.

The remodeling of chromatin promotes the binding of 53BP1, BRCA1, and RPA which are necessary for DNA repair.

5.4.2 BPTF

BPTF is the largest subunit of the NURF complex. BPTF resides on 17q and is frequently duplicated in several cancer types, with the greatest frequency in neuroblastoma (Bown et al., 1999). A nonreciprocal translocation of BPTF on 17q was characterized in a human lung cell line following continuous culturing (Buganim et al., 2010). This translocation resulted in increased BPTF expression and correlated with increased cell proliferation. Knockdown of BPTF in these cells reduced proliferation. Consistent with the frequent duplication of 17q in human tumors, the BPTF gene was found to be amplified in various human tumors, especially neuroblastomas and lung cancers and many human cancer cell lines. Whether BPTF duplication provides an advantage to cancer cells, or is just a benign consequence of the 17q duplication, needs to be specifically addressed.

5.4.3 RSF1

RSF1, located on chromosome 11q13.5, is also frequently duplicated in human tumors. Several studies have documented that RSF1 is overexpressed in human cancers. In oral squamous cell cancers, RSF1 was found to be overexpressed relative to normal tissue, which correlates with decreased survival (Fang et al., 2011). Knockdown of RSF1 in these cancer cells results in decreased proliferation and increased cell death. In addition to oral cancer, the duplication of this region is also associated with paclitaxel resistance in ovarian cancers (Choi et al., 2009). A shRNA knockdown screen showed that the RSF1 gene located in the duplicated region is essential for paclitaxel resistance. The knockdown of either subunit of the RSF complex, RSF1 or the ATPase SNF2H, results in increased sensitivity to paclitaxel. Interestingly, the expression of a RSF1 mutant with SNF2H-binding capacity, but not remodeling activity, resulted in the same sensitivity to paclitaxel. These results suggest that the loss of RSF remodeling activity, rather than the redistribution of SNF2H to other remodeling complexes, is what causes sensitivity to paclitaxel. Candidate pathways involved in RSF mediated paclitaxel resistance include NF-κB, ERK, Akt, and EGR1. How RSF1 regulates these pathways is unknown. Pull down and mass spectrometry experiments from an ovarian carcinoma cell line showed that RSF1 physically interacts with Cyclin E1, an important G1/S-phase regulator (Sheu et al., 2013). Overexpression of both RSF1 and Cyclin E1 promotes cellular proliferation but only in a p53 null background. Ectopic expression of a dominant negative truncation of RSF1 in mouse xenograft tumor models results in reduced cancer cell proliferation and tumor size. In combination, these results suggest that a RSF1–Cyclin E1 interaction promotes p53 null ovarian tumors. Similar RSF1 overexpression studies in nontransformed ovarian cells results in increased DNA damage, activation of DNA damage repair pathways, and growth arrest (Sheu et al., 2010). The growth arrest accompanies upregulation of p53 and p21 and depends on an active p53 allele. shRNA knockdown experiments showed that the increase in DNA damage observed with RSF1 overexpression requires SNF2H suggesting that it is dependent on the formation of the RSF complex. In lung cancers, RSF1 is also frequently overexpressed, and knockdown of RSF1 in lung cancer cell lines resulted in reduced proliferation and increased apoptosis [222]. In addition to defects in proliferation and apoptosis, there was an accompanying decrease in Cyclin D1 and phospho-ERK levels in these cells suggesting that RSF1 overexpression promotes G1/S-phase transition and mitogenic ERK signaling.

6. THERAPEUTIC POTENTIAL OF CHROMATIN REMODELING COMPLEXES IN HUMAN CANCER

Targeting epigenetic regulators for the treatment of human cancers has proved successful for histone deacetylases and DNA methyltransferases. In each case, the therapeutics developed are approved for difficult to treat cancers providing valuable assets to oncologists (Arrowsmith, Bountra, Fish, Lee, & Schapira, 2012; Dawson & Kouzarides, 2012). Examples include: vorinostat and romidepsin, FDA-approved HDAC inhibitors used to treat cutaneous T cell lymphoma; and azacitidine and decitabine, FDA-approved DNMT inhibitors for the treatment of myelodysplastic syndromes (Dawson, Kouzarides, & Huntly, 2012; Nebbioso, Carafa, Benedetti, & Altucci, 2012). Next generation, more specific HDAC and DNA methyltransferase inhibitors are currently being developed to improve on these successes. Because of these successes, many other small molecules are actively being pursued which target posttranslational modifications on chromatin for therapeutic benefit. These therapies primarily target a large number of histone modifying enzymes and small RNA regulatory pathways. Histone modifying enzymes currently being targeted include deacetylases, acetyltransferases, and demethylases. In some cases, small molecules targeting these regulators have made it into Phase II clinical trials (Arrowsmith et al., 2012; Dawson et al., 2012;

Nebbioso et al., 2012; Plass et al., 2013). These molecules are tested alone and in combination chemotherapy and include HDAC inhibitors (entinostat, mocetinostat, panobinostat, bellinostat, valproic acid, givinostat, CHR-3996, CHR-2845, and SB939), the KDM1A demethylase inhibitor tranylcypromine, and the HAT inhbitior curcumin (Arrowsmith et al., 2012; Dawson et al., 2012; Nebbioso et al., 2012; Plass et al., 2013). Phase II trials revealed promising results for mocetinostat and bellinostat. The considerable efforts invested into developing small molecules targeting these mechanisms highlights their widely perceived therapeutic potential. This potential can also be realized by developing small regulators to chromatin remodeling complexes, another major class of chromatin regulator.

The discovery that chromatin remodeling complexes regulate gene expression important for cancer progression has stimulated interest in screening them for small molecule regulators. Using completely different approaches, two successful screens have been performed for the BRG1 and RUVBL1 ATPases (Dykhuizen, Carmody, Tolliday, Crabtree, & Palmer, 2012; Elkaim et al., 2012). To discover small molecule regulators of BRG1, a live cell screen was performed by measuring the expression of Bmi, a known target of BRG1 in embryonic stem cells, by high-throughput qRT-PCR. Follow-up assays were performed using embryonic stem cell lines with a knock-in Bmi-luciferase reporter allele. Several compounds (which were not described) were discovered with this approach and are being investigated further. In contrast to the live cell approach used for BRG1, a structure-based biochemical screen was performed to discover small molecule regulators for the RUVBL1 ATPase. To identify inhibitors of the RUVBL1 ATPase activity, the authors used its known structure to model molecules from several libraries to the ATP-binding site. Promising compounds that could bind the ATP-binding site were assayed for the ability to inhibit a DNA-dependent ATPase activity for recombinant RUVBL1. This approach yielded several compounds, some of which inhibited the proliferation of cancer cells.

In addition to discovering small molecules targeting the ATPase activity as was performed for RUVBL1, resources can be invested to discover molecules to inhibit key protein–protein interactions essential for the function of chromatin remodeling complexes. The inhibition of the BAF57–AR interaction is a successful example of this strategy. BAF57 is known to physically interact with AR, and is required for AR binding to chromatin and AR-regulated gene expression in prostate cancer cells. The ectopic expression of a peptide mimic to the sequence on BAF57 that interacts with

AR resulted in inhibited AR binding to chromatin, reduced AR-stimulated transcription, and inhibited AR-dependent prostate cancer cell proliferation (Link et al., 2008). These results are consistent with the possibility that the peptide is inhibiting the BAF57–AR interaction surface and preventing BAF57-dependent AR binding and transactivation. Additional means to inhibit chromatin remodeling complexes with this approach can also include targeting essential subunit interactions within a remodeling complex. While this approach benefits from the focus of a structure-based design, it will suffer from the difficulty of delivering peptides to intracellular targets in a patient's tumor.

Another strategy to therapeutically regulate chromatin remodeling is to identify small molecules that can regulate the expression of genes encoding essential subunits of chromatin remodeling complexes. This has successfully been done to reexpress the BRM ATPase in cancer cell lines that have silenced its expression without a deletion (Gramling & Reisman, 2011; Gramling, Rogers, Liu, & Reisman, 2011). This screen used a glucocorticoid receptor-regulated luciferase reporter that is induced in the presence of BRM. As a result of this screen, two compounds were identified which increased the expression of BRM, inducing the expression of BRM target genes, and decreased the growth of cancer cell lines. Each of these compounds regulates BRM expression through an unknown mechanism.

These different strategies provide excellent models on how to approach small molecule discovery screens for chromatin remodeling complexes. Biochemical screens can be performed for individual ATPase subunits or chromatin remodeling complexes which can be produced recombinantly. Conversely, live cell assays can be used for chromatin remodeling complexes with many subunits, which are not feasible to produce recombinantly. Once candidate compounds are identified, they can be then characterized using biochemical assays requiring small amounts of purified complex.

7. CONCLUDING REMARKS

Many changes occur to the cell during the transition from normal tissue to advanced stage cancer. In many cases, the changes result in abnormal gene expression patterns, which in turn establish and maintain key aspects of cancer cell biology. Over the past decade, chromatin remodeling complexes have been shown to establish and maintain abnormal gene expression in cancer cells. In addition, the recent explosion in sequencing cancer cell genomes has documented that subunits of chromatin remodeling complexes are frequently mutated in several cancer cell types suggesting that some of these mutations can drive the cancer cell phenotype. The sum of these results demonstrates that chromatin remodeling complexes have the potential to be a novel class of targets for developing cancer therapeutics.

To rigorously explore if chromatin remodeling complexes are viable therapeutic targets, several avenues of research should be pursued. The sequence data from cancer cell genomes suggests that subunits of chromatin remodeling complexes might be drivers of cancer development. To investigate the potential driver status of these mutations, conditional mouse alleles must be developed and used in longevity studies. These studies should be done in combination with common oncogene transgenes (MYC and RasG12V) or the knockout of other tumor suppressor pathways (p53 or pRB) to study their tumor-promoting potential in context with other common tumor-promoting mutations. Proof of principle studies have been performed with the INI1 mutation showing that its tumor-promoting activities require a functional BRG1, polycomb, and Cyclin D1. A BRG1 requirement for tumor development with INI mutations suggests that they result from deregulated SWI/SNF complexes lacking the INI1 subunit. Similar studies should be performed for mutations in other SWI/SNF accessory subunits including ARID1A and BAF180. If tumor-promoting activities for these mutations require an active ATPase, then small molecules could be designed to BRG1 for therapeutic advantage for these types of tumors.

Chromatin remodeling activities are perceived to be essential for cell viability. While it is almost universal that they are required for embryonic development, many reports document that at least a handful of remodeling complexes are not cell essential, and therefore may not be required for adult organisms. In order to justify the design of small molecule regulators for these complexes, it will be important to know the function of these complexes in adult cells. These studies are relevant from a therapeutic standpoint because in a majority of cases the targeting of chromatin remodeling complexes with therapies will be performed in an adult human. The use of tissue-specific knockout alleles to deplete subunits in specific tissues should promote these studies. In combination, conditional organism-wide knockouts should be pursued using ligand-inducible knockout alleles in adults. These experiments are critical for determining the possible consequences to the adult animal from targeting these complexes with small molecule regulators.

In addition to oncogenes and tumor suppressors, which have important functions for establishing cancer, many genes are necessary to maintain progrowth signals in cancer cells. These genes are not essential for normal cells but are essential for the cancer cell to maintain its phenotype. This class of nononcogenes has been actively investigated by several groups using shRNA knockdown screening strategies (Luo, Solimini, & Elledge, 2009). Relevant to chromatin remodeling, preliminary evidence shows that SNF2L containing chromatin remodeling complexes like NURF, RSF, and CERF could participate in these nononcogene pathways. It would be useful to the field to conduct shRNA knockdown screens specifically looking for nononcogene function for chromatin remodeling complexes in multiple cancer types. From a practical sense, these studies are only useful in combination with studies showing that the remodeling complexes are not essential for normal cell survival in adult organisms (see above).

In order to rationally approach the design of small molecule regulators for these complexes, their atomic structures must be solved. Much success has been realized over several years solving the low-resolution structure of chromatin remodeling complexes by electron microscopy (Leschziner, 2011). While these structures are a major step forward, their resolution must be improved. The size and heterogeneity of chromatin remodeling complexes, particularly in complex with a nucleosome, will likely prevent their crystallization. However, atomic resolution structures can be assembled using integrated structural techniques (Ward, Sali, & Wilson, 2013). In this approach, a low-resolution EM structure serves as a scaffold with which to model atomic resolution structures of individual subunits or domains. Recent success using this approach has been realized for the INO80 and SWR1 complexes (Nguyen et al., 2013; Tosi et al., 2013). With these structures, we can begin to study the molecular contacts these complexes make between subunits and the nucleosome, with the hope of rationally designing small molecules to disrupt these interactions. Significant progress into rationally designing small molecules to the ATPase-binding sites can also be realized by solving the crystal structures of the ATPase containing subunit.

Chromatin remodeling complexes should be feasible drug targets because, like histone modifying enzymes and DNA methyltransferases, they are enzymes. Druggable active sites which would have the most profound effect on the activity of the complexes include the ATP and nucleosome-binding sites. To discover small molecules for these sites, rational approaches would be preferred which will require the atomic resolution structure (see above) but can also be approached with live cell or biochemical screens for small molecule inhibitors. These approaches have been used successfully in the past (see Section 6). Off target effects for small molecules targeting the nucleosome-binding site are less of a concern than the ATP-binding site because significant nucleosome contacts are established by subunits unique to each complex (Leschziner, 2011; Nguyen et al., 2013; Tosi et al., 2013). Beyond targeting the ATPase and nucleosome-binding sites, these complexes have many binding sites for histone modifications including chromo, PHD, and bromodomains. Targeting histone-binding domains with small molecules might provide a means to inhibit their ability to bind chromatin (Dawson et al., 2012).

The explosion of research into the field of epigenetics has yielded much understanding on how these mechanisms operate under normal circumstances to regulate gene expression. While much more work needs to be done to further our understanding of these mechanisms, we are in a position to begin to seriously explore whether epigenetic mechanisms can be viable drug targets for the treatment of human disease. As outlined in this review, there is potential for these complexes to participate in cancer as driving mutations, regulating oncogenes and tumor suppressor pathways, and participating as nononcogenes. As such, chromatin remodeling complexes have a great potential to be successfully targeted for therapeutic benefit. The next decade of research on chromatin remodeling complexes should yield many discoveries, some of which may have value in the clinic.

ACKNOWLEDGMENTS

The authors would like to thank the V Foundation for Cancer Research, the Jeffress Memorial Fund, Virginia Commonwealth University School of Medicine, and the Massey Cancer Center for financial support. The authors would also like to thank Kevin Hogan for editorial review of the manuscript.

Conflict of Interest: The authors do not declare any conflicts of interest.

REFERENCES

Aalfs, J. D., & Kingston, R. E. (2000). What does 'chromatin remodeling' mean? *Trends in Biochemical Sciences*, *25*(11), 548–555.

Adams, M. E., Hurd, E. A., Beyer, L. A., Swiderski, D. L., Raphael, Y., & Martin, D. M. (2007). Defects in vestibular sensory epithelia and innervation in mice with loss of Chd7 function: Implications for human CHARGE syndrome. *The Journal of Comparative Neurology*, *504*(5), 519–532.

Allis, C. D., Jenuwein, T., & Reinberg, D. (2007). *Epigenetics*. Cold Spring Harbor, N.Y.: Cold Spring Harbor Laboratory Press.

Ard, P. G., Chatterjee, C., Kunjibettu, S., Adside, L. R., Gralinski, L. E., & McMahon, S. B. (2002). Transcriptional regulation of the mdm2 oncogene by p53 requires TRRAP acetyltransferase complexes. *Molecular and Cellular Biology*, *22*(16), 5650–5661.

Arrowsmith, C. H., Bountra, C., Fish, P. V., Lee, K., & Schapira, M. (2012). Epigenetic protein families: A new frontier for drug discovery. *Nature Reviews Drug Discovery*, *11*(5), 384–400.

Ayllon, V., & Rebollo, A. (2000). Ras-induced cellular events (review). *Molecular Membrane Biology, 17*(2), 65–73.

Badenhorst, P., Xiao, H., Cherbas, L., Kwon, S. Y., Voas, M., Rebay, I., et al. (2005). The Drosophila nucleosome remodeling factor NURF is required for Ecdysteroid signaling and metamorphosis. *Genes & Development, 19*(21), 2540–2545.

Bagchi, A., Papazoglu, C., Wu, Y., Capurso, D., Brodt, M., Francis, D., et al. (2007). CHD5 is a tumor suppressor at human 1p36. *Cell, 128*(3), 459–475.

Bagheri-Yarmand, R., Balasenthil, S., Gururaj, A. E., Talukder, A. H., Wang, Y. H., Lee, J. H., et al. (2007). Metastasis-associated protein 1 transgenic mice: A new model of spontaneous B-cell lymphomas. *Cancer Research, 67*(15), 7062–7067.

Bagheri-Yarmand, R., Talukder, A. H., Wang, R. A., Vadlamudi, R. K., & Kumar, R. (2004). Metastasis-associated protein 1 deregulation causes inappropriate mammary gland development and tumorigenesis. *Development, 131*(14), 3469–3479.

Balasubramaniam, S., Comstock, C. E., Ertel, A., Jeong, K. W., Stallcup, M. R., Addya, S., et al. (2013). Aberrant BAF57 signaling facilitates prometastatic phenotypes. *Clinical Cancer Research, 19*(10), 2657–2667.

Banting, G. S., Barak, O., Ames, T. M., Burnham, A. C., Kardel, M. D., Cooch, N. S., et al. (2005). CECR2, a protein involved in neurulation, forms a novel chromatin remodeling complex with SNF2L. *Human Molecular Genetics, 14*(4), 513–524.

Barak, O., Lazzaro, M. A., Lane, W. S., Speicher, D. W., Picketts, D. J., & Shiekhattar, R. (2003). Isolation of human NURF: A regulator of Engrailed gene expression. *The EMBO Journal, 22*(22), 6089–6100.

Barlev, N. A., Liu, L., Chehab, N. H., Mansfield, K., Harris, K. G., Halazonetis, T. D., et al. (2001). Acetylation of p53 activates transcription through recruitment of coactivators/histone acetyltransferases. *Molecular Cell, 8*(6), 1243–1254.

Bartke, T., Vermeulen, M., Xhemalce, B., Robson, S. C., Mann, M., & Kouzarides, T. (2010). Nucleosome-interacting proteins regulated by DNA and histone methylation. *Cell, 143*(3), 470–484.

Bartlett, C., Orvis, T. J., Rosson, G. S., & Weissman, B. E. (2011). BRG1 mutations found in human cancer cell lines inactivate Rb-mediated cell-cycle arrest. *Journal of Cellular Physiology, 226*(8), 1989–1997.

Bauer, A., Chauvet, S., Huber, O., Usseglio, F., Rothbacher, U., Aragnol, D., et al. (2000). Pontin52 and reptin52 function as antagonistic regulators of beta-catenin signalling activity. *The EMBO Journal, 19*(22), 6121–6130.

Becker, P. B., & Workman, J. L. (2013). Nucleosome remodeling and epigenetics. *Cold Spring Harbor Perspectives in Biology, 5*(9).

Biegel, J. A., Fogelgren, B., Wainwright, L. M., Zhou, J. Y., Bevan, H., & Rorke, L. B. (2000). Germline INI1 mutation in a patient with a central nervous system atypical teratoid tumor and renal rhabdoid tumor. *Genes, Chromosomes & Cancer, 28*(1), 31–37.

Biegel, J. A., Zhou, J. Y., Rorke, L. B., Stenstrom, C., Wainwright, L. M., & Fogelgren, B. (1999). Germ-line and acquired mutations of INI1 in atypical teratoid and rhabdoid tumors. *Cancer Research, 59*(1), 74–79.

Bochar, D. A., Wang, L., Beniya, H., Kinev, A., Xue, Y., Lane, W. S., et al. (2000). BRCA1 is associated with a human SWI/SNF-related complex: Linking chromatin remodeling to breast cancer. *Cell, 102*(2), 257–265.

Bowman, G. D. (2010). Mechanisms of ATP-dependent nucleosome sliding. *Current Opinion in Structural Biology, 20*(1), 73–81.

Bown, N., Cotterill, S., Lastowska, M., O'Neill, S., Pearson, A. D., Plantaz, D., et al. (1999). Gain of chromosome arm 17q and adverse outcome in patients with neuroblastoma. *The New England Journal of Medicine, 340*(25), 1954–1961.

Buganim, Y., Goldstein, I., Lipson, D., Milyavsky, M., Polak-Charcon, S., Mardoukh, C., et al. (2010). A novel translocation breakpoint within the BPTF gene is associated with a pre-malignant phenotype. *PLoS One, 5*(3), e9657.

Bultman, S., Gebuhr, T., Yee, D., La Mantia, C., Nicholson, J., Gilliam, A., et al. (2000). A Brg1 null mutation in the mouse reveals functional differences among mammalian SWI/SNF complexes. *Molecular Cell*, *6*(6), 1287–1295.

Bultman, S. J., Herschkowitz, J. I., Godfrey, V., Gebuhr, T. C., Yaniv, M., Perou, C. M., et al. (2008). Characterization of mammary tumors from Brg1 heterozygous mice. *Oncogene*, *27*(4), 460–468.

Burkhardt, L., Fuchs, S., Krohn, A., Masser, S., Mader, M., Kluth, M., et al. (2013). CHD1 is a 5q21 tumor suppressor required for ERG rearrangement in prostate cancer. *Cancer Research*, *73*(9), 2795–2805.

Burrows, A. E., Smogorzewska, A., & Elledge, S. J. (2010). Polybromo-associated BRG1-associated factor components BRD7 and BAF180 are critical regulators of p53 required for induction of replicative senescence. *Proceedings of the National Academy of Sciences of the United States of America*, *107*(32), 14280–14285.

Cai, C., Ashktorab, H., Pang, X., Zhao, Y., Sha, W., Liu, Y., et al. (2012). MicroRNA-211 expression promotes colorectal cancer cell growth in vitro and in vivo by targeting tumor suppressor CHD5. *PLoS One*, 7(1), e29750.

Cai, Y., Jin, J., Tomomori-Sato, C., Sato, S., Sorokina, I., Parmely, T. J., et al. (2003). Identification of new subunits of the multiprotein mammalian TRRAP/TIP60-containing histone acetyltransferase complex. *The Journal of Biological Chemistry*, *278*(44), 42733–42736.

Cancer Genome Atlas Research Network (2013). Comprehensive molecular characterization of clear cell renal cell carcinoma. *Nature*, *499*(7456), 43–49.

Cavellan, E., Asp, P., Percipalle, P., & Farrants, A. K. (2006). The WSTF-SNF2h chromatin remodeling complex interacts with several nuclear proteins in transcription. *The Journal of Biological Chemistry*, *281*(24), 16264–16271.

Chandler, R. L., Brennan, J., Schisler, J. C., Serber, D., Patterson, C., & Magnuson, T. (2013). ARID1a-DNA interactions are required for promoter occupancy by SWI/SNF. *Molecular and Cellular Biology*, *33*(2), 265–280.

Chang, K., Creighton, C. J., Davis, C., Donehower, L., Drummond, J., Wheeler, D., et al. (2013). The Cancer Genome Atlas Pan-Cancer analysis project. *Nature Genetics*, *45*(10), 1113–1120.

Chen, L., Cai, Y., Jin, J., Florens, L., Swanson, S. K., Washburn, M. P., et al. (2011). Subunit organization of the human INO80 chromatin remodeling complex: An evolutionarily conserved core complex catalyzes ATP-dependent nucleosome remodeling. *The Journal of Biological Chemistry*, *286*(13), 11283–11289.

Cheng, S. W., Davies, K. P., Yung, E., Beltran, R. J., Yu, J., & Kalpana, G. V. (1999). c-MYC interacts with INI1/hSNF5 and requires the SWI/SNF complex for transactivation function. *Nature Genetics*, *22*(1), 102–105.

Choi, J. H., Sheu, J. J., Guan, B., Jinawath, N., Markowski, P., Wang, T. L., et al. (2009). Functional analysis of 11q13.5 amplicon identifies Rsf-1 (HBXAP) as a gene involved in paclitaxel resistance in ovarian cancer. *Cancer Research*, *69*(4), 1407–1415.

Cismasiu, V. B., Adamo, K., Gecewicz, J., Duque, J., Lin, Q., & Avram, D. (2005). BCL11B functionally associates with the NuRD complex in T lymphocytes to repress targeted promoter. *Oncogene*, *24*(45), 6753–6764.

Clevers, H. (2006). Wnt/beta-catenin signaling in development and disease. *Cell*, *127*(3), 469–480.

Dawson, M. A., & Kouzarides, T. (2012). Cancer epigenetics: From mechanism to therapy. *Cell*, *150*(1), 12–27.

Dawson, M. A., Kouzarides, T., & Huntly, B. J. (2012). Targeting epigenetic readers in cancer. *The New England Journal of Medicine*, *367*(7), 647–657.

Decristofaro, M. F., Betz, B. L., Rorie, C. J., Reisman, D. N., Wang, W., & Weissman, B. E. (2001). Characterization of SWI/SNF protein expression in human breast cancer cell lines and other malignancies. *Journal of Cellular Physiology*, *186*(1), 136–145.

De Koning, L., Corpet, A., Haber, J. E., & Almouzni, G. (2007). Histone chaperones: An escort network regulating histone traffic. *Nature Structural & Molecular Biology*, *14*(11), 997–1007. [Research Support, N.I.H., Extramural Research Support, Non-U.S. Gov't Review].

Deleu, L., Shellard, S., Alevizopoulos, K., Amati, B., & Land, H. (2001). Recruitment of TRRAP required for oncogenic transformation by E1A. *Oncogene*, *20*(57), 8270–8275.

Dugan, K. A., Wood, M. A., & Cole, M. D. (2002). TIP49, but not TRRAP, modulates c-Myc and E2F1 dependent apoptosis. *Oncogene*, *21*(38), 5835–5843.

Dunaief, J. L., Strober, B. E., Guha, S., Khavari, P. A., Alin, K., Luban, J., et al. (1994). The retinoblastoma protein and BRG1 form a complex and cooperate to induce cell cycle arrest. *Cell*, *79*(1), 119–130.

Dykhuizen, E. C., Carmody, L. C., Tolliday, N., Crabtree, G. R., & Palmer, M. A. (2012). Screening for inhibitors of an essential chromatin remodeler in mouse embryonic stem cells by monitoring transcriptional regulation. *Journal of Biomolecular Screening*, *17*(9), 1221–1230.

Dykhuizen, E. C., Hargreaves, D. C., Miller, E. L., Cui, K., Korshunov, A., Kool, M., et al. (2013). BAF complexes facilitate decatenation of DNA by topoisomerase IIalpha. *Nature*, *497*(7451), 624–627.

Eckey, M., Kuphal, S., Straub, T., Rummele, P., Kremmer, E., Bosserhoff, A. K., et al. (2012). Nucleosome remodeler SNF2L suppresses cell proliferation and migration and attenuates Wnt signaling. *Molecular and Cellular Biology*, *32*(13), 2359–2371.

Elkaim, J., Castroviejo, M., Bennani, D., Taouji, S., Allain, N., Laguerre, M., et al. (2012). First identification of small-molecule inhibitors of Pontin by combining virtual screening and enzymatic assay. *The Biochemical Journal*, *443*(2), 549–559.

Endo, M., Yasui, K., Zen, Y., Gen, Y., Zen, K., Tsuji, K., et al. (2013). Alterations of the SWI/SNF chromatin remodelling subunit-BRG1 and BRM in hepatocellular carcinoma. *Liver International*, *33*(1), 105–117.

Erdel, F., & Rippe, K. (2011). Chromatin remodelling in mammalian cells by ISWI-type complexes—Where, when and why? *The FEBS Journal*, *278*(19), 3608–3618.

Erikson, J., ar-Rushdi, A., Drwinga, H. L., Nowell, P. C., & Croce, C. M. (1983). Transcriptional activation of the translocated c-myc oncogene in burkitt lymphoma. *Proceedings of the National Academy of Sciences of the United States of America*, *80*(3), 820–824.

Fang, F. M., Li, C. F., Huang, H. Y., Lai, M. T., Chen, C. M., Chiu, I. W., et al. (2011). Overexpression of a chromatin remodeling factor, RSF-1/HBXAP, correlates with aggressive oral squamous cell carcinoma. *The American Journal of Pathology*, *178*(5), 2407–2415.

Feng, Y., Lee, N., & Fearon, E. R. (2003). TIP49 regulates beta-catenin-mediated neoplastic transformation and T-cell factor target gene induction via effects on chromatin remodeling. *Cancer Research*, *63*(24), 8726–8734.

Finkbeiner, M. G., Sawan, C., Ouzounova, M., Murr, R., & Herceg, Z. (2008). HAT cofactor TRRAP mediates beta-catenin ubiquitination on the chromatin and the regulation of the canonical Wnt pathway. *Cell Cycle*, 7(24), 3908–3914.

Forbes, S. A., Bindal, N., Bamford, S., Cole, C., Kok, C. Y., Beare, D., et al. (2011). COSMIC: Mining complete cancer genomes in the Catalogue of Somatic Mutations in Cancer. *Nucleic Acids Research*, *39*(Database issue), D945–D950.

Francis, N. J., Saurin, A. J., Shao, Z., & Kingston, R. E. (2001). Reconstitution of a functional core polycomb repressive complex. *Molecular Cell*, *8*(3), 545–556.

Fu, J., Qin, L., He, T., Qin, J., Hong, J., Wong, J., et al. (2011). The TWIST/Mi2/NuRD protein complex and its essential role in cancer metastasis. *Cell Research*, *21*(2), 275–289.

Fu, J., Zhang, L., He, T., Xiao, X., Liu, X., Wang, L., et al. (2012). TWIST represses estrogen receptor-alpha expression by recruiting the NuRD protein complex in breast cancer cells. *International Journal of Biological Sciences*, *8*(4), 522–532.

Fuchs, M., Gerber, J., Drapkin, R., Sif, S., Ikura, T., Ogryzko, V., et al. (2001). The p400 complex is an essential E1A transformation target. *Cell*, *106*(3), 297–307.
Fujii, T., Ueda, T., Nagata, S., & Fukunaga, R. (2010). Essential role of p400/mDomino chromatin-remodeling ATPase in bone marrow hematopoiesis and cell-cycle progression. *The Journal of Biological Chemistry*, *285*(39), 30214–30223.
Fujiki, R., Kim, M. S., Sasaki, Y., Yoshimura, K., Kitagawa, H., & Kato, S. (2005). Ligand-induced transrepression by VDR through association of WSTF with acetylated histones. *The EMBO Journal*, *24*(22), 3881–3894.
Gao, X., Tate, P., Hu, P., Tjian, R., Skarnes, W. C., & Wang, Z. (2008). ES cell pluripotency and germ-layer formation require the SWI/SNF chromatin remodeling component BAF250a. *Proceedings of the National Academy of Sciences of the United States of America*, *105*(18), 6656–6661.
Garcia-Pedrero, J. M., Kiskinis, E., Parker, M. G., & Belandia, B. (2006). The SWI/SNF chromatin remodeling subunit BAF57 is a critical regulator of estrogen receptor function in breast cancer cells. *The Journal of Biological Chemistry*, *281*(32), 22656–22664.
Gevry, N., Chan, H. M., Laflamme, L., Livingston, D. M., & Gaudreau, L. (2007). p21 transcription is regulated by differential localization of histone H2A.Z. *Genes & Development*, *21*(15), 1869–1881.
Gitenay, D., & Baron, V. T. (2009). Is EGR1 a potential target for prostate cancer therapy? *Future Oncology*, *5*(7), 993–1003.
Gospodinov, A., Vaissiere, T., Krastev, D. B., Legube, G., Anachkova, B., & Herceg, Z. (2011). Mammalian Ino80 mediates double-strand break repair through its role in DNA end strand resection. *Molecular and Cellular Biology*, *31*(23), 4735–4745.
Gramling, S., & Reisman, D. (2011). Discovery of BRM targeted therapies: Novel reactivation of an anti-cancer gene. *Letters in Drug Design & Discovery*, *8*(1), 93–99.
Gramling, S., Rogers, C., Liu, G., & Reisman, D. (2011). Pharmacologic reversal of epigenetic silencing of the anticancer protein BRM: A novel targeted treatment strategy. *Oncogene*, *30*(29), 3289–3294.
Grigoletto, A., Lestienne, P., & Rosenbaum, J. (2011). The multifaceted proteins Reptin and Pontin as major players in cancer. *Biochimica et Biophysica Acta*, *1815*(2), 147–157.
Guan, B., Wang, T. L., & Shih Ie, M. (2011). ARID1A, a factor that promotes formation of SWI/SNF-mediated chromatin remodeling, is a tumor suppressor in gynecologic cancers. *Cancer Research*, *71*(21), 6718–6727.
Gunduz, E., Gunduz, M., Ouchida, M., Nagatsuka, H., Beder, L., Tsujigiwa, H., et al. (2005). Genetic and epigenetic alterations of BRG1 promote oral cancer development. *International Journal of Oncology*, *26*(1), 201–210.
Gunther, K., Rust, M., Leers, J., Boettger, T., Scharfe, M., Jarek, M., et al. (2013). Differential roles for MBD2 and MBD3 at methylated CpG islands, active promoters and binding to exon sequences. *Nucleic Acids Research*, *41*(5), 3010–3021.
Haas, J. E., Palmer, N. F., Weinberg, A. G., & Beckwith, J. B. (1981). Ultrastructure of malignant rhabdoid tumor of the kidney. A distinctive renal tumor of children. *Human Pathology*, *12*(7), 646–657.
Hagelkruys, A., Sawicka, A., Rennmayr, M., & Seiser, C. (2011). The biology of HDAC in cancer: The nuclear and epigenetic components. *Handbook of Experimental Pharmacology*, *206*, 13–37.
Hah, N., Kolkman, A., Ruhl, D. D., Pijnappel, W. W., Heck, A. J., Timmers, H. T., et al. (2010). A role for BAF57 in cell cycle-dependent transcriptional regulation by the SWI/SNF chromatin remodeling complex. *Cancer Research*, *70*(11), 4402–4411.
Hall, J. A., & Georgel, P. T. (2007). CHD proteins: A diverse family with strong ties. *Biochemistry and Cell Biology*, *85*(4), 463–476.
Hanahan, D., & Weinberg, R. A. (2011). Hallmarks of cancer: The next generation. *Cell*, *144*(5), 646–674.

Hendrich, B., & Bird, A. (1998). Identification and characterization of a family of mammalian methyl-CpG binding proteins. *Molecular and Cellular Biology*, *18*(11), 6538–6547.

Henrich, K. O., Schwab, M., & Westermann, F. (2012). 1p36 tumor suppression—A matter of dosage? *Cancer Research*, *72*(23), 6079–6088.

Hill, D. A., de la Serna, I. L., Veal, T. M., & Imbalzano, A. N. (2004). BRCA1 interacts with dominant negative SWI/SNF enzymes without affecting homologous recombination or radiation-induced gene activation of p21 or Mdm2. *Journal of Cellular Biochemistry*, *91*(5), 987–998.

Ho, L., & Crabtree, G. R. (2010). Chromatin remodelling during development. *Nature*, *463*(7280), 474–484.

Huang, S., Gulzar, Z. G., Salari, K., Lapointe, J., Brooks, J. D., & Pollack, J. R. (2012). Recurrent deletion of CHD1 in prostate cancer with relevance to cell invasiveness. *Oncogene*, *31*(37), 4164–4170.

Huen, J., Kakihara, Y., Ugwu, F., Cheung, K. L., Ortega, J., & Houry, W. A. (2010). Rvb1-Rvb2: Essential ATP-dependent helicases for critical complexes. *Biochemistry and Cell Biology*, *88*(1), 29–40.

Ibragimova, I., Maradeo, M. E., Dulaimi, E., & Cairns, P. (2013). Aberrant promoter hypermethylation of PBRM1, BAP1, SETD2, KDM6A and other chromatin-modifying genes is absent or rare in clear cell RCC. *Epigenetics*, *8*(5), 486–493.

Ikura, T., Ogryzko, V. V., Grigoriev, M., Groisman, R., Wang, J., Horikoshi, M., et al. (2000). Involvement of the TIP60 histone acetylase complex in DNA repair and apoptosis. *Cell*, *102*(4), 463–473.

Isakoff, M. S., Sansam, C. G., Tamayo, P., Subramanian, A., Evans, J. A., Fillmore, C. M., et al. (2005). Inactivation of the Snf5 tumor suppressor stimulates cell cycle progression and cooperates with p53 loss in oncogenic transformation. *Proceedings of the National Academy of Sciences of the United States of America*, *102*(49), 17745–17750.

Jin, J., Cai, Y., Yao, T., Gottschalk, A. J., Florens, L., Swanson, S. K., et al. (2005). A mammalian chromatin remodeling complex with similarities to the yeast INO80 complex. *The Journal of Biological Chemistry*, *280*(50), 41207–41212.

Jones, S., Li, M., Parsons, D. W., Zhang, X., Wesseling, J., Kristel, P., et al. (2012). Somatic mutations in the chromatin remodeling gene ARID1A occur in several tumor types. *Human Mutation*, *33*(1), 100–103.

Jones, S., Wang, T. L., Shih Ie, M., Mao, T. L., Nakayama, K., Roden, R., et al. (2010). Frequent mutations of chromatin remodeling gene ARID1A in ovarian clear cell carcinoma. *Science*, *330*(6001), 228–231.

Kadoch, C., Hargreaves, D. C., Hodges, C., Elias, L., Ho, L., Ranish, J., et al. (2013). Proteomic and bioinformatic analysis of mammalian SWI/SNF complexes identifies extensive roles in human malignancy. *Nature Genetics*, *45*(6), 592–601.

Kandoth, C., McLellan, M. D., Vandin, F., Ye, K., Niu, B., Lu, C., et al. (2013). Mutational landscape and significance across 12 major cancer types. *Nature*, *502*(7471), 333–339.

Kennison, J. A., & Tamkun, J. W. (1988). Dosage-dependent modifiers of polycomb and antennapedia mutations in Drosophila. *Proceedings of the National Academy of Sciences of the United States of America*, *85*(21), 8136–8140.

Kia, S. K., Gorski, M. M., Giannakopoulos, S., & Verrijzer, C. P. (2008). SWI/SNF mediates polycomb eviction and epigenetic reprogramming of the INK4b-ARF-INK4a locus. *Molecular and Cellular Biology*, *28*(10), 3457–3464.

Klochendler-Yeivin, A., Fiette, L., Barra, J., Muchardt, C., Babinet, C., & Yaniv, M. (2000). The murine SNF5/INI1 chromatin remodeling factor is essential for embryonic development and tumor suppression. *EMBO Reports*, *1*(6), 500–506.

Kong, L., Yu, X. P., Bai, X. H., Zhang, W. F., Zhang, Y., Zhao, W. M., et al. (2007). RbAp48 is a critical mediator controlling the transforming activity of human papillomavirus type 16 in cervical cancer. *The Journal of Biological Chemistry*, *282*(36), 26381–26391.

Kornberg, R. D. (1974). Chromatin structure: A repeating unit of histones and DNA. *Science, 184*(4139), 868–871.

Kouzarides, T. (2007). Chromatin modifications and their function. *Cell, 128*(4), 693–705.

Kumar, R., Balasenthil, S., Manavathi, B., Rayala, S. K., & Pakala, S. B. (2010). Metastasis-associated protein 1 and its short form variant stimulates Wnt1 transcription through promoting its derepression from Six3 corepressor. *Cancer Research, 70*(16), 6649–6658.

Kumar, R., Balasenthil, S., Pakala, S. B., Rayala, S. K., Sahin, A. A., & Ohshiro, K. (2010). Metastasis-associated protein 1 short form stimulates Wnt1 pathway in mammary epithelial and cancer cells. *Cancer Research, 70*(16), 6598–6608.

Kuwahara, Y., Charboneau, A., Knudsen, E. S., & Weissman, B. E. (2010). Reexpression of hSNF5 in malignant rhabdoid tumor cell lines causes cell cycle arrest through a p21 (CIP1/WAF1)-dependent mechanism. *Cancer Research, 70*(5), 1854–1865.

Kuwahara, Y., Wei, D., Durand, J., & Weissman, B. E. (2013). SNF5 reexpression in malignant rhabdoid tumors regulates transcription of target genes by recruitment of SWI/SNF complexes and RNAPII to the transcription start site of their promoters. *Molecular Cancer Research, 11*(3), 251–260.

Landry, J., Sharov, A. A., Piao, Y., Sharova, L. V., Xiao, H., Southon, E., et al. (2008). Essential role of chromatin remodeling protein Bptf in early mouse embryos and embryonic stem cells. *PLoS Genetics, 4*(10), e1000241.

Lange, M., Demajo, S., Jain, P., & Di Croce, L. (2011). Combinatorial assembly and function of chromatin regulatory complexes. *Epigenomics, 3*(5), 567–580.

Larsen, D. H., Poinsignon, C., Gudjonsson, T., Dinant, C., Payne, M. R., Hari, F. J., et al. (2010). The chromatin-remodeling factor CHD4 coordinates signaling and repair after DNA damage. *The Journal of Cell Biology, 190*(5), 731–740.

Laybourn, P. J., & Kadonaga, J. T. (1991). Role of nucleosomal cores and histone H1 in regulation of transcription by RNA polymerase II. *Science, 254*(5029), 238–245.

Lee, J. Y., & Lee, T. H. (2012). Effects of histone acetylation and CpG methylation on the structure of nucleosomes. *Biochimica et Biophysica Acta, 1824*(8), 974–982.

Lee, K., Lee, A. Y., Kwon, Y. K., & Kwon, H. (2011). Suppression of HPV E6 and E7 expression by BAF53 depletion in cervical cancer cells. *Biochemical and Biophysical Research Communications, 412*(2), 328–333.

Lee, R. S., Stewart, C., Carter, S. L., Ambrogio, L., Cibulskis, K., Sougnez, C., et al. (2012). A remarkably simple genome underlies highly malignant pediatric rhabdoid cancers. *The Journal of Clinical Investigation, 122*(8), 2983–2988.

LeRoy, G., Loyola, A., Lane, W. S., & Reinberg, D. (2000). Purification and characterization of a human factor that assembles and remodels chromatin. *The Journal of Biological Chemistry, 275*(20), 14787–14790.

LeRoy, G., Orphanides, G., Lane, W. S., & Reinberg, D. (1998). Requirement of RSF and FACT for transcription of chromatin templates in vitro. *Science, 282*(5395), 1900–1904.

Leschziner, A. E. (2011). Electron microscopy studies of nucleosome remodelers. *Current Opinion in Structural Biology, 21*(6), 709–718.

Li, D. Q., Divijendra Natha Reddy, S., Pakala, S. B., Wu, X., Zhang, Y., Rayala, S. K., et al. (2009). MTA1 coregulator regulates p53 stability and function. *The Journal of Biological Chemistry, 284*(50), 34545–34552.

Li, G. C., Guan, L. S., & Wang, Z. Y. (2003). Overexpression of RbAp46 facilitates stress-induced apoptosis and suppresses tumorigenicity of neoplastigenic breast epithelial cells. *International Journal of Cancer, 105*(6), 762–768.

Li, D. Q., Ohshiro, K., Reddy, S. D., Pakala, S. B., Lee, M. H., Zhang, Y., et al. (2009). E3 ubiquitin ligase COP1 regulates the stability and functions of MTA1. *Proceedings of the National Academy of Sciences of the United States of America, 106*(41), 17493–17498.

Li, D. Q., Pakala, S. B., Nair, S. S., Eswaran, J., & Kumar, R. (2012). Metastasis-associated protein 1/nucleosome remodeling and histone deacetylase complex in cancer. *Cancer Research*, *72*(2), 387–394.

Li, D. Q., Pakala, S. B., Reddy, S. D., Ohshiro, K., Peng, S. H., Lian, Y., et al. (2010). Revelation of p53-independent function of MTA1 in DNA damage response via modulation of the p21 WAF1-proliferating cell nuclear antigen pathway. *The Journal of Biological Chemistry*, *285*(13), 10044–10052.

Li, D. Q., Pakala, S. B., Reddy, S. D., Ohshiro, K., Zhang, J. X., Wang, L., et al. (2011). Bidirectional autoregulatory mechanism of metastasis-associated protein 1-alternative reading frame pathway in oncogenesis. *Proceedings of the National Academy of Sciences of the United States of America*, *108*(21), 8791–8796.

Li, R., Zhang, H., Yu, W., Chen, Y., Gui, B., Liang, J., et al. (2009). ZIP: A novel transcription repressor, represses EGFR oncogene and suppresses breast carcinogenesis. *The EMBO Journal*, *28*(18), 2763–2776.

Link, K. A., Balasubramaniam, S., Sharma, A., Comstock, C. E., Godoy-Tundidor, S., Powers, N., et al. (2008). Targeting the BAF57 SWI/SNF subunit in prostate cancer: A novel platform to control androgen receptor activity. *Cancer Research*, *68*(12), 4551–4558.

Link, K. A., Burd, C. J., Williams, E., Marshall, T., Rosson, G., Henry, E., et al. (2005). BAF57 governs androgen receptor action and androgen-dependent proliferation through SWI/SNF. *Molecular and Cellular Biology*, *25*(6), 2200–2215.

Lister, R., Pelizzola, M., Dowen, R. H., Hawkins, R. D., Hon, G., Tonti-Filippini, J., et al. (2009). Human DNA methylomes at base resolution show widespread epigenomic differences. *Nature*, *462*(7271), 315–322.

Liu, W., Lindberg, J., Sui, G., Luo, J., Egevad, L., Li, T., et al. (2012). Identification of novel CHD1-associated collaborative alterations of genomic structure and functional assessment of CHD1 in prostate cancer. *Oncogene*, *31*(35), 3939–3948.

Luger, K., Mader, A. W., Richmond, R. K., Sargent, D. F., & Richmond, T. J. (1997). Crystal structure of the nucleosome core particle at 2.8 A resolution. *Nature*, *389*(6648), 251–260.

Luijsterburg, M. S., & van Attikum, H. (2011). Chromatin and the DNA damage response: The cancer connection. *Molecular Oncology*, *5*(4), 349–367.

Luo, J., Solimini, N. L., & Elledge, S. J. (2009). Principles of cancer therapy: Oncogene and non-oncogene addiction. *Cell*, *136*(5), 823–837.

Luo, J., Su, F., Chen, D., Shiloh, A., & Gu, W. (2000). Deacetylation of p53 modulates its effect on cell growth and apoptosis. *Nature*, *408*(6810), 377–381.

Magdinier, F., & Wolffe, A. P. (2001). Selective association of the methyl-CpG binding protein MBD2 with the silent p14/p16 locus in human neoplasia. *Proceedings of the National Academy of Sciences of the United States of America*, *98*(9), 4990–4995.

Majid, S., Dar, A. A., Ahmad, A. E., Hirata, H., Kawakami, K., Shahryari, V., et al. (2009). BTG3 tumor suppressor gene promoter demethylation, histone modification and cell cycle arrest by genistein in renal cancer. *Carcinogenesis*, *30*(4), 662–670.

Marignani, P. A., Kanai, F., & Carpenter, C. L. (2001). LKB1 associates with Brg1 and is necessary for Brg1-induced growth arrest. *The Journal of Biological Chemistry*, *276*(35), 32415–32418.

Marzook, H., Li, D. Q., Nair, V. S., Mudvari, P., Reddy, S. D., Pakala, S. B., et al. (2012). Metastasis-associated protein 1 drives tumor cell migration and invasion through transcriptional repression of RING finger protein 144A. *The Journal of Biological Chemistry*, *287*(8), 5615–5626.

Mazumdar, A., Wang, R. A., Mishra, S. K., Adam, L., Bagheri-Yarmand, R., Mandal, M., et al. (2001). Transcriptional repression of oestrogen receptor by metastasis-associated protein 1 corepressor. *Nature Cell Biology*, *3*(1), 30–37.

McMahon, S. B., Van Buskirk, H. A., Dugan, K. A., Copeland, T. D., & Cole, M. D. (1998). The novel ATM-related protein TRRAP is an essential cofactor for the c-Myc and E2F oncoproteins. *Cell*, *94*(3), 363–374.

Medina, P. P., Carretero, J., Fraga, M. F., Esteller, M., Sidransky, D., & Sanchez-Cespedes, M. (2004). Genetic and epigenetic screening for gene alterations of the chromatin-remodeling factor, SMARCA4/BRG1, in lung tumors. *Genes, Chromosomes & Cancer*, *41*(2), 170–177.

Medina, P. P., Romero, O. A., Kohno, T., Montuenga, L. M., Pio, R., Yokota, J., et al. (2008). Frequent BRG1/SMARCA4-inactivating mutations in human lung cancer cell lines. *Human Mutation*, *29*(5), 617–622.

Menard, L., Taras, D., Grigoletto, A., Haurie, V., Nicou, A., Dugot-Senant, N., et al. (2010). In vivo silencing of Reptin blocks the progression of human hepatocellular carcinoma in xenografts and is associated with replicative senescence. *Journal of Hepatology*, *52*(5), 681–689.

Migliori, V., Mapelli, M., & Guccione, E. (2012). On WD40 proteins: Propelling our knowledge of transcriptional control? *Epigenetics*, 7(8), 815–822.

Min, J. N., Tian, Y., Xiao, Y., Wu, L., Li, L., & Chang, S. (2013). The mINO80 chromatin remodeling complex is required for efficient telomere replication and maintenance of genome stability. *Cell Research*, *23*, 1396–1413.

Molli, P. R., Singh, R. R., Lee, S. W., & Kumar, R. (2008). MTA1-mediated transcriptional repression of BRCA1 tumor suppressor gene. *Oncogene*, *27*(14), 1971–1980.

Moon, H. E., Cheon, H., Chun, K. H., Lee, S. K., Kim, Y. S., Jung, B. K., et al. (2006). Metastasis-associated protein 1 enhances angiogenesis by stabilization of HIF-1alpha. *Oncology Reports*, *16*(4), 929–935.

Moon, H. E., Cheon, H., & Lee, M. S. (2007). Metastasis-associated protein 1 inhibits p53-induced apoptosis. *Oncology Reports*, *18*(5), 1311–1314.

Morey, L., Brenner, C., Fazi, F., Villa, R., Gutierrez, A., Buschbeck, M., et al. (2008). MBD3, a component of the NuRD complex, facilitates chromatin alteration and deposition of epigenetic marks. *Molecular and Cellular Biology*, *28*(19), 5912–5923.

Morrison, A. J., & Shen, X. (2009). Chromatin remodelling beyond transcription: The INO80 and SWR1 complexes. *Nature Reviews Molecular Cell Biology*, *10*(6), 373–384.

Murr, R., Loizou, J. I., Yang, Y. G., Cuenin, C., Li, H., Wang, Z. Q., et al. (2006). Histone acetylation by Trrap-Tip60 modulates loading of repair proteins and repair of DNA double-strand breaks. *Nature Cell Biology*, *8*(1), 91–99.

Murr, R., Vaissiere, T., Sawan, C., Shukla, V., & Herceg, Z. (2007). Orchestration of chromatin-based processes: Mind the TRRAP. *Oncogene*, *26*(37), 5358–5372.

Nagarajan, P., Onami, T. M., Rajagopalan, S., Kania, S., Donnell, R., & Venkatachalam, S. (2009). Role of chromodomain helicase DNA-binding protein 2 in DNA damage response signaling and tumorigenesis. *Oncogene*, *28*(8), 1053–1062.

Nagl, N. G., Jr., Patsialou, A., Haines, D. S., Dallas, P. B., Beck, G. R., Jr., & Moran, E. (2005). The p270 (ARID1A/SMARCF1) subunit of mammalian SWI/SNF-related complexes is essential for normal cell cycle arrest. *Cancer Research*, *65*(20), 9236–9244.

Nagl, N. G., Jr., Wang, X., Patsialou, A., Van Scoy, M., & Moran, E. (2007). Distinct mammalian SWI/SNF chromatin remodeling complexes with opposing roles in cell-cycle control. *The EMBO Journal*, *26*(3), 752–763.

Naidu, S. R., Love, I. M., Imbalzano, A. N., Grossman, S. R., & Androphy, E. J. (2009). The SWI/SNF chromatin remodeling subunit BRG1 is a critical regulator of p53 necessary for proliferation of malignant cells. *Oncogene*, *28*(27), 2492–2501.

Nair, S. S., Li, D. Q., & Kumar, R. (2013). A core chromatin remodeling factor instructs global chromatin signaling through multivalent reading of nucleosome codes. *Molecular Cell*, *49*(4), 704–718.

Nebbioso, A., Carafa, V., Benedetti, R., & Altucci, L. (2012). Trials with 'epigenetic' drugs: An update. *Molecular Oncology*, *6*(6), 657–682.

Nguyen, V. Q., Ranjan, A., Stengel, F., Wei, D., Aebersold, R., Wu, C., et al. (2013). Molecular architecture of the ATP-dependent chromatin-remodeling complex SWR1. *Cell*, *154*(6), 1220–1231.

Niimi, A., Chambers, A. L., Downs, J. A., & Lehmann, A. R. (2012). A role for chromatin remodellers in replication of damaged DNA. *Nucleic Acids Research*, *40*(15), 7393–7403.

Nowell, P. C. (1976). The clonal evolution of tumor cell populations. *Science*, *194*(4260), 23–28.

Ohshiro, K., Rayala, S. K., Wigerup, C., Pakala, S. B., Natha, R. S., Gururaj, A. E., et al. (2010). Acetylation-dependent oncogenic activity of metastasis-associated protein 1 co-regulator. *EMBO Reports*, *11*(9), 691–697.

Oya, H., Yokoyama, A., Yamaoka, I., Fujiki, R., Yonezawa, M., Youn, M. Y., et al. (2009). Phosphorylation of Williams syndrome transcription factor by MAPK induces a switching between two distinct chromatin remodeling complexes. *The Journal of Biological Chemistry*, *284*(47), 32472–32482.

Pan, M. R., Hsieh, H. J., Dai, H., Hung, W. C., Li, K., Peng, G., et al. (2012). Chromodomain helicase DNA-binding protein 4 (CHD4) regulates homologous recombination DNA repair, and its deficiency sensitizes cells to poly(ADP-ribose) polymerase (PARP) inhibitor treatment. *The Journal of Biological Chemistry*, *287*(9), 6764–6772.

Papamichos-Chronakis, M., Watanabe, S., Rando, O. J., & Peterson, C. L. (2011). Global regulation of H2A.Z localization by the INO80 chromatin-remodeling enzyme is essential for genome integrity. *Cell*, *144*(2), 200–213.

Park, E. J., Hur, S. K., & Kwon, J. (2010). Human INO80 chromatin-remodelling complex contributes to DNA double-strand break repair via the expression of Rad54B and XRCC3 genes. *The Biochemical Journal*, *431*(2), 179–187.

Park, H. Y., Jeon, Y. K., Shin, H. J., Kim, I. J., Kang, H. C., Jeong, S. J., et al. (2007). Differential promoter methylation may be a key molecular mechanism in regulating BubR1 expression in cancer cells. *Experimental & Molecular Medicine*, *39*(2), 195–204.

Park, J., Wood, M. A., & Cole, M. D. (2002). BAF53 forms distinct nuclear complexes and functions as a critical c-Myc-interacting nuclear cofactor for oncogenic transformation. *Molecular and Cellular Biology*, *22*(5), 1307–1316.

Paul, S., Kuo, A., Schalch, T., Vogel, H., Joshua-Tor, L., McCombie, W. R., et al. (2013). Chd5 requires PHD-mediated histone 3 binding for tumor suppression. *Cell Reports*, *3*(1), 92–102.

Pawlowski, R., Muhl, S. M., Sulser, T., Krek, W., Moch, H., & Schraml, P. (2013). Loss of PBRM1 expression is associated with renal cell carcinoma progression. *International Journal of Cancer*, *132*(2), E11–E17.

Perez-Pinera, P., Ousterout, D. G., & Gersbach, C. A. (2012). Advances in targeted genome editing. *Current Opinion in Chemical Biology*, *16*(3–4), 268–277.

Plass, C., Pfister, S. M., Lindroth, A. M., Bogatyrova, O., Claus, R., & Lichter, P. (2013). Mutations in regulators of the epigenome and their connections to global chromatin patterns in cancer. *Nature Reviews Genetics*, *14*(11), 765–780.

Pleasance, E. D., Stephens, P. J., O'Meara, S., McBride, D. J., Meynert, A., Jones, D., et al. (2010). A small-cell lung cancer genome with complex signatures of tobacco exposure. *Nature*, *463*(7278), 184–190.

Polakis, P. (2000). Wnt signaling and cancer. *Genes & Development*, *14*(15), 1837–1851.

Polo, S. E., Kaidi, A., Baskcomb, L., Galanty, Y., & Jackson, S. P. (2010). Regulation of DNA-damage responses and cell-cycle progression by the chromatin remodelling factor CHD4. *The EMBO Journal*, *29*(18), 3130–3139.

Poot, R. A., Bozhenok, L., van den Berg, D. L., Hawkes, N., & Varga-Weisz, P. D. (2005). Chromatin remodeling by WSTF-ISWI at the replication site: Opening a window of opportunity for epigenetic inheritance? *Cell Cycle*, *4*(4), 543–546. [Research Support, Non-U.S. Gov't Review].

Poot, R. A., Bozhenok, L., van den Berg, D. L., Steffensen, S., Ferreira, F., Grimaldi, M., et al. (2004). The Williams syndrome transcription factor interacts with PCNA to target chromatin remodelling by ISWI to replication foci. *Nature Cell Biology*, *6*(12), 1236–1244.

Poot, R. A., Dellaire, G., Hulsmann, B. B., Grimaldi, M. A., Corona, D. F., Becker, P. B., et al. (2000). HuCHRAC, a human ISWI chromatin remodelling complex contains hACF1 and two novel histone-fold proteins. *The EMBO Journal*, *19*(13), 3377–3387.

Potts, R. C., Zhang, P., Wurster, A. L., Precht, P., Mughal, M. R., Wood, W. H., 3rd., et al. (2011). CHD5, a brain-specific paralog of Mi2 chromatin remodeling enzymes, regulates expression of neuronal genes. *PLoS One*, *6*(9), e24515.

Ptashne, M. (2013). Epigenetics: Core misconcept. *Proceedings of the National Academy of Sciences of the United States of America*, *110*(18), 7101–7103.

Pulukuri, S. M., & Rao, J. S. (2006). CpG island promoter methylation and silencing of 14-3-3sigma gene expression in LNCaP and Tramp-C1 prostate cancer cell lines is associated with methyl-CpG-binding protein MBD2. *Oncogene*, *25*(33), 4559–4572.

Qu, Y., Dang, S., & Hou, P. (2013). Gene methylation in gastric cancer. *Clinica Chimica Acta*, *424*, 53–65.

Rao, M., Casimiro, M. C., Lisanti, M. P., D'Amico, M., Wang, C., Shirley, L. A., et al. (2008). Inhibition of cyclin D1 gene transcription by Brg-1. *Cell Cycle*, *7*(5), 647–655.

Reisman, D. N., Sciarrotta, J., Wang, W., Funkhouser, W. K., & Weissman, B. E. (2003). Loss of BRG1/BRM in human lung cancer cell lines and primary lung cancers: Correlation with poor prognosis. *Cancer Research*, *63*(3), 560–566.

Reyes, J. C., Barra, J., Muchardt, C., Camus, A., Babinet, C., & Yaniv, M. (1998). Altered control of cellular proliferation in the absence of mammalian brahma (SNF2alpha). *The EMBO Journal*, *17*(23), 6979–6991.

Rhodes, D. R., Yu, J., Shanker, K., Deshpande, N., Varambally, R., Ghosh, D., et al. (2004). ONCOMINE: A cancer microarray database and integrated data-mining platform. *Neoplasia*, *6*(1), 1–6.

Roberts, C. W., Galusha, S. A., McMenamin, M. E., Fletcher, C. D., & Orkin, S. H. (2000). Haploinsufficiency of Snf5 (integrase interactor 1) predisposes to malignant rhabdoid tumors in mice. *Proceedings of the National Academy of Sciences of the United States of America*, *97*(25), 13796–13800.

Roberts, C. W., Leroux, M. M., Fleming, M. D., & Orkin, S. H. (2002). Highly penetrant, rapid tumorigenesis through conditional inversion of the tumor suppressor gene Snf5. *Cancer Cell*, *2*(5), 415–425.

Rodriguez-Nieto, S., Canada, A., Pros, E., Pinto, A. I., Torres-Lanzas, J., Lopez-Rios, F., et al. (2011). Massive parallel DNA pyrosequencing analysis of the tumor suppressor BRG1/SMARCA4 in lung primary tumors. *Human Mutation*, *32*(2), E1999–E2017.

Romero, O. A., Setien, F., John, S., Gimenez-Xavier, P., Gomez-Lopez, G., Pisano, D., et al. (2012). The tumour suppressor and chromatin-remodelling factor BRG1 antagonizes Myc activity and promotes cell differentiation in human cancer. *EMBO Molecular Medicine*, *4*(7), 603–616.

Rosenbaum, J., Baek, S. H., Dutta, A., Houry, W. A., Huber, O., Hupp, T. R., et al. (2013). The emergence of the conserved AAA+ ATPases Pontin and Reptin on the signaling landscape. *Science Signaling*, *6*(266), mr1.

Ruhl, D. D., Jin, J., Cai, Y., Swanson, S., Florens, L., Washburn, M. P., et al. (2006). Purification of a human SRCAP complex that remodels chromatin by incorporating the histone variant H2A.Z into nucleosomes. *Biochemistry*, *45*(17), 5671–5677.

Sanchez-Tillo, E., Lazaro, A., Torrent, R., Cuatrecasas, M., Vaquero, E. C., Castells, A., et al. (2010). ZEB1 represses E-cadherin and induces an EMT by recruiting the SWI/SNF chromatin-remodeling protein BRG1. *Oncogene*, *29*(24), 3490–3500.

Schlabach, M. R., Luo, J., Solimini, N. L., Hu, G., Xu, Q., Li, M. Z., et al. (2008). Cancer proliferation gene discovery through functional genomics. *Science*, *319*(5863), 620–624.

Sevenet, N., Sheridan, E., Amram, D., Schneider, P., Handgretinger, R., & Delattre, O. (1999). Constitutional mutations of the hSNF5/INI1 gene predispose to a variety of cancers. *American Journal of Human Genetics*, *65*(5), 1342–1348.

Shain, A. H., & Pollack, J. R. (2013). The spectrum of SWI/SNF mutations, ubiquitous in human cancers. *PLoS One*, *8*(1), e55119.

Shen, L., & Zhang, Y. (2013). 5-Hydroxymethylcytosine: Generation, fate, and genomic distribution. *Current Opinion in Cell Biology*, *25*(3), 289–296.

Sheu, J. J., Choi, J. H., Guan, B., Tsai, F. J., Hua, C. H., Lai, M. T., et al. (2013). Rsf-1, a chromatin remodelling protein, interacts with cyclin E1 and promotes tumour development. *The Journal of Pathology*, *229*(4), 559–568.

Sheu, J. J., Guan, B., Choi, J. H., Lin, A., Lee, C. H., Hsiao, Y. T., et al. (2010). Rsf-1, a chromatin remodeling protein, induces DNA damage and promotes genomic instability. *The Journal of Biological Chemistry*, *285*(49), 38260–38269.

Shukeir, N., Pakneshan, P., Chen, G., Szyf, M., & Rabbani, S. A. (2006). Alteration of the methylation status of tumor-promoting genes decreases prostate cancer cell invasiveness and tumorigenesis in vitro and in vivo. *Cancer Research*, *66*(18), 9202–9210.

Skene, P. J., & Henikoff, S. (2013). Histone variants in pluripotency and disease. *Development*, *140*(12), 2513–2524.

Smeenk, G., Wiegant, W. W., Vrolijk, H., Solari, A. P., Pastink, A., & van Attikum, H. (2010). The NuRD chromatin-remodeling complex regulates signaling and repair of DNA damage. *The Journal of Cell Biology*, *190*(5), 741–749.

Srinivasan, R., Mager, G. M., Ward, R. M., Mayer, J., & Svaren, J. (2006). NAB2 represses transcription by interacting with the CHD4 subunit of the nucleosome remodeling and deacetylase (NuRD) complex. *The Journal of Biological Chemistry*, *281*(22), 15129–15137.

Stopka, T., & Skoultchi, A. I. (2003). The ISWI ATPase Snf2h is required for early mouse development. *Proceedings of the National Academy of Sciences of the United States of America*, *100*(24), 14097–14102.

Strobeck, M. W., DeCristofaro, M. F., Banine, F., Weissman, B. E., Sherman, L. S., & Knudsen, E. S. (2001). The BRG-1 subunit of the SWI/SNF complex regulates CD44 expression. *The Journal of Biological Chemistry*, *276*(12), 9273–9278.

Strohner, R., Nemeth, A., Jansa, P., Hofmann-Rohrer, U., Santoro, R., Langst, G., et al. (2001). NoRC—A novel member of mammalian ISWI-containing chromatin remodeling machines. *The EMBO Journal*, *20*(17), 4892–4900.

Tamborero, D., Gonzalez-Perez, A., Perez-Llamas, C., Deu-Pons, J., Kandoth, C., Reimand, J., et al. (2013). Comprehensive identification of mutational cancer driver genes across 12 tumor types. *Scientific Reports*, *3*, 2650.

Taub, R., Kirsch, I., Morton, C., Lenoir, G., Swan, D., Tronick, S., et al. (1982). Translocation of the c-myc gene into the immunoglobulin heavy chain locus in human Burkitt lymphoma and murine plasmacytoma cells. *Proceedings of the National Academy of Sciences of the United States of America*, *79*(24), 7837–7841.

Toiber, D., Erdel, F., Bouazoune, K., Silberman, D. M., Zhong, L., Mulligan, P., et al. (2013). SIRT6 recruits SNF2H to DNA break sites, preventing genomic instability through chromatin remodeling. *Molecular Cell*, *51*(4), 454–468.

Tong, J. K., Hassig, C. A., Schnitzler, G. R., Kingston, R. E., & Schreiber, S. L. (1998). Chromatin deacetylation by an ATP-dependent nucleosome remodelling complex. *Nature*, *395*(6705), 917–921.

Tosi, A., Haas, C., Herzog, F., Gilmozzi, A., Berninghausen, O., Ungewickell, C., et al. (2013). Structure and subunit topology of the INO80 chromatin remodeler and its nucleosome complex. *Cell*, *154*(6), 1207–1219.

Toyoshima, M., Howie, H. L., Imakura, M., Walsh, R. M., Annis, J. E., Chang, A. N., et al. (2012). Functional genomics identifies therapeutic targets for MYC-driven cancer. *Proceedings of the National Academy of Sciences of the United States of America, 109*(24), 9545–9550.

Tsikitis, M., Zhang, Z., Edelman, W., Zagzag, D., & Kalpana, G. V. (2005). Genetic ablation of Cyclin D1 abrogates genesis of rhabdoid tumors resulting from Ini1 loss. *Proceedings of the National Academy of Sciences of the United States of America, 102*(34), 12129–12134.

Udugama, M., Sabri, A., & Bartholomew, B. (2011). The INO80 ATP-dependent chromatin remodeling complex is a nucleosome spacing factor. *Molecular and Cellular Biology, 31*(4), 662–673.

Valdman, A., Nordenskjold, A., Fang, X., Naito, A., Al-Shukri, S., Larsson, C., et al. (2003). Mutation analysis of the BRG1 gene in prostate cancer clinical samples. *International Journal of Oncology, 22*(5), 1003–1007.

Van Rechem, C., Boulay, G., Pinte, S., Stankovic-Valentin, N., Guerardel, C., & Leprince, D. (2010). Differential regulation of HIC1 target genes by CtBP and NuRD, via an acetylation/SUMOylation switch, in quiescent versus proliferating cells. *Molecular and Cellular Biology, 30*(16), 4045–4059.

Villaronga, M. A., Lopez-Mateo, I., Markert, L., Espinosa, E., Fresno Vara, J. A., & Belandia, B. (2011). Identification and characterization of novel potentially oncogenic mutations in the human BAF57 gene in a breast cancer patient. *Breast Cancer Research and Treatment, 128*(3), 891–898.

Vries, R. G., Bezrookove, V., Zuijderduijn, L. M., Kia, S. K., Houweling, A., Oruetxebarria, I., et al. (2005). Cancer-associated mutations in chromatin remodeler hSNF5 promote chromosomal instability by compromising the mitotic checkpoint. *Genes & Development, 19*(6), 665–670.

Wang, J., Chen, H., Fu, S., Xu, Z. M., Sun, K. L., & Fu, W. N. (2011). The involvement of CHD5 hypermethylation in laryngeal squamous cell carcinoma. *Oral Oncology, 47*(7), 601–608.

Wang, W., Cote, J., Xue, Y., Zhou, S., Khavari, P. A., Biggar, S. R., et al. (1996). Purification and biochemical heterogeneity of the mammalian SWI-SNF complex. *The EMBO Journal, 15*(19), 5370–5382.

Wang, Z., Zang, C., Rosenfeld, J. A., Schones, D. E., Barski, A., Cuddapah, S., et al. (2008). Combinatorial patterns of histone acetylations and methylations in the human genome. *Nature Genetics, 40*(7), 897–903.

Wang, S., Zhang, B., & Faller, D. V. (2002). Prohibitin requires Brg-1 and Brm for the repression of E2F and cell growth. *The EMBO Journal, 21*(12), 3019–3028.

Ward, A. B., Sali, A., & Wilson, I. A. (2013). Biochemistry. Integrative structural biology. *Science, 339*(6122), 913–915.

Weinstein, I. B. (2002). Cancer. Addiction to oncogenes—The Achilles heal of cancer. *Science, 297*(5578), 63–64.

Wiegand, K. C., Shah, S. P., Al-Agha, O. M., Zhao, Y., Tse, K., Zeng, T., et al. (2010). ARID1A mutations in endometriosis-associated ovarian carcinomas. *The New England Journal of Medicine, 363*(16), 1532–1543.

Wilson, B. G., Wang, X., Shen, X., McKenna, E. S., Lemieux, M. E., Cho, Y. J., et al. (2010). Epigenetic antagonism between polycomb and SWI/SNF complexes during oncogenic transformation. *Cancer Cell, 18*(4), 316–328.

Wong, M. M., Cox, L. K., & Chrivia, J. C. (2007). The chromatin remodeling protein, SRCAP, is critical for deposition of the histone variant H2A.Z at promoters. *The Journal of Biological Chemistry, 282*(36), 26132–26139.

Wong, A. K., Shanahan, F., Chen, Y., Lian, L., Ha, P., Hendricks, K., et al. (2000). BRG1, a component of the SWI-SNF complex, is mutated in multiple human tumor cell lines. *Cancer Research, 60*(21), 6171–6177.

Wong-Staal, F., Dalla-Favera, R., Franchini, G., Gelmann, E. P., & Gallo, R. C. (1981). Three distinct genes in human DNA related to the transforming genes of mammalian sarcoma retroviruses. *Science*, *213*(4504), 226–228.

Wood, M. A., McMahon, S. B., & Cole, M. D. (2000). An ATPase/helicase complex is an essential cofactor for oncogenic transformation by c-Myc. *Molecular Cell*, *5*(2), 321–330.

Woodcock, C. L., & Ghosh, R. P. (2010). Chromatin higher-order structure and dynamics. *Cold Spring Harbor Perspectives in Biology*, *2*(5), a000596.

Workman, J. L., & Kingston, R. E. (1992). Nucleosome core displacement in vitro via a metastable transcription factor-nucleosome complex. *Science*, *258*(5089), 1780–1784.

Wu, S., Shi, Y., Mulligan, P., Gay, F., Landry, J., Liu, H., et al. (2007). A YY1-INO80 complex regulates genomic stability through homologous recombination-based repair. *Nature Structural & Molecular Biology*, *14*(12), 1165–1172.

Wurdak, H., Zhu, S., Romero, A., Lorger, M., Watson, J., Chiang, C. Y., et al. (2010). An RNAi screen identifies TRRAP as a regulator of brain tumor-initiating cell differentiation. *Cell Stem Cell*, *6*(1), 37–47.

Xia, W., Nagase, S., Montia, A. G., Kalachikov, S. M., Keniry, M., Su, T., et al. (2008). BAF180 is a critical regulator of p21 induction and a tumor suppressor mutated in breast cancer. *Cancer Research*, *68*(6), 1667–1674.

Xu, Y., & Price, B. D. (2011). Chromatin dynamics and the repair of DNA double strand breaks. *Cell Cycle*, *10*(2), 261–267.

Xu, Y., Sun, Y., Jiang, X., Ayrapetov, M. K., Moskwa, P., Yang, S., et al. (2010). The p400 ATPase regulates nucleosome stability and chromatin ubiquitination during DNA repair. *The Journal of Cell Biology*, *191*(1), 31–43.

Ye, Y., Xiao, Y., Wang, W., Gao, J. X., Yearsley, K., Yan, Q., et al. (2012). Singular v dual inhibition of SNF2L and its isoform, SNF2LT, have similar effects on DNA damage but opposite effects on the DNA damage response, cancer cell growth arrest and apoptosis. *Oncotarget*, *3*(4), 475–489.

Ye, Y., Xiao, Y., Wang, W., Wang, Q., Yearsley, K., Wani, A. A., et al. (2009). Inhibition of expression of the chromatin remodeling gene, SNF2L, selectively leads to DNA damage, growth inhibition, and cancer cell death. *Molecular Cancer Research*, 7(12), 1984–1999.

Yeh, J. E., Toniolo, P. A., & Frank, D. A. (2013). Targeting transcription factors: Promising new strategies for cancer therapy. *Current Opinion in Oncology*, *25*(6), 652–658.

Yip, D. J., Corcoran, C. P., Alvarez-Saavedra, M., DeMaria, A., Rennick, S., Mears, A. J., et al. (2012). Snf2l regulates Foxg1-dependent progenitor cell expansion in the developing brain. *Developmental Cell*, *22*(4), 871–878.

Yoo, Y. G., Kong, G., & Lee, M. O. (2006). Metastasis-associated protein 1 enhances stability of hypoxia-inducible factor-1alpha protein by recruiting histone deacetylase 1. *The EMBO Journal*, *25*(6), 1231–1241.

Yoshida, T., Hazan, I., Zhang, J., Ng, S. Y., Naito, T., Snippert, H. J., et al. (2008). The role of the chromatin remodeler Mi-2beta in hematopoietic stem cell self-renewal and multilineage differentiation. *Genes & Development*, *22*(9), 1174–1189.

Yun, M., Wu, J., Workman, J. L., & Li, B. (2011). Readers of histone modifications. *Cell Research*, *21*(4), 564–578.

Zang, Z. J., Cutcutache, I., Poon, S. L., Zhang, S. L., McPherson, J. R., Tao, J., et al. (2012). Exome sequencing of gastric adenocarcinoma identifies recurrent somatic mutations in cell adhesion and chromatin remodeling genes. *Nature Genetics*, *44*(5), 570–574.

Zhang, X. Y., DeSalle, L. M., Patel, J. H., Capobianco, A. J., Yu, D., Thomas-Tikhonenko, A., et al. (2005). Metastasis-associated protein 1 (MTA1) is an essential downstream effector of the c-MYC oncoprotein. *Proceedings of the National Academy of Sciences of the United States of America*, *102*(39), 13968–13973.

Zhang, Z., Wippo, C. J., Wal, M., Ward, E., Korber, P., & Pugh, B. F. (2011). A packing mechanism for nucleosome organization reconstituted across a eukaryotic genome. *Science*, *332*(6032), 977–980.

Zhao, R., Yan, Q., Lv, J., Huang, H., Zheng, W., Zhang, B., et al. (2012). CHD5, a tumor suppressor that is epigenetically silenced in lung cancer. *Lung Cancer*, *76*(3), 324–331.

Zhuravel, E., Shestakova, T., Glushko, N., Soldatkina, M., & Pogrebnoy, P. (2008). Expression patterns of murine beta-defensin-2 mRNA in Lewis lung carcinoma cells in vitro and in vivo. *Experimental Oncology*, *30*(3), 206–211.

Ziller, M. J., Muller, F., Liao, J., Zhang, Y., Gu, H., Bock, C., et al. (2011). Genomic distribution and inter-sample variation of non-CpG methylation across human cell types. *PLoS Genetics*, 7(12), e1002389.

Zupkovitz, G., Grausenburger, R., Brunmeir, R., Senese, S., Tischler, J., Jurkin, J., et al. (2010). The cyclin-dependent kinase inhibitor p21 is a crucial target for histone deacetylase 1 as a regulator of cellular proliferation. *Molecular and Cellular Biology*, *30*(5), 1171–1181.

CHAPTER SIX

Diffuse Intrinsic Pontine Gliomas: Treatments and Controversies

Amy Lee Bredlau[*,†,1], **David N. Korones**[‡,§]
[*]Department of Pediatrics, Medical University of South Carolina, Charleston, South Carolina, USA
[†]Department of Neurosciences, Medical University of South Carolina, Charleston, South Carolina, USA
[‡]Department of Pediatrics, University of Rochester, Rochester, New York, USA
[§]Department of Palliative Care, University of Rochester, Rochester, New York, USA
[1]Corresponding author: e-mail address: bredlau@musc.edu

Contents

Abstract

Diffuse intrinsic pontine gliomas (DIPGs) are a fairly common pediatric brain tumor, and children with these tumors have a dismal prognosis. They generally are diagnosed within the first decade of life, and due to their location within the pons, these tumors are not surgically resectable. The median survival for children with DIPGs is less than 1 year, in spite of decades of clinical trial development of unique approaches to radiation therapy and chemotherapy. Novel therapies are under investigation for these deadly tumors. As clinicians and researchers make a concerted effort to obtain tumor tissue, the molecular signals of these tumors are being investigated in an attempt to uncover targetable therapies for DIPGs. In addition, direct application of chemotherapies into the tumor (convection-enhanced delivery) is being investigated as a novel delivery system for treatment of DIPGs. Overall, DIPGs require creative thinking and a disciplined approach for development of a therapy that can improve the prognosis for these unfortunate children.

Advances in Cancer Research, Volume 121
ISSN 0065-230X
http://dx.doi.org/10.1016/B978-0-12-800249-0.00006-8

1. INTRODUCTION

Diffuse intrinsic pontine gliomas (DIPGs) (also known as pontine gliomas and brain stem gliomas) make up around 10% of all pediatric brain tumors (Guillamo, Doz, & Delattre, 2001). In spite of decades of investigation, these tumors remain refractory to therapy and result in a mean life expectancy of 9–12 months from diagnosis (Broniscer & Gajjar, 2004; Guillamo et al., 2001; Hargrave, Bartels, & Bouffet, 2006; Korones, 2007). DIPGs appear to be astrocytomas of World Health Organization grades II–IV (Farmer et al., 2001), though they do not respond to therapies that effectively treat some of these astrocytomas in other areas of the brain (Fig. 6.1). The differences in the efficacy of therapies between DIPGs and other astrocytomas are under investigation, as design of effective therapies will hinge upon deeper understanding of the unique challenges and molecular differences present in DIPGs.

2. DIAGNOSIS

Children with DIPGs are diagnosed on the basis of neurologic symptoms and magnetic resonance imaging (MRI) findings and do not traditionally

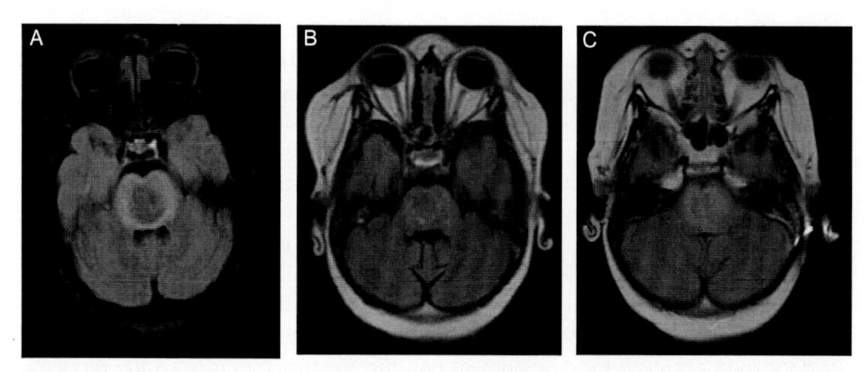

Figure 6.1 MRI images of a child with DIPG through the disease course. (A) T2 FLAIR image at diagnosis with heterogeneity and edema of a large pontine mass, (B) FLAIR image after radiation therapy completed, 3 months after diagnosis (note the decrease in size of pontine lesion), and (C) FLAIR image at radiographic recurrence, with increased signal intensity and size of lesion, 6 months from diagnosis. The patient remained asymptomatic with recurrent disease, off therapy, for one more month but succumbed to disease, 9 months from diagnosis.

require a biopsy, a relatively unique approach among children with brain tumors. The diagnostic criteria for a DIPG include fewer than 6 months of symptoms, two of three major neurological abnormalities (cranial nerve deficits, long-tract signs, and/or ataxia), and classic imaging findings on MRI (Donaldson, Laningham, & Fisher, 2006; Hargrave, Bartels, et al., 2006; Korones, 2007) (Fig. 6.2). Although the diagnostic criteria include symptoms up to 6 months, children with more than 3 months of symptoms at presentation are not likely to follow the expected clinical course for a DIPG (Jackson et al., 2013; Shuper et al., 1998) and, arguably, should not be diagnosed with a DIPG without biopsy confirmation.

Younger children with DIPGs can present with unusual signs. Mood changes and irritability can be found in some children with DIPGs (Guillamo et al., 2001). Other children can have "pathological laughter" (gelastic seizures) at presentation. This presents as laughter during sleep or at other inappropriate times. Additionally, separation anxiety that is inappropriate for age (and of a sudden onset) can be found in some patients with DIPG, both at presentation and at progression (Hargrave, Mabbott, & Bouffet, 2006).

The classic major MRI findings of DIPG include hypointensity on T1- and hyperintensity on T2-weighted images (Kornreich et al., 2005). Also, minor findings in support of DIPGs include engulfment of the basilar artery and/or extension into the cerebellopontine angle (Zimmerman, 1996). Most DIPGs are not contrast enhancing at diagnosis. When present, contrast enhancement is a poor prognostic indicator (Poussaint et al., 2011) (Fig. 6.3).

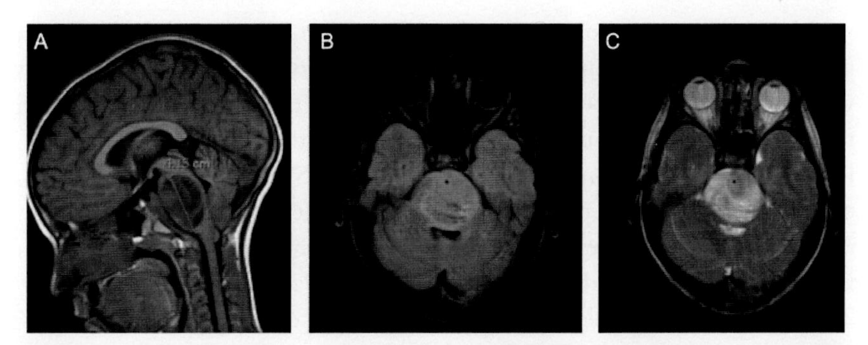

Figure 6.2 MRI images of a young child with newly diagnosed DIPG. (A) T1 sagittal FLAIR with typical hypointensity of DIPG, (B) T2 axial FLAIR with large, heterogeneous, mass in the pons, and (C) T2 axial MRI with typical hyperintensity of pontine glioma.

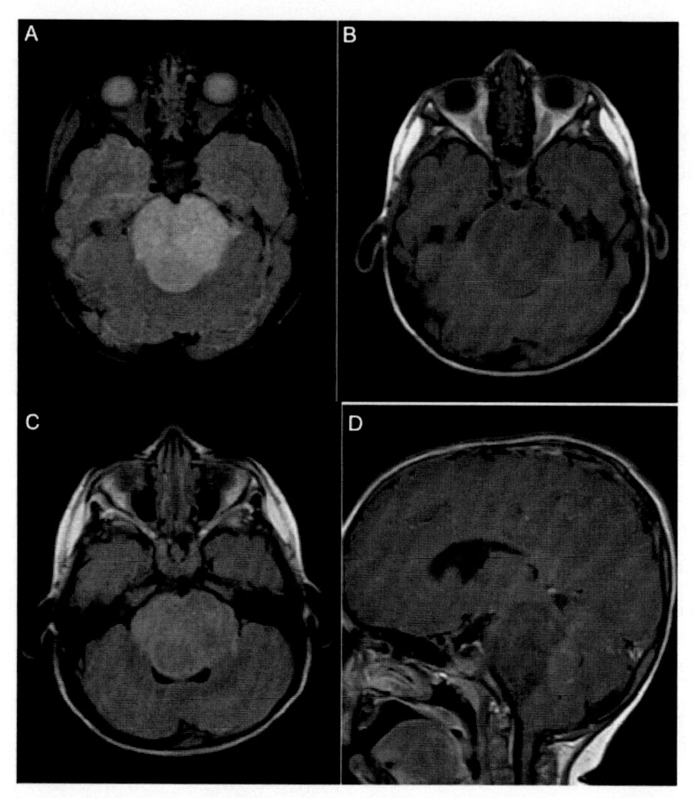

Figure 6.3 MRI images of a toddler with newly diagnosed DIPG. (A) T2-weighted axial image with heterogeneous hyperintensity in the pons, (B) T1-weighted axial image with heterogeneous hypointensity in the pons, (C) T1 FLAIR image with mild/moderate edema in the surrounding cerebellum, and (D) T1 contrast enhanced sagittal image with heterogeneous contrast enhancement that is sometimes seen in DIPGs.

Many different radiographic approaches for prognostication of children with DIPGs have been investigated, as there are some children with DIPGs who survive far longer than the median, though their clinical presentation may not be different from those who succumb to their disease more rapidly. In 2002, magnetic resonance spectroscopy (MRS) was first used to determine the malignancy of a brainstem tumor at the time of diagnosis. A change in the choline:creatine ratio was useful in determining if a lesion was aggressive versus indolent (Curless, Bowen, Pattany, Gonik, & Kramer, 2002). This study revealed that increased choline:creatine peaks are consistent with more aggressive DIPGs and shorter overall survival. Subsequently, MRS was noted to measure spectral abnormalities prior to MRI deterioration or at diagnosis, giving further prognostic information to patients and their families (Helton et al., 2006; Laprie et al., 2005; Thakur, Karimi, Dunkel, Koutcher, & Huang, 2006). Since that time, diffusion tensor imaging (DTI) has been investigated for DIPGs. DTI allows description of the anterior transverse pontine fibers, medial lemnisci, and coritcospinal tracts and can be used to determine early tumor progression (Prabhu et al., 2011). Similarly, PET scanning along with MRI can be prognostic because DIPGs that are ^{18}F-FDG avid in more than 50% of the tumor are likely to progress earlier than those with less PET avidity (Zukotynski et al., 2011).

Of note, age at diagnosis may also be prognostic. Children <3 years old at diagnosis can have a prolonged survival, living years after their diagnosis, often with temporary disease stabilization (Broniscer et al., 2008; Jackson et al., 2013; Thompson & Kosnik, 2005).

While primary dissemination of DIPGs is rare (Benesch, Wagner, Berthold, & Wolff, 2005), dissemination of DIPGs at recurrence is common (Gururangan et al., 2006; Sethi et al., 2011). The pattern of disease spread at progression is not prognostic (Wagner et al., 2006).

3. HISTORICAL PERSPECTIVES

Previous to the "MRI era," and even before computed tomography (CT) scan era, brain stem tumors were diagnosed with angiograms and air encephalograms. With these imprecise tools, tumors found in the brainstem required biopsy for diagnosis. Tumors found in the area were generally glioblastoma, astrocytoma, or ependymoma (Reigel, Scarff, & Woodford, 1979). Surgery is no longer considered standard of care in the diagnosis of DIPGs, due in large part to the rarity of misdiagnosis when the clinical and MRI diagnostic criteria are met. This new standard of care was established after an article by Albright in 1993 (Albright et al., 1993), wherein he argued that symptomalogic and radiographic criteria for diagnosis were adequate for diagnosis and perhaps superior to pathologic diagnosis due to the potential morbidity of a brain biopsy. This was motivated, in part, by the safety and reliability of diagnosis as above, the widespread availability of MRI, and, in part, by evidence that the outcome of patients with DIPGs were not improved by surgery (Epstein & McCleary, 1986; Pierre-Kahn et al., 1993), though stereotaxic biopsies of pontine tumors were being done safely at that time (Hood, Gebarski, McKeever, & Venes, 1986).

However, more recently, this noninvasive approach is being reevaluated as the lack of tissue is limiting the ability to research the biology of this vexing tumor and ultimately to develop targeted therapies for it. In the 1990s, molecular diagnosis of tumors was uncommon, whereas today it is beginning to guide therapy. Therefore, with 20 years of noninvasive diagnosis of DIPG, the lack of data regarding the molecular makeup of this tumor at diagnosis is keenly felt (MacDonald, 2012). Clinicians are now more willing to consider biopsy at the time of progression (if not diagnosis) to increase understanding of the molecular makeup of these tumors, though it is not yet standard of care in most regions to do so (Cage et al., 2013).

The ethical dilemma inherent in this debate is not to be underestimated. Though stereotactic biopsies are safe to perform (Cartmill & Punt, 1999; Chico-Ponce de Leon et al., 2003; Goncalves-Ferreira, Herculano-Carvalho, & Pimentel, 2003; Phi, Chung, Wang, Ryu, & Kim, 2013), it is difficult to argue that the outcome of any single patient will be improved by undergoing such a procedure when clinical and radiographic data are typically all that is required to support an accurate diagnosis. However, one could also argue that the same single patient is harmed by the lack of progress in the past 20 years in understanding the molecular makeup of DIPGs because biopsies were not done. In fact, a European consensus statement has recently been released, recommending biopsies for presumed DIPGs "to ascertain biological characteristics to enhance understanding and targeting of treatments, especially in clinical trials" (Walker et al., 2013). Nevertheless, because there are risks to biopsy (albeit not high) and no direct benefit to patients, biopsy outside the context of a clinical trial cannot be recommended. This could certainly change in an era of molecularly guided therapy, which is a promising thought in treatment of DIPGs (see below). However, this is not yet standard of care therapy for any brain tumors, let alone DIPGs.

Brainstem (and other brain) tumors have been treated with radiation since the 1940s. Previously, radiation was often performed after surgical intervention for resection or biopsy for pathological diagnosis (Peirce & Bouchard, 1950). In 1975, Lee examined different radiation doses for children with infratentorial tumors. This group found that radiation of at least 50 Gy was needed "since the goal of treatment is cure" (Lee, 1975). In 1979, 19 patients with brainstem gliomas were treated with 36–50 Gy of radiation, with a mean survival of 8.7 months (Atac & Blaauw, 1979). In 1980, Kim et al. treated 63 patients with radiation therapy, noting that survival was slightly better, again, when patients were treated with more than 50 Gy (Kim, Chin, Pollan, Hazel, & Webster, 1980). Littman et al. demonstrated a survival of about 30% at 5 years with 24–64 Gy of radiation therapy. Of the patients in this study, 19 of 62 had biopsy proven DIPGs (Littman et al., 1980). It should be noted that these reports were made prior to the CT and MRI eras, and so the initial clinicoradiographic diagnoses are not necessarily those of DIPGs.

In the 1980s and 1990s, hyperfractionation of radiation was investigated in a series of clinical trials for treatment of DIPGs. Hyperfractionation includes delivery of a high dose of radiation (>70 Gy) over a few weeks by delivering two doses of radiation daily, 5 days a week. This approach was attempted a number of times, but no hyperfractionated dose of radiation was demonstrated to cure DIPGs nor even to improve upon the median survival time for these children (Edwards et al., 1989; Freeman et al., 1988; Mandell et al., 1999; Packer et al., 1994). It was the radiation therapy trials of this period that established the median survival of DIPGs in the range of 8–12 months (Atac & Blaauw, 1979; Edwards et al., 1989; Freeman et al., 1988; Hibi et al., 1992; Lewis, Lucraft, & Gholkar, 1997; Packer et al., 1994).

4. CURRENT TREATMENTS

DIPGs are currently treated with radiation therapy shortly after diagnosis. As noted above, biopsy is generally not performed at the time of diagnosis. Symptoms are controlled with glucocorticoid therapy until radiation therapy enables weaning of the glucocorticoid (which have significant short-term morbidities, including insomnia, hyperphagia, hyperglycemia, hypertension, and severe mood swings, all of which severely impact quality of life). In most patients with DIPGs, the glucocorticoids can be weaned off completely during or shortly after radiation therapy is complete.

Current radiation therapy given to children with DIPGs is based upon a randomized clinical trial in which hyperfractionated radiotherapy was compared to once daily radiation, and no survival benefit was seen with the hyperfractionated approach (Mandell et al., 1999). Radiation therapy is given in one daily treatment of 1.8 Gy per fraction to a total dose of 54–55.8 Gy delivered over approximately 6 weeks (Mandell et al., 1999). Children are often treated with either opposed lateral field radiation or, more commonly, 3D conformal radiation therapy or intensity modulated radiation therapy, which enables the radiation oncology team to minimize the toxicities of radiation to the uninvolved brain. These radiation fields are given to the tumor as well as a 1–1.5 cm margin beyond the fluid-attenuated inversion recovery (FLAIR) extent of tumor (Halperin, Constine, Tarbell, & Kun, 2011).

No form of adjuvant chemotherapy has yet been proved in randomized controlled trials to be superior to radiation alone. Table 6.1 lists Phase I and II

Table 6.1 Phase I/II trials of chemotherapy for treatment of DIPGs

Year	Author	Therapy	Outcome	Comments
1987	Jenkin (Jenkin et al., 1987)	Randomized between RT alone or with lomustine, vincristine, and prednisone	No increase in median survival in chemotherapy arm	Increased risk of infection on chemotherapy arm
1993	Kretschmar (Kretschmar et al., 1993)	Cisplatin and cyclophosphamide prior to RT	No increase in median survival	
1996	Packer (Packer et al., 1996)	Beta-interferon with RT and as maintenance	No increase in survival	
1997	Needle (Needle et al., 1997)	Etoposide for recurrent brain tumors	No tumor response in three treated patients	
1997	Bouffet (Bouffet et al., 1997)	High-dose carmustine with ABMT	No increase in survival	
1998	Walter (Walter et al., 1998)	Carboplatin and etoposide with HRT	No increase in median survival	
1998	Dunkel (Dunkel et al., 1998)	High-dose chemotherapy with thiotepa and etoposide, with and without carmustine and carboplatin prior to ABMT	No increase in median survival	
1999	Jakacki (Jakacki, Siffert, Jamison, Velasquez, & Allen, 1999)	Dose-intensive procarbazine, lomustine, and vincristine with ABMT before, during, and after RT	No increase in median survival	
1999	Allen (Allen et al., 1999)	Carboplatin with HRT	No increase in median survival	Three long-term survivors (46–104 MO) with typical DIPG

Table 6.1 Phase I/II trials of chemotherapy for treatment of DIPGs—cont'd

Year	Author	Therapy	Outcome	Comments
2000	Freeman (Freeman et al., 2000)	Cisplatin and HRT	Lower 1 year survival in cisplatin group	Cisplatin was investigated as a radiosensitizer
2000	Bouffet (Bouffet et al., 2000)	Radiation therapy followed by busulfan and thiotepa followed by ABMT	No increase in median survival	
2000	Broniscer (Broniscer, Leite, Lanchote, Machado, & Cristofani, 2000)	Tamoxifen with and after RT	No increase in median survival	
2001	Benesch (Benesch et al., 2001)	Ifosfamide, etoposide, methotrexate, cisplatin, and cytarabine with RT, followed by carmustine, carboplatin, and vincristine maintenance	36% 22 MO or more survival (median and overall survival not reported)	Toxic therapy, Only one 22 MO survivor met current DIPG diagnostic criteria
2001	Hurwitz (Hurwitz et al., 2001)	Paclitaxel alone in progressive or recurrent tumors	No increase in median survival	
2002	Doz (Doz et al., 2002)	Carboplatin before and after RT	No increase in median survival	
2002	Jennings (Jennings et al., 2002)	Carboplatin, etoposide, vincristine, or cisplatin, etoposide, cyclophosphamide, and vincristine before HRT	No increase in median survival in either regimen	
2002	Lashford (Lashford et al., 2002)	Temozolomide for progressive or recurrent tumors	No increase in median survival	

Continued

Table 6.1 Phase I/II trials of chemotherapy for treatment of DIPGs—cont'd

Year	Author	Therapy	Outcome	Comments
2002	Wolff (Wolff et al., 2002)	Trophosphamide and etoposide with RT	No increase in median survival	
2003	Dreyer (Dreyer et al., 2003)	Idarubicin for progressive or recurrent tumors	No increase in median survival	
2003	Marcus (Marcus et al., 2003)	Etanidazole with HRT	No increase in median survival	
2003	Sanghavi (Sanghavi et al., 2003)	Topotecan as radiosensitizer for RT, Phase I study	Median survival 15 MO (compared to 8–12 MO)	Not designed to detect a survival advantage
2005	Bernier-Chastagner (Bernier-Chastagner et al., 2005)	Topotecan as radiosensitizer for RT, Phase II study	No increase in median survival	
2005	Broniscer (Broniscer et al., 2005)	Temozolomide after RT	No increase in median survival	
2005	Greenberg (Greenberg et al., 2005)	Etoposide, vincristine, cyclosporine with RT	No increase in median survival	Excessively toxic regimen
2005	Packer (Packer et al., 2005)	RMP-7 and carboplatin with RT	No increase in median survival	Drug development stopped for other reasons
2006	Warren (Warren et al., 2006)	Lobradimil and carboplatin for progressive or recurrent tumors	No increase in median survival	
2006	Burzynksi (Burzynski, Janicki, Weaver, & Burzynski, 2006)	Antineoplaston therapy for progressive, recurrent, or high-grade tumors	No increase in median survival	

Table 6.1 Phase I/II trials of chemotherapy for treatment of DIPGs—cont'd

Year	Author	Therapy	Outcome	Comments
2007	Turner (Turner et al., 2007)	Thalidomide and RT	No increase in median survival	
2008	Korones (Korones et al., 2008)	Vincristine and oral etoposide with RT	No increase in median survival	
2008	Sirachainan (Sirachainan et al., 2008)	Temozolomide with RT, followed by temozolomide and *cis*-retinoic acid	Slightly increased median survival	12 patients, nonrandomized
2010	Broniscer (Broniscer et al., 2010)	Vandetanib during and after RT	No increase in median survival	
2010	Jalali (Jalali et al., 2010)	Temozolomide with RT	No increase in median survival	
2010	Gururangan (Gururangan et al., 2010)	Bevacizumab and irinotecan for progressive or recurrent tumors	No increase in median survival	
2010	Wolff (Wolff et al., 2010)	Intensive chemotherapy with cisplatin, etoposide, vincristine followed by cisplatin, etoposide, ifosfamide with RT, followed by maintenance with cisplatin, etoposide, ifosfamide	No increase in median survival for patients with DIPGs	
2011	Pollack (Pollack et al., 2011)	Gefitinib during and after RT	No increase in median survival	
2011	Haas-Kogan (Haas-Kogan et al., 2011)	Tipifarnib with RT	No increase in median survival	
2011	Cohen (Cohen et al., 2011)	Temozolomide during and after RT	No increase in median survival	

Continued

Table 6.1 Phase I/II trials of chemotherapy for treatment of DIPGs—cont'd

Year	Author	Therapy	Outcome	Comments
2012	Warren (Warren et al., 2012)	Pegylated interferon α-2b after RT	No increase in median survival	
2013	Bradley (Bradley et al., 2013)	Motexafin gadolinium and RT	No increase in median survival	
2013	Cohen (Cohen et al., 2013)	Arsenic trioxide and RT	Safe dosage of arsenic for future studies identified	Phase I clinical trial

RT, radiation therapy; HRT, hyperfractionated radiation therapy; ABMT, autologous bone marrow transplant; CED, convection-enhanced delivery; MO, month.
Median survival, for comparison, is 8–12 months from diagnosis. Note that regimens for progressive and recurrent brain tumors do not include radiation therapy, as reirradiating tumors is generally avoided due to concerns for local toxicity.

clinical trials, evaluating various systemic therapies for DIPGs that have been performed over the past 20 years. Various other therapies that have had minimal success in case reports, but have not been confirmed in prospective or randomized trials, include gefitinib (Daw et al., 2005), interferon beta, and ranimustine with radiation (Ohno, Natsume, Fujii, Ito, & Wakabayashi, 2009), bevacizumab and temozolomide with radiation (Aguilera et al., 2013), and bevacizumab and irinotecan (Torcuator, Zuniga, Loutfi, & Mikkelsen, 2009).

5. RECENT DEVELOPMENTS

5.1. Diagnosis and prognosis

In the evaluation of DIPGs, MRIs are expensive and require sedation, with its inherent health risks, for many DIPG patients. There continues to be a need for better and more affordable data collection in the assessment of DIPG patients. Saratsis et al. (2012) noted that cyclophillin A (CypA) and dimethylarginase-1 are both found preferentially in the cerebrospinal fluid of patients with DIPGs, not in patients with supratentorial gliomas or normal brains. Also, CypA is found in the urine of DIPG patients, both before and after radiation therapy. This could be a useful, affordable screening mechanism for DIPGs.

Other biomarkers that have been investigated for DIPGs include urinary vascular endothelial growth factor, serum basic fibroblast factor, and matrix metalloproteinase 9. These have been noted in a small group of patients to be lower in those children who live for more than 2 years after diagnosis (Warren et al., 2012).

5.2. Molecular analysis

There is a good deal of emphasis placed recently on the understanding of genetic mutations found in certain tumors. This is thought to be important in determining molecularly guided therapy for tumors and has shown promising results in the treatment of certain cancers over the past decade or so. For example, chronic myelogenous leukemia (CML) is a disease that was once very difficult to treat and was often deadly. It was known that a specific gene (BCR–ABL) caused the myeloid proliferation of CML by increasing tyrosine kinase in the cell precursors (Druker et al., 1996). Molecular drugs (tyrosine kinase inhibitors) were developed that have transformed CML from a fatal disease to a chronic, easily controlled condition (Lyseng-Williamson & Jarvis, 2001; Mauro & Druker, 2001).

The understanding of these molecular markers and genetic mutations lags behind in the study of DIPGs, as mentioned in Section 3. Another deficiency in the ability to analyze and test therapies for DIPGs has been the lack of an animal model of DIPG. However, there are some advances being made in this field, both in understanding of the molecular dysfunction present in DIPGs and in the creation of murine DIPG models.

In 2003, Gilbertson et al. determined that the ERBB1 inhibitor of EGFR tyrosine kinase is increased in the higher grade DIPGs at both diagnosis and autopsy (Gilbertson et al., 2003). ERBB1 inhibitor mutations are generally termed EGFR mutations, due to their position and activity in the molecular pathway. EGFR mutations have been noted in many cancers and are also being utilized to design molecularly targeted therapies against lung (Roengvoraphoj, Tsongalis, Dragnev, & Rigas, 2013), colorectal (Strimpakos et al., 2013), ovarian (Harter et al., 2013), pancreatic, and breast cancers (Yewale, Baradia, Vhora, Patil, & Misra, 2013). In general, these molecularly targeted agents are monoclonal antibodies against EGFR and tyrosine kinase inhibitors (Yewale et al., 2013).

Gilbertson's group also demonstrated that few tumors had increased or mutated tumor protein 53 (TP 53) (Gilbertson et al., 2003), though Badhe et al. demonstrated that p53 mutations are found in higher grade astrocytomas in the brainstem (Badhe, Chauhan, & Mehta, 2004). Mutations in p53 are known to increase many cancers, such as testicular (Koster, van

Vugt, Timmer-Bosscha, Gietema, & de Jong, 2013), breast (Turner et al., 2013), ovarian, lung, head and neck, esophageal, liver, other brain, and bladder (Tazawa, Kagawa, & Fujiwara, 2013). Current therapeutic strategies for targeted p53 therapies include viral vectors of wild-type p53 (Tazawa et al., 2013) as well as nanoparticle delivery of p53 (Zhao, Yin, & Li, 2014).

In other studies, it has been noted that there are increases in platelet-derived growth factor receptor alpha (PDGFRAα) (Zarghooni et al., 2010), MET (Paugh et al., 2011), phosphatidylinositol-4,5-bisphosphanate 3-kinase catalytic subunit alpha (PI3KCA), and ataxia telangiectasia mutated/monophosphoryl lipid A (ATM/MPL) (Grill et al., 2012) in DIPGs. Interestingly, the mutations noted in PDGFRAα in DIPGs (and other pediatric high-grade gliomas) are different from those found in adult high-grade gliomas (Paugh et al., 2013). However, these molecular abnormalities are not universally present in all DIPGs. Grill et al. found only 15% of DIPGs had an increase in PI3KCA and 5% of tumors had increased ATM/MPL (Grill et al., 2012). Nonetheless, these targetable molecules are of interest for development of DIPG-specific agents. For example, Paugh et al. (2013) noted that these unique pediatric PDGFRAα molecules are inhibited by both dasatinib and crenolanib, *in vitro*.

It has recently been discovered that histone mutations are involved in the development of DIPGs. Wu et al. discovered *K27M* mutations in H3F3A and HIST1H2B proteins in 78% of DIPGs they studied. However, in non-brainstem gliomas, there were guanine-to-adenine transitions in H3F3A, and these mutations were mutually exclusive. In the DIPG tumors studied, the germline DNA was found to be without these mutations (i.e., wild type), which supports that these mutations are somatic. It is believed that these mutations are involved in the development of DIPGs because they were noted in tissue samples taken before therapy was begun for the DIPG patients. Similar mutations are not found in other pediatric tumors (low-grade gliomas of the brainstem, anaplastic astrocytomas, medulloblastomas, ependymomas, or non-central nervous system tumors) (Wu et al., 2012).

Further investigations of these H3 mutations demonstrate that the *H3K27M* mutation is a gain-of-function mutation. The abnormal histone inhibits polycomb repressive complex 2 (PRC2) activity. In normal cells, PRC2 helps to maintain epigenetic gene silencing and X chromosome inactivation by methylating K27 on histone 3. These mutations in PRC2 are associated with tumor progression (Lewis et al., 2013). This increasing understanding of the development of DIPGs is exciting, although it remains to be seen whether our current understanding will translate into effective, targeted therapies. Along those lines, the Children's Oncology Group (COG) is currently performing a clinical trial of vorinostat (suberoylanilide hydroxamic acid), which is a histone deacetylase inhibitor.

Another molecule currently under investigation is the WEE1 kinase, which is a main regulator of the G_2 cell cycle checkpoint. This molecule has been demonstrated to increase radiosensitization of glioma cells. Recently, inhibitors of WEE1 have been investigated and are being tested in murine xenograft models with *in vivo* evidence of efficacy against DIPG cells (Caretti et al., 2013). MK-1775, a promising WEE1 inhibitor, is currently in a Phase I clinical trial being run by COG.

In addition, mouse models of DIPG tumors have recently been developed. Caretti et al. developed a representative DIPG mouse model by injecting human E98 glioma cells into the murine pons (Caretti et al., 2011). Thereafter, development of a DIPG cell line and xenograft mouse model was established (Aoki et al., 2012; Hashizume et al., 2012). This will facilitate drug development via the traditional routes of laboratory medicine preceding clinical trial assessments in DIPG and should allow for greater understanding of DIPG tumors through orthotopic techniques.

A final challenge for molecularly guided therapy against DIPGs is the blood–brain barrier. There is question whether these molecular therapies are able to cross the blood–brain barrier, and whether the current delivery mechanisms (i.e., intravenous injection or enteral delivery) are sufficient for delivery of these medications past the blood–brain barrier and, specifically, into the pons.

5.3. Therapeutic approaches

One theory regarding the resistance of DIPGs to traditional chemotherapy is that the blood–brain barrier is more durable in the pons (Bradley et al., 2008). Hence, therapeutic approaches to circumvent the blood–brain barrier have been in development. In 2007, Murad et al. demonstrated the safety of delivering gemcitabine safely into the healthy brainstems of six primates by convection-enhanced delivery (CED) (Murad et al., 2007). CED of chemotherapies involves stereotactic placement of catheters into the tumor itself, followed by very slow administration of chemotherapy through these small catheters.

In 2013, two children with DIPGs were treated with CED of topotecan over 100 h (Anderson et al., 2013). Both of these children progressed and died of their disease, which is not surprising given topotecan's lack of therapeutic efficacy in clinical trials for DIPGs. However, it is worth noting that these two children tolerated the CED delivery of chemotherapy, though they did have worsening neurologic status, which did not resolve. In another case, a 5-year-old child was administered 5 days of intratumoral carboplatin via CED with reversible neurological changes (Barua et al., 2013). This child had an improvement in some symptoms but worsening of others in the months following his CED therapy. The follow-up MRI 1 month after therapy demonstrated signal changes consistent with tumor cell death in the treatment area and progression outside of the carboplatin delivery area. These reports demonstrate the feasibility of this approach, though more research will need to be done before efficacy will be evaluable.

Immunotherapy is another popular field of investigation for treatment of many cancers, including DIPGs. The basic theory behind these treatments is that, if there is a molecular signal (outlined above), then perhaps an antibody to these molecular markers would direct the innate immune system toward the tumor, increasing apoptosis and autophagy of the tumor cells. Immunotherapy is thus given either in isolation or in combination with known chemotherapeutic agents. For example, a recent case report demonstrates the utilization of nimotuzumab in combination with temozolomide and radiation therapy for a child with leptomeningeal dissemination of a DIPG (Muller et al., 2013). This child's disease did progress as is typically seen in treatment of DIPGs, however, with death approximately 9 months after diagnosis.

Of interest, a group in Amsterdam recently tested high-grade gliomas and DIPGs *in vitro* with a number of classic chemotherapeutic agents (e.g., melphalan, doxorubicin, mitoxantrone, and BCNU) as well as targeted agents (vandetanib and bortezomib). They demonstrated *in vitro* tumor kill with these agents and identified three major drug efflux transporters that may be involved in the observed chemotherapy resistance of and/or diminished influx past the blood–brain barrier into DIPGs (Veringa et al., 2013). This group identified an increase in P-glycoprotein/MDR1 (P-gp), multidrug-resistant protein 1 (MRP1), and breast cancer resistance protein 1 (BCRP1) in the vasculature endothelial cells in some DIPGs (which are theorized to decrease the amount of chemotherapy that passes through the blood–brain barrier). These early *in vitro* data are intriguing and certainly bear further investigation.

Another approach in treatment of DIPGs that is currently being investigated is "molecular guided therapy" (Grill et al., 2012). There are clinical trials ongoing, currently, which require biopsy of the DIPG. This biopsy is analyzed and any molecular signals identified within the tumor are then utilized for formulation of the chemotherapeutic (or immunotherapeutic) treatment plan.

6. CONCLUSION

DIPGs are a challenging group of tumors to treat and for which to design novel therapies. DIPGs are most likely a diverse group of tumors, at least at the molecular level, and they are in an anatomic location into which delivery of chemotherapy is a challenge. Increasing the number of tumor biopsies (taken at diagnosis, progression, or autopsy) is likely to be an imperative step to increasing the scientific community's understanding of the molecular characterization of DIPGs. To date, no chemotherapy has been developed that can prolong survival past that of radiation therapy, alone. Nevertheless, research into treatment of DIPGs is an active area, and there are many exciting avenues of investigation open to researchers interested in tackling this enigmatic tumor.

REFERENCES

Aguilera, D. G., Mazewski, C., Hayes, L., Jordan, C., Esiashivilli, N., Janns, A., et al. (2013). Prolonged survival after treatment of diffuse intrinsic pontine glioma with radiation, temozolamide, and bevacizumab: Report of 2 cases. *Journal of Pediatric Hematology/Oncology, 35*(1), e42–e46. http://dx.doi.org/10.1097/MPH.0b013e318279aed8.

Albright, A. L., Packer, R. J., Zimmerman, R., Rorke, L. B., Boyett, J., & Hammond, G. D. (1993). Magnetic resonance scans should replace biopsies for the diagnosis of diffuse brain stem gliomas: A report from the Children's Cancer Group. *Neurosurgery, 33*(6), 1026–1029.

Allen, J., Siffert, J., Donahue, B., Nirenberg, A., Jakacki, R., Robertson, P., et al. (1999). A phase I/II study of carboplatin combined with hyperfractionated radiotherapy for brainstem gliomas. *Cancer, 86*(6), 1064–1069.

Anderson, R. C., Kennedy, B., Yanes, C. L., Garvin, J., Needle, M., Canoll, P., et al. (2013). Convection-enhanced delivery of topotecan into diffuse intrinsic brainstem tumors in children. *Journal of Neurosurgery. Pediatrics, 11*(3), 289–295. http://dx.doi.org/10.3171/2012.10.PEDS12142.

Aoki, Y., Hashizume, R., Ozawa, T., Banerjee, A., Prados, M., James, C. D., et al. (2012). An experimental xenograft mouse model of diffuse pontine glioma designed for therapeutic testing. *Journal of Neuro-Oncology, 108*(1), 29–35. http://dx.doi.org/10.1007/s11060-011-0796-x.

Atac, M. S., & Blaauw, G. (1979). Radiotherapy in brain-stem gliomas in children. *Clinical Neurology and Neurosurgery, 81*(4), 281–290.

Badhe, P. B., Chauhan, P. P., & Mehta, N. K. (2004). Brainstem gliomas—A clinicopathological study of 45 cases with p53 immunohistochemistry. *Indian Journal of Cancer, 41*(4), 170–174.

Barua, N. U., Lowis, S. P., Woolley, M., O'Sullivan, S., Harrison, R., & Gill, S. S. (2013). Robot-guided convection-enhanced delivery of carboplatin for advanced brainstem glioma. *Acta Neurochirurgica (Wien), 155*(8), 1459–1465. http://dx.doi.org/10.1007/s00701-013-1700-6.

Benesch, M., Lackner, H., Moser, A., Kerbl, R., Schwinger, W., Oberbauer, R., et al. (2001). Outcome and long-term side effects after synchronous radiochemotherapy for childhood brain stem gliomas. *Pediatric Neurosurgery, 35*(4), 173–180. http://dx.doi.org/10.1159/000050418.

Benesch, M., Wagner, S., Berthold, F., & Wolff, J. E. (2005). Primary dissemination of high-grade gliomas in children: Experiences from four studies of the Pediatric Oncology and Hematology Society of the German Language Group (GPOH). *Journal of Neuro-Oncology, 72*(2), 179–183. http://dx.doi.org/10.1007/s11060-004-3546-5.

Bernier-Chastagner, V., Grill, J., Doz, F., Bracard, S., Gentet, J. C., Marie-Cardine, A., et al. (2005). Topotecan as a radiosensitizer in the treatment of children with malignant diffuse brainstem gliomas: Results of a French Society of Paediatric Oncology Phase II Study. *Cancer, 104*(12), 2792–2797. http://dx.doi.org/10.1002/cncr.21534.

Bouffet, E., Khelfaoui, F., Philip, I., Biron, P., Brunat-Mentigny, M., & Philip, T. (1997). High-dose carmustine for high-grade gliomas in childhood. *Cancer Chemotherapy and Pharmacology, 39*(4), 376–379.

Bouffet, E., Raquin, M., Doz, F., Gentet, J. C., Rodary, C., Demeocq, F., et al. (2000). Radiotherapy followed by high dose busulfan and thiotepa: A prospective assessment of high dose chemotherapy in children with diffuse pontine gliomas. *Cancer, 88*(3), 685–692.

Bradley, K. A., Pollack, I. F., Reid, J. M., Adamson, P. C., Ames, M. M., Vezina, G., et al. (2008). Motexafin gadolinium and involved field radiation therapy for intrinsic pontine glioma of childhood: A Children's Oncology Group phase I study. *Neuro-Oncology, 10*(5), 752–758. http://dx.doi.org/10.1215/15228517-2008-043.

Bradley, K. A., Zhou, T., McNall-Knapp, R. Y., Jakacki, R. I., Levy, A. S., Vezina, G., et al. (2013). Motexafin-gadolinium and involved field radiation therapy for intrinsic pontine glioma of childhood: A children's oncology group phase 2 study. *International Journal of Radiation Oncology, Biology, Physics, 85*(1), e55–e60. http://dx.doi.org/10.1016/j.ijrobp.2012.09.004.

Broniscer, A., Baker, J. N., Tagen, M., Onar-Thomas, A., Gilbertson, R. J., Davidoff, A. M., et al. (2010). Phase I study of vandetanib during and after radiotherapy in children with diffuse intrinsic pontine glioma. *Journal of Clinical Oncology, 28*(31), 4762–4768. http://dx.doi.org/10.1200/JCO.2010.30.3545.

Broniscer, A., & Gajjar, A. (2004). Supratentorial high-grade astrocytoma and diffuse brainstem glioma: Two challenges for the pediatric oncologist. *The Oncologist, 9*(2), 197–206.

Broniscer, A., Iacono, L., Chintagumpala, M., Fouladi, M., Wallace, D., Bowers, D. C., et al. (2005). Role of temozolomide after radiotherapy for newly diagnosed diffuse brainstem glioma in children: Results of a multiinstitutional study (SJHG-98). *Cancer, 103*(1), 133–139. http://dx.doi.org/10.1002/cncr.20741.

Broniscer, A., Laningham, F. H., Sanders, R. P., Kun, L. E., Ellison, D. W., & Gajjar, A. (2008). Young age may predict a better outcome for children with diffuse pontine glioma. *Cancer, 113*(3), 566–572. http://dx.doi.org/10.1002/cncr.23584.

Broniscer, A., Leite, C. C., Lanchote, V. L., Machado, T. M., & Cristofani, L. M. (2000). Radiation therapy and high-dose tamoxifen in the treatment of patients with diffuse brainstem gliomas: Results of a Brazilian cooperative study. Brainstem Glioma Cooperative Group. *Journal of Clinical Oncology, 18*(6), 1246–1253.

Burzynski, S. R., Janicki, T. J., Weaver, R. A., & Burzynski, B. (2006). Targeted therapy with antineoplastons A10 and AS2-1 of high-grade, recurrent, and progressive brainstem glioma. *Integrative Cancer Therapies*, *5*(1), 40–47. http://dx.doi.org/10.1177/1534735405285380.

Cage, T. A., Samagh, S. P., Mueller, S., Nicolaides, T., Haas-Kogan, D., Prados, M., et al. (2013). Feasiblity, safety, and indications for surgical biopsy of intrinsic brainstem tumors in children. *Child's Nervous System*, *29*(8), 1314–1319.

Caretti, V., Hiddingh, L., Lagerweij, T., Schellen, P., Koken, P. W., Hulleman, E., et al. (2013). WEE1 kinase inhibition enhances the radiation response of diffuse intrinsic pontine gliomas. *Molecular Cancer Therapeutics*, *12*(2), 141–150. http://dx.doi.org/10.1158/1535-7163.MCT-12-0735.

Caretti, V., Zondervan, I., Meijer, D. H., Idema, S., Vos, W., Hamans, B., et al. (2011). Monitoring of tumor growth and post-irradiation recurrence in a diffuse intrinsic pontine glioma mouse model. *Brain Pathology*, *21*(4), 441–451. http://dx.doi.org/10.1111/j.1750-3639.2010.00468.x.

Cartmill, M., & Punt, J. (1999). Diffuse brain stem glioma. A review of stereotactic biopsies. *Child's Nervous System*, *15*(5), 235–237, discussion 238.

Chico-Ponce de Leon, F., Perezpena-Diazconti, M., Castro-Sierra, E., Guerrero-Jazo, F. J., Gordillo-Dominguez, L. F., Gutierrez-Guerra, R., et al. (2003). Stereotactically-guided biopsies of brainstem tumors. *Child's Nervous System*, *19*(5–6), 305–310.

Cohen, K. J., Gibbs, I. C., Fisher, P. G., Hayashi, R. J., Macy, M. E., & Gore, L. (2013). A phase I trial of arsenic trioxide chemoradiotherapy for infiltrating astrocytomas of childhood. *Neuro-Oncology*, *15*(6), 783–787. http://dx.doi.org/10.1093/neuonc/not021.

Cohen, K. J., Heideman, R. L., Zhou, T., Holmes, E. J., Lavey, R. S., Bouffet, E., et al. (2011). Temozolomide in the treatment of children with newly diagnosed diffuse intrinsic pontine gliomas: A report from the Children's Oncology Group. *Neuro-Oncology*, *13*(4), 410–416. http://dx.doi.org/10.1093/neuonc/noq205.

Curless, R. G., Bowen, B. C., Pattany, P. M., Gonik, R., & Kramer, D. L. (2002). Magnetic resonance spectroscopy in childhood brainstem tumors. *Pediatric Neurology*, *26*(5), 374–378.

Daw, N. C., Furman, W. L., Stewart, C. F., Iacono, L. C., Krailo, M., Bernstein, M. L., et al. (2005). Phase I and pharmacokinetic study of gefitinib in children with refractory solid tumors: A Children's Oncology Group Study. *Journal of Clinical Oncology*, *23*(25), 6172–6180. http://dx.doi.org/10.1200/JCO.2005.11.429.

Donaldson, S. S., Laningham, F., & Fisher, P. G. (2006). Advances toward an understanding of brainstem gliomas. *Journal of Clinical Oncology*, *24*(8), 1266–1272. http://dx.doi.org/10.1200/JCO.2005.04.6599.

Doz, F., Neuenschwander, S., Bouffet, E., Gentet, J. C., Schneider, P., Kalifa, C., et al. (2002). Carboplatin before and during radiation therapy for the treatment of malignant brain stem tumours: A study by the Societe Francaise d'Oncologie Pediatrique. *European Journal of Cancer*, *38*(6), 815–819.

Dreyer, Z. E., Kadota, R. P., Stewart, C. F., Friedman, H. S., Mahoney, D. H., Kun, L. E., et al. (2003). Phase 2 study of idarubicin in pediatric brain tumors: Pediatric Oncology Group study POG 9237. *Neuro-Oncology*, *5*(4), 261–267. http://dx.doi.org/10.1215/S115285170200056X.

Druker, B. J., Tamura, S., Buchdunger, E., Ohno, S., Segal, G. M., Fanning, S., et al. (1996). Effects of a selective inhibitor of the Abl tyrosine kinase on the growth of Bcr-Abl positive cells. *Nature Medicine*, *2*(5), 561–566.

Dunkel, I. J., Garvin, J. H., Jr., Goldman, S., Ettinger, L. J., Kaplan, A. M., Cairo, M., et al. (1998). High dose chemotherapy with autologous bone marrow rescue for children with diffuse pontine brain stem tumors. Children's Cancer Group. *Journal of Neuro-Oncology*, *37*(1), 67–73.

Edwards, M. S., Wara, W. M., Urtasun, R. C., Prados, M., Levin, V. A., Fulton, D., et al. (1989). Hyperfractionated radiation therapy for brain-stem glioma: A phase I-II trial. *Journal of Neurosurgery*, *70*(5), 691–700. http://dx.doi.org/10.3171/jns.1989.70.5.0691.

Epstein, F., & McCleary, E. L. (1986). Intrinsic brain-stem tumors of childhood: Surgical indications. *Journal of Neurosurgery*, *64*(1), 11–15. http://dx.doi.org/10.3171/jns.1986.64.1.0011.

Farmer, J. P., Montes, J. L., Freeman, C. R., Meagher-Villemure, K., Bond, M. C., & O'Gorman, A. M. (2001). Brainstem gliomas. A 10-year institutional review. *Pediatric Neurosurgery*, *34*(4), 206–214. http://dx.doi.org/10.1159/000056021.

Freeman, C. R., Kepner, J., Kun, L. E., Sanford, R. A., Kadota, R., Mandell, L., et al. (2000). A detrimental effect of a combined chemotherapy-radiotherapy approach in children with diffuse intrinsic brain stem gliomas? *International Journal of Radiation Oncology, Biology, Physics*, *47*(3), 561–564.

Freeman, C. R., Krischer, J., Sanford, R. A., Burger, P. C., Cohen, M., & Norris, D. (1988). Hyperfractionated radiotherapy in brain stem tumors: Results of a Pediatric Oncology Group study. *International Journal of Radiation Oncology, Biology, Physics*, *15*(2), 311–318.

Gilbertson, R. J., Hill, D. A., Hernan, R., Kocak, M., Geyer, R., Olson, J., et al. (2003). ERBB1 is amplified and overexpressed in high-grade diffusely infiltrative pediatric brain stem glioma. *Clinical Cancer Research*, *9*(10 Pt 1), 3620–3624.

Goncalves-Ferreira, A. J., Herculano-Carvalho, M., & Pimentel, J. (2003). Stereotactic biopsies of focal brainstem lesions. *Surgical Neurology*, *60*(4), 311–320, discussion 320.

Greenberg, M. L., Fisher, P. G., Freeman, C., Korones, D. N., Bernstein, M., Friedman, H., et al. (2005). Etoposide, vincristine, and cyclosporin A with standard-dose radiation therapy in newly diagnosed diffuse intrinsic brainstem gliomas: A pediatric oncology group phase I study. *Pediatric Blood & Cancer*, *45*(5), 644–648. http://dx.doi.org/10.1002/pbc.20382.

Grill, J., Puget, S., Andreiuolo, F., Philippe, C., MacConaill, L., & Kieran, M. W. (2012). Critical oncogenic mutations in newly diagnosed pediatric diffuse intrinsic pontine glioma. *Pediatric Blood & Cancer*, *58*(4), 489–491. http://dx.doi.org/10.1002/pbc.24060.

Guillamo, J. S., Doz, F., & Delattre, J. Y. (2001). Brain stem gliomas. *Current Opinion in Neurology*, *14*(6), 711–715.

Gururangan, S., Chi, S. N., Young Poussaint, T., Onar-Thomas, A., Gilbertson, R. J., Vajapeyam, S., et al. (2010). Lack of efficacy of bevacizumab plus irinotecan in children with recurrent malignant glioma and diffuse brainstem glioma: A Pediatric Brain Tumor Consortium study. *Journal of Clinical Oncology*, *28*(18), 3069–3075. http://dx.doi.org/10.1200/JCO.2009.26.8789.

Gururangan, S., McLaughlin, C. A., Brashears, J., Watral, M. A., Provenzale, J., Coleman, R. E., et al. (2006). Incidence and patterns of neuraxis metastases in children with diffuse pontine glioma. *Journal of Neuro-Oncology*, 77(2), 207–212. http://dx.doi.org/10.1007/s11060-005-9029-5.

Haas-Kogan, D. A., Banerjee, A., Poussaint, T. Y., Kocak, M., Prados, M. D., Geyer, J. R., et al. (2011). Phase II trial of tipifarnib and radiation in children with newly diagnosed diffuse intrinsic pontine gliomas. *Neuro-Oncology*, *13*(3), 298–306. http://dx.doi.org/10.1093/neuonc/noq202.

Halperin, E. C., Constine, L. S., Tarbell, N. J., & Kun, L. (2011). *Pediatric radiation oncology* (5th ed.). Philadelphia: Wolters Kluwer, Lippincott Williams & Wilkins.

Hargrave, D., Bartels, U., & Bouffet, E. (2006). Diffuse brainstem glioma in children: Critical review of clinical trials. *The Lancet Oncology*, 7(3), 241–248. http://dx.doi.org/10.1016/S1470-2045(06)70615-5.

Hargrave, D. R., Mabbott, D. J., & Bouffet, E. (2006). Pathological laughter and behavioural change in childhood pontine glioma. *Journal of Neuro-Oncology*, 77(3), 267–271. http://dx.doi.org/10.1007/s11060-005-9034-8.

Harter, P., Sehouli, J., Kimmig, R., Rau, J., Hilpert, F., Kurzeder, C., et al. (2013). Addition of vandetanib to pegylated liposomal doxorubicin (PLD) in patients with recurrent ovarian cancer. A randomized phase I/II study of the AGO Study Group (AGO-OVAR 2.13). *Investigational New Drugs*, *31*(6), 1499–1504. http://dx.doi.org/10.1007/s10637-013-0011-3.

Hashizume, R., Smirnov, I., Liu, S., Phillips, J. J., Hyer, J., McKnight, T. R., et al. (2012). Characterization of a diffuse intrinsic pontine glioma cell line: Implications for future investigations and treatment. *Journal of Neuro-Oncology*, *110*(3), 305–313. http://dx.doi.org/10.1007/s11060-012-0973-6.

Helton, K. J., Phillips, N. S., Khan, R. B., Boop, F. A., Sanford, R. A., Zou, P., et al. (2006). Diffusion tensor imaging of tract involvement in children with pontine tumors. *AJNR American Journal of Neuroradiology*, *27*(4), 786–793.

Hibi, T., Shitara, N., Genka, S., Fuchinoue, T., Hayakawa, I., Tsuchida, T., et al. (1992). Radiotherapy for pediatric brain stem glioma: Radiation dose, response, and survival. *Neurosurgery*, *31*(4), 643–650, discussion 650–641.

Hood, T. W., Gebarski, S. S., McKeever, P. E., & Venes, J. L. (1986). Stereotaxic biopsy of intrinsic lesions of the brain stem. *Journal of Neurosurgery*, *65*(2), 172–176. http://dx.doi.org/10.3171/jns.1986.65.2.0172.

Hurwitz, C. A., Strauss, L. C., Kepner, J., Kretschmar, C., Harris, M. B., Friedman, H., et al. (2001). Paclitaxel for the treatment of progressive or recurrent childhood brain tumors: A pediatric oncology phase II study. *Journal of Pediatric Hematology/Oncology*, *23*(5), 277–281.

Jackson, S., Patay, Z., Howarth, R., Pai Panandiker, A. S., Onar-Thomas, A., Gajjar, A., et al. (2013). Clinico-radiologic characteristics of long-term survivors of diffuse intrinsic pontine glioma. *Journal of Neuro-Oncology*, *114*(3), 339–344. http://dx.doi.org/10.1007/s11060-013-1189-0.

Jakacki, R. I., Siffert, J., Jamison, C., Velasquez, L., & Allen, J. C. (1999). Dose-intensive, time-compressed procarbazine, CCNU, vincristine (PCV) with peripheral blood stem cell support and concurrent radiation in patients with newly diagnosed high-grade gliomas. *Journal of Neuro-Oncology*, *44*(1), 77–83.

Jalali, R., Raut, N., Arora, B., Gupta, T., Dutta, D., Munshi, A., et al. (2010). Prospective evaluation of radiotherapy with concurrent and adjuvant temozolomide in children with newly diagnosed diffuse intrinsic pontine glioma. *International Journal of Radiation Oncology, Biology, Physics*, 77(1), 113–118. http://dx.doi.org/10.1016/j.ijrobp.2009.04.031.

Jenkin, R. D., Boesel, C., Ertel, I., Evans, A., Hittle, R., Ortega, J., et al. (1987). Brain-stem tumors in childhood: A prospective randomized trial of irradiation with and without adjuvant CCNU, VCR, and prednisone. A report of the Children's Cancer Study Group. *Journal of Neurosurgery*, *66*(2), 227–233. http://dx.doi.org/10.3171/jns.1987.66.2.0227.

Jennings, M. T., Sposto, R., Boyett, J. M., Vezina, L. G., Holmes, E., Berger, M. S., et al. (2002). Preradiation chemotherapy in primary high-risk brainstem tumors: Phase II study CCG-9941 of the Children's Cancer Group. *Journal of Clinical Oncology*, *20*(16), 3431–3437.

Kim, T. H., Chin, H. W., Pollan, S., Hazel, J. H., & Webster, J. H. (1980). Radiotherapy of primary brain stem tumors. *International Journal of Radiation Oncology, Biology, Physics*, *6*(1), 51–57.

Kornreich, L., Schwarz, M., Karmazyn, B., Cohen, I. J., Shuper, A., Michovitz, S., et al. (2005). Role of MRI in the management of children with diffuse pontine tumors: A study of 15 patients and review of the literature. *Pediatric Radiology*, *35*(9), 872–879. http://dx.doi.org/10.1007/s00247-005-1502-y.

Korones, D. N. (2007). Treatment of newly diagnosed diffuse brain stem gliomas in children: In search of the holy grail. *Expert Review of Anticancer Therapy*, 7(5), 663–674. http://dx.doi.org/10.1586/14737140.7.5.663.

Korones, D. N., Fisher, P. G., Kretschmar, C., Zhou, T., Chen, Z., Kepner, J., et al. (2008). Treatment of children with diffuse intrinsic brain stem glioma with radiotherapy, vincristine and oral VP-16: A Children's Oncology Group phase II study. *Pediatric Blood & Cancer, 50*(2), 227–230. http://dx.doi.org/10.1002/pbc.21154.
Koster, R., van Vugt, M. A., Timmer-Bosscha, H., Gietema, J. A., & de Jong, S. (2013). Unravelling mechanisms of cisplatin sensitivity and resistance in testicular cancer. *Expert Reviews in Molecular Medicine, 15*, e12. http://dx.doi.org/10.1017/erm.2013.13.
Kretschmar, C. S., Tarbell, N. J., Barnes, P. D., Krischer, J. P., Burger, P. C., & Kun, L. (1993). Pre-irradiation chemotherapy and hyperfractionated radiation therapy 66 Gy for children with brain stem tumors. A phase II study of the Pediatric Oncology Group, Protocol 8833. *Cancer, 72*(4), 1404–1413.
Laprie, A., Pirzkall, A., Haas-Kogan, D. A., Cha, S., Banerjee, A., Le, T. P., et al. (2005). Longitudinal multivoxel MR spectroscopy study of pediatric diffuse brainstem gliomas treated with radiotherapy. *International Journal of Radiation Oncology, Biology, Physics, 62*(1), 20–31. http://dx.doi.org/10.1016/j.ijrobp.2004.09.027.
Lashford, L. S., Thiesse, P., Jouvet, A., Jaspan, T., Couanet, D., Griffiths, P. D., et al. (2002). Temozolomide in malignant gliomas of childhood: A United Kingdom Children's Cancer Study Group and French Society for Pediatric Oncology Intergroup Study. *Journal of Clinical Oncology, 20*(24), 4684–4691.
Lee, F. (1975). Radiation of infratentorial and supratentorial brain-stem tumors. *Journal of Neurosurgery, 43*(1), 65–68. http://dx.doi.org/10.3171/jns.1975.43.1.0065.
Lewis, J., Lucraft, H., & Gholkar, A. (1997). UKCCSG study of accelerated radiotherapy for pediatric brain stem gliomas. United Kingdom Childhood Cancer Study Group. *International Journal of Radiation Oncology, Biology, Physics, 38*(5), 925–929.
Lewis, P. W., Muller, M. M., Koletsky, M. S., Cordero, F., Lin, S., Banaszynski, L. A., et al. (2013). Inhibition of PRC2 activity by a gain-of-function H3 mutation found in pediatric glioblastoma. *Science, 340*(6134), 857–861. http://dx.doi.org/10.1126/science.1232245.
Littman, P., Jarrett, P., Bilaniuk, L. T., Rorke, L. B., Zimmerman, R. A., Bruce, D. A., et al. (1980). Pediatric brain stem gliomas. *Cancer, 45*(11), 2787–2792.
Lyseng-Williamson, K., & Jarvis, B. (2001). Imatinib. *Drugs, 61*(12), 1765–1774, discussion 1775–1766.
MacDonald, T. J. (2012). Diffuse intrinsic pontine glioma (DIPG): Time to biopsy again? *Pediatric Blood & Cancer, 58*(4), 487–488. http://dx.doi.org/10.1002/pbc.24090.
Mandell, L. R., Kadota, R., Freeman, C., Douglass, E. C., Fontanesi, J., Cohen, M. E., et al. (1999). There is no role for hyperfractionated radiotherapy in the management of children with newly diagnosed diffuse intrinsic brainstem tumors: Results of a Pediatric Oncology Group phase III trial comparing conventional vs. hyperfractionated radiotherapy. *International Journal of Radiation Oncology, Biology, Physics, 43*(5), 959–964.
Marcus, K. J., Dutton, S. C., Barnes, P., Coleman, C. N., Pomeroy, S. L., Goumnerova, L., et al. (2003). A phase I trial of etanidazole and hyperfractionated radiotherapy in children with diffuse brainstem glioma. *International Journal of Radiation Oncology, Biology, Physics, 55*(5), 1182–1185.
Mauro, M. J., & Druker, B. J. (2001). STI571: Targeting BCR-ABL as therapy for CML. *The Oncologist, 6*(3), 233–238.
Muller, K., Schlamann, A., Seidel, C., Warmuth-Metz, M., Christiansen, H., Vordermark, D., et al. (2013). Craniospinal irradiation with concurrent temozolomide and nimotuzumab in a child with primary metastatic diffuse intrinsic pontine glioma. A compassionate use treatment. *Strahlentherapie und Onkologie, 189*(8), 693–696. http://dx.doi.org/10.1007/s00066-013-0370-x.
Murad, G. J., Walbridge, S., Morrison, P. F., Szerlip, N., Butman, J. A., Oldfield, E. H., et al. (2007). Image-guided convection-enhanced delivery of gemcitabine to the brainstem. *Journal of Neurosurgery, 106*(2), 351–356. http://dx.doi.org/10.3171/jns.2007.106.2.351.

Needle, M. N., Molloy, P. T., Geyer, J. R., Herman-Liu, A., Belasco, J. B., Goldwein, J. W., et al. (1997). Phase II study of daily oral etoposide in children with recurrent brain tumors and other solid tumors. *Medical and Pediatric Oncology*, *29*(1), 28–32.

Ohno, M., Natsume, A., Fujii, M., Ito, M., & Wakabayashi, T. (2009). Interferon-beta, MCNU, and conventional radiotherapy for pediatric patients with brainstem glioma. *Pediatric Blood & Cancer*, *53*(1), 37–41. http://dx.doi.org/10.1002/pbc.21987.

Packer, R. J., Boyett, J. M., Zimmerman, R. A., Albright, A. L., Kaplan, A. M., Rorke, L. B., et al. (1994). Outcome of children with brain stem gliomas after treatment with 7800 cGy of hyperfractionated radiotherapy. A Children's Cancer Group Phase I/II Trial. *Cancer*, *74*(6), 1827–1834.

Packer, R. J., Krailo, M., Mehta, M., Warren, K., Allen, J., Jakacki, R., et al. (2005). Phase 1 study of concurrent RMP-7 and carboplatin with radiotherapy for children with newly diagnosed brainstem gliomas. *Cancer*, *104*(6), 1281–1287. http://dx.doi.org/10.1002/cncr.21301.

Packer, R. J., Prados, M., Phillips, P., Nicholson, H. S., Boyett, J. M., Goldwein, J., et al. (1996). Treatment of children with newly diagnosed brain stem gliomas with intravenous recombinant beta-interferon and hyperfractionated radiation therapy: A children's cancer group phase I/II study. *Cancer*, 77(10), 2150–2156. http://dx.doi.org/10.1002/(SICI)1097-0142(19960515)77:10<2150::AID-CNCR28>3.0.CO;2-T.

Paugh, B. S., Broniscer, A., Qu, C., Miller, C. P., Zhang, J., Tatevossian, R. G., et al. (2011). Genome-wide analyses identify recurrent amplifications of receptor tyrosine kinases and cell-cycle regulatory genes in diffuse intrinsic pontine glioma. *Journal of Clinical Oncology*, *29*(30), 3999–4006. http://dx.doi.org/10.1200/JCO.2011.35.5677.

Paugh, B. S., Zhu, X., Qu, C., Endersby, R., Diaz, A. K., Zhang, J., et al. (2013). Novel oncogenic PDGFRA mutations in pediatric high-grade gliomas. *Cancer Research*, *73*, 6219–6229. http://dx.doi.org/10.1158/0008-5472.CAN-13-1491.

Peirce, C. B., & Bouchard, J. (1950). Role of radiation therapy in the control of malignant neoplasms of the brain and brain stem. *Radiology*, *55*(3), 337–343.

Phi, J. H., Chung, H. T., Wang, K. C., Ryu, S. K., & Kim, S. K. (2013). Transcerebellar biopsy of diffuse pontine gliomas in children: A technical note. *Child's Nervous System*, *29*(3), 489–493. http://dx.doi.org/10.1007/s00381-012-1933-3.

Pierre-Kahn, A., Hirsch, J. F., Vinchon, M., Payan, C., Sainte-Rose, C., Renier, D., et al. (1993). Surgical management of brain-stem tumors in children: Results and statistical analysis of 75 cases. *Journal of Neurosurgery*, *79*(6), 845–852. http://dx.doi.org/10.3171/jns.1993.79.6.0845.

Pollack, I. F., Stewart, C. F., Kocak, M., Poussaint, T. Y., Broniscer, A., Banerjee, A., et al. (2011). A phase II study of gefitinib and irradiation in children with newly diagnosed brainstem gliomas: A report from the Pediatric Brain Tumor Consortium. *Neuro-Oncology*, *13*(3), 290–297. http://dx.doi.org/10.1093/neuonc/noq199.

Poussaint, T. Y., Kocak, M., Vajapeyam, S., Packer, R. I., Robertson, R. L., Geyer, R., et al. (2011). MRI as a central component of clinical trials analysis in brainstem glioma: A report from the Pediatric Brain Tumor Consortium (PBTC). *Neuro-Oncology*, *13*(4), 417–427. http://dx.doi.org/10.1093/neuonc/noq200.

Prabhu, S. P., Ng, S., Vajapeyam, S., Kieran, M. W., Pollack, I. F., Geyer, R., et al. (2011). DTI assessment of the brainstem white matter tracts in pediatric BSG before and after therapy: A report from the Pediatric Brain Tumor Consortium. *Child's Nervous System*, *27*(1), 11–18. http://dx.doi.org/10.1007/s00381-010-1323-7.

Reigel, D. H., Scarff, T. B., & Woodford, J. E. (1979). Biopsy of pediatric brain stem tumors. *Child's Brain*, *5*(3), 329–340.

Roengvoraphoj, M., Tsongalis, G. J., Dragnev, K. H., & Rigas, J. R. (2013). Epidermal growth factor receptor tyrosine kinase inhibitors as initial therapy for non-small cell lung cancer: Focus on epidermal growth factor receptor mutation testing and mutation-positive patients. *Cancer Treatment Reviews*, *39*(8), 839–850. http://dx.doi.org/10.1016/j.ctrv.2013.05.001.

Sanghavi, S. N., Needle, M. N., Krailo, M. D., Geyer, J. R., Ater, J., & Mehta, M. P. (2003). A phase I study of topotecan as a radiosensitizer for brainstem glioma of childhood: First report of the Children's Cancer Group-0952. *Neuro-Oncology*, *5*(1), 8–13.

Saratsis, A. M., Yadavilli, S., Magge, S., Rood, B. R., Perez, J., Hill, D. A., et al. (2012). Insights into pediatric diffuse intrinsic pontine glioma through proteomic analysis of cerebrospinal fluid. *Neuro-Oncology*, *14*(5), 547–560. http://dx.doi.org/10.1093/neuonc/nos067.

Sethi, R., Allen, J., Donahue, B., Karajannis, M., Gardner, S., Wisoff, J., et al. (2011). Prospective neuraxis MRI surveillance reveals a high risk of leptomeningeal dissemination in diffuse intrinsic pontine glioma. *Journal of Neuro-Oncology*, *102*(1), 121–127. http://dx.doi.org/10.1007/s11060-010-0301-y.

Shuper, A., Kornreich, L., Loven, D., Michowitz, S., Schwartz, M., & Cohen, I. J. (1998). Diffuse brain stem gliomas. Are we improving outcome? *Child's Nervous System*, *14*(10), 578–581.

Sirachainan, N., Pakakasama, S., Visudithbhan, A., Chiamchanya, S., Tuntiyatorn, L., Dhanachai, M., et al. (2008). Concurrent radiotherapy with temozolomide followed by adjuvant temozolomide and cis-retinoic acid in children with diffuse intrinsic pontine glioma. *Neuro-Oncology*, *10*(4), 577–582. http://dx.doi.org/10.1215/15228517-2008-025.

Strimpakos, A., Pentheroudakis, G., Kotoula, V., De Roock, W., Kouvatseas, G., Papakostas, P., et al. (2013). The prognostic role of ephrin A2 and endothelial growth factor receptor pathway mediators in patients with advanced colorectal cancer treated with cetuximab. *Clinical Colorectal Cancer*, *12*(4), 267.e2–274.e2. http://dx.doi.org/10.1016/j.clcc.2013.07.001, e262.

Tazawa, H., Kagawa, S., & Fujiwara, T. (2013). Advances in adenovirus-mediated p53 cancer gene therapy. *Expert Opinion on Biological Therapy*, *13*(11), 1569–1583. http://dx.doi.org/10.1517/14712598.2013.845662.

Thakur, S. B., Karimi, S., Dunkel, I. J., Koutcher, J. A., & Huang, W. (2006). Longitudinal MR spectroscopic imaging of pediatric diffuse pontine tumors to assess tumor aggression and progression. *AJNR American Journal of Neuroradiology*, *27*(4), 806–809.

Thompson, W. D., Jr., & Kosnik, E. J. (2005). Spontaneous regression of a diffuse brainstem lesion in the neonate. Report of two cases and review of the literature. *Journal of Neurosurgery*, *102*(1 Suppl.), 65–71. http://dx.doi.org/10.3171/ped.2005.102.1.0065.

Torcuator, R., Zuniga, R., Loutfi, R., & Mikkelsen, T. (2009). Bevacizumab and irinotecan treatment for progressive diffuse brainstem glioma: Case report. *Journal of Neuro-Oncology*, *93*(3), 409–412. http://dx.doi.org/10.1007/s11060-008-9782-3.

Turner, C. D., Chi, S., Marcus, K. J., MacDonald, T., Packer, R. J., Poussaint, T. Y., et al. (2007). Phase II study of thalidomide and radiation in children with newly diagnosed brain stem gliomas and glioblastoma multiforme. *Journal of Neuro-Oncology*, *82*(1), 95–101. http://dx.doi.org/10.1007/s11060-006-9251-9.

Turner, N., Moretti, E., Siclari, O., Migliaccio, I., Santarpia, L., D'Incalci, M., et al. (2013). Targeting triple negative breast cancer: Is p53 the answer? *Cancer Treatment Reviews*, *39*(5), 541–550. http://dx.doi.org/10.1016/j.ctrv.2012.12.001.

Veringa, S. J., Biesmans, D., van Vuurden, D. G., Jansen, M. H., Wedekind, L. E., Horsman, I., et al. (2013). In vitro drug response and efflux transporters associated with drug resistance in pediatric high grade glioma and diffuse intrinsic pontine glioma. *PLoS One*, *8*(4), e61512. http://dx.doi.org/10.1371/journal.pone.0061512.

Wagner, S., Benesch, M., Berthold, F., Gnekow, A. K., Rutkowski, S., Strater, R., et al. (2006). Secondary dissemination in children with high-grade malignant gliomas and diffuse intrinsic pontine gliomas. *British Journal of Cancer*, *95*(8), 991–997. http://dx.doi.org/10.1038/sj.bjc.6603402.

Walker, D. A., Liu, J., Kieran, M., Jabado, N., Picton, S., Packer, R., et al. (2013). A multidisciplinary consensus statement concerning surgical approaches to low-grade, high-grade astrocytomas and diffuse intrinsic pontine gliomas in childhood (CPN Paris 2011) using the Delphi method. *Neuro-Oncology*, *15*(4), 462–468. http://dx.doi.org/10.1093/neuonc/nos330.

Walter, A. W., Gajjar, A., Ochs, J. S., Langston, J. W., Sanford, R. A., Kun, L. E., et al. (1998). Carboplatin and etoposide with hyperfractionated radiotherapy in children with newly diagnosed diffuse pontine gliomas: A phase I/II study. *Medical and Pediatric Oncology*, *30*(1), 28–33.

Warren, K., Bent, R., Wolters, P. L., Prager, A., Hanson, R., Packer, R., et al. (2012). A phase 2 study of pegylated interferon alpha-2b (PEG-Intron((R))) in children with diffuse intrinsic pontine glioma. *Cancer*, *118*(14), 3607–3613. http://dx.doi.org/10.1002/cncr.26659.

Warren, K., Jakacki, R., Widemann, B., Aikin, A., Libucha, M., Packer, R., et al. (2006). Phase II trial of intravenous lobradimil and carboplatin in childhood brain tumors: A report from the Children's Oncology Group. *Cancer Chemotherapy and Pharmacology*, *58*(3), 343–347. http://dx.doi.org/10.1007/s00280-005-0172-7.

Wolff, J. E., Driever, P. H., Erdlenbruch, B., Kortmann, R. D., Rutkowski, S., Pietsch, T., et al. (2010). Intensive chemotherapy improves survival in pediatric high-grade glioma after gross total resection: Results of the HIT-GBM-C protocol. *Cancer*, *116*(3), 705–712. http://dx.doi.org/10.1002/cncr.24730.

Wolff, J. E., Westphal, S., Molenkamp, G., Gnekow, A., Warmuth-Metz, M., Rating, D., et al. (2002). Treatment of paediatric pontine glioma with oral trophosphamide and etoposide. *British Journal of Cancer*, *87*(9), 945–949. http://dx.doi.org/10.1038/sj.bjc.6600552.

Wu, G., Broniscer, A., McEachron, T. A., Lu, C., Paugh, B. S., Becksfort, J., et al. (2012). Somatic histone H3 alterations in pediatric diffuse intrinsic pontine gliomas and non-brainstem glioblastomas. *Nature Genetics*, *44*(3), 251–253. http://dx.doi.org/10.1038/ng.1102.

Yewale, C., Baradia, D., Vhora, I., Patil, S., & Misra, A. (2013). Epidermal growth factor receptor targeting in cancer: A review of trends and strategies. *Biomaterials*, *34*(34), 8690–8707. http://dx.doi.org/10.1016/j.biomaterials.2013.07.100.

Zarghooni, M., Bartels, U., Lee, E., Buczkowicz, P., Morrison, A., Huang, A., et al. (2010). Whole-genome profiling of pediatric diffuse intrinsic pontine gliomas highlights platelet-derived growth factor receptor alpha and poly (ADP-ribose) polymerase as potential therapeutic targets. *Journal of Clinical Oncology*, *28*(8), 1337–1344. http://dx.doi.org/10.1200/JCO.2009.25.5463.

Zhao, F., Yin, H., & Li, J. (2014). Supramolecular self-assembly forming a multifunctional synergistic system for targeted co-delivery of gene and drug. *Biomaterials*, *35*, 1050–1062. http://dx.doi.org/10.1016/j.biomaterials.2013.10.044.

Zimmerman, R. A. (1996). Neuroimaging of primary brainstem gliomas: Diagnosis and course. *Pediatric Neurosurgery*, *25*(1), 45–53.

Zukotynski, K. A., Fahey, F. H., Kocak, M., Alavi, A., Wong, T. Z., Treves, S. T., et al. (2011). Evaluation of 18F-FDG PET and MRI associations in pediatric diffuse intrinsic brain stem glioma: A report from the Pediatric Brain Tumor Consortium. *Journal of Nuclear Medicine*, *52*(2), 188–195. http://dx.doi.org/10.2967/jnumed.110.081463.

CHAPTER SEVEN

In Vivo Modeling of Malignant Glioma: The Road to Effective Therapy

Timothy P. Kegelman[*], Bin Hu[*], Luni Emdad[*,†,‡], Swadesh K. Das[*,†], Devanand Sarkar[*,†,‡], Paul B. Fisher[*,†,‡,1]

[*]Department of Human and Molecular Genetics, Virginia Commonwealth University, School of Medicine, Richmond, Virginia, USA
[†]VCU Institute of Molecular Medicine, Virginia Commonwealth University, School of Medicine, Richmond, Virginia, USA
[‡]VCU Massey Cancer Center, Virginia Commonwealth University, School of Medicine, Richmond, Virginia, USA
[1]Corresponding author: e-mail address: pbfisher@vcu.edu

Contents

Abstract

Despite an increased emphasis on developing new therapies for malignant gliomas, they remain among the most intractable tumors faced today as they demonstrate a remarkable ability to evade current treatment strategies. Numerous candidate

Advances in Cancer Research, Volume 121
ISSN 0065-230X
http://dx.doi.org/10.1016/B978-0-12-800249-0.00007-X

treatments fail at late stages, often after showing promising preclinical results. This disconnect highlights the continued need for improved animal models of glioma, which can be used to both screen potential targets and authentically recapitulate the human condition. This review examines recent developments in the animal modeling of glioma, from more established rat models to intriguing new systems using *Drosophila* and zebrafish that set the stage for higher throughput studies of potentially useful targets. It also addresses the versatility of mouse modeling using newly developed techniques recreating human protocols and sophisticated genetically engineered approaches that aim to characterize the biology of gliomagenesis. The use of these and future models will elucidate both new targets and effective combination therapies that will impact on disease management.

1. INTRODUCTION

The nearly 70,000 patients estimated to be diagnosed with glioma in 2013 would look to treatments developed using the latest insights into the pathology of this devastating form of cancer (Dolecek, Propp, Stroup, & Kruchko, 2012). While expected survival for glioblastoma (GBM) patients remains vanishingly small, increases in our understanding of the biology of this tumor have been made in the last decade. GBM has emerged as the most extensively described cancer through genomic profiling (Dunn et al., 2012), uncovering subtleties that may prove highly useful in development of novel therapies. To support this, the lag time from discovering a possible target to implementation of therapy has decreased (Chabner, 2011). Dedication to a deep understanding of the molecular biology of GBM has been shown through marked progress in characterizing the genome (Beroukhim et al., 2007; Cancer Genome Atlas Research Network, 2008) and the transcriptome (Phillips et al., 2006; Verhaak et al., 2010) of a wide sampling of patient tumors. This process has provided a far more detailed topography of the GBM landscape as well as unmasked what seemed to be a collection of highly related tumors to show a heterogeneous cast of distinct subtypes with unique tendencies and characteristics.

2. MALIGNANT GLIOMA

The World Health Organization guidelines for classification of gliomas are still the most widely used categorization method for clinical grading (Louis et al., 2007). While this method remains rooted in morphological and histopathological findings, insights into molecular signatures of tumor grade have been supplemented and advocated for incorporation into updated standards (Dunn et al., 2012; Huse & Holland, 2010). Glioma can include tumors defined as astrocytomas, oligodendrogliomas, or oligoastrocytomas, among others (Louis et al., 2007). Grade I gliomas have a limited proliferation profile and have the possibility of cure with surgical resection alone (Louis et al., 2007). While still considered low grade, grade II astrocytomas are known as "diffuse astrocytomas," show cytological atypia, and invade into surrounding tissue as the name suggests. Gliomas are considered as high grade at grade III, noted for high mitotic activity and anaplasia. Once microvascular proliferation and/or necrosis are observed, gliomas are categorized as grade IV, or GBM (Louis et al., 2007). Ninety to ninety-five percent of GBMs arise as "primary GBM," while 5–10% develop from lower grade gliomas, predominately in younger patients, and are known as "secondary GBMs" (Biernat, Huang, Yokoo, Kleihues, & Ohgaki, 2004; Ohgaki & Kleihues, 2005). These tumors show distinct molecular profiles as discussed below. Using just this classification system without molecular profiling, stratification for survival can be observed based on tumor grade (Dolecek et al., 2012; Louis et al., 2007). As molecular characterization and diagnostics become more prevalent, future classification systems will undoubtedly include genomic profiling to delineate tumor subclasses (von Deimling, Korshunov, & Hartmann, 2011).

Cancers of the central nervous system (CNS), like all other cancer types, have characteristic genetic instability with alterations in chromosome structure and/or copy number (Dunn et al., 2012). The systemic analysis of genetic data from tumors has identified a number of new, common somatic copy number alterations (SCNAs) present in cancer (Chin, Hahn, Getz, & Meyerson, 2011). Many of the most recent advanced methods used to identify new, important targets in tumors were first used to investigate GBM, and later other cancer types, through the probing of numerous primary patient samples (Beroukhim et al., 2007; Dunn et al., 2012). Such investigation into SCNAs resulted in a compilation of presumed targets of amplification (which includes *EGFR*, *MET*, *PDGFRA*, *MDM4*, *MDM2*, *CCND2*, *PIK3CA*, *MYC*, *CDK4*, and *CDK6*) and deletion (*CDKN2A/B*, *CDKN2C*, *PTEN*, and *RB1*) (Bredel et al., 2005; Kotliarov et al., 2006). This was a precursor to the first of a number of consortium-based analyses of the cancer genome, The Cancer Genome Atlas pilot study, dedicated to resolve the molecular picture of GBM through multiplatform profiling (Dolecek et al., 2012; Parsons et al., 2008). This project both confirmed earlier results and identified new molecular abnormalities, including amplification of *AKT3* and homozygous deletion of *PARK2* (Dolecek et al., 2012; Parsons et al., 2008). This project was able to quantify the dysregulation of major signaling pathways known to be important in GBM, while confirming previous studies (Ekstrand et al., 1991; Henson et al., 1994; Louis, 1994, 2006; Reifenberger, Reifenberger, Ichimura, Meltzer, & Collins, 1994; Schmidt, Ichimura, Reifenberger, & Collins, 1994; Ueki et al., 1996). Although *TP53* was known to be a commonly deleted gene, its signaling was impaired in 87% of analyzed samples. This occurred through mechanisms such as *CDKN2A* deletion and *MDM2* or *MDM4* amplification. Similarly, Rb signaling was inhibited in nearly 80% of samples, through either direct mutation or deletion of *RB1* or *CDK2NA*, or alternatively amplification of *CDK4*, *CDK6*, and *CCND2*. Oncogenic signaling through drivers such as receptor tyrosine kinases (RTKs), RAS signaling, or PI3K activation was present in almost 90% of tumors.

3. NOTABLE ABERRANT SIGNALING PATHWAYS IN MALIGNANT GLIOMA

Alterations in important checkpoint and driver pathways have been shown to contribute to development of gliomas. Enhanced signaling by RTKs (especially EGFR, PDGFR, and MET), PI3K pathway activation, signaling pathways left unchecked by decreased PTEN or NF1 activity, and effects of mutant isocitrate dehydrogenase (IDH) proteins play critical roles in the molecular biology of GBM (Cancer Genome Atlas Research Network, 2008; Parsons et al., 2008; Phillips et al., 2006; Verhaak et al., 2010).

3.1. Activating kinases

3.1.1 EGFR/EGFRvIII

EGFR is the flagship protein of the EGFR/ErbB receptor family, consisting of a ligand-binding extracellular domain connected by a hydrophobic transmembrane region to a cytoplasmic domain with tyrosine kinase activity (Huang, Xu, & White, 2009; Wells, 1999). It is amplified in about half of primary GBMs and is associated with a poor prognosis (Hurtt, Moossy, Donovan-Peluso, & Locker, 1992; Jaros et al., 1992; Schlegel et al., 1994). In cells with *EGFR* amplification, about half have a truncated variant, *EGFRvIII*, resulting from an intragenic gene rearrangement produced by the in-frame deletion of extracellular region exons 2–7. EGFRvIII expression portends a worse prognosis than wild-type EGFR expression alone (Heimberger et al., 2005; Shinojima et al., 2003). EGFR can initiate its signaling through the Shc–Grb2–Ras pathway as well as through the activation of PI3K (Huang et al., 1997, 2009; Narita et al., 2002). When overexpressed *in vitro*, EGFRvIII has been shown to display constitutive phosphorylation, enhanced cell proliferation, and superior tumorigenicity (Huang et al., 1997; Narita et al., 2002), as well as modulation of prosurvival protein Bcl–xL expression leading to resistance to apoptosis by DNA-damaging agents (Nagane, Levitzki, Gazit, Cavenee, & Huang, 1998). Unless EGF ligand is present at high concentration, wild-type EGFR overexpression cannot simulate the downstream effects of EGFRvIII (Bachoo et al., 2002). EGFR has been shown to translocate to both the mitochondria (Boerner, Demory, Silva, & Parsons, 2004) and the nucleus (Wang & Hung, 2009). In the nucleus, EGFR and EGFRvIII are possibly acting through both transcriptional and signaling mechanisms to drive proliferation as well as induce DNA damage repair (Wang & Hung, 2009).

Expression of EGFRvIII in human tumors is in fact heterogeneous, often observed in only a subpopulation of cells (Nishikawa et al., 2004). This subset of cells likely promotes its own growth and potentiates the proliferation of neighbor, wild-type EGFR cells, through the transmembrane glycoprotein gp130 (Inda et al., 2010). Downstream EGFR signaling has been shown to induce IL-8 production through NF-κB (Bonavia et al., 2012). Moreover, heterozygous deletion of *NFKBIA* (encoding for endogenous NF-κB inhibitor IκBα) displays a mutually exclusive pattern to *EGFR* amplification (Bredel et al., 2010).

When cultured *in vitro*, most lines fail to maintain faithful *EGFR* amplifications or EGFRvIII expression. However, passage of cells using stem cell conditions or serial *in vivo* xenograft harvesting has been shown to result in preservation of *EGFR* status (Stockhausen et al., 2011; Yacoub et al., 2008). While *EGFR* alterations are known to be common in other cancer types, the types of genetic alterations in GBM have been shown to be distinct from others such as non-small cell lung cancer (NSCLC). For example, *EGFR* amplifications in glioma tend to be focal and at very high copy numbers (>20). Mutations in nonglioma cancers are often found in the intracellular domains of EGFR (Janne, Engelman, & Johnson, 2005), while the majority of EGFR mutations in glioma, including the vIII mutation, are found in the extracellular domain (Cancer Genome Atlas Research Network, 2008; Lee, Vivanco, et al., 2006).

3.1.2 Platelet-derived growth factor receptor

Platelet-derived growth factor receptor (PDGFR) is another RTK with demonstrated importance in GBM signaling. *PDGFRA* is amplified in 15% of all samples, but enriched in the proneural subtype (discussed below) (Phillips et al., 2006; Verhaak et al., 2010). Of all samples with gene amplification, 40% harbor an intragenic deletion, $PDGFRA^{\partial 8,9}$, an in-frame deletion of exons 8 and 9, leading to a truncated extracellular domain (Ozawa et al., 2010). Additionally, an in-frame gene fusion of the extracellular domain of VEGFR-2 and the intracellular/kinase domains of PDGFRA has been identified. Both these mutations lead to constitutively active, transforming mutant proteins (Cancer Genome Atlas Research Network, 2008). *PDGFRB* expression appears to be restricted to proliferating endothelial cells within GBM tumors (Di Rocco, Carroll, Zhang, & Black, 1998; Fleming et al., 1992; Hermanson et al., 1992; Smith et al., 2000).

Other methods of activating the PDGFR signaling pathway include the production of endogenous ligands, PDGF(A–D), which are overexpressed in ~30% of glioma tissue samples and cell lines (Lokker, Sullivan, Hollenbach, Israel, & Giese, 2002). Both autocrine and paracrine mechanisms are likely candidates responsible for potentiating PDGF/PDGFR signaling within a tumor. Work in animal models has shown that, much like EGFR/EGFRvIII, tumors can be comprised of diverse populations of cells, some of which overexpress PDGF, and recruit cells which do not, leading to heterogeneous malignant gliomas (Assanah et al., 2006; Fomchenko et al., 2011).

3.1.3 c-Met

A third RTK that is amplified in 5% of GBMs, though rarely mutated, is *c-Met* (Cancer Genome Atlas Research Network, 2008). It is also coactivated in cells with increased levels of EGFR/EGFRvIII (Huang et al., 2007; Pillay et al., 2009; Stommel et al., 2007). Activated EGFR can associate with c-Met, resulting in ligand-free activation of c-Met (Jo et al., 2000). The c-Met–EGFR relationship is further intertwined in that ligand for c-Met, HGF, can transcriptionally activate EGFR ligands TGF-α and EGF, leading to activation of EGFR (Reznik et al., 2008). Additionally, blocking EGFRvIII activity through monoclonal antibody delivery (panitumumab) can lead to a switch to c-Met activation through HGF binding. This is a demonstration of interpathway cross talk that leads to obvious means for intrinsic drug resistance. Indeed, this relationship holds in lung cancer expressing mutated EGFR, where c-Met overexpression can lead to resistance to gefitinib, an EGFR-tyrosine kinase-specific inhibitor (Engelman et al., 2007). Therefore, it is imperative to apply a multifocal approach to RTK-specific inhibitors. Treatment targeted against EGFR (erlotinib), PDGFR (imatinib), and c-Met (SU11274) showed a significant improvement of GBM cell growth inhibition *in vitro* (Stommel et al., 2007). However, when c-Met is amplified alone, crizotinib treatment, directed against ALK and c-Met activity (Christensen et al., 2007), shows significant radiographic and clinical improvement in a case report (Chi et al., 2012). A phase I clinical trial is ongoing (A8081001).

3.1.4 Src-family kinases

Nonreceptor kinases belonging to the Src family often mediate signaling from growth factor receptors and are widely expressed in GBM (Lu et al., 2009). Src has been implicated in numerous cellular processes and can mediate pro-oncogenic processes such as cell viability, migration, invasion, and metastasis (Boukerche et al., 2010; Boukerche, Su, Prvot, Sarkar, & Fisher, 2008; Du et al., 2009). These proteins are often activated in GBM tumors and cell lines (Du et al., 2009; Stettner et al., 2005) and are effectors of oncogenic EGFR signaling (Lu et al., 2009). In 31 samples of GBM tumors, SRC was significantly activated in 61%. Inhibition via dasatinib inhibits viability and migration *in vitro* as well as tumor growth *in vivo* (Du et al., 2009). Src family kinases are among the most frequently observed tyrosine kinase activations in a study of 130 human cancer cells, along with EGFR, fibroblast growth factor receptor 3 (FGFR3), and focal adhesion kinase (FAK) (Du et al., 2009). Additionally, Src has been shown to be a crucial component of signaling related to novel scaffolding protein, melanoma differentiation-associated protein-9 (MDA-9)/Syntenin, demonstrated to be a promoter of malignant phenotypes in numerous cancers, including GBM (Das et al., 2012; Kegelman et al., 2014).

3.1.5 PI3K-related signaling

One of the most extensively studied and important dysregulated pathways in glioma signaling is the PI3K signaling cascade; driving survival, proliferation, migration, and invasion (Engelman, Luo, & Cantley, 2006). PI3K signaling can be activated through GTP-bound Ras as well as RTK activity, recruiting PI3K to the plasma membrane where it catalyzes the phosphorylation of phosphatidylinositol (4,5)-bisphosphate (PIP_2) to phosphatidylinositol

(3,4,5)-trisphosphate (PIP_3). The major tumor-suppressor PTEN reverses this reaction. PIP_3 recruits serine/threonine kinase AKT to the plasma membrane, where it is fully activated via phosphorylation at T308 by PDK1 and S473 by mTORC2 (Sarbassov, Guertin, Ali, & Sabatini, 2005). Elevated AKT phosphorylation is noted in up to 85% of GBM patient samples and cell lines (Wang et al., 2004). Another important RTK, Ephrin A2 (EphA2), highly expressed in GBM (Wykosky, Gibo, Stanton, & Debinski, 2005) binds PI3K after ligand stimulation and induces GBM cell migration by activating AKT (Miao et al., 2009).

Aside from RTK-dependent activation, PI3K signaling can be activated via mutation or amplification of catalytic subunit isoform *PIK3CA* (Cancer Genome Atlas Research Network, 2008; Gallia et al., 2006; Kita, Yonekawa, Weller, & Ohgaki, 2007) or overexpression of *PIK3CD* (Mizoguchi, Nutt, Mohapatra, & Louis, 2004). Important interactions with other proteins can also govern PI3K signaling. PI3K can interact with Src family kinase Yes, promoting CD95-driven invasion through activation of GSK3β and matrix metalloproteinase (MMP) expression (Kleber et al., 2008). In GBM that lacks EGFR amplification, insulin-like growth factor-2 (IGF2) promotes aggressive growth via insulin-like growth factor-1 receptor (IGF-R1) and phosphoinositide-3-kinase regulatory subunit 3 (PI3K-R3) (Soroceanu et al., 2007). In the absence of functioning PTEN, resulting AKT activation can contribute to RTK inhibitor sensitivity (Mellinghoff et al., 2005; Mellinghoff, Cloughesy, & Mischel, 2007). Inhibitors directed against PI3K have proved to be cytostatic as opposed to cytotoxic, which may be due to arrest at different phases of the cell cycle (Fan et al., 2010; Paternot & Roger, 2009). Therefore, numerous combination studies have been undertaken, simultaneously inhibiting other targets such as mTORC1, mTORC2, and MEK G1 (Fan et al., 2006, 2010; Liu, Wang, et al., 2011; Paternot & Roger, 2009). Due to the variety of mechanisms by which downstream targets can be activated, inhibition of the PI3K pathway has proved difficult. RTK signaling can circumvent AKT through PKC signaling, activating mTORC1 independent of AKT (Fan et al., 2009). Additionally, PI3K activation can modulate phosphorylation and inactivation of the pro-apoptotic protein Bad through PDK1 and PKC, increasing cell survival (Desai, Pillai, Win-Piazza, & Acevedo-Duncan, 2011). Overall, the PI3K cascade and feedback mechanisms, though widely studied, have proved complex and are not yet completely elucidated. Thus, inhibition strategies must be implemented as part of a combination treatment plan (Akhavan, Cloughesy, & Mischel, 2010).

3.2. Tumor suppressor pathways

3.2.1 PTEN

Loss of *PTEN* function leads to high levels of PI3K activity and downstream signaling effects. *PTEN* is lost, mutated, or epigenetically inactivated in up to 50% of gliomas and is one of the most frequently lost tumor suppressors in all cancers (Koul, 2008). Under normal conditions, PTEN stability is governed posttranscriptionally by GSK3β-mediated phosphorylation at T366 (Maccario, Perera, Davidson, Downes, & Leslie, 2007) and NEDD4-1-mediated ubiquitination leading to proteasomal degradation (Wang et al., 2007). Thus, PTEN function can be impeded without being genetically mutated or deleted. Since PTEN is part of a heterotrimeric tumor suppressor complex with NHERF-1 (Na^+/H^+ exchanger regulatory factor-1) and PHLPP-1 (pleckstrin homology domain leucine-rich repeat protein phosphatase-1), upregulation of NEDD4-1 or a similar alteration to the stability of other members of the complex can lead to disrupted PTEN function (Molina et al., 2012). Overexpression of forkhead transcription factor, FoxM1B, frequently enhanced in glioma samples, is associated with NEDD4-1 overexpression (Dai et al., 2007, 2010). It follows that the estimate of 50% of glioma samples showing *PTEN* mutation or deletion is an underestimate of the true number of tumors affected by altered PTEN function (Dunn et al., 2012). PTEN is involved in additional, disparate functions in the cell, including drug resistance, protein stability, and metabolism. *PTEN* loss was shown to increase the transporter protein ABCG2, implicated in drug efflux mechanisms (Bleau et al., 2009). In other cancer types, increases in ENTPD-5 (ectonucleoside triphosphate diphosphohydrolase-5) in *PTEN*-null cells lead to an increase in growth factor levels, enhanced AKT-mediated anabolism, and higher aerobic glycolysis (Warburg effect) (Fang et al., 2010). While *PTEN* loss and mutation often affect the phosphatase activity of the protein, these distinctions are not necessarily functionally equal in all respects. PTEN has been shown to localize to the nucleus to enhance the tumor-suppressor activity of the APC–CDH1 complex in a phosphatase-independent manner (Song et al., 2011). PTEN remains a crucial brake for diverse cellular functions and an especially crucial participant in gliomagenesis.

3.2.2 NF1

Neurofibromin, the protein product of *NF1*, is a Ras-GTPase-activating protein (Ras-GAP) and is known to inhibit Ras–mTOR signaling in astrocytes (Scheffzek et al., 1998). In gliomas, inactivation of NF1 can occur through deletion, mutation (Parsons et al., 2008), or enhanced proteasomal degradation by PKC hyperactivation (Cancer Genome Atlas Research Network, 2008; McGillicuddy et al., 2009). *NF1* loss can act through Ras-mediated overactivation of mTOR, leading to increases in proliferation and greater migratory abilities in primary murine astrocytes (Sandsmark et al., 2007). Other downstream effects include the activation of Stat3 through mTORC1 and Rac1 signaling, thereby resulting in transcription of cyclinD1 in *NF1*-null cells (Banerjee et al., 2010). In total, RTK/RAS/PI3K activation is found in 88% of tumors, which can occur through either activating mutations or loss of tumor suppressor function (Cancer Genome Atlas Research Network, 2008).

3.3. IDH mutations

IDH proteins have emerged as important new targets in glioma, particularly in grade II/III gliomas and in secondary GBM. While only 3–7% of primary GBMs show evidence of *IDH1* mutation, 60–80% of grade II and III gliomas and up to 80% of secondary GBM harbor mutant *IDH1*, with the most common mutation being an R132H substitution (Balss et al., 2008; Hartmann et al., 2009; Ichimura et al., 2009; Yan et al., 2009). As an enzyme in the tricarboxylic acid cycle, IDH1 catalyzes the $NADP^+$-dependent reduction of isocitrate to 2-oxoglutarate (α-ketoglutarate, α-KG), resulting in production of NADPH (Raimundo, Baysal, & Shadel, 2011). However, mutant IDH1 catalyzes α-KG to an *R*-enantiomer of 2-hydroxyglutarate (2-HG), the same chemical built up in D-2-hydroxyglutaric aciduria (D-2-HGA), a rare autosomal recessive organic aciduria (Dunn et al., 2012). It is possible that this and similar conditions could lead to increased risk for malignancies (Aghili, Zahedi, & Rafiee, 2009; Bhagwat & Levine, 2010). Both the wild-type and mutant IDH1 proteins can catalyze this reaction, but the mutant protein does so with much greater efficiency, leading to an excess of 2-HG (Dang et al., 2010). 2-HG can impair function of α-KG-dependent proteins, affecting a wide range of cellular functions including DNA demethylation, histone demethylation, fatty acid metabolism, hypoxic state detection, and collagen modification among others (Loenarz & Schofield, 2008). A global DNA hypermethylated state is observed in mutant IDH1 gliomas, as well as IDH1/2 mutant AML samples (Figueroa et al., 2010; Noushmehr et al., 2010). In normal astrocytes expressing mutant IDH1, this hypermethylated state can be replicated, possibly due to the TET2 enzyme, an α-KG-dependent enzyme (Turcan et al., 2012; Xu et al., 2011). The persistent hypermethylated state could favor a dedifferentiated state (Figueroa et al., 2010), supported by additional dysfunction in histone demethylases in IDH1 mutant cells (Chowdhury et al., 2011; Lu et al., 2012; Xu et al., 2011). Presence of repressive histone methylation in IDH1 mutant cells precedes observed DNA hypermethylation and impairs differentiation (Lu et al., 2012). Additionally, 2-HG can alter the cell's ability to regulate physiological reactions to hypoxia, possibly by stabilizing hypoxia-inducible factor-1 (HIF-1) (Zhao et al., 2009), leading to a less robust HIF-1 response to hypoxia in mutant IDH1 cells (Koivunen et al., 2012).

When comparing patients with varying grades of glioma who have mutant IDH1 expression to those with wild-type IDH1, those with mutant IDH1 tend to present at a significantly younger age and have notably longer survival times compared to those with wild-type expression (Hartmann et al., 2010; Parsons et al., 2008; Sanson et al., 2009; Yan et al., 2009). Such mutations could be a leading event in glioma progression, as a subset of patients having only IDH1 mutations at a preliminary biopsy acquired TP53 or 1p19q loss at subsequent biopsies (Dunn et al., 2012). The IDH1 status of tumors has a clear clinical relevance, and efforts have been made to develop noninvasive tests specific to this mutation. Although 2-HG is easily monitored in the serum of AML patients, its presence may be less specific in glioma (Capper et al., 2012). Other studies point to the possibility of imaging the brain to monitor 2-HG using magnetic resonance spectroscopy (Pope et al., 2012). The morphological characteristics and histological findings are often indistinguishable. However, it has become clear that IDH1 mutant and wild-type GBM undergo disparate disease progressions.

4. MOLECULAR CLASSIFICATION OF GBM

The heterogeneous nature of GBM has been revealed transcriptional profiling efforts to discover multiple subtypes with unique expression patterns through mRNA analysis of tumor samples (Golub et al., 1999). While these tumors may share histopathological and morphological characteristics as well as WHO tumor grade, recent studies have elucidated solid evidence for subtypes of GBM (Phillips et al., 2006; Verhaak et al., 2010). Phillips et al. (2006) analyzed expression of a panel of genes linked to differences in survival outcomes to define three subtypes of GBM: proneural, mesenchymal, and proliferative. Alternatively, using unsupervised hierarchical clustering analysis, Verhaak et al. (2010) classified 200 TCGA samples of GBM into four different subtypes: proneural, mesenchymal, classical, and neural. Each of these subtypes had at least 210 genes defining it, and three of the four subtypes showed unique molecular alterations. Amplification of *PDGFRA*, *CDK6*, *CDK4*, and *MET* was linked to the proneural subtype, as well as *PI3KCA/PIK3R1* mutation and mutation or loss of heterozygosity of *TP53*. Importantly, *IDH1* mutations were observed frequently in the proneural group, and it contained the highest number of young patients, corresponding with other *IDH1* data showing its prevalence in grade II and III gliomas and secondary GBM. The signature alteration in the mesenchymal group was *NF1* mutation, and other changes included mutation or loss of *TP53* and *CDK2NA*. *EGFR* amplification along with *PTEN* and *CDK2NA* loss were common in the classical group, while the neural group so far has no defining mutations.

Since the proneural and mesenchymal subtypes were created using different methods and sample sets and thus are considered the most compelling and definitive (Huse, Phillips, & Brennan, 2011). Using both methods, expression of *DLL3* and *OLIG2* was found to be strong in the proneural subtype. Expression of *CD40* and *CHI3L1/YKL-40*, a potential target for serum protein monitoring of GBM progression (Iwamoto et al., 2011), was robust in the mesenchymal subtype in both studies. A subset of genes that are associated with the various subtypes of GBM are represented in a 9-gene panel shown to predict patient outcome, with expression related to a mesenchymal phenotype showing poor prognosis (Colman et al., 2010). However, not all biologically relevant data can be distilled from purely genomic data. Using unsupervised clustering of proteomic analysis, three groups defined by signaling pathways were developed: EGFR related, PDGFR related, or signaling coordinated with *NF1* loss (Brennan, 2011). Analysis indicated that these pathways were nonoverlapping, as activation of each signaling pathway was mutually exclusive. These results demonstrated a need for protein-based data to complement genomic profiling as tumors with high PDGF expression often had low PDGF mRNA and no amplifications. Therefore, a given signaling pathway could be undervalued based solely on transcription-based approaches. Further extensions in these efforts are being undertaken to uncover networks and signaling programs that lead to a particular transcriptional state. Computational network analysis is reverse engineering the identity of critical regulatory molecules and transcription factors unique to each subtype. For example, six transcription factors were identified as being largely responsible for the transcriptional state of the mesenchymal subtype, and validation of STAT3 and C/EBPβ showed that they were each required for successful growth of orthotopic xenografts (Carro et al., 2010). Further studies using similar methods aimed to discover other important regulators of the mesenchymal state. A group of important molecules were defined including YAP, MAFB, HCLS1, and TAZ, a transcriptional coactivator. TAZ was hypermethylated in proneural subtype tumors, and found to drive the mesenchymal phenotype when overexpressed, cooperating with PDGFB to induce malignant progression (Bhat et al., 2011). Continued investigation into the control of GBM subtype selection will assist in focusing the identification of novel treatment targets for enhanced clinical therapy.

5. PROGRESSION OF GLIOBLASTOMA

5.1. Invasion

The majority of GBMs treated with current standard-of-care therapy recur within centimeters of the primary tumor mass (Chamberlain, 2011; Hochberg & Pruitt, 1980). Even in cases of grade II diffuse gliomas, evidence can be found of cells up to 2 cm from the primary tumor mass (Yordanova, Moritz-Gasser, & Duffau, 2011). Up to 20% of patients show evidence of macroscopic invasion upon presentation. These observations can be of multifocal disease, bihemispheric invasion along white matter tracts (also known as "butterfly pattern" glioma), and invasion along the subependymal and subarachnoid spaces (Chamberlain, 2011; Parsa et al., 2005). Early studies of glioma show up to 50% of untreated patients with histological findings demonstrating bilateral hemispheric involvement (Matsukado, Maccarty, & Kernohan, 1961; Scherer, 1940). While magnetic resonance imaging (MRI) is more sensitive than computed tomography (CT) scanning in detecting small tumors, and can identify extratumoral involvement such as peritumoral edema, it is accepted that radiographic study is sure to miss the full extent of the disease area (Burger, Heinz, Shibata, & Kleihues, 1988; Hochberg, Fischman, & Metz, 1994).

Spread through lymphatic or vascular systems is exceptionally limited, unlike other cancer indications (Hoffman & Duffner, 1985). Thus, the invasive nature of glioma is imposed on surrounding brain tissue. To successfully invade, tumor cells must detach from the primary tumor mass, adhere to the extracellular matrix (ECM), degrade the ECM, and activate motility programs to instigate movement (Onishi, Ichikawa, Kurozumi, & Date, 2011).

The composition of the ECM makes the brain parenchyma unique compared to extra CNS tissue sites. Glia externa limitans, which coats the cortical surface and encloses the cerebral blood vessels, is a conventional collagen-rich basement membrane (Gritsenko, Ilina, & Friedl, 2012; Louis, 2006; Rutka, Apodaca, Stern, & Rosenblum, 1988). Unique to the brain parenchyma is the perineuronal network, a lattice of predominately hyaluronan sulfate proteoglycans as well as chondroitin sulfate proteoglycans, link proteins, and tenascins (Gritsenko et al., 2012; Kwok, Dick, Wang, & Fawcett, 2011). The subventricular zone (SVZ), and area of neurogenesis, is more enriched for chondroitin and heparan sulfate proteoglycans (Sirko, von Holst, Wizenmann, Gotz, & Faissner, 2007).

Receptors for hyaluronan, immunoglobulin superfamily member CD44 and hyaluronan-mediated motility receptor (RHAMM) are overexpressed by GBM cells (Akiyama et al., 2001). These proteins are repressed by p53, pointing toward a linkage between cell checkpoints and migration/invasion abilities (Dunn et al., 2012; Godar et al., 2008; Sohr & Engeland, 2008). The actions of integrins, especially $\alpha v\beta 3$ and $\alpha v\beta 5$ heterodimers, contribute to adherence to the ECM through cytoskeletal rearrangement. This proceeds through intracellular proteins including FAK (Riemenschneider, Mueller, Betensky, Mohapatra, & Louis, 2005; Rutka et al., 1999) and Pyk2 (Lipinski et al., 2008). Testing of a $\alpha v\beta 5$ inhibitor, cilengitide, together with TMZ and radiation is underway in a phase III clinical trial (Reardon et al., 2011).

After adhering to the ECM, cells must clear a path through degradation of the ECM meshwork. MMPs are heavily involved in this process. MMP2 and MMP9 have been widely studied for their involvement in GBM invasion (Forsyth et al., 1999). CD95/FasR, the expression of which is enhanced at the leading edge of invasion, can activate AKT1, recruiting SFK, Yes and p85, upregulating both MMP2 and MMP9 (Kleber et al., 2008). Migration has been shown to be increased by LRP1 (low-density lipoprotein receptor-related protein-1), act through ERK to upregulate both MMP2 and MMP9 (Song, Li, Lee, Schwartz, & Bu, 2009). FoxM1B is an important transcription factor, overexpressed in GBM and shown to transform immortalized human astrocytes into invasive cells. It can assert its activity through PTEN degradation and AKT activation, leading to MMP2 upregulation (Dai et al., 2010). Among the growing list of additional molecules involved in GBM invasion (Nakada et al., 2007; Teodorczyk & Martin-Villalba, 2010), MMP1 has been implicated through an EGFR-dependent mechanism in which its expression is increased through EGFR-induced guanylate-binding protein 1 (GBP1) (Li et al., 2011). Cultured cell lines usually do not replicate the invasive growth seen in GBM patients. However, tumor-initiating cells grown under neurosphere conditions have demonstrated the ability to recapitulate an invasion patterns similar to the tumors from which they were derived (Wakimoto et al., 2012).

After gaining a foothold and beginning to degrade the surrounding ECM, invading tumor cells must engage cellular programs to induce locomotion. These programs will largely involve the rearrangement of the actin cytoskeleton. CD44 is cleaved by ADAM proteases and MMP9 leading to cytoskeletal reorganization and promotes motility (Bourguignon, 2008; Chetty et al., 2011; Murai et al., 2004). Girdin, an actin binding protein known to be involved in directing neural cell migration, is regulated by AKT and is important in the invasiveness of tumor-initiating cells (Natsume et al., 2012). Myosin II is also an important protein for invasion in GBM as a predominant initiator of cellular force. It also enables GBM cells to invade through pores less than the width of their nuclear diameter, a vital advantage in the narrow ECM of the brain parenchyma (Beadle et al., 2008). It will be important to test future inhibitors of invasion in models that accurately recreate the infiltrating nature of human GBM.

5.2. Angiogenesis

Angiogenesis is a hallmark of GBM, and microvascular proliferation, along with necrosis, is a major characteristic that delineates WHO grade III from grade IV tumors (Louis et al., 2007). Glioma cells need blood vessels not only to replenish oxygen, deliver nutrients, and remove waste, but also to create a vascular niche to selectively support glioma stem cells (GSCs) (Calabrese et al., 2007; Gilbertson & Rich, 2007). Several mechanisms are used to accomplish this including angiogenesis, vasculogenesis, and tumor cell repurposing. Angiogenesis involves using the existing vasculature to create new blood vessels (Folkman, 1971; Kerbel, 2008). Alternatively, bone marrow-derived endothelial progenitor cells can be recruited to form new blood vessels in the absence of preexisting ones in a process known as vasculogenesis. The newly formed vasculature can then expand and be pruned through angiogenesis (Patenaude, Parker, & Karsan, 2010). Tumor cells themselves can be recruited directly into the vascular wall or may directly differentiate into vascular endothelium (Ricci-Vitiani et al., 2010; Wang et al., 2010). It is well known that the intratumoral vasculature formed through these processes is incomplete, highly aberrant, and tortuous, which leads to sections of acidosis, hypoxia, and the development of peritumoral edema (Long, 1970).

The constant balance between pro- and antiangiogenic signaling molecules determines the outcome of blood vessel formation, and these molecules can be highly dysregulated in the tumor microenvironment (Bergers & Benjamin, 2003; Onishi et al., 2011). Vascular endothelial growth factor (VEGF)-related signaling is thought to be a crucial proangiogenic factor in tumor angiogenesis and can be induced by HIF-1α. VEGF largely acts through the VEGFR-2/KDR pathway, stimulating signaling in endothelial cells to promote proliferation, survival, migration, and permeability (Hicklin & Ellis, 2005). Numerous other signaling factors have been implicated in stimulating angiogenesis in GBM including PDGF, FGF, angiopoetin signaling (ANG/TIE system), Notch signaling, Integrins, Ephrins, and SDF-1/CXCL12 (Carmeliet & Jain, 2011a, 2011b; Weis & Cheresh, 2011). Additionally, IL-8 has been shown to be an important proangiogenic molecule in GBM, most recently through the action of MDA-9/Syntenin (Kegelman et al., 2014). Angiogenesis can be inhibited by endogenous factors such as angiostatin, interferons, thrombospondins, endostatin, and tumstatin (Nyberg, Xie, & Kalluri, 2005). Blood vessel formation is favored when proangiogenic signaling outweighs antiangiogenic signaling.

Pharmaceutical angiogenesis inhibitors have been shown to be successful in treating GBM in patients, with Bevacizumab, an anti-VEGF antibody, undergoing accelerated approval for use in GBM (Cohen, Shen, Keegan, & Pazdur, 2009). Most angiogenesis inhibitors have similar approaches, targeting the VEGF ligand (Vredenburgh, Desjardins, Herndon, Marcello, et al., 2007), the VEGFR-2/KDR receptor (Batchelor et al., 2010), or the downstream signaling molecules. Bevacizumab has been the most successful inhibitory antibody used thus far and is widely used in patients. Two initial phase II trials combining bevacizumab with irinotecan, a topoisomerase I inhibitor, showed remarkable increases of the 6-month, progression-free survival percentage from 9–15% to 38–46%, median survival gains from 22–26 weeks increased to 40–42 weeks, as well as a 60% radiographic response rate (Vredenburgh, Desjardins, Herndon, Dowell, et al., 2007; Vredenburgh, Desjardins, Herndon, Marcello, et al., 2007). Two subsequent studies showed more modest, but still significant, treatment improvements (Friedman et al., 2009; Kreisl et al., 2009), with a phase III study in progress. However, inhibition of angiogenesis is known to decrease the peritumoral edema caused by permeable tumor vasculature, dramatically altering the appearance on MRI (Gerstner et al., 2009; Wen et al., 2010). Moreover, tumors that recur after antiangiogenic therapy tend to progress rapidly and rarely respond to additional chemotherapy (Ellis & Hicklin, 2008; Quant et al., 2009). We need to continue to understand these phenomena, as there is growing evidence that angiogenic inhibitors instigate invasion (Iwamoto et al., 2009; Narayana et al., 2012; Norden et al., 2008). The use of relevant animal models will aid in understanding the complex interplay between treatment approaches.

6. INTRODUCTION TO ANIMAL MODELING IN GLIOMA

To be useful in the modeling of human cancer, any particular model must strike a balance of faithfulness in recapitulating the human disease and convenience of use to the researcher. The consummate model would display authentic histopathology observed in human disease, including the infiltration, angiogenesis, and distant spread so often seen in the human as tumors progress through similar pathological stages. Signaling pathways found to be significant in patient screens would be prominent, while the heterogeneity of human tumors, both cellular and molecular, also would be preserved. Stromal interactions should remain as faithful to human counterparts as possible, including recapitulating immune system influences. Once these stipulations are met, this ultimate model should develop tumors with high penetrance, preferably with short latencies and within a predictable window of time, all while possessing the means to reliably monitor tumor initiation and progression noninvasively. Of course, human tumors themselves do not adhere to these requirements, as wide heterogeneity and differences in progression are common. Each *in vivo* system has advantages and drawbacks whether in authenticity or convenience (Fig. 7.1). Nonetheless, useful data can be generated from diverse models, and progression of new treatments through fruitful preclinical experimentation will lead to advances in treatment, especially for patients with aggressive disease that portends a grim prognosis.

While the catalog of genetic abnormalities observed from sampling end-stage tumors has led to numerous advances in the understanding of glioma biology, modeling using an *in vivo* system allows us to scrutinize the role of specific alterations and their effects on gliomagenesis. Likewise, genetic profiling that has identified subtypes of GBM based on expression patterns has revealed intriguing differences about prognosis and resistance to standard therapies. By using animal modeling systems, we can further address which

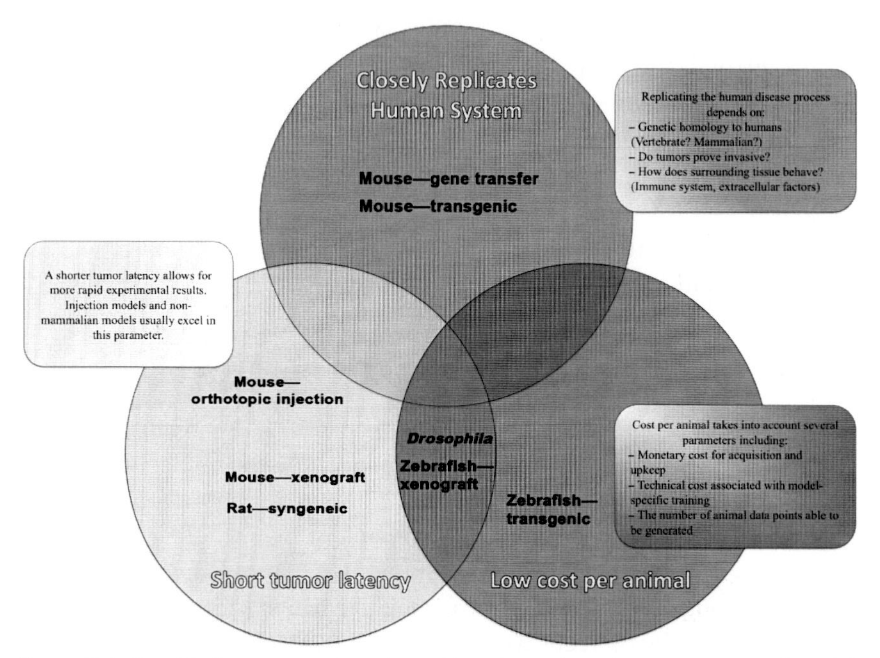

Figure 7.1 Graphical representation of considerations when choosing an animal model. Three areas are represented: Human tumor recapitulation (top), tumor latency (bottom left), and cost on a per animal basis (bottom right). (See the color plate.)

therapies will be most effective for each subtype and devise new, more targeted therapeutic strategies. The natural progression of human glioma often leads to advanced disease before symptoms arise, and due to this late presentation, there is a scarcity of information on early stage lesions. Animal modeling of glioma is built to confront these challenges in various ways. (1) *Gliomagenesis*: Animal modeling can identify which genetic aberrations are necessary to initiate a tumor. Additionally, models can help categorize specific activating lesions as necessary and others that may not be sufficient on their own, but are effective in accelerating the process. (2) *Tumor progression*: During the course of disease, tumors acquire mutations that endow advantageous properties such as enhancements in growth rate, angiogenic potential, or invasive ability. A tumor must adapt to what may be considered hostile conditions as it grows. It must evade immune surveillance, deal with inflammatory conditions, and grow despite hypoxic surroundings. Exploring the signaling pathways involved in these features gives insight into how the glioma cells survive and thrive in the brain. (3) *Treatments*: The ultimate goal of animal modeling is to improve and test treatments, efficiently moving effective new options into the clinical arena. Animal models must be able to select appropriate molecular targets. The knowledge acquired in uncovering essential tumor signaling pathways and aberrations that lead to an enhancement in tumor progression will yield novel therapeutic strategies that can be tested in animal models. The time to chemically develop a new compound for testing is considerably shorter than the time it takes to verify its efficacy *in vivo*. Therefore, the animal models used must faithfully recapitulate the human disease in a manner that will permit effective testing of new compounds and the evaluation of drug toxicity. When time is a dominating factor in experimental planning, tumor latency must be carefully considered when choosing an appropriate animal model (Fig. 7.2). Ideally, an animal model would help identify biomarkers that are able to demonstrate the efficacy of a tested drug. Moreover, addressing how tumors become resistant to therapy is a valuable aspect of *in vivo* disease modeling. This opens up the possibility of exploring combinations of treatments, from refining existing therapies to devising new, more effective fusions of targeted treatments. Modeling glioma *in vivo* provides the opportunity to examine the relative contributions of factors outside the neoplasm itself. One can address the

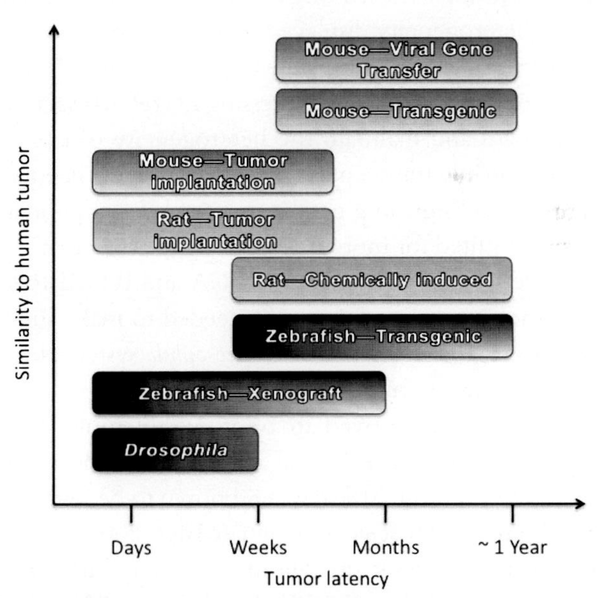

Figure 7.2 Tumor latency can vary widely between, but also within, animal models of glioma. Pictured is a representation of comparative tumor latencies for *Drosophila*, zebrafish, rat, and mouse models. (See the color plate.)

significance of ECM composition, supporting roles of normal brain tissues, and immune modulation, each of which may help or hinder glioma growth.

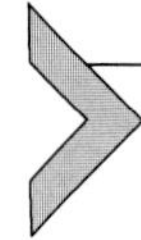

7. NON-MAMMALIAN MODELS OF GLIOMA

7.1. *Drosophila*

While not as widely used as mammalian models, a system for the simulation of invasive glioma in the fruit fly has recently been developed. A *Drosophila* model could have significant advantages over *in vitro* experiments without the time investment of genetically engineered mouse (GEM) models. Often, cultured GBM cells lose key genetic and phenotypic features that drove their *in vivo* pathogenesis (Lee, Kotliarova, et al., 2006). *EGFR* amplification and invasive growth upon reimplantation are commonly not retained after extended *in vitro* maintenance. Neural stem cell (NSC) culture can propagate tumor cells in neurosphere form, thought to retain the most aggressive tumor-initiating cells' characteristics, yet these can only be established from a subset of GBM tumors (Chen et al., 2010; Laks et al., 2009). *In vitro* models are further restricted due to the lack of surrounding environmental cues normally present within the CNS stroma, including signals from reactive astrocytes, microglia, and vasculature (Pong & Gutmann, 2011). Even within the same tumor, the heterogeneity displayed *in vivo* is a crucial driver of proliferation and survival. In tumors with large populations of cells with amplified EGFR, a smaller population of cells expressing EGFRvIII secretes cytokines that promote growth and maintain the heterogeneity of the entire tumor (Inda et al., 2010). While transgenic mouse models have been indispensible in elucidating roles and signaling mechanisms of known glioma mutations, they are not as well suited for more high-throughput screening of the many mutations revealed by databases such as TCGA and REMBRANDT. The investment of time, money, and expertise needed to make such large-scale studies routine is prohibitive. Therefore, *Drosophila* systems may be able to find an important niche within GBM *in vivo* modeling.

Drosophila models have proved to be powerful tools in the study of human disease, including neurological disease and cancer (Hirth, 2010; Reiter & Bier, 2002). Almost 70% of genes known to be involved in human disease have orthologs in *Drosophila* (Reiter & Bier, 2002). Important genes with established involvement in glioma signaling are among highly conserved pathways including EGFR/RTK/Ras, PI3K, Notch, Wnt, Jak-STAT, Hedgehog, and TGF-β. Numerous components of these crucial signaling pathways were uncovered in studies of neural development and tumorigenesis in fruit flies (Wilson & Chuang, 2010). Observations made in fruit fly models on human pathogenesis have been shown to translate to mammalian models of disease (Choi et al., 2011; McBride et al., 2005; Read et al., 2005). The ability of targeting individual cell types is crucial in the modeling of human disease and is a strength of *Drosophila* systems.

Precise cell-specific alterations in gene expression have been demonstrated in fruit flies, including those in the CNS (reviewed in St Johnston, 2002). Knockdown models are available for almost all genes with a sequenced and well-annotated genome. This allows for the exploration and identification of novel gene functions as well as the ability to delve into more complex, multigene interactions (Read, 2011). Notably, *Drosophila* NSCs and neurons share many similarities with their vertebrate counterparts (Doherty, Logan, Tasdemir, & Freeman, 2009; Freeman & Doherty, 2006). The remaining question is, would a fly model of glioma capture the crucial aspects of the human pathology, both genetically and phenotypically?

The loss of tumor suppressor-like genes has been shown to yield disruptions in the divisions and cell fate determination of neural progenitor cells (Neumuller & Knoblich, 2009). Alternatively, the study of oncogenic drivers has made recent advances. Since signaling pathways found to be important in glioma signaling are intact in the fruit fly, such as EGFR-Ras and PI3K pathways, development of a *Drosophila* glioma model has focused on these as targets for alteration. Using the repo-Gal4 glial-specific driver of transcription, orthologs of activated EGFR and PI3K (dEGFR and dp110 in *Drosophila*, respectively) were overexpressed in the brains of *Drosophila*, leading to glial neoplasia in developing larva (Read, Cavenee, Furnari, & Thomas, 2009; Witte, Jeibmann, Klambt, & Paulus, 2009). This was accomplished through the concurrent expression of constitutively active variants, $dEGFR^{\lambda}$ and $dp110^{CAAX}$, at 50- to 100-fold above normal observations in glia (Read et al., 2009). The neoplastic glia, first observed in post-embryonic stages, lose their normal glial morphology while becoming robustly proliferative and form multilayered aggregations in the brain, disrupting normal architecture (Read et al., 2009). Likewise, overexpression of *Drosophila* orthologs of genes known to be important in glioma produced similar results. Overexpression of dAkt (or knockdown of dPTEN via siRNA), $dRas^{V12}$ (activated Ras), as well as orthologs for activated PDGFRα and FGFR led to neoplastic growth and patterns of inappropriate migration (Read et al., 2009; Witte et al., 2009). Cells from $dEGFR^{\lambda}$ and $dp110^{CAAX}$ mutant glia formed sizeable, invasive tumors when injected into the abdomen of host fruit flies, an established test of tumorigenicity.

Notably, these tumors were adept at stimulating new trachea or commandeering existing trachea, or oxygen delivery tubules in *Drosophila* (Read et al., 2009).

One advantage of fruit fly models is the ability to engage in forward genetic screens, uncovering the identity of other genes that suppress or enhance disease phenotypes when mutated or inhibited. Fruit fly models have proved successful in these screens and have identified novel genes and gene functions within cell processes and signaling pathways (Karim et al., 1996; Read et al., 2005; Yoshikawa, McKinnon, Kokel, & Thomas, 2003). This valuable approach to *in vivo* screening, utilized in "enhancer–suppressor screens," has uncovered core components and cell type-specific regulators of signaling pathways (Karim et al., 1996; Simon, 2000; Voas & Rebay, 2004). Signaling transduction pathways in fruit flies, especially the RTK pathways, are highly conserved to the degree that mouse and human components can functionally replace *Drosophila* homologues (Halder, Callaerts, & Gehring, 1995; Read, 2011). One recent study demonstrating the advantages of such an approach used RNAi to interrogate almost the entire *Drosophila* kinome, the kinases within the genome (Read et al., 2013). Two hundred twenty-three of the 243 kinases in the fruit fly genome were knocked down and 45 were found to modify the development of neoplasia in the $dEGFR^{\lambda}$;$dp110^{CAAX}$ model of glioma. Knockdown targeting dRIOK1 and dRIOK2, orthologs of mammalian RIO (right open reading frame) atypical kinases, resulted in especially robust inhibition of tumorigenesis. This was in turn validated via database and human cell line experimentation, showing that RIOK1 and RIOK2 were highly expressed in GBM and upregulated in cell lines expressing EGFRvIII compared to wtEGFR. Further, mouse $Pten^{-/-}$;$Ink4a/arf^{-/-}$ astrocytes overexpressing RIOK2 formed invasive, high-grade gliomas (HGGs) in 70% of intracranial injections, while control $Pten^{-/-}$;$Ink4a/arf^{-/-}$ astrocytes formed none. Knockdown of RIOK1 and RIOK2 led to apoptosis and chemosensitization in the presence of wild-type p53 (Read et al., 2013). Such a study exemplifies the strength of an *in vivo* system capable of undergoing a large genetic screen and still results in relevant output.

Overall, *Drosophila* modeling is a robust system with a high level of conserved cellular signaling in the nervous system. It has already been used to test known and potential treatments for GBM (Witte et al., 2009) as compounds can be directly fed to the fruit flies. Live imaging is possible in rapidly developing tumors, and successful drug and drug targets can be verified in corresponding mammalian systems. The complexity and heterogeneity of

GBM signaling are becoming apparent as interactions and cross talk between pathways are being discovered. Therefore, the ability for this model to engage in larger scale genetic screens is a major advantage in decoding crucial targets for future therapies.

7.2. Zebrafish

Zebrafish (*Danio rerio*) systems have emerged as viable hosts for modeling human cancer pathology in recent years. In addition to the added physiologic and genetic homology to mammals as a vertebrate model, a number of advantages are inherent in using zebrafish. Genetic manipulation is easily accomplished and animal maintenance is relatively inexpensive. Additionally, zebrafish have large numbers of offspring, embryos undergo rapid development, and their translucent nature readily allows for structure visualization (Geiger, Fu, & Kao, 2008). Studies of human tumor cells injected into zebrafish embryos have shown that these cells can survive and interact with zebrafish tissues, highlighting a potential for new modeling approaches to glioma. Geiger and colleagues injected U251 cells into the yolk sac at 2 days postfertilization (dpf) to successfully form tumors. There are important differences in zebrafish and human tissue culture. Zebrafish embryos are usually maintained at 28 °C, and do not survive well at 37 °C, a typical temperature for cell line maintenance (Yang, Chen, et al., 2013). However, U251 cells showed no significant growth changes at temperatures above 28 °C, and zebrafish fare well at temperatures of 30–35 °C, so xenograft maintenance in this range is acceptable (Geiger et al., 2008; Yang, Chen, et al., 2013).

The transgenic zebrafish model *fli1:EGFP* is frequently used to visualize zebrafish blood vessels and has been combined with xenograft approaches to study several aspects of glioma in recent years. Temozolomide and radiation therapy were used in combination showing that TMZ has a radiosensitizing effect on xenograft tumors in zebrafish, similar to what has been shown in mammals. Moreover, TMZ treatment alone showed no developmental effects to the embryo (Geiger et al., 2008). Tumor angiogenic and invasive studies were also undertaken using the same *fli1:EGFP* model. The angiogenic potential of U87 tumors was evaluated after injecting cells into the yolk sacs of zebrafish embryos and measuring vessel formation and length in the ventrolateral aspect of the yolk sac. Inhibiting JNK through pharmacological means decreased the observed angiogenic vessel growth (Yang, Cui, et al., 2013). Invasion of GSCs, as designated by CD133 expression, was also evaluated using this model. Cells isolated from tumorspheres and injected into the yolk sac invaded via vessels away from the injection site. Injections enriched for CD133 expression proved to be more invasive, invading to the edges of the yolk sac, and had higher expression of MMP9. However, after cells migrated away from the primary injection site, only 42.3% of the cells retained CD133 expression compared to 74% at the point of injection (Yang, Cui, et al., 2013).

Transgenic models of zebrafish are also being used to explore *NF1* mutations, important in glioma formation. Knockouts of both *NF1* orthologs, *nf1a* and *nf1b*, were found to be developmentally lethal with aberrant proliferation and differentiation of OPCs (Shin et al., 2012). However, if one intact allele remained, as in $nf1a^{+/-};nf1b^{-/-}$, loss of tumor suppressor function cooperated with p53 mutation to form tumors with 62% penetrance at 45 weeks postfertilization (wpf). This is significantly higher than the 28% penetrance at 66 wpf observed in *p53*-null zebrafish (Berghmans et al., 2005). Both brain and peripheral nerve sheath tumors developed, but brain tumors developed early (31–33 wpf) and showed features of human HGG, likely grade III due to lack of necrosis or vascular proliferation (Shin et al., 2012). Another recent study demonstrated that when a dominant active form of Akt1, DAAkt1, is overexpressed in zebrafish, gliomas of grades varying from I to IV develop in over 1/3 of fish by 6 months and almost 50% by 9 months. When an active form of Rac1, DARac1, is coexpressed in this model, the penetrance increases to 62% and 73%, with enhanced invasion, higher grade of tumors observed, and shorter survival times (Jung et al., 2013).

8. MAMMALIAN MODELS OF GLIOMA

8.1. Rat

Until relatively recently, rat models were used more widely than mouse models for the study of glioma (Fomchenko & Holland, 2006). Xenograft transplantation of tumor cells into immunocompromised rats has been used extensively (Candolfi et al., 2007), but the popularity of the syngeneic orthotopic approach is unique to rat modeling of glioma. Development began in the 1970s using carcinogenic exposure to MNU (*N*-methylnitrosourea) or ENU (*N*-ethyl-*N*-nitrosourea), which led to the production of reproducible tumors. These tumors could then be transplanted into syngeneic hosts through stereotactic injection (Barker et al., 1973). Advantages of using rats include the size of the brain, roughly three times larger than that of a mouse, which tolerates larger volumes when injecting tumor cells (Barth & Kaur, 2009). Additionally, thicker rat skulls can tolerate screws to implant tumor cells or deliver therapeutics, something not possible in the thinner skulls of mice (Saini, Roser, Samii, & Bellinzona, 2004). However, most rat models are not genetically engineered; therefore, specific targeting of pathways is not as prevalent as with mouse models. Likewise, exploration into tumor-initiating cell types or the effect of stromal factors on tumor development is more easily achieved with the genetic pliability of mouse models (Reilly et al., 2008). Nonetheless, rat models have great utility and have been used to test numerous treatment modalities and imaging techniques over the years.

The C6 model was originally produced in outbred Wistar rats over 8 months through the repeated administration of MNU and eventually characterized as a glial tumor. Features include pleomorphic cells and occasional points of invasion into the surrounding brain. Mutations were shown in *Ink4a*, while *p53* remained unaltered (Asai et al., 1994; Schlegel et al., 1999). Increased expression of notable genes involved in human brain tumors were demonstrated, such as *PDGFβ*, *EGFR*, *IGF-1*, and *Erb3* (Guo et al., 2003; Morford, Boghaert, Brooks, & Roszman, 1997). Although numerous therapies have been tested using this model, including antiangiogenic therapy, radiation, and oncolytic viral therapy (Barth & Kaur, 2009), no syngeneic host exists in which these cells can be maintained. The tumor line was created using outbred Wistar rats; therefore, even Wistar rats develop an immune response to the cells (Parsa et al., 2000).

The most prevalent rat brain tumor model has been the 9L gliosarcoma, produced through MNU exposure over 26 weeks in Fischer 344 rats (Schmidek, Nielsen, Schiller, & Messer, 1971). They have a spindle-like, sarcomatoid appearance and grow rapidly when implanted. *p53* is mutated in this line, but *Ink4a* and *Arf* remain normal (Asai et al., 1994; Schlegel et al., 1999). *EGFR* is overexpressed in these cells as is *TGFα*, one of its ligands (Sibenaller et al., 2005). 9L tumors have been shown to contain cancer stem cell (CSC)s, which can grow as neurospheres in culture and express NSC markers Nestin and Sox2 (Ghods et al., 2007). Studies using this cell line have yielded insight into transporting therapeutics across the blood–brain barrier (BBB) (Khan, Jallo, Liu, Carson, & Guarnieri, 2005) as well as imaging techniques that utilize MRI and PET (Bansal, Shuyan, Hara, Harris, & Degrado, 2008). Despite the fact that 9L cells can form tumors in allogeneic Wistar rats (Stojiljkovic et al., 2003), the tumors they form in Fischer rats have been shown to be significantly immunogenic (Barth & Kaur, 2009).

T9 rat glioma cells may in fact be 9L gliosarcoma cells under a different name and show strikingly similar phenotypes (Barth, 1998; Denlinger, Axler, Koestner, & Liss, 1975). Repeated MNU injections over 6 months also led to the development of CNS-1 glioma in an inbred Lewis rat (Kruse et al., 1994). When implanted, these cells formed tumors that had many features of human HGG: invasive growth in a periventricular and perivascular pattern, nuclear atypia, necrotic foci, and some evidence of pseudopalisading arrangements, but not to the level observed in human samples (Candolfi et al., 2007). These features, coupled with the observation of infiltrating macrophages and T cells, led to this model being useful for studying glioma invasion as well as tumor–stroma interactions (Matthews et al., 2000; Nutt, Zerillo, Kelly, & Hockfield, 2001; Owens et al., 1998).

RG2 and F98 glioma cells were produced concurrently in the same lab and originate from progeny of Fischer 344 rats treated with ENU on gestation day 20. They have a spindle-like appearance with fusiform nuclei present and are highly invasive *in vivo*, which make them more authentic representatives of human GBM (Weizsacker, Nagamune, Winkelstroter, Vieten, & Wechsler, 1982). Both lines overexpress *PDGFβ*, *Ras*, and *EGFR*, and RG2 has demonstrated a loss in the *Ink4a* locus (Schlegel et al., 1999; Sibenaller et al., 2005). Importantly, both RG2 and F98 tumors are weakly immunogenic (Tzeng, Barth, Orosz, & James, 1991; Weizsacker et al., 1982), making them useful models to study pathways relevant to glioma resistance to immunotherapy. The BT4C glioma was also developed through administration of ENU to a pregnant rat, in this case a Lewis inbred line, and subsequently isolated *in vitro* after 200 days in culture (Laerum & Rajewsky, 1975). A somewhat heterogeneous population of cells resulted with some appearing flattened, others showing glial-like multipolarity, and sporadic giant cells (Laerum et al., 1977). Tumors demonstrate signs of dilated, nonuniform blood vessels, irregular nuclei, and areas of high proliferation and dense cellularity (Stuhr et al., 2007). The periphery of the tumor shows marked increases in VEGF, tPA, uPA, and microvessel density (Barth & Kaur, 2009). Numerous studies have employed this tumor model, including investigations into combining VEGF inhibition with temozolomide and radiation (Sandstrom, Johansson, Bergstrom, Bergenheim, & Henriksson, 2008).

While previously discussed models have been created through the use of carcinogen exposure, the RT-2 glioma line was established through the intracranial injection of Rous sarcoma virus in neonatal Fischer rats (Copeland, Talley, & Bigner, 1976). These tumors appear to evoke a

$CD8^+$ T-cell-mediated immune response, likely from virally encoded antigens (Shah & Ramsey, 2003). These cells have been used in wide-ranging studies, including cytotoxic gene therapy and its effect on radiosensitization (Valerie et al., 2000, 2001) as well as quantification of invasion into the surrounding brain through the use of GFP-tagged cells (Mourad et al., 2003).

One transgenic model in rats was developed expressing a viral form of EGFR (v-erbB) under the control of the S100β promoter (Ohgaki et al., 2006). Over 60% were found to develop tumors at a mean latency of 59 weeks (Yokoo, Tanaka, Nobusawa, Nakazato, & Ohgaki, 2008). These were classified as malignant glioma, anaplastic oligodendroglioma, and low-grade oligodendroglioma, though there were reports of occasional lung metastases (Ohgaki et al., 2006; Yokoo et al., 2008). Most recently, this model was used to characterize the tumor-associated macrophages that infiltrate gliomas (Sasaki et al., 2013). Though the long latency and variability in tumor grade may not be ideal, future studies could be aided by adding transgenic rat models to glioma research, particularly future imaging investigations.

8.2. Mouse

Mouse models of human cancers remain the most widely used *in vivo* model for the study of tumor pathogenesis. The mouse genome has a high level of similarity to humans and is well characterized. A strong foundation of transgenic animals has been built and thoroughly described to establish a wide availability of tumor models of all kinds. Glioma modeling using murine systems varies widely in sophistication, speed, and authenticity of human tumor recreation. Each has unique advantages and disadvantages.

8.2.1 Tumor implantation techniques

The simplest approach to growing tumors *in vivo* is a subcutaneous xenograft in immunocompromised mice. These are the least challenging technically and this can result in quick tumors that can be monitored noninvasively through a caliper method. Additionally, one can inject tumors into both flanks on the same mouse to test for "bystander" effects of treatment (Bhutia et al., 2013; Sarkar et al., 2005). Other advantages include the ease of using a variety of cell lines. Well-described, commercially available established lines are often used in subcutaneous xenografts, but cells isolated from patient samples are also used in injection techniques (Pandita, Aldape, Zadeh, Guha, & James, 2004; Yacoub et al., 2008). Drawbacks include the obvious: a CNS tumor growing under the skin will not encounter the same microenvironment as it would within the brain. Furthermore, a lack of BBB precludes reliable analysis of therapeutics as the pharmaceutical kinetics are altered. Consequently, orthotopic implantation of tumor cells is a superior model for CNS tumors.

While technically more challenging, direct injection of tumor cells into the brain of immunocompromised mice leads to fast-forming tumors within a predictable timeframe and narrow window (Radaelli et al., 2009). This method carries all the advantages of using genetically altered cultured cells to test signaling pathways *in vivo*, whether using established cell lines, or primary tumor cells at low passage (Yacoub et al., 2004). Disadvantages of this model include failure of some cells that grow robustly in culture to engraft tumors *in vivo* (de Ridder, Laerum, Mork, & Bigner, 1987; Liu et al., 2006). Cells that are successful in growing tumors can fail to replicate pathology often seen in human HGG. Specifically, human cell lines can be prone to forming tumors that are well circumscribed, compact, and have little infiltration of the surrounding brain parenchyma (Finkelstein et al., 1994; Radaelli et al., 2009). Multiple passages *in vitro* can lead to an alteration in genetic profile due to different selective pressures experienced by the cells (Li et al., 2008). Brief culture under neurosphere conditions or harvesting the cells from a flank xenograft before intracranial injection has been shown to maintain the expression patterns of the primary tumor (Claes et al., 2008; Galli et al., 2004; Giannini et al., 2005; Stockhausen et al., 2011). Cells exposed less to conditions outside of their normal *milieu* will more faithfully simulate the parental tumor in histological presentation, invasion ability, $CD133^+$ expression, and expression profile, thus yielding reproducible tumors (Shu et al., 2008; Yi, Hua, & Lin, 2011) (Fig. 7.3).

8.2.2 Syngeneic mouse models

An alternative to using human tumors for intracranial injection is to use mouse tumors harvested from spontaneous or induced methods. Immunocompromised mouse models do not completely recapitulate the clinical histopathology of human tumors and remain unable to study tumor-specific immune responses (Waziri, 2010). Therefore a syngeneic model of glioma in mice provides the technical control of an injection model, while retaining the authenticity of an immunocompetent host. Binello and colleagues (2012) use the CT-2A cell line in such a model. These cells were derived in the same fashion as the C1261 cell line, harvested from induced gliomas resulting from the intracranial implant of 3-methylcholanthrene pellets into C57/B6 mice (Szatmari et al., 2006). CT-2A cells were collected from

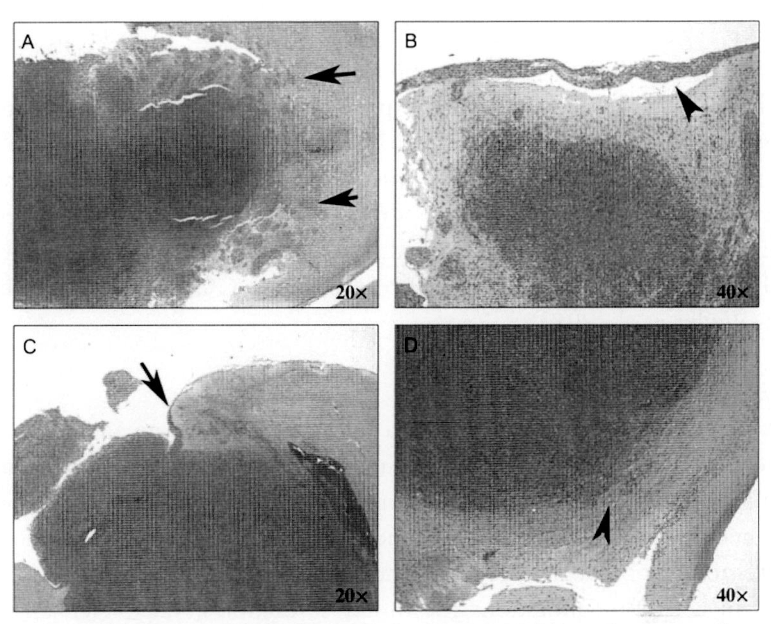

Figure 7.3 Orthotopic xenograft placement in mice can yield invasive tumors that recapitulate the phenotype of human tumors. This also allows for preimplantation modification of the tumor cells. (A, B) In an unaltered GBM6 cell line, we note tumor invasion surrounding brain parenchyma, with finger-like projections and tumor nests apart from the main tumor mass showing infiltration along Virchow–Robin spaces (arrows) and leptomeningeal invasion (arrowhead). (C, D) In this sample, a proinvasive gene, *mda-9/syntenin*, is knocked down, yielding less invasive tumors. Note the very focal leptomeningeal invasion compared to control (arrow) and minimal focal tumor invasion into surrounding brain parenchyma (arrowhead) with a largely demarcated margin. *Reproduced with permission from Kegelman et al. (2014).* (See the color plate.)

malignant astrocytoma formed after implantation of 20-methylcholanthrene pellets in C57/B6 mice (Zimmerman & Arnold, 1941). Genetically, these cells have wild-type p53, are PTEN deficient, and capture a number of features of HGG: a high cell density and mitotic index, nuclear polymorphism, hemorrhage, pseudopalisading necrosis, and microvascular proliferation (Martinez-Murillo & Martinez, 2007). CT-2A cells were shown to readily form neurospheres and could be used for BTSC studies in the future (Binello, Qadeer, Kothari, Emdad, & Germano, 2012).

Another example of syngeneic mouse modeling was utilized to investigate the inhibition of mTOR in combination with radiotherapy. The SMA (spontaneous murine astrocytoma)-560 cell line was used orthotopically to study temsirolimus (CCI-779) and radiation treatment. Since sublethal radiotherapy leads to increased invasiveness in glioma (Tabatabai, Frank, Mohle, Weller, & Wick, 2006; Wild-Bode, Weller, Rimner, Dichgans, & Wick, 2001), and relapsed GBM tends to evade treatment targeting EGFR and mTOR signaling, an initial treatment that combats the proinvasive effects of radiation is needed. Weiler and colleagues showed that CCI-779 targeted RGS4, a driver of invasion in GBM, thereby inhibiting RT-induced invasion and leading to a survival benefit in mice with injected syngeneic tumors that received both CCI-779 and radiotherapy (Weiler et al., 2013).

8.2.3 GEM models

The well-described and often engineered murine genome allows for the generation of tumor models in immunocompetent mice, extensively used in glioma modeling (Table 7.1). Germline deletion of a tumor suppressor can determine its importance in glioma formation. Similarly, enhanced expression of potential oncogenes under a designated promoter can reveal roles in activating mutations for tumor formation (Nagy, Gertsenstein, Vintersten, & Behringer, 2003). Yet, global approaches lead to simulation of tumor predisposition syndromes throughout a target organ, or significant developmental aberrations, some of which can be lethal to the animal even before a tumor forms. Some targeting is achieved by using specific promoters; however, these can depend on stage of development and may express in multiple types of cells (Orban, Chui, & Marth, 1992).

Efforts to target expression of a mutation are often achieved through a Cre–lox system, wherein expression of Cre recombinase aids in the knockout or expression of a gene of interest. To develop a transgenic mouse using this system, loxP is inserted, flanking a strong transcriptional stop site followed by the gene of interest. When Cre is present, the STOP cassette is removed, leading to gene expression in these cells. Alternatively, a conditional knockout mouse can be created by flanking a tumor suppressor with loxP sites, leading to its loss in cells expressing Cre (Orban et al., 1992; Sauer & Henderson, 1988). Most commonly, transgenic Cre mice express the recombinase under the control of glial fibrillary acidic protein (GFAP), observed in mature astrocytes and SVZ glial progenitor cells, or Nestin, expressed in the neural progenitor compartment (Rankin, Zhu, & Baker, 2012). Refined control over gene expression can be executed with an inducible Cre system using fusion Cre-estrogen receptor (ER) protein. After exposure to tamoxifen, Cre-ER translocates to the nucleus and initiates recombination (Feil et al., 1996).

Table 7.1 Genetically engineered mouse models (GEMs) of glioma

Promoter	Genetics	Incidence/latency	Morphology	Mechanism	References
Global	$Nf1^{+/-};p53^{+/-}$	90% at 6 months	Variable grade astrocytoma	Conventional KO	Reilly, Loisel, Bronson, McLaughlin, and Jacks (2000)
GFAP	$Nf1^{+/-};p53^{+/-}$	100% at 5–10 months	Variable grade astrocytoma	Conditional KO	Zhu et al. (2005)
GFAP	$Nf1^{+/-};p53^{+/-}; Pten^{-/-}$	100% at 5–8 months	High-grade astrocytoma	Conditional KO	Kwon et al. (2008)
GFAP	$GFAP\text{-}T_{121}$	100% at 10–12 months	Low-grade astrocytoma	TG	Xiao, Wu, Pandolfi, Louis, and Van Dyke (2002)
GFAP	$GFAP\text{-}T_{121}; Pten^{-/-}$	100% at 6 months	High-grade astrocytoma	TG/conditional KO	Xiao et al. (2005)
GFAP	$H\text{-}Ras\ (V^{12})$	100% at 3 months	High-grade astrocytoma	TG	Ding et al. (2001)
GFAP	$H\text{-}Ras\ (V^{12})$; *EGFRvIII*	100% at 3 months	High-grade ODG	TG; AdV	Ding et al. (2003)
GFAP	$H\text{-}Ras\ (V^{12}); Pten^{-/-}$	100% at 6 weeks	High-grade astrocytoma	TG; conventional KO	Wei et al. (2006)
GFAP	$p53^{-/-};Pten^{+/-}$	70% at 9 months	High-grade astrocytoma/ GBM	Conditional KO	Zheng et al. (2008)

Continued

Table 7.1 Genetically engineered mouse models (GEMs) of glioma—cont'd

Promoter	Genetics	Incidence/latency	Morphology	Mechanism	References
GFAP	*PDGFB*; $p53^{+/-}$	40–70% at 6 months	High-grade ODG/GBM	TG	Hede et al. (2009)
GFAP	*c-myc*	50% at 1.5 months	High-grade astrocytoma	TG	Jensen et al. (2003)
GFAP	$K\text{-}Ras^{(G12D)}$	100% at 3 months	Intermediate-grade astrocytoma	TG	Abel et al. (2009)
GFAP	$PDGFA_L$	80% at 1 year	High-grade OA	TG	Nazarenko et al. (2011)
GFAP	$K\text{-}Ras^{(G12D)}$; $Ink4a/Arf^{-/-}$	33% at 3 months	Variable grade lioma/sarcoma	RCAS; conventional KO	Uhrbom et al. (2002)
GFAP	$K\text{-}Ras^{(G12D)}$; *Akt*; $Ink4a/Arf^{-/-}$	40% at 3 months	Variable grade astrocytoma	RCAS; conventional KO	Uhrbom et al. (2002)
GFAP	*Akt*; *H-Ras* (V^{12}); $p53^{+/-}$	75% (SVZ)–100% (HP) at 6 months	High-grade glioma	LV; conventional KO	Marumoto et al. (2009)
GFAP	$p53^{+/-}$;$Pten^{-/-}$	30% at 8 months	Variable grade astrocytoma	AdV-Cre in SVZ	Jacques et al. (2010)
GFAP	*PDGFB*; $Ink4a/Arf^{-/-}$	40% at 3 months	Variable grade ODG/OA	RCAS; conventional KO	Dai et al. (2001)

Nestin	*EGFR*; *Ink4a/Arf*$^{-/-}$	40% at 2.5 months	Variable grade glioma	RCAS; conventional KO	Holland, Hively, DePinho, and Varmus (1998)
Nestin	*PDGFB*; *Ink4a/Arf*$^{-/-}$	60–95% at 3 months	Variable grade ODG	RCAS; conventional KO	Dai et al. (2001), Shih et al. (2004)
Nestin	*PDGFB*; *Akt*	45% at 3 months	Variable grade astrocytoma	RCAS	Dai et al. (2005)
Nestin	*K-Ras*$^{(G12D)}$; *Pten*$^{-/-}$	60% at 3 months	Variable grade astrocytoma	RCAS; conditional KO	Hu et al. (2005)
Nestin	*K-Ras*$^{(G12D)}$; *Akt*	25% at 3 months	Variable grade astrocytoma	RCAS	Uhrbom et al. (2002)
Nestin	*K-Ras*$^{(G12D)}$; *Akt*; *Ink4a/Arf*$^{-/-}$	50% at 3 months	Variable grade astrocytoma	RCAS; conventional KO	Uhrbom et al. (2002)
Nestin	*Nf1*$^{+/-}$;*p53*$^{-/-}$; *Pten*$^{+/-}$	100% at 60 weeks	High-grade astrocytoma/GBM	Conditional KO (Nestin-Cre/AdV-Cre to SVZ)	Alcantara Llaguno et al. (2009)
Nestin	*Nf1*$^{-/-}$;*p53*$^{-/-}$	100% at 60 weeks	High-grade astrocytoma/GBM	Conditional KO (Nestin-Cre/AdV-Cre to SVZ)	Alcantara Llaguno et al. (2009)
Nestin/NG2/GFAP	*Nf1*$^{-/-}$;*p53*$^{-/-}$	100% at 5 months	High-grade astrocytoma/GBM	MADM	Liu, Sage, et al. (2011)

Continued

Table 7.1 Genetically engineered mouse models (GEMs) of glioma—cont'd

Promoter	Genetics	Incidence/latency	Morphology	Mechanism	References
S100β	*v-ErbB*; *Ink4a/Arf*$^{+/-}$	90% at 6 months	High-grade ODG	TG; conventional KO	Weiss et al. (2003)
S100β	*v-ErbB*; *Ink4a/Arf*$^{+/-}$	90% at 6 months	High-grade ODG	TG; conventional KO	Persson et al. (2010)
CNP	*PDGFB*	33% at 3 months	Low-grade ODG/glioma	RCAS	Lindberg, Kastemar, Olofsson, Smits, and Uhrbom (2009)
Rostral subcortical white matter	*PDGFB*; *Pten*$^{-/-}$	90% at 3 months	High-grade Glioma/GBM	RV-Cre injection to WM	Lei et al. (2011)
Rostral subcortical white matter	*PDGFB*; *Pten*$^{-/-}$; *p53*$^{-/-}$	90% at 1.5 months	High-grade Glioma/GBM	RV-Cre injection to WM	Lei et al. (2011)

GBM, glioblastoma; TG, transgenic; KO, knockout; OA, oligoastrocytoma; ODG, oligodendroglioma; SVZ, subventricular zone; WM, white matter; TG, transgenic; AdV, adenovirus; RCAS, RCAS/TVA system; LV, lentivirus; RV, retrovirus; MADM, mosaic analysis with double markers.

As with any GEM model, multiple mutations within the same cell type can be achieved with appropriate breeding strategies. Tumors developed in this fashion show a range of latencies and tumor grades depending on the genes altered in each model (Huse & Holland, 2009). The downside to this approach can be the long times involved in tumor formation as well as the investment in money, time, expertise, and resources to design and develop genetically engineered mice.

8.2.4 Approaches to mouse modeling of glioma

8.2.4.1 Loss or inhibition of key tumor suppressors

Tumor suppressors that are typically targeted in the development of mouse models of glioma include p53, PTEN, Nf1, and Rb. Targeting multiple tumor suppressors under different settings has yielded variable tumorigenesis and penetrance. The mutation of one allele of Nf1 and p53 has led to the development of a wide range in grade of astrocytomas at a rate of 92% by 6 months. The majority of these tumors have lost heterozygosity at the remaining Nf1 and p53 alleles, which reside close to each other in the mouse genome. Of note, this model may be well suited to study more slowly developing astrocytic tumors such as secondary GBM, as older mice tend to have tumors of a higher grade. Outside the CNS, sarcomas have been observed in this model (Reilly et al., 2000). This extracranial tumor incidence is reduced by utilizing a floxed *Nf1* allele on a GFAP-Cre background. p53 remains mutated, but astrocytes now have two mutations. This leads to tumors occurring with 100% penetrance, but on a somewhat longer timescale: 5–10 months (Zhu et al., 2005). Mutation to PTEN leads to a decreased latency in this model as well as an enhancement in the grade of tumors formed (Kwon et al., 2008). When p53 is lost in astrocytes along with one copy of PTEN, grade III astrocytomas and GBMs develop by 15–40 weeks at a rate of over 70% (Zheng et al., 2008).

Additionally, an effort was made to disentangle the respective contributions of *Ink4a* and its splice variant, *Arf*. Commonly, both may be deleted in glioma models; however, when using a PDGF-driven model of glioma, it was found that *Arf* loss contributed more to tumor development and enhancement of tumor grade than *Ink4a*, especially in glial progenitor cells (Tchougounova et al., 2007). Expression of genes that often drive other cancer models also can initiate glioma in mice. For example, when mouse astrocytes express an activated, truncated form of SV40 T antigen, T_{121}, which inhibits both the p53 and Rb signaling pathways, low-grade astrocytomas result within 300 days (Xiao et al., 2002). Once again, additional mutation to PTEN leads to shorter tumor latency, more frequent higher grade tumors, as well as increases in cellularity and mitotic activity (Xiao et al., 2005).

8.2.4.2 Expression of oncogenes

Genetic analysis of human HGG reveals a wide range of activating mutations. Therefore, mouse modeling has attempted to address the relative importance of particular mutations and the combinations that consistently yield tumors *in vivo*. Constitutively active and oncogenic v-Src induces growth and survival signaling, promoting glioma formation when overexpressed in mice (Smilowitz et al., 2007; Weissenberger et al., 1997). Since the additional loss of p53 or Rb in this model did not enhance tumor formation, it can be concluded that these pathways were also impaired by v-Src expression (Maddalena, Hainfellner, Hegi, Glatzel, & Aguzzi, 1999), making it a difficult model in which to differentiate mutations that are necessary or sufficient for tumorigenesis.

Commonly, HGG exhibits strong RAS activation upon examination of human tumors. The transgenic expression of activated Ras, V^{12}Ha-Ras, is a common tactic to produce glioma in mice. Tumor formation and grade corresponds to Ras dosage, with homozygous animals reported to develop high-grade tumors with grade IV characteristics at 2 weeks of age (Ding et al., 2001; Shannon et al., 2005). However, these tumors show mutations in other pathways as they develop. Consistent formation of high-grade lesions requires combination with other mutations, such as combination with EGFRvIII (Ding et al., 2003), which can both decrease latency and demonstrate oligodendroglial phenotypes, or PTEN knockout (Wei et al., 2006), which can increase tumor grade while hastening formation (Robinson et al., 2010, 2011; Uhrbom et al., 2002). While EGFR mutation is commonly observed in human GBM, Ras mutation is rare, possibly also due to its activation under RTK amplified signaling. Nonetheless, it must be a consideration when choosing a Ras-activated model for preclinical testing of a targeted inhibitor in GBM (Rankin et al., 2012).

EGFR amplification, overexpression, and mutation are common in GBM and can often be present along with INK4A/ARF deletion, which impairs both p53 and Rb (Holland et al., 1998; Liu, Feng, et al., 2011; Zhu et al., 2009). Conforming to data found in humans, the combination of PI3K signaling with Rb and p53 disruption leads to enhancement of tumor formation (Cancer Genome Atlas Research Network, 2008; Parsons et al., 2008). A transforming variant of EGFR, v-erbB, can be targeted to a subset of astrocytes under the S100β promoter, which is expressed in astrocytes that surround blood vessels and NG2-expressing cells (Wang & Bordey, 2008). This results in oligodendrogliomas of low grade at 60% of mice at 1 year. When either INK4a/ARF or p53 is knocked out, higher grade lesions are observed and nearly 100% of mice develop tumors in 12 months (Weiss et al., 2003). Monoallelic knockout of either tumor suppressor locus yields intermediate effects.

8.2.4.3 Somatic gene transfer by viral vector

Even when gene alterations in GEM are controlled under cell-specific promoters, tumor initiation can occur in multiple locations, as opposed to the isolated focal nature that normally occurs in humans. To produce a model that more closely recapitulates this aspect of human tumorigenesis, transfer of genes to somatic cells can be accomplished using a viral vector, which only infects a limited number of cells in a targeted region of the brain. Genes can be activating oncogenes, or Cre can be delivered to excise a tumor suppressor with loxP flanking sites. Advantages of this approach are that it more closely represents what is observed in humans, one can deliver multiple genes of interest, as well as imaging markers, by implementing more than one virus, and one can circumvent some of the difficulties of creating transgenic or knockout mouse lines. At the same time, there can be some limits on the size of genes packaged within a virus, the technical aspect of intracranial injection can be challenging, and inflammatory conditions are created by the nature of the injection technique.

Retroviruses are a useful vector in which to package relevant genes of interest. A murine retrovirus, MoMuLV, was injected into the forebrain of newborn pups and carried PDGFB chain (PDGFB), a growth factor found to be expressed in GBM (Uhrbom, Hesselager, Nister, & Westermark, 1998). Forty percent of mice developed tumors, which had a wide range of pathological features, including GBM characteristics, but also primitive neuroectodermal tumor features. Another approach utilizes an avian virus coupled with transgenic mice expressing the receptor for the virus expressed under specific promoter control, GFAP or Nestin. The replication-competent avian sarcoma–leukosis virus long terminal repeat with splice acceptor (RCAS), which exclusively recognizes the tv-a receptor, is normally foreign to mammalian cells. The tv-a receptor was expressed in mice under the control of GFAP (Gtv-a), Nestin (Ntv-a), or CNPase (present in oligodendrocytes progenitors) (Ctv-a) promoters (Holland et al., 1998; Holland & Varmus, 1998; Lindberg et al., 2009). This targets the RCAS virus containing genes of interest to specific cells in a desired location of the brain, leading to focused tumor formation. Purified virus or infected chicken fibroblasts that shed RCAS can be injected intracranially. When constitutively active variants of Akt and KRas are delivered in combination, Ntv-a mice yield astrocytic tumors at a rate of 25% by 12 weeks (Holland et al., 2000). KRas delivery alone is sufficient when PTEN is deleted (Hu et al., 2005). PDGFB delivered by RCAS in either Ntv-a or Gtv-a leads to a dosage-dependent formation of tumors with oligodendroglial or a mixed oligoastrocytic phenotype in up to 100% of mice (Shih et al., 2004). Much like other models, the loss of tumor suppressors like *INK4a/ARF* or *PTEN* yields more frequent tumors of a higher grade. Examples of this include improvement from 25% to 50% incidence with KRas and Akt delivery to mice with homozygous deletions of *INK4a/ARF* (Uhrbom et al., 2002). Additionally, *INK4a/ARF* or *PTEN* loss decreases latency and enhances high-grade histology in RCAS-PDGFB models (Dai et al., 2001). While the RCAS–tv-a system has shown the ability to produce tumors in adult cells, the virus only infects dividing cells and infection efficiency declines as the animal ages (Hambardzumyan, Amankulor, Helmy, Becher, & Holland, 2009).

Lentiviral vectors can infect both dividing and nondividing cells and have been gaining in popularity as a somatic gene transfer vector. One model utilized lentiviral vectors to deliver activated H-RasV12 and Akt to GFAP-Cre mice (Marumoto et al., 2009). The oncogenes followed a cytomegalovirus (CMV) immediate-early promoter-loxP-red fluorescent protein-loxP cassette, which, in the presence of Cre, would lead to transcription of the oncogene as well as a downstream GFP reporter. The injections were targeted to specific areas of the brain, and differential tumorigenesis was observed based on region. While lentiviral vectors with H-RasV12 or Akt alone did not produce tumors, injecting both vectors simultaneously to the hippocampus led to 40% of mice developing high-grade tumors by 5 months. The same injections to the cortex produced no tumors, and the SVZ yielded only one tumor out of nine mice. When GFAP-Cre;p53$^{+/-}$ mice were used, H-RasV12 on its own produced tumors in 60% of mice injected in the hippocampus. The combination of H-RasV12 and Akt yielded tumors in 100% of mice injected in the hippocampus, 75% in the SVZ, and only 7% in the cortex. Tumors displayed many of the characteristics of GBM and occurred with much shorter latency in GFAP-Cre;p53$^{+/-}$ mice (Marumoto et al., 2009).

In another model, the lentiviruses themselves carried Cre driven by either a GFAP or a CMV promoter. A combination of *INK4a/ARF*,

PTEN, and *p53* was floxed in the mice, while mutated *K-Ras*^{V12} was under the control of a STOP-lox cassette, leading to KRAS production in the presence of Cre. Loss of *INK4a/ARF* and *K-Ras*^{V12} expression with and without concomitant *PTEN* loss led to 30–35% of mice developing tumors by 150 days when injected with a GFAP-Cre lentivirus. Additional *p53* loss raised the incidence to 80%, with over half categorized as grade IV lesions. When CMV-Cre lentiviruses were used, 100% of mice developed tumors with much shorter latencies, although less than 1/3 of them were grade IV tumors (de Vries et al., 2010).

Through the use of the RCAS/Ntv-a model, TAZ altered the typically proneural phenotype observed in PDGFB-driven tumors. TAZ is a transcriptional coactivator, shown to regulate mesenchymal differentiation *in vitro* through binding to the TEAD transcription factor (Bhat et al., 2011). PDGFB expression in Nestin^{+} NPCs yielded grade II tumors and a median survival of about 11 weeks, similar to data shown in other studies (Dai et al., 2001). When TAZ or constitutively nuclear TAZ (4SA) was used alone in NPCs, no tumors developed in 90 days. In combination with PDGFB, survival was reduced to less than 5 weeks with TAZ or 4SA expression. These tumors were consistently grade III or IV and showed a more mesenchymal profile compared to a proneural profile of tumors from PDGFB alone (Bhat et al., 2011).

Discovery of a mutation resulting in a fusion protein present in ~3% of human GBM samples was validated through lentiviral infection in immunocompetent mice (Singh et al., 2012). The FGFR3–TACC3 fusion protein was found to be present in just 3 of 97 samples, but was shown to induce GBM formation, and more importantly, identified a potentially treatable subset of human GBM. FGFR3 is a member of the FGFR RTK family, and TACC3 is a member of the TACC family, which mediates localization to the mitotic spindle and was shown to be oncogenic in several human tumors (Duncan et al., 2010; Yao et al., 2012). The expression of this fusion protein induced high rates of chromosome instability, aneuploidy, and enhanced proliferation in primary astrocytes. Injection of lentiviruses containing *FGFR3-TACC3* and sh*p53* led to HGG formation in 87.5% of mice within 240 days. Further study using orthotopic injection of *INK4a/ARF*$^{-/-}$ astrocytes expressing FGFR3–TACC3 showed that the resulting tumors could be inhibited by an FGFR inhibitor, AZD4547, prolonging survival by 28 days (Singh et al., 2012). This potential for treatment in a subset of GBM demonstrates the value of validating a rare mutation in relevant animal models.

Additionally, viral injection strategies can couple vectors with bioluminescent tracking. This method can provide insight on cooperative mutations and information on cells of origin for glioma. Drawbacks in injecting somatic genes include the inflammatory response to the trauma of injection, with subsequent immune cell recruitment possibly affecting results.

8.2.5 Investigating glioma pathology

8.2.5.1 Stromal interactions

The numerous genetically modified mice available and the relative ease with which the mouse genome can be engineered lead to a multitude of options in exploring the interactions between glioma and stromal cells. Specifically, progression can be shaped by the makeup of ECM proteins, cytokines, and other stromal factors encountered by neoplastic cells (Hoelzinger, Demuth, & Berens, 2007; van Kempen, Ruiter, van Muijen, & Coussens, 2003). By selecting genetic backgrounds that alter the paracrine microenvironment, interactions involving a tumor and its surroundings can be investigated. While biallelic deletion of Nf1 in astrocytes did not produce glioma in the presence of normal stroma (Bajenaru et al., 2002), mice with an $Nf^{+/-}$ genotype produce optic pathway gliomas when the second Nf1 gene is knocked out specifically in astrocytes (Bajenaru et al., 2003). Additionally, by using *CD38-null* mice, Levy and colleagues demonstrated that CD38 was important in the tumor-supporting function of infiltrating microglia and macrophages (Levy et al., 2012). PDGF and its receptor PDGFRα, altered in GBM, have been shown to be involved in stromal autocrine or paracrine signaling promoting tumorigenesis (Shih & Holland, 2006) as well as in the recruitment of distant cells affecting the progression of high-grade tumors (Assanah et al., 2006; Fomchenko et al., 2011). Notably, aberrant PDGF signaling introduced in neural progenitors can lead to development of oligodendrogliomas (Calzolari & Malatesta, 2010; Dai et al., 2001) and astrocytomas of the proneural subtype (Becher et al., 2010; Hambardzumyan et al., 2009; Hede et al., 2009; Hesselager, Uhrbom, Westermark, & Nister, 2003; Lei et al., 2011).

Xenograft modeling has recently been utilized to understand why oncolytic virotherapy has been underwhelming in clinical trials. Conditionally replicative oncolytic herpes simplex virus (oHSV) has been shown to effectively lyse tumor cells in preclinical studies (Chiocca, 2002), but seems to fall short in early-phase clinical trials (Markert et al., 2009). Using orthotopic models of GBM in SCID-γc^{null} mice, it was determined that natural killer (NK) cells impede on the efficacy of oHSV therapy through rapid viral clearance via cell-mediated killing and macrophage activation. This was dependent on the NK cell receptor NKp46, as sublethally radiated mice transplanted with *Ncr1*$^{-/-}$ NK cells showed more tumor inhibition after oHSV therapy than those transplanted with WT NK cells (Alvarez-Breckenridge et al., 2012).

8.2.5.2 Cell of origin

Compelling evidence suggests that HGGs do not arise from only one cell type. The NSC has the ability to self-renew, proliferating through adulthood, and retaining multipotentiality. Due to its continued proliferation in adults, it is more susceptible to oncogenic mutations that could lead to glioma formation. The SVZ harbors numerous types of progenitor cells, and this heterogeneity could help to explain the heterogeneous nature of human glioma (Canoll & Goldman, 2008). Nonetheless, gliomas are not necessarily found adjacent to these proliferative niches within the brain. Often, they are found in the hemispheres, suggesting that transformed cells have enhanced migratory abilities on top of an already mobile phenotype (Alcantara Llaguno et al., 2009; Kwon et al., 2008). In the context of mouse models, however, mutations introduced in areas of neural progenitors (e.g., the hippocampus and SVZ) are more efficient at transforming cells than those introduced to the cortex (Alcantara Llaguno et al., 2009; Hambardzumyan et al., 2009; Marumoto et al., 2009). Additionally, oligodendrocyte progenitor cells (OPCs) can generate gliomas in mice as evidenced by PDGFB expression in OPCs driving oligodendroglioma formation in mice (Lindberg et al., 2009; Persson et al., 2010). A viral oncogenic form of EGFR, v-erbB, can also drive oligodendrogliomas with expression signatures similar to OPCs rather than NSCs. This particular model shows authentic oligodendroglioma formation along white matter tracts, much like in humans (Persson et al., 2010). OPCs were suggested to be the cell of origin in a model that develops proneural type GBM with PDGF stimulation combined with *p53* and *PTEN* deletion (Lei et al., 2011). The OPC-proneural link was further demonstrated in a study that revealed that only the OPCs derived from NSCs with mutations developed aberrant growth patterns, even when compared to the NSCs themselves, and eventually formed tumors with proneural features (Liu, Sage, et al., 2011).

Glioma cells often express NSC-specific proteins such as Nestin and Sox2. However, this may be the result of defects in differentiation control rather than markers of a cell of origin, as more differentiated cells can generate gliomas in mice (Rankin et al., 2012). Substantial evidence exists that supports astrocytes as tumorigenic cells for glioma formation. In the absence of INK4a/ARF, expression of mutated EGFR transformed astrocytes and NSCs at a similar rate (Bachoo et al., 2002). EGFRvIII expression combined with p53 and PTEN deletion also leads to the transformation of astrocytes (Endersby, Zhu, Hay, Ellison, & Baker, 2011). However, in the absence of a driving mutation, but with PTEN, p53, and Rb deletion, only NSCs became tumorigenic (Jacques et al., 2010). Finally, recent data show that even mature neurons can give rise to GBM tumors in mice in addition to NSCs and astrocytes (Friedmann-Morvinski et al., 2012). These inconsistent results point toward the difficulty in ascertaining a definitive cell of origin for glioma, and even the sources for oligodendroglioma versus astrocytoma. In light of the ability to reprogram fully differentiated cells into pluripotent stem cells, it is completely plausible that multiple cell types can lead to a spectrum of glioma tumors if given the necessary mutations.

One study in transgenic mice concurrently investigated identical mutations in different cells within the mouse brain. Oncogenic Ras ($Kras^{G12D}$) along with p53 loss ($p53^{fl/fl}$) was targeted to GFAP-expressing cells, resulting in multifocal tumors arising in different parts of the brain. Depending on the origin of the tumor, the resulting genetic signature followed distinct patterns. Cells arising from the SVZ expressed a hallmark of neural stem/progenitor cells (NPSCs), while those that arose from the cortex or leptomeninges lacked this signature. Differences in resulting tumors were shown by harvesting and separating astrocytes and NPSCs from neonatal pups. Astrocytic tumors tended to be more aggressive, had strong GFAP expression, and matched a mesenchymal pattern, while NPSC-derived tumors retained their NPSC markers, were prone to differentiate, and had proneural expression patterns (Ghazi et al., 2012).

Intricate transgenic models can be of great use for defining cells responsible for tumor recurrence and exploring the existence of cancer stem cells in glioma. The model used by Chen et al. (2010) had conditional deletions of *Nf1*, *p53*, and *Pten*, resulting in 100% penetrance of malignant gliomas likely deriving from NSCs of the SVZ (Alcantara Llaguno et al., 2009). Building on this model, a transgene was added that expresses modified thymidine kinase (ΔTK) from herpes simplex virus (HSV) under the control of a Nestin promoter. This allows for the destruction of Nestin-expressing cells through systemic ganciclovir (GCV) administration. One to three days following TMZ treatment, most proliferating cells expressed Nestin. However, a loss of stem cell properties occurred around 7 days post-TMZ treatment. Subsequent GCV administration prolonged survival significantly compared to

TMZ treatment alone, but not any longer than sole GCV treatment. This was postulated to be due to lack of a ΔTK transgene present in ventrally located tumors, separate from the dorsally growing tumors usually observed in this model (Chen et al., 2012). These findings show strong support for a quiescent CSC that avoids traditional chemotherapy, only to differentiate and self-renew, seeding recurring tumors.

Astrocytes, glial and oligodendrocyte precursors, and NSCs have been suggested to be the candidate cells for glioma origin. Recently, differentiated neurons were transduced via lentivirus to become oncogenic *in vivo* (Friedmann-Morvinski et al., 2012). Using mice expressing Cre under a Synapsin I promoter (SynI-Cre), specific to neurons, injection of a lentivirus that expresses shp53 and either H-Ras-V12 or shNF1 in the presence of Cre led to tumor formation in 20/20 mice within 6–10 weeks. A similar experiment with CamK2a-Cre mice, which targets mature neurons, led to tumors with a much longer latency of 9–12 months (Friedmann-Morvinski et al., 2012). The molecular signature of SynI-Cre tumors strongly correlated with a mesenchymal molecular subtype and was very similar to tumors derived from GFAP-Cre mice. Alternatively, Nestin-Cre mice produced a neural subtype with these lentiviruses. These findings highlight the potential of mature cells to dedifferentiate into a more stem-like state and, when in a supportive microenvironment, can both sustain the renewal of stem-like cells and differentiation of progeny cells (Friedmann-Morvinski et al., 2012).

8.2.5.3 Modeling interplay between standard and novel therapies

All animal models aim to advance the treatment of human disease, though some approaches skillfully combine standard treatments with novel approaches. This inclusion of conventional therapies more closely models typical clinical trial design, which evaluates the efficacy of new therapies against, or in combination with, standard of practice.

An interesting approach for advancing GBM treatment is the usage of therapeutic, engineered stem cells in combination with traditional treatments. A recent study utilized modified mouse NSCs expressing S-TRAIL, known to induce apoptosis in 50% of GBM (Panner, James, Berger, & Pieper, 2005; Rieger, Naumann, Glaser, Ashkenazi, & Weller, 1998), in combination with surgical resection (Kauer, Figueiredo, Hingtgen, & Shah, 2011). U87 cells were injected into immunocompromised mice, formed tumors, and were resected 21 days after injection. A profound increase in survival was shown when therapeutic S-TRAIL NSCs encapsulated in a synthetic ECM were applied to the resection cavity. Hundred percent of these mice were alive 42 days posttreatment, while mice with resection along with control NSCs had a median survival of 14.5 days posttreatment (Kauer et al., 2011).

Along with surgical resection, radiation treatment for GBM is among standard practices. Another stem cell approach utilized the combination of radiation and umbilical cord blood-derived mesenchymal stem cells (MSCs) expressing TRAIL. Radiation was shown to enhance the homing potential of stem cells by upregulating IL-8 production. Orthotopic tumor models in nude mice showed a radiation dose-dependent survival benefit of the combination of MSC-TRAIL treatment plus radiation with 80% of mice surviving more than 80 days posttreatment, but only at the highest (10 Gy) dose of radiation (Kim et al., 2010). This was significantly longer than mock-treated mice (median survival <30 days) and MSC treatment alone or in combination with 5 Gy radiation (median survival <50 days) (Kim et al., 2010). Often, chemotherapy accompanies radiation in standard treatment of human gliomas. One such study in mice employed a model in which GBM was induced using Nestin-tv-a/*INK4a*/*ARF*$^{-/-}$/*PTEN*$^{loxp/loxp}$ mice injected with DF-1 (chicken fibroblast cells), which were infected with RCAS-PDGFB and RCAS-Cre. The resulting model was used with a combination of chemotherapy and radiation to show gemcitabine can be a radiosensitizer in a proneural like PDGF-driven GBM subtype (Galban et al., 2012).

One study used a combination of viral introduction of oncogenes and syngeneic transplant to expose potential combination toxicity. The PDGF-IRES-Cre retrovirus was injected into mice with conditional deletions of *Pten* and *p53* and resulted in 100% tumor penetrance and a median survival of 27 days postinjection (Lei et al., 2011). The resulting tumors were used in a syngeneic model to test sunitinib, a small molecule RTK inhibitor, and radiation. Alone, sunitinib or high-dose radiation provided a modest survival, and the combination of low-dose radiation with sunitinib provided a delay in tumor growth. However, combining high-dose radiation with sunitinib resulted in a fatal toxicity to the animals (D'Amico et al., 2012).

Another intriguing new field encompasses the development of small molecules that can act as therapeutics as well as diagnostic markers. This can be especially difficult considering the challenge of the BBB in treating GBM. Using predictive computer modeling, the antitumor cytokine MDA-7/IL-24 was modified to add a diagnostic luciferase domain while maintaining the ability to induce its effects through IL-20Rα/β receptors

(Hingtgen et al., 2012). This modified protein, SM7L, was then delivered by mouse NSCs to an *in vivo* orthotopic GBM model using implanted U87 cells. Treatment with the mNSC-SM7L cells resulted in localized expression, specific to the tumor tissue, and inhibition of tumor progression. When combined with S-TRAIL treatment through double-secreting mNSC-SM7L/S-TRAIL cells, the antitumor efficacy was further augmented.

9. CONCLUSIONS AND FUTURE PERSPECTIVES

The molecular characterization of genetic alterations and creation of publically available information has accelerated our understanding of GBM within recent years. Notably, the view of GBM as a heterogeneous disease with subpopulations of cells within tumors emphasizes the difficulty in treating this neoplasm. While similar processes undoubtedly occur in other cancer types, the nature of GBM presents an added challenge. A propensity for invasion, the invaluable nature of the surrounding tissue to a patient's quality of life, and the difficulties delivering treatment across the BBB are a few of the obstacles unique to GBM. Looking ahead, we must create a tangible connection from retrospective, population-based genomic characterization to the planning of treatment in individual patients by using profiling of clinical samples (MacConaill et al., 2009; Sequist et al., 2011). Animal modeling will be an indispensible tool for reaching those goals.

Ample opportunities exist to improve upon the already targeted treatments developed to combat GBM. To date, EGFR and PDGFRα inhibitors have failed to show significant clinical benefit to the outcomes in patient trials (De Witt Hamer, 2010). Although considerable effort has been put forth to produce targeted agents that negate the oncogenic effect of these signaling pathways, the high rates of proliferation and mutation within GBM tumors lead to subpopulations that evade such treatments. Additional RTK signaling cascades may cooperate and compensate to maintain a minimum signaling threshold that is not adequately attenuated by inhibition of a single RTK (Huang et al., 2007; Stommel et al., 2007). Often, a pattern emerges in which a tumor is comprised of subpopulations that have discrete RTK amplifications. In fact, 13% of GBMs with EGFR, PDGFRα, or MET amplifications have additional amplifications in another RTK (Szerlip et al., 2012), with three or more RTK signaling pathways activated in the majority of GBM cell lines and patient samples (Stommel et al., 2007). Therefore, RTK profiling will be an important step in developing a personalized GBM treatment regimen and will be accomplished through the use of preclinical animal models of glioma. RTK signaling can often lead to the production of MMPs and enhanced invasive ability for GBM cells. However, even MMP inhibition through marimastat treatment did not show efficacy in a clinical trial setting. Much more work remains to investigate possible compensatory methods with which GBM cells evade current treatments. Potential roles for other signaling motifs including STAT3 and the Hippo pathways have been uncovered and are beginning to be explored (Bhat et al., 2011; Carro et al., 2010).

Further animal modeling of the transformation from low-grade glioma into malignant glioma will prove useful in understanding this relatively little known transition. Up to 70% of patients with low-grade glioma will progress to malignant disease within 5 years (Ohgaki & Kleihues, 2005). These typically younger patients with secondary GBM comprise less than 10% of all diagnosed GBMs, but the opportunity to explore the changes associated with progression is unique to this population. Further efforts to enhance subclassification of low-grade tumors such as oligodendrogliomas with notable mutations such as 1p/19q codeletion/translocation or *CIC* mutations, and each mutation's role in the context of wild-type or mutant p53 will ultimately lead to more targeted and personalized treatment (Dunn et al., 2012; Snuderl et al., 2009; Yip et al., 2012). The continued use of animal models to explore the progression of *IDH* mutant glioma from low grade to secondary GBM could lead to significant progress in attacking this subset of tumor. Such advances coupled with the earlier age of onset of these tumors could translate to large gains in progression-free years for these patients.

A defining feature of GBM is its propensity to develop strong angiogenic signaling within the tumor microenvironment. Targeting angiogenic signaling has yielded some success with the use of Bevacizumab in GBM. However, evidence exists showing that angiogenic inhibitors can promote invasive growth patterns and could explain the highly refractive nature of residual tumor cells that survive Bevacizumab treatment (Iwamoto et al., 2009; Narayana et al., 2012). The combination of nonredundant treatment options that address multiple pathways involved in invasion and angiogenesis will be necessary. Animal modeling will be an invaluable resource to investigate the cellular mechanisms involved in the interplay between angiogenesis and invasion. *In vivo* signaling within a heterogeneous tumor, interacting with ECM factors, and often hijacking local noncancerous cells within the tumor *milieu* can only be addressed through sophisticated animal modeling. Continued study into the role of infiltrating microglia and astrocytic involvement will uncover interactions that may prove to be key targets in combination therapy strategies.

A number of recent studies have been aimed at defining the importance of cells of origin in gliomas. In what is a recurring trend for this tumor, a heterogeneous population appears to be responsible. As we progress toward a more complete understanding of developmental gliomagenesis, comprehensive genetic characterization and deep sequencing efforts will provide insight into the underlying programs that govern the most aggressive GBM phenotypes. The ability to take inventory of differences in epigenetic signatures and transcriptional expression profiles will result in the identification of novel targets for therapy. This next generation of therapeutics will have to attack the capability of cells to maintain multipotentiality, self-renewal, and dedifferentiation, which all contribute to the overall aggressive nature of the tumor.

Initial screens for novel compounds that may be active in inhibiting multiple signaling pathways will be aided by the ability to perform more high-throughput type analysis *in vivo*. The development of glioma models in *Drosophila* and zebrafish will aid in achieving these goals. Ultimately, screens of compound libraries in whole organism settings could be used to analyze potential targets for GBM treatment. In zebrafish, such a method for performing high-throughput screening for antiangiogenic compounds has already been described, identifying and validating the involvement of phosphorylase kinase subunit G1 (PhKG1) (Camus et al., 2012). Further, this target was shown to be upregulated and copy number alterations demonstrated in human tumor samples. In *Drosophila*, global *in vivo* RNAi screening has been defined in the context of heart function regulators. The critical function of CCR4-Not components was discovered and subsequently validated in a *not3* heterozygous knockout mouse (Neely et al., 2010). Similar approaches utilizing organisms conducive to high-throughput analysis, such as *Drosophila* and zebrafish, could be highly valuable in the screening for novel genetic targets and compound library analysis for therapeutic development in GBM. Follow-up studies investigating prospective genes or novel molecules in mouse models could lead to higher rates of success in developing candidate therapies for clinical trials.

As we develop animal models to more closely reflect the disease progression observed in humans, the possibility of combining advanced techniques with cutting edge imaging techniques will be particularly useful. As described previously, a xenograft model utilizing resection techniques recapitulates the surgical intervention commonly incorporated into management of human GBM (Kauer et al., 2011). One could envision the combination of this approach in a transgenic mouse model, especially when employing viral gene transfer. While traditional transgenic mouse models often show multifocal tumor development throughout the brain, through the use of viral gene transfer, the likelihood of more focused tumorigenesis is achieved. Monitoring of tumor development through imaging techniques could be coupled with a resection protocol, ultimately leading to the development of a model that imitates the local recurrence after surgical resection observed in humans. Novel approaches to reduce this phenomenon could lead to methods to combat the most lethal aspect of GBM.

Expanding the utilization of veterinary clinical studies opens another avenue to gain insight into novel treatments and protocols for human patients. A closer evolutionary relationship exists between the canine genome and humans than the mouse, and dogs have four times the rate of spontaneous intracranial neoplasia (15–20 per 100,000 canines per year) when compared to humans (Dobson, Samuel, Milstein, Rogers, & Wood, 2002; Gavin, Fike, & Hoopes, 1995; Heidner, Kornegay, Page, Dodge, & Thrall, 1991; Kirkness et al., 2003). Additionally, the typically advanced age of 8.6 years at the time of astrocytoma diagnosis mirrors the generally older patient typical in human GBM (Keller & Madewell, 1992). GSCs have been identified (Stoica et al., 2009), and overexpression of EGFR, PDGFRα, and IGFBP2 was demonstrated (Higgins et al., 2010) in spontaneous canine GBM. No gold standard of care is established for canines undergoing treatment for GBM, which allows for the opportunity to introduce earlier testing of novel treatments in veterinary clinical trials (Hansen & Khanna, 2004). As with any veterinary clinical trial, the number of subjects available for testing would be dependent on pet owners' consent. The relatively low numbers of subjects balanced by the flexibility of the treatment options puts the use of spontaneous cases of canine GBM in a niche between mouse modeling and human clinical trials.

While no animal model completely replicates all the characteristics of human disease, each is useful for investigating both glioma pathogenesis and preclinical screening of novel therapeutic strategies. The continued use of each animal model type will advance promising treatments to clinical studies in a timely manner and be able to translate more therapy successes seen in animal trials to humans. We need to continue to understand the strengths and idiosyncrasies of individual models to produce the maximal useful information in furthering GBM treatment development.

ACKNOWLEDGMENTS

This study was supported by National Institutes of Health, National Cancer Institute Grant, 1R01 CA134721, and a VCU Massey Cancer Center Developmental Grant to P. B. F., the James S. McDonnell Foundation (D. S.). D. S. is a Harrison Scholar in the MCC and P. B. F. holds the Thelma Newmeyer Corman Chair in Cancer Research in the MCC.

REFERENCES

Abel, T. W., Clark, C., Bierie, B., Chytil, A., Aakre, M., Gorska, A., et al. (2009). GFAP-cre-mediated activation of oncogenic K-ras results in expansion of the subventricular zone and infiltrating glioma. *Molecular Cancer Research: MCR*, 7(5), 645–653.

Aghili, M., Zahedi, F., & Rafiee, E. (2009). Hydroxyglutaric aciduria and malignant brain tumor: A case report and literature review. *Journal of Neuro-Oncology*, *91*(2), 233–236.

Akhavan, D., Cloughesy, T. F., & Mischel, P. S. (2010). mTOR signaling in glioblastoma: Lessons learned from bench to bedside. *Neuro-Oncology*, *12*(8), 882–889.

Akiyama, Y., Jung, S., Salhia, B., Lee, S., Hubbard, S., Taylor, M., et al. (2001). Hyaluronate receptors mediating glioma cell migration and proliferation. *Journal of Neuro-Oncology*, *53*(2), 115–127.

Alcantara Llaguno, S., Chen, J., Kwon, C. H., Jackson, E. L., Li, Y., Burns, D. K., et al. (2009). Malignant astrocytomas originate from neural stem/progenitor cells in a somatic tumor suppressor mouse model. *Cancer Cell*, *15*(1), 45–56.

Alvarez-Breckenridge, C. A., Yu, J., Price, R., Wojton, J., Pradarelli, J., Mao, H., et al. (2012). NK cells impede glioblastoma virotherapy through NKp30 and NKp46 natural cytotoxicity receptors. *Nature Medicine*, *18*(12), 1827–1834.

Asai, A., Miyagi, Y., Sugiyama, A., Gamanuma, M., Hong, S. H., Takamoto, S., et al. (1994). Negative effects of wild-type p53 and s-myc on cellular growth and tumorigenicity of glioma cells. Implication of the tumor suppressor genes for gene therapy. *Journal of Neuro-Oncology*, *19*(3), 259–268.

Assanah, M., Lochhead, R., Ogden, A., Bruce, J., Goldman, J., & Canoll, P. (2006). Glial progenitors in adult white matter are driven to form malignant gliomas by platelet-derived growth factor-expressing retroviruses. *The Journal of Neuroscience: The Official Journal of the Society for Neuroscience*, *26*(25), 6781–6790.

Bachoo, R. M., Maher, E. A., Ligon, K. L., Sharpless, N. E., Chan, S. S., You, M. J., et al. (2002). Epidermal growth factor receptor and Ink4a/arf: Convergent mechanisms governing terminal differentiation and transformation along the neural stem cell to astrocyte axis. *Cancer Cell*, *1*(3), 269–277.

Bajenaru, M. L., Hernandez, M. R., Perry, A., Zhu, Y., Parada, L. F., Garbow, J. R., et al. (2003). Optic nerve glioma in mice requires astrocyte Nf1 gene inactivation and Nf1 brain heterozygosity. *Cancer Research*, *63*(24), 8573–8577.

Bajenaru, M. L., Zhu, Y., Hedrick, N. M., Donahoe, J., Parada, L. F., & Gutmann, D. H. (2002). Astrocyte-specific inactivation of the neurofibromatosis 1 gene (NF1) is insufficient for astrocytoma formation. *Molecular and Cellular Biology*, *22*(14), 5100–5113.

Balss, J., Meyer, J., Mueller, W., Korshunov, A., Hartmann, C., & von Deimling, A. (2008). Analysis of the IDH1 codon 132 mutation in brain tumors. *Acta Neuropathologica*, *116*(6), 597–602.

Banerjee, S., Byrd, J. N., Gianino, S. M., Harpstrite, S. E., Rodriguez, F. J., Tuskan, R. G., et al. (2010). The neurofibromatosis type 1 tumor suppressor controls cell growth by regulating signal transducer and activator of transcription-3 activity in vitro and in vivo. *Cancer Research*, *70*(4), 1356–1366.

Bansal, A., Shuyan, W., Hara, T., Harris, R. A., & Degrado, T. R. (2008). Biodisposition and metabolism of [(18)F]fluorocholine in 9L glioma cells and 9L glioma-bearing fisher rats. *European Journal of Nuclear Medicine and Molecular Imaging*, *35*(6), 1192–1203.

Barker, M., Hoshino, T., Gurcay, O., Wilson, C. B., Nielsen, S. L., Downie, R., et al. (1973). Development of an animal brain tumor model and its response to therapy with 1,3-bis(2-chloroethyl)-1-nitrosourea. *Cancer Research*, *33*(5), 976–986.

Barth, R. F. (1998). Rat brain tumor models in experimental neuro-oncology: The 9L, C6, T9, F98, RG2 (D74), RT-2 and CNS-1 gliomas. *Journal of Neuro-Oncology*, *36*(1), 91–102.

Barth, R. F., & Kaur, B. (2009). Rat brain tumor models in experimental neuro-oncology: The C6, 9L, T9, RG2, F98, BT4C, RT-2 and CNS-1 gliomas. *Journal of Neuro-Oncology*, *94*(3), 299–312.

Batchelor, T. T., Duda, D. G., di Tomaso, E., Ancukiewicz, M., Plotkin, S. R., Gerstner, E., et al. (2010). Phase II study of cediranib, an oral pan-vascular endothelial growth factor receptor tyrosine kinase inhibitor, in patients with recurrent glioblastoma. *Journal of Clinical Oncology: Official Journal of the American Society of Clinical Oncology*, *28*(17), 2817–2823.

Beadle, C., Assanah, M., Monzo, P., Vallee, R., Rosenfeld, S., & Canoll, P. (2008). The role of myosin II in glioma invasion of the brain. *Molecular Biology of the Cell*, *19*(8), 3357–3368.

Becher, O. J., Hambardzumyan, D., Walker, T. R., Helmy, K., Nazarian, J., Albrecht, S., et al. (2010). Preclinical evaluation of radiation and perifosine in a genetically and histologically accurate model of brainstem glioma. *Cancer Research*, *70*(6), 2548–2557.

Bergers, G., & Benjamin, L. E. (2003). Tumorigenesis and the angiogenic switch. *Nature Reviews Cancer*, *3*(6), 401–410.

Berghmans, S., Murphey, R. D., Wienholds, E., Neuberg, D., Kutok, J. L., Fletcher, C. D., et al. (2005). Tp53 mutant zebrafish develop malignant peripheral nerve sheath tumors. *Proceedings of the National Academy of Sciences of the United States of America*, *102*(2), 407–412.

Beroukhim, R., Getz, G., Nghiemphu, L., Barretina, J., Hsueh, T., Linhart, D., et al. (2007). Assessing the significance of chromosomal aberrations in cancer: Methodology and application to glioma. *Proceedings of the National Academy of Sciences of the United States of America*, *104*(50), 20007–20012.

Bhagwat, N., & Levine, R. L. (2010). Metabolic syndromes and malignant transformation: Where the twain shall meet. *Science Translational Medicine*, *2*(54), 54ps50.

Bhat, K. P., Salazar, K. L., Balasubramaniyan, V., Wani, K., Heathcock, L., Hollingsworth, F., et al. (2011). The transcriptional coactivator TAZ regulates mesenchymal differentiation in malignant glioma. *Genes & Development*, *25*(24), 2594–2609.

Bhutia, S. K., Das, S. K., Azab, B., Menezes, M. E., Dent, P., Wang, X. Y., et al. (2013). Targeting breast cancer-initiating/stem cells with melanoma differentiation-associated gene-7/interleukin-24. *International Journal of Cancer*, *133*, 2726–2736.

Biernat, W., Huang, H., Yokoo, H., Kleihues, P., & Ohgaki, H. (2004). Predominant expression of mutant EGFR (EGFRvIII) is rare in primary glioblastomas. *Brain Pathology (Zurich, Switzerland)*, *14*(2), 131–136.

Binello, E., Qadeer, Z. A., Kothari, H. P., Emdad, L., & Germano, I. M. (2012). Stemness of the CT-2A immunocompetent mouse brain tumor model: Characterization in vitro. *Journal of Cancer*, *3*, 166–174.

Bleau, A. M., Hambardzumyan, D., Ozawa, T., Fomchenko, E. I., Huse, J. T., Brennan, C. W., et al. (2009). PTEN/PI3K/akt pathway regulates the side population phenotype and ABCG2 activity in glioma tumor stem-like cells. *Cell Stem Cell*, *4*(3), 226–235.

Boerner, J. L., Demory, M. L., Silva, C., & Parsons, S. J. (2004). Phosphorylation of Y845 on the epidermal growth factor receptor mediates binding to the mitochondrial protein cytochrome c oxidase subunit II. *Molecular and Cellular Biology, 24*(16), 7059–7071.

Bonavia, R., Inda, M. M., Vandenberg, S., Cheng, S., Nagane, M., Hadwiger, P., et al. (2012). EGFRvIII promotes glioma angiogenesis and growth through the NF-Î°B, interleukin-8 pathway. *Oncogene, 31*(36), 4054–4066.

Boukerche, H., Aissaoui, H., Prvost, C., Hirbec, H., Das, S. K., Su, Z., et al. (2010). Src kinase activation is mandatory for MDA-9/syntenin-mediated activation of nuclear factor-kappaB. *Oncogene, 29*(21), 3054–3066.

Boukerche, H., Su, Z., Prvot, C., Sarkar, D., & Fisher, P. (2008). Mda-9/syntenin promotes metastasis in human melanoma cells by activating c-src. *Proceedings of the National Academy of Sciences of the United States of America, 105*(41), 15914–15919.

Bourguignon, L. Y. (2008). Hyaluronan-mediated CD44 activation of RhoGTPase signaling and cytoskeleton function promotes tumor progression. *Seminars in Cancer Biology, 18*(4), 251–259.

Bredel, M., Bredel, C., Juric, D., Harsh, G. R., Vogel, H., Recht, L. D., et al. (2005). High-resolution genome-wide mapping of genetic alterations in human glial brain tumors. *Cancer Research, 65*(10), 4088–4096.

Bredel, M., Scholtens, D. M., Yadav, A. K., Alvarez, A. A., Renfrow, J. J., Chandler, J. P., et al. (2010). NFKBIA deletion in glioblastomas. *The New England Journal of Medicine, 364*, 627–637. http://dx.doi.org/10.1056/NEJMoa1006312.

Brennan, C. (2011). Genomic profiles of glioma. *Current Neurology and Neuroscience Reports, 11*(3), 291–297.

Burger, P. C., Heinz, E. R., Shibata, T., & Kleihues, P. (1988). Topographic anatomy and CT correlations in the untreated glioblastoma multiforme. *Journal of Neurosurgery, 68*(5), 698–704.

Calabrese, C., Poppleton, H., Kocak, M., Hogg, T. L., Fuller, C., Hamner, B., et al. (2007). A perivascular niche for brain tumor stem cells. *Cancer Cell, 11*(1), 69–82.

Calzolari, F., & Malatesta, P. (2010). Recent insights into PDGF-induced gliomagenesis. *Brain Pathology (Zurich, Switzerland), 20*(3), 527–538.

Camus, S., Quevedo, C., Menendez, S., Paramonov, I., Stouten, P. F. W., Janssen, R. A. J., et al. (2012). Identification of phosphorylase kinase as a novel therapeutic target through high-throughput screening for anti-angiogenesis compounds in zebrafish. *Oncogene, 31*(39), 4333–4342.

Cancer Genome Atlas Research Network (2008). Comprehensive genomic characterization defines human glioblastoma genes and core pathways. *Nature, 455*(7216), 1061–1068.

Candolfi, M., Curtin, J. F., Nichols, W. S., Muhammad, A. G., King, G. D., Pluhar, G. E., et al. (2007). Intracranial glioblastoma models in preclinical neuro-oncology: Neuropathological characterization and tumor progression. *Journal of Neuro-Oncology, 85*(2), 133–148.

Canoll, P., & Goldman, J. E. (2008). The interface between glial progenitors and gliomas. *Acta Neuropathologica, 116*(5), 465–477.

Capper, D., Simon, M., Langhans, C. D., Okun, J. G., Tonn, J. C., Weller, M., et al. (2012). 2-Hydroxyglutarate concentration in serum from patients with gliomas does not correlate with IDH1/2 mutation status or tumor size. *International Journal of Cancer, 131*(3), 766–768.

Carmeliet, P., & Jain, R. K. (2011a). Molecular mechanisms and clinical applications of angiogenesis. *Nature, 473*(7347), 298–307.

Carmeliet, P., & Jain, R. K. (2011b). Principles and mechanisms of vessel normalization for cancer and other angiogenic diseases. *Nature Reviews Drug Discovery, 10*(6), 417–427.

Carro, M. S., Lim, W. K., Alvarez, M. J., Bollo, R. J., Zhao, X., Snyder, E. Y., et al. (2010). The transcriptional network for mesenchymal transformation of brain tumours. *Nature, 463*(7279), 318–325.

Chabner, B. A. (2011). Early accelerated approval for highly targeted cancer drugs. *The New England Journal of Medicine, 364*(12), 1087–1089. http://dx.doi.org/10.1056/NEJMp1100548.

Chamberlain, M. C. (2011). Radiographic patterns of relapse in glioblastoma. *Journal of Neuro-Oncology, 101*(2), 319–323.

Chen, J., Li, Y., Yu, T. S., McKay, R. M., Burns, D. K., Kernie, S. G., et al. (2012). A restricted cell population propagates glioblastoma growth after chemotherapy. *Nature, 488*(7412), 522–526.

Chen, R., Nishimura, M. C., Bumbaca, S. M., Kharbanda, S., Forrest, W. F., Kasman, I. M., et al. (2010). A hierarchy of self-renewing tumor-initiating cell types in glioblastoma. *Cancer Cell, 17*(4), 362–375.

Chetty, C., Vanamala, S. K., Gondi, C. S., Dinh, D. H., Gujrati, M., & Rao, J. S. (2011). MMP-9 induces CD44 cleavage and CD44 mediated cell migration in glioblastoma xenograft cells. *Cellular Signalling, 24*, 549–559. http://dx.doi.org/10.1016/j.cellsig.2011.10.008.

Chi, A. S., Batchelor, T. T., Kwak, E. L., Clark, J. W., Wang, D. L., Wilner, K. D., et al. (2012). Rapid radiographic and clinical improvement after treatment of a MET-amplified recurrent glioblastoma with a mesenchymal-epithelial transition inhibitor. *Journal of Clinical Oncology: Official Journal of the American Society of Clinical Oncology, 30*(3), e30–e33.

Chin, L., Hahn, W. C., Getz, G., & Meyerson, M. (2011). Making sense of cancer genomic data [reviews]. *Genes & Development, 25*(6), 534–555. http://dx.doi.org/10.1101/gad.2017311.

Chiocca, E. A. (2002). Oncolytic viruses. *Nature Reviews Cancer, 2*(12), 938–950.

Choi, C. H., Schoenfeld, B. P., Bell, A. J., Hinchey, P., Kollaros, M., Gertner, M. J., et al. (2011). Pharmacological reversal of synaptic plasticity deficits in the mouse model of fragile X syndrome by group II mGluR antagonist or lithium treatment. *Brain Research, 1380*, 106–119.

Chowdhury, R., Yeoh, K. K., Tian, Y. M., Hillringhaus, L., Bagg, E. A., Rose, N. R., et al. (2011). The oncometabolite 2-hydroxyglutarate inhibits histone lysine demethylases. *EMBO Reports, 12*(5), 463–469.

Christensen, J. G., Zou, H. Y., Arango, M. E., Li, Q., Lee, J. H., McDonnell, S. R., et al. (2007). Cytoreductive antitumor activity of PF-2341066, a novel inhibitor of anaplastic lymphoma kinase and c-met, in experimental models of anaplastic large-cell lymphoma. *Molecular Cancer Therapeutics, 6*(12 Pt 1), 3314–3322.

Claes, A., Schuuring, J., Boots-Sprenger, S., Hendriks-Cornelissen, S., Dekkers, M., van der Kogel, A. J., et al. (2008). Phenotypic and genotypic characterization of orthotopic human glioma models and its relevance for the study of anti-glioma therapy. *Brain Pathology (Zurich, Switzerland), 18*(3), 423–433.

Cohen, M. H., Shen, Y. L., Keegan, P., & Pazdur, R. (2009). FDA drug approval summary: Bevacizumab (avastin) as treatment of recurrent glioblastoma multiforme. *The Oncologist, 14*(11), 1131–1138.

Colman, H., Zhang, L., Sulman, E. P., McDonald, J. M., Shooshtari, N. L., Rivera, A., et al. (2010). A multigene predictor of outcome in glioblastoma. *Neuro-Oncology, 12*(1), 49–57.

Copeland, D. D., Talley, F. A., & Bigner, D. D. (1976). The fine structure of intracranial neoplasms induced by the inoculation of avian sarcoma virus in neonatal and adult rats. *The American Journal of Pathology, 83*(1), 149–176.

Dai, C., Celestino, J. C., Okada, Y., Louis, D. N., Fuller, G. N., & Holland, E. C. (2001). PDGF autocrine stimulation dedifferentiates cultured astrocytes and induces oligodendrogliomas and oligoastrocytomas from neural progenitors and astrocytes in vivo. *Genes & Development*, *15*(15), 1913–1925.

Dai, B., Kang, S. H., Gong, W., Liu, M., Aldape, K. D., Sawaya, R., et al. (2007). Aberrant FoxM1B expression increases matrix metalloproteinase-2 transcription and enhances the invasion of glioma cells. *Oncogene*, *26*(42), 6212–6219.

Dai, C., Lyustikman, Y., Shih, A., Hu, X., Fuller, G. N., Rosenblum, M., et al. (2005). The characteristics of astrocytomas and oligodendrogliomas are caused by two distinct and interchangeable signaling formats. *Neoplasia (New York, N.Y.)*, 7(4), 397–406.

Dai, B., Pieper, R. O., Li, D., Wei, P., Liu, M., Woo, S. Y., et al. (2010). FoxM1B regulates NEDD4-1 expression, leading to cellular transformation and full malignant phenotype in immortalized human astrocytes. *Cancer Research*, *70*(7), 2951–2961.

D'Amico, R., Lei, L., Kennedy, B. C., Sisti, J., Ebiana, V., Crisman, C., et al. (2012). The addition of sunitinib to radiation delays tumor growth in a murine model of glioblastoma. *Neurological Research*, *34*(3), 252–261.

Dang, L., White, D. W., Gross, S., Bennett, B. D., Bittinger, M. A., Driggers, E. M., et al. (2010). Cancer-associated IDH1 mutations produce 2-hydroxyglutarate. *Nature*, *465*(7300), 966.

Das, S. K., Bhutia, S. K., Azab, B., Kegelman, T. P., Peachy, L., Santhekadur, P. K., et al. (2012). MDA-9/syntenin and IGFBP-2 promote angiogenesis in human melanoma. *Cancer Research*, *73*, 844–854. http://dx.doi.org/10.1158/0008-5472.CAN-12-1681.

de Ridder, L. I., Laerum, O. D., Mork, S. J., & Bigner, D. D. (1987). Invasiveness of human glioma cell lines in vitro: Relation to tumorigenicity in athymic mice. *Acta Neuropathologica*, *72*(3), 207–213.

de Vries, N. A., Bruggeman, S. W., Hulsman, D., de Vries, H. I., Zevenhoven, J., Buckle, T., et al. (2010). Rapid and robust transgenic high-grade glioma mouse models for therapy intervention studies. *Clinical Cancer Research: An Official Journal of the American Association for Cancer Research*, *16*(13), 3431–3441.

De Witt Hamer, P. C. (2010). Small molecule kinase inhibitors in glioblastoma: A systematic review of clinical studies. *Neuro-Oncology*, *12*(3), 304–316.

Denlinger, R. H., Axler, D. A., Koestner, A., & Liss, L. (1975). Tumor-specific transplantation immunity to intracerebral challenge with cells from a methylnitrosourea-induced brain tumor. *Journal of Medicine*, *6*(3–4), 249–259.

Desai, S., Pillai, P., Win-Piazza, H., & Acevedo-Duncan, M. (2011). PKC-iota promotes glioblastoma cell survival by phosphorylating and inhibiting BAD through a phosphatidylinositol 3-kinase pathway. *Biochimica Et Biophysica Acta*, *1813*(6), 1190–1197.

Di Rocco, F., Carroll, R. S., Zhang, J., & Black, P. M. (1998). Platelet-derived growth factor and its receptor expression in human oligodendrogliomas. *Neurosurgery*, *42*(2), 341–346.

Ding, H., Roncari, L., Shannon, P., Wu, X., Lau, N., Karaskova, J., et al. (2001). Astrocyte-specific expression of activated p21-ras results in malignant astrocytoma formation in a transgenic mouse model of human gliomas. *Cancer Research*, *61*(9), 3826–3836.

Ding, H., Shannon, P., Lau, N., Wu, X., Roncari, L., Baldwin, R. L., et al. (2003). Oligodendrogliomas result from the expression of an activated mutant epidermal growth factor receptor in a RAS transgenic mouse astrocytoma model. *Cancer Research*, *63*(5), 1106–1113.

Dobson, J. M., Samuel, S., Milstein, H., Rogers, K., & Wood, J. L. N. (2002). Canine neoplasia in the UK: Estimates of incidence rates from a population of insured dogs. *Journal of Small Animal Practice*, *43*(6), 240–246.

Doherty, J., Logan, M. A., Tasdemir, O. E., & Freeman, M. R. (2009). Ensheathing glia function as phagocytes in the adult drosophila brain. *The Journal of Neuroscience: The Official Journal of the Society for Neuroscience*, *29*(15), 4768–4781.

Dolecek, T. A., Propp, J. M., Stroup, N. E., & Kruchko, C. (2012). CBTRUS statistical report: Primary brain and central nervous system tumors diagnosed in the united states in 2005-2009. *Neuro-Oncology*, *14*(Suppl. 5), v1–v49.

Du, J., Bernasconi, P., Clauser, K. R., Mani, D. R., Finn, S. P., Beroukhim, R., et al. (2009). Bead-based profiling of tyrosine kinase phosphorylation identifies SRC as a potential target for glioblastoma therapy. *Nature Biotechnology*, *27*(1), 77–83.

Duncan, C. G., Killela, P. J., Payne, C. A., Lampson, B., Chen, W. C., Liu, J., et al. (2010). Integrated genomic analyses identify ERRFI1 and TACC3 as glioblastoma-targeted genes. *Oncotarget*, *1*(4), 265–277.

Dunn, G. P., Rinne, M. L., Wykosky, J., Genovese, G., Quayle, S. N., Dunn, I. F., et al. (2012). Emerging insights into the molecular and cellular basis of glioblastoma. *Genes & Development*, *26*(8), 756–784.

Ekstrand, A. J., James, C. D., Cavenee, W. K., Seliger, B., Pettersson, R. F., & Collins, V. P. (1991). Genes for epidermal growth factor receptor, transforming growth factor alpha, and epidermal growth factor and their expression in human gliomas in vivo. *Cancer Research*, *51*(8), 2164–2172.

Ellis, L. M., & Hicklin, D. J. (2008). Pathways mediating resistance to vascular endothelial growth factor-targeted therapy. *Clinical Cancer Research: An Official Journal of the American Association for Cancer Research*, *14*(20), 6371–6375.

Endersby, R., Zhu, X., Hay, N., Ellison, D. W., & Baker, S. J. (2011). Nonredundant functions for akt isoforms in astrocyte growth and gliomagenesis in an orthotopic transplantation model. *Cancer Research*, *71*(12), 4106–4116.

Engelman, J., Luo, J., & Cantley, L. (2006). The evolution of phosphatidylinositol 3-kinases as regulators of growth and metabolism. *Nature Reviews. Genetics*, 7(8), 606–619.

Engelman, J. A., Zejnullahu, K., Mitsudomi, T., Song, Y., Hyland, C., Park, J. O., et al. (2007). MET amplification leads to gefitinib resistance in lung cancer by activating ERBB3 signaling. *Science (New York, NY)*, *316*(5827), 1039–1043.

Fan, Q. W., Cheng, C., Hackett, C., Feldman, M., Houseman, B. T., Nicolaides, T., et al. (2010). Akt and autophagy cooperate to promote survival of drug-resistant glioma. *Science Signaling*, *3*(147), ra81.

Fan, Q. W., Cheng, C., Knight, Z. A., Haas-Kogan, D., Stokoe, D., James, C. D., et al. (2009). EGFR signals to mTOR through PKC and independently of akt in glioma. *Science Signaling*, *2*(55), ra4.

Fan, Q. W., Knight, Z. A., Goldenberg, D. D., Yu, W., Mostov, K. E., Stokoe, D., et al. (2006). A dual PI3 kinase/mTOR inhibitor reveals emergent efficacy in glioma. *Cancer Cell*, *9*(5), 341–349.

Fang, M., Shen, Z., Huang, S., Zhao, L., Chen, S., Mak, T. W., et al. (2010). The ER UDPase ENTPD5 promotes protein N-glycosylation, the warburg effect, and proliferation in the PTEN pathway. *Cell*, *143*(5), 711–724.

Feil, R., Brocard, J., Mascrez, B., LeMeur, M., Metzger, D., & Chambon, P. (1996). Ligand-activated site-specific recombination in mice. *Proceedings of the National Academy of Sciences of the United States of America*, *93*(20), 10887–10890.

Figueroa, M. E., Abdel-Wahab, O., Lu, C., Ward, P. S., Patel, J., Shih, A., et al. (2010). Leukemic IDH1 and IDH2 mutations result in a hypermethylation phenotype, disrupt TET2 function, and impair hematopoietic differentiation. *Cancer Cell*, *18*(6), 553–567.

Finkelstein, S. D., Black, P., Nowak, T. P., Hand, C. M., Christensen, S., & Finch, P. W. (1994). Histological characteristics and expression of acidic and basic fibroblast growth factor genes in intracerebral xenogeneic transplants of human glioma cells. *Neurosurgery*, *34*(1), 136–143.

Fleming, T. P., Saxena, A., Clark, W. C., Robertson, J. T., Oldfield, E. H., Aaronson, S. A., et al. (1992). Amplification and/or overexpression of platelet-derived growth factor receptors and epidermal growth factor receptor in human glial tumors. *Cancer Research, 52*(16), 4550–4553.

Folkman, J. (1971). Tumor angiogenesis: Therapeutic implications. *The New England Journal of Medicine, 285*(21), 1182–1186.

Fomchenko, E. I., Dougherty, J. D., Helmy, K. Y., Katz, A. M., Pietras, A., Brennan, C., et al. (2011). Recruited cells can become transformed and overtake PDGF-induced murine gliomas in vivo during tumor progression. *PLoS One, 6*(7), e20605.

Fomchenko, E., & Holland, E. (2006). Mouse models of brain tumors and their applications in preclinical trials. *Clinical Cancer Research, 12*(18), 5288–5297.

Forsyth, P. A., Wong, H., Laing, T. D., Rewcastle, N. B., Morris, D. G., Muzik, H., et al. (1999). Gelatinase-A (MMP-2), gelatinase-B (MMP-9) and membrane type matrix metalloproteinase-1 (MT1-MMP) are involved in different aspects of the pathophysiology of malignant gliomas. *British Journal of Cancer, 79*(11–12), 1828–1835.

Freeman, M. R., & Doherty, J. (2006). Glial cell biology in drosophila and vertebrates. *Trends in Neurosciences, 29*(2), 82–90.

Friedman, H. S., Prados, M. D., Wen, P. Y., Mikkelsen, T., Schiff, D., Abrey, L. E., et al. (2009). Bevacizumab alone and in combination with irinotecan in recurrent glioblastoma. *Journal of Clinical Oncology: Official Journal of the American Society of Clinical Oncology, 27*(28), 4733–4740.

Friedmann-Morvinski, D., Bushong, E. A., Ke, E., Soda, Y., Marumoto, T., Singer, O., et al. (2012). Dedifferentiation of neurons and astrocytes by oncogenes can induce gliomas in mice. *Science (New York, NY), 338*(6110), 1080–1084.

Galban, S., Lemasson, B., Williams, T. M., Li, F., Heist, K. A., Johnson, T. D., et al. (2012). DW-MRI as a biomarker to compare therapeutic outcomes in radiotherapy regimens incorporating temozolomide or gemcitabine in glioblastoma. *PLoS One*, 7(4), e35857. http://dx.doi.org/10.1371/journal.pone.0035857.

Galli, R., Binda, E., Orfanelli, U., Cipelletti, B., Gritti, A., De Vitis, S., et al. (2004). Isolation and characterization of tumorigenic, stem-like neural precursors from human glioblastoma. *Cancer Research, 64*(19), 7011–7021.

Gallia, G. L., Rand, V., Siu, I. M., Eberhart, C. G., James, C. D., Marie, S. K., et al. (2006). PIK3CA gene mutations in pediatric and adult glioblastoma multiforme. *Molecular Cancer Research: MCR, 4*(10), 709–714.

Gavin, P. R., Fike, J. R., & Hoopes, P. J. (1995). Central nervous system tumors. *Seminars in Veterinary Medicine and Surgery (Small Animal), 10*(3), 180–189.

Geiger, G. A., Fu, W., & Kao, G. D. (2008). Temozolomide-mediated radiosensitization of human glioma cells in a zebrafish embryonic system. *Cancer Research, 68*(9), 3396–3404.

Gerstner, E. R., Duda, D. G., di Tomaso, E., Ryg, P. A., Loeffler, J. S., Sorensen, A. G., et al. (2009). VEGF inhibitors in the treatment of cerebral edema in patients with brain cancer. *Nature Reviews Clinical Oncology, 6*(4), 229–236.

Ghazi, S. O., Stark, M., Zhao, Z., Mobley, B. C., Munden, A., Hover, L., et al. (2012). Cell of origin determines tumor phenotype in an oncogenic ras/p53 knockout transgenic model of high-grade glioma. *Journal of Neuropathology and Experimental Neurology, 71*(8), 729–740.

Ghods, A. J., Irvin, D., Liu, G., Yuan, X., Abdulkadir, I. R., Tunici, P., et al. (2007). Spheres isolated from 9L gliosarcoma rat cell line possess chemoresistant and aggressive cancer stem-like cells. *Stem Cells (Dayton, Ohio), 25*(7), 1645–1653.

Giannini, C., Sarkaria, J. N., Saito, A., Uhm, J. H., Galanis, E., Carlson, B. L., et al. (2005). Patient tumor EGFR and PDGFRA gene amplifications retained in an invasive intracranial xenograft model of glioblastoma multiforme. *Neuro-Oncology*, 7(2), 164–176.

Gilbertson, R. J., & Rich, J. N. (2007). Making a tumour's bed: Glioblastoma stem cells and the vascular niche. *Nature Reviews. Cancer, 7*(10), 733–736.
Godar, S., Ince, T. A., Bell, G. W., Feldser, D., Donaher, J. L., Bergh, J., et al. (2008). Growth-inhibitory and tumor-suppressive functions of p53 depend on its repression of CD44 expression. *Cell, 134*(1), 62–73.
Golub, T. R., Slonim, D. K., Tamayo, P., Huard, C., Gaasenbeek, M., Mesirov, J. P., et al. (1999). Molecular classification of cancer: Class discovery and class prediction by gene expression monitoring. *Science (New York, NY), 286*(5439), 531–537.
Gritsenko, P. G., Ilina, O., & Friedl, P. (2012). Interstitial guidance of cancer invasion. *The Journal of Pathology, 226*(2), 185–199.
Guo, P., Hu, B., Gu, W., Xu, L., Wang, D., Huang, H. J., et al. (2003). Platelet-derived growth factor-B enhances glioma angiogenesis by stimulating vascular endothelial growth factor expression in tumor endothelia and by promoting pericyte recruitment. *The American Journal of Pathology, 162*(4), 1083–1093.
Halder, G., Callaerts, P., & Gehring, W. J. (1995). Induction of ectopic eyes by targeted expression of the eyeless gene in drosophila. *Science (New York, NY), 267*(5205), 1788–1792.
Hambardzumyan, D., Amankulor, N. M., Helmy, K. Y., Becher, O. J., & Holland, E. C. (2009). Modeling adult gliomas using RCAS/t-va technology. *Translational Oncology, 2*(2), 89–95.
Hansen, K., & Khanna, C. (2004). Spontaneous and genetically engineered animal models; use in preclinical cancer drug development. *European Journal of Cancer (Oxford, England 1990), 40*(6), 858–880.
Hartmann, C., Hentschel, B., Wick, W., Capper, D., Felsberg, J., Simon, M., et al. (2010). Patients with IDH1 wild type anaplastic astrocytomas exhibit worse prognosis than IDH1-mutated glioblastomas, and IDH1 mutation status accounts for the unfavorable prognostic effect of higher age: Implications for classification of gliomas. *Acta Neuropathologica, 120*(6), 707–718.
Hartmann, C., Meyer, J., Balss, J., Capper, D., Mueller, W., Christians, A., et al. (2009). Type and frequency of IDH1 and IDH2 mutations are related to astrocytic and oligodendroglial differentiation and age: A study of 1,010 diffuse gliomas. *Acta Neuropathologica, 118*(4), 469–474.
Hede, S. M., Hansson, I., Afink, G. B., Eriksson, A., Nazarenko, I., Andrae, J., et al. (2009). GFAP promoter driven transgenic expression of PDGFB in the mouse brain leads to glioblastoma in a Trp53 null background. *Glia, 57*(11), 1143–1153.
Heidner, G. L., Kornegay, J. N., Page, R. L., Dodge, R. K., & Thrall, D. E. (1991). Analysis of survival in a retrospective study of 86 dogs with brain tumors. *Journal of Veterinary Internal Medicine/American College of Veterinary Internal Medicine, 5*(4), 219–226.
Heimberger, A. B., Hlatky, R., Suki, D., Yang, D., Weinberg, J., Gilbert, M., et al. (2005). Prognostic effect of epidermal growth factor receptor and EGFRvIII in glioblastoma multiforme patients. *Clinical Cancer Research: An Official Journal of the American Association for Cancer Research, 11*(4), 1462–1466.
Henson, J. W., Schnitker, B. L., Correa, K. M., von Deimling, A., Fassbender, F., Xu, H. J., et al. (1994). The retinoblastoma gene is involved in malignant progression of astrocytomas. *Annals of Neurology, 36*(5), 714–721.
Hermanson, M., Funa, K., Hartman, M., Claesson-Welsh, L., Heldin, C. H., Westermark, B., et al. (1992). Platelet-derived growth factor and its receptors in human glioma tissue: Expression of messenger RNA and protein suggests the presence of autocrine and paracrine loops. *Cancer Research, 52*(11), 3213–3219.
Hesselager, G., Uhrbom, L., Westermark, B., & Nister, M. (2003). Complementary effects of platelet-derived growth factor autocrine stimulation and p53 or Ink4a-arf deletion in a mouse glioma model. *Cancer Research, 63*(15), 4305–4309.

Hicklin, D. J., & Ellis, L. M. (2005). Role of the vascular endothelial growth factor pathway in tumor growth and angiogenesis. *Journal of Clinical Oncology: Official Journal of the American Society of Clinical Oncology, 23*(5), 1011–1027.

Higgins, R. J., Dickinson, P. J., LeCouteur, R. A., Bollen, A. W., Wang, H., Wang, H., et al. (2010). Spontaneous canine gliomas: Overexpression of EGFR, PDGFRalpha and IGFBP2 demonstrated by tissue microarray immunophenotyping. *Journal of Neuro-Oncology, 98*(1), 49–55.

Hingtgen, S., Kasmieh, R., Elbayly, E., Nesterenko, I., Figueiredo, J. L., Dash, R., et al. (2012). A first-generation multi-functional cytokine for simultaneous optical tracking and tumor therapy. *PLoS One,* 7(7), e40234.

Hirth, F. (2010). Drosophila melanogaster in the study of human neurodegeneration. *CNS & Neurological Disorders Drug Targets, 9*(4), 504–523.

Hochberg, F. H., Fischman, A. J., & Metz, R. (1994). Imaging of brain tumors. *Cancer, 74*(12), 3080–3082.

Hochberg, F. H., & Pruitt, A. (1980). Assumptions in the radiotherapy of glioblastoma. *Neurology, 30*(9), 907–911.

Hoelzinger, D. B., Demuth, T., & Berens, M. E. (2007). Autocrine factors that sustain glioma invasion and paracrine biology in the brain microenvironment. *Journal of the National Cancer Institute, 99*(21), 1583–1593.

Hoffman, H. J., & Duffner, P. K. (1985). Extraneural metastases of central nervous system tumors. *Cancer, 56*(7 Suppl.), 1778–1782.

Holland, E. C., Celestino, J., Dai, C., Schaefer, L., Sawaya, R. E., & Fuller, G. N. (2000). Combined activation of ras and akt in neural progenitors induces glioblastoma formation in mice. *Nature Genetics, 25*(1), 55–57.

Holland, E. C., Hively, W. P., DePinho, R. A., & Varmus, H. E. (1998). A constitutively active epidermal growth factor receptor cooperates with disruption of G1 cell-cycle arrest pathways to induce glioma-like lesions in mice. *Genes & Development, 12*(23), 3675–3685.

Holland, E. C., & Varmus, H. E. (1998). Basic fibroblast growth factor induces cell migration and proliferation after glia-specific gene transfer in mice. *Proceedings of the National Academy of Sciences of the United States of America, 95*(3), 1218–1223.

Hu, X., Pandolfi, P. P., Li, Y., Koutcher, J. A., Rosenblum, M., & Holland, E. C. (2005). mTOR promotes survival and astrocytic characteristics induced by pten/AKT signaling in glioblastoma. *Neoplasia (New York, NY),* 7(4), 356–368.

Huang, P. H., Mukasa, A., Bonavia, R., Flynn, R. A., Brewer, Z. E., Cavenee, W. K., et al. (2007). Quantitative analysis of EGFRvIII cellular signaling networks reveals a combinatorial therapeutic strategy for glioblastoma. *Proceedings of the National Academy of Sciences of the United States of America, 104*(31), 12867–12872.

Huang, H. S., Nagane, M., Klingbeil, C. K., Lin, H., Nishikawa, R., Ji, X. D., et al. (1997). The enhanced tumorigenic activity of a mutant epidermal growth factor receptor common in human cancers is mediated by threshold levels of constitutive tyrosine phosphorylation and unattenuated signaling. *The Journal of Biological Chemistry, 272*(5), 2927–2935.

Huang, P. H., Xu, A. M., & White, F. M. (2009). Oncogenic EGFR signaling networks in glioma. *Science Signaling, 2*(87), re6.

Hurtt, M. R., Moossy, J., Donovan-Peluso, M., & Locker, J. (1992). Amplification of epidermal growth factor receptor gene in gliomas: Histopathology and prognosis. *Journal of Neuropathology and Experimental Neurology, 51*(1), 84–90.

Huse, J., & Holland, E. (2009). Genetically engineered mouse models of brain cancer and the promise of preclinical testing. *Brain Pathology, 19*(1), 132–143.

Huse, J., & Holland, E. (2010). Targeting brain cancer: Advances in the molecular pathology of malignant glioma and medulloblastoma. *Nature Reviews Cancer, 10*(5), 319–331.

Huse, J. T., Phillips, H. S., & Brennan, C. W. (2011). Molecular subclassification of diffuse gliomas: Seeing order in the chaos. *Glia, 59*(8), 1190–1199.
Ichimura, K., Pearson, D. M., Kocialkowski, S., Backlund, L. M., Chan, R., Jones, D. T., et al. (2009). IDH1 mutations are present in the majority of common adult gliomas but rare in primary glioblastomas. *Neuro-Oncology, 11*(4), 341–347.
Inda, M. M., Bonavia, R., Mukasa, A., Narita, Y., Sah, D. W., Vandenberg, S., et al. (2010). Tumor heterogeneity is an active process maintained by a mutant EGFR-induced cytokine circuit in glioblastoma. *Genes & Development, 24*(16), 1731–1745.
Iwamoto, F. M., Abrey, L. E., Beal, K., Gutin, P. H., Rosenblum, M. K., Reuter, V. E., et al. (2009). Patterns of relapse and prognosis after bevacizumab failure in recurrent glioblastoma. *Neurology, 73*(15), 1200–1206.
Iwamoto, F. M., Hottinger, A. F., Karimi, S., Riedel, E., Dantis, J., Jahdi, M., et al. (2011). Serum YKL-40 is a marker of prognosis and disease status in high-grade gliomas. *Neuro-Oncology, 13*(11), 1244–1251.
Jacques, T. S., Swales, A., Brzozowski, M. J., Henriquez, N. V., Linehan, J. M., Mirzadeh, Z., et al. (2010). Combinations of genetic mutations in the adult neural stem cell compartment determine brain tumour phenotypes. *The EMBO Journal, 29*(1), 222–235.
Janne, P. A., Engelman, J. A., & Johnson, B. E. (2005). Epidermal growth factor receptor mutations in non-small-cell lung cancer: Implications for treatment and tumor biology. *Journal of Clinical Oncology: Official Journal of the American Society of Clinical Oncology, 23*(14), 3227–3234.
Jaros, E., Perry, R. H., Adam, L., Kelly, P. J., Crawford, P. J., Kalbag, R. M., et al. (1992). Prognostic implications of p53 protein, epidermal growth factor receptor, and ki-67 labelling in brain tumours. *British Journal of Cancer, 66*(2), 373–385.
Jensen, N. A., Pedersen, K. M., Lihme, F., Rask, L., Nielsen, J. V., Rasmussen, T. E., et al. (2003). Astroglial c-myc overexpression predisposes mice to primary malignant gliomas. *The Journal of Biological Chemistry, 278*(10), 8300–8308.
Jo, M., Stolz, D. B., Esplen, J. E., Dorko, K., Michalopoulos, G. K., & Strom, S. C. (2000). Cross-talk between epidermal growth factor receptor and c-met signal pathways in transformed cells. *The Journal of Biological Chemistry, 275*(12), 8806–8811.
Jung, I. H., Leem, G. L., Jung, D. E., Kim, M. H., Kim, E. Y., Kim, S. H., et al. (2013). Glioma is formed by active Akt1 alone and promoted by active Rac1 in transgenic zebrafish. *Neuro-Oncology, 15*, 290–304. http://dx.doi.org/10.1093/neuonc/nos387.
Karim, F. D., Chang, H. C., Therrien, M., Wassarman, D. A., Laverty, T., & Rubin, G. M. (1996). A screen for genes that function downstream of Ras1 during drosophila eye development. *Genetics, 143*(1), 315–329.
Kauer, T. M., Figueiredo, J. L., Hingtgen, S., & Shah, K. (2011). Encapsulated therapeutic stem cells implanted in the tumor resection cavity induce cell death in gliomas. *Nature Neuroscience, 15*(2), 197–204.
Kegelman, T. P., Das, S. K., Hu, B., Menezes, M. E., Emdad, L., Dasgupta, S., et al. (2014). MDA-9/syntenin is a key regulator of glioma pathogenesis. *Neuro-Oncology, 16*, 50–61.
Keller, E. T., & Madewell, B. R. (1992). Locations and types of neoplasms in immature dogs: 69 cases (1964-1989). *Journal of the American Veterinary Medical Association, 200*(10), 1530–1532.
Kerbel, R. S. (2008). Tumor angiogenesis. *The New England Journal of Medicine, 358*(19), 2039–2049.
Khan, A., Jallo, G. I., Liu, Y. J., Carson, B. S., Sr., & Guarnieri, M. (2005). Infusion rates and drug distribution in brain tumor models in rats. *Journal of Neurosurgery, 102*(1 Suppl), 53–58.
Kim, S. M., Oh, J. H., Park, S. A., Ryu, C. H., Lim, J. Y., Kim, D. S., et al. (2010). Irradiation enhances the tumor tropism and therapeutic potential of TRAIL-secreting

human umbilical cord blood-derived mesenchymal stem cells in glioma therapy. *Stem Cells (Dayton, Ohio)*, *28*, 2217–2228. http://dx.doi.org/10.1002/stem.543.

Kirkness, E. F., Bafna, V., Halpern, A. L., Levy, S., Remington, K., Rusch, D. B., et al. (2003). The dog genome: Survey sequencing and comparative analysis. *Science (New York, NY)*, *301*(5641), 1898–1903.

Kita, D., Yonekawa, Y., Weller, M., & Ohgaki, H. (2007). PIK3CA alterations in primary (de novo) and secondary glioblastomas. *Acta Neuropathologica*, *113*(3), 295–302.

Kleber, S., Sancho-Martinez, I., Wiestler, B., Beisel, A., Gieffers, C., Hill, O., et al. (2008). Yes and PI3K bind CD95 to signal invasion of glioblastoma. *Cancer Cell*, *13*(3), 235–248.

Koivunen, P., Lee, S., Duncan, C. G., Lopez, G., Lu, G., Ramkissoon, S., et al. (2012). Transformation by the (R)-enantiomer of 2-hydroxyglutarate linked to EGLN activation. *Nature*, *483*(7390), 484–488.

Kotliarov, Y., Steed, M. E., Christopher, N., Walling, J., Su, Q., Center, A., et al. (2006). High-resolution global genomic survey of 178 gliomas reveals novel regions of copy number alteration and allelic imbalances. *Cancer Research*, *66*(19), 9428–9436.

Koul, D. (2008). PTEN signaling pathways in glioblastoma. *Cancer Biology & Therapy*, *7*(9), 1321–1325.

Kreisl, T. N., Kim, L., Moore, K., Duic, P., Royce, C., Stroud, I., et al. (2009). Phase II trial of single-agent bevacizumab followed by bevacizumab plus irinotecan at tumor progression in recurrent glioblastoma. *Journal of Clinical Oncology: Official Journal of the American Society of Clinical Oncology*, *27*(5), 740–745.

Kruse, C. A., Molleston, M. C., Parks, E. P., Schiltz, P. M., Kleinschmidt-DeMasters, B. K., & Hickey, W. F. (1994). A rat glioma model, CNS-1, with invasive characteristics similar to those of human gliomas: A comparison to 9L gliosarcoma. *Journal of Neuro-Oncology*, *22*(3), 191–200.

Kwok, J. C., Dick, G., Wang, D., & Fawcett, J. W. (2011). Extracellular matrix and perineuronal nets in CNS repair. *Developmental Neurobiology*, *71*(11), 1073–1089.

Kwon, C. H., Zhao, D., Chen, J., Alcantara, S., Li, Y., Burns, D. K., et al. (2008). Pten haploinsufficiency accelerates formation of high-grade astrocytomas. *Cancer Research*, *68*(9), 3286–3294.

Laerum, O. D., & Rajewsky, M. F. (1975). Neoplastic transformation of fetal rat brain cells in culture after exposure to ethylnitrosourea in vivo. *Journal of the National Cancer Institute*, *55*(5), 1177–1187.

Laerum, O. D., Rajewsky, M. F., Schachner, M., Stavrou, D., Haglid, K. G., & Haugen, A. (1977). Phenotypic properties of neoplastic cell lines developed from fetal rat brain cells in culture after exposure to ethylnitrosourea in vivo. *Zeitschrift Fur Krebsforschung Und Klinische Onkologie Cancer Research and Clinical Oncology*, *89*(3), 273–295.

Laks, D. R., Masterman-Smith, M., Visnyei, K., Angenieux, B., Orozco, N. M., Foran, I., et al. (2009). Neurosphere formation is an independent predictor of clinical outcome in malignant glioma. *Stem Cells (Dayton, Ohio)*, *27*(4), 980–987.

Lee, J., Kotliarova, S., Kotliarov, Y., Li, A., Su, Q., Donin, N. M., et al. (2006). Tumor stem cells derived from glioblastomas cultured in bFGF and EGF more closely mirror the phenotype and genotype of primary tumors than do serum-cultured cell lines. *Cancer Cell*, *9*(5), 391–403.

Lee, J. C., Vivanco, I., Beroukhim, R., Huang, J. H., Feng, W. L., DeBiasi, R. M., et al. (2006). Epidermal growth factor receptor activation in glioblastoma through novel missense mutations in the extracellular domain. *PLoS Medicine*, *3*(12), e485.

Lei, L., Sonabend, A. M., Guarnieri, P., Soderquist, C., Ludwig, T., Rosenfeld, S., et al. (2011). Glioblastoma models reveal the connection between adult glial progenitors and the proneural phenotype. *PLoS One*, *6*(5), e20041.

Levy, A., Blacher, E., Vaknine, H., Lund, F. E., Stein, R., & Mayo, L. (2012). CD38 deficiency in the tumor microenvironment attenuates glioma progression and modulates features of tumor-associated microglia/macrophages. *Neuro-Oncology*, *14*(8), 1037–1049.

Li, M., Mukasa, A., Inda, M. M., Zhang, J., Chin, L., Cavenee, W., et al. (2011). Guanylate binding protein 1 is a novel effector of EGFR-driven invasion in glioblastoma. *The Journal of Experimental Medicine*, *208*(13), 2657–2673.

Li, A., Walling, J., Kotliarov, Y., Center, A., Steed, M. E., Ahn, S. J., et al. (2008). Genomic changes and gene expression profiles reveal that established glioma cell lines are poorly representative of primary human gliomas. *Molecular Cancer Research: MCR*, *6*(1), 21–30.

Lindberg, N., Kastemar, M., Olofsson, T., Smits, A., & Uhrbom, L. (2009). Oligodendrocyte progenitor cells can act as cell of origin for experimental glioma. *Oncogene*, *28*(23), 2266–2275.

Lipinski, C. A., Tran, N. L., Viso, C., Kloss, J., Yang, Z., Berens, M. E., et al. (2008). Extended survival of Pyk2 or FAK deficient orthotopic glioma xenografts. *Journal of Neuro-Oncology*, *90*(2), 181–189.

Liu, M., Dai, B., Kang, S. H., Ban, K., Huang, F. J., Lang, F. F., et al. (2006). FoxM1B is overexpressed in human glioblastomas and critically regulates the tumorigenicity of glioma cells. *Cancer Research*, *66*(7), 3593–3602.

Liu, K. W., Feng, H., Bachoo, R., Kazlauskas, A., Smith, E. M., Symes, K., et al. (2011). SHP-2/PTPN11 mediates gliomagenesis driven by PDGFRA and INK4A/ARF aberrations in mice and humans. *The Journal of Clinical Investigation*, *121*(3), 905–917.

Liu, C., Sage, J. C., Miller, M. R., Verhaak, R. G., Hippenmeyer, S., Vogel, H., et al. (2011). Mosaic analysis with double markers reveals tumor cell of origin in glioma. *Cell*, *146*(2), 209–221.

Liu, Q., Wang, J., Kang, S. A., Thoreen, C. C., Hur, W., Ahmed, T., et al. (2011). Discovery of 9-(6-aminopyridin-3-yl)-1-(3-(trifluoromethyl)phenyl)benzo[h][1,6]naphthyridin-2(1H)-one (Torin2) as a potent, selective, and orally available mammalian target of rapamycin (mTOR) inhibitor for treatment of cancer. *Journal of Medicinal Chemistry*, *54*(5), 1473–1480.

Loenarz, C., & Schofield, C. J. (2008). Expanding chemical biology of 2-oxoglutarate oxygenases. *Nature Chemical Biology*, *4*(3), 152–156.

Lokker, N. A., Sullivan, C. M., Hollenbach, S. J., Israel, M. A., & Giese, N. A. (2002). Platelet-derived growth factor (PDGF) autocrine signaling regulates survival and mitogenic pathways in glioblastoma cells: Evidence that the novel PDGF-C and PDGF-D ligands may play a role in the development of brain tumors. *Cancer Research*, *62*(13), 3729–3735.

Long, D. M. (1970). Capillary ultrastructure and the blood-brain barrier in human malignant brain tumors. *Journal of Neurosurgery*, *32*(2), 127–144.

Louis, D. N. (1994). The p53 gene and protein in human brain tumors. *Journal of Neuropathology and Experimental Neurology*, *53*(1), 11–21.

Louis, D. N. (2006). Molecular pathology of malignant gliomas. *Annual Review of Pathology*, *1*, 97–117.

Louis, D. N., Ohgaki, H., Wiestler, O. D., Cavenee, W. K., Burger, P. C., Jouvet, A., et al. (2007). The 2007 WHO classification of tumours of the central nervous system. *Acta Neuropathologica*, *114*(2), 97–109.

Lu, C., Ward, P. S., Kapoor, G. S., Rohle, D., Turcan, S., Abdel-Wahab, O., et al. (2012). IDH mutation impairs histone demethylation and results in a block to cell differentiation. *Nature*, *483*(7390), 474–478.

Lu, K., Zhu, S., Cvrljevic, A., Huang, T., Sarkaria, S., Ahkavan, D., et al. (2009). Fyn and SRC are effectors of oncogenic epidermal growth factor receptor signaling in glioblastoma patients. *Cancer Research*, *69*(17), 6889–6898.

Maccario, H., Perera, N. M., Davidson, L., Downes, C. P., & Leslie, N. R. (2007). PTEN is destabilized by phosphorylation on Thr366. *The Biochemical Journal*, *405*(3), 439–444.

MacConaill, L. E., Campbell, C. D., Kehoe, S. M., Bass, A. J., Hatton, C., Niu, L., et al. (2009). Profiling critical cancer gene mutations in clinical tumor samples. *PLoS One*, *4*(11), e7887.

Maddalena, A. S., Hainfellner, J. A., Hegi, M. E., Glatzel, M., & Aguzzi, A. (1999). No complementation between TP53 or RB-1 and v-src in astrocytomas of GFAP-v-src transgenic mice. *Brain Pathology (Zurich, Switzerland)*, *9*(4), 627–637.

Markert, J. M., Liechty, P. G., Wang, W., Gaston, S., Braz, E., Karrasch, M., et al. (2009). Phase ib trial of mutant herpes simplex virus G207 inoculated pre-and post-tumor resection for recurrent GBM. *Molecular Therapy: The Journal of the American Society of Gene Therapy*, *17*(1), 199–207.

Martinez-Murillo, R., & Martinez, A. (2007). Standardization of an orthotopic mouse brain tumor model following transplantation of CT-2A astrocytoma cells. *Histology and Histopathology*, *22*(12), 1309–1326.

Marumoto, T., Tashiro, A., Friedmann-Morvinski, D., Scadeng, M., Soda, Y., Gage, F., et al. (2009). Development of a novel mouse glioma model using lentiviral vectors. *Nature Medicine*, *15*(1), 110–116.

Matsukado, Y., Maccarty, C. S., & Kernohan, J. W. (1961). The growth of glioblastoma multiforme (astrocytomas, grades 3 and 4) in neurosurgical practice. *Journal of Neurosurgery*, *18*, 636–644.

Matthews, R. T., Gary, S. C., Zerillo, C., Pratta, M., Solomon, K., Arner, E. C., et al. (2000). Brain-enriched hyaluronan binding (BEHAB)/brevican cleavage in a glioma cell line is mediated by a disintegrin and metalloproteinase with thrombospondin motifs (ADAMTS) family member. *The Journal of Biological Chemistry*, *275*(30), 22695–22703.

McBride, S. M., Choi, C. H., Wang, Y., Liebelt, D., Braunstein, E., Ferreiro, D., et al. (2005). Pharmacological rescue of synaptic plasticity, courtship behavior, and mushroom body defects in a drosophila model of fragile X syndrome. *Neuron*, *45*(5), 753–764.

McGillicuddy, L. T., Fromm, J. A., Hollstein, P. E., Kubek, S., Beroukhim, R., De Raedt, T., et al. (2009). Proteasomal and genetic inactivation of the NF1 tumor suppressor in gliomagenesis. *Cancer Cell*, *16*(1), 44–54.

Mellinghoff, I. K., Cloughesy, T. F., & Mischel, P. S. (2007). PTEN-mediated resistance to epidermal growth factor receptor kinase inhibitors. *Clinical Cancer Research: An Official Journal of the American Association for Cancer Research*, *13*(2 Pt 1), 378–381.

Mellinghoff, I. K., Wang, M. Y., Vivanco, I., Haas-Kogan, D. A., Zhu, S., Dia, E. Q., et al. (2005). Molecular determinants of the response of glioblastomas to EGFR kinase inhibitors. *The New England Journal of Medicine*, *353*(19), 2012–2024.

Miao, H., Li, D. Q., Mukherjee, A., Guo, H., Petty, A., Cutter, J., et al. (2009). EphA2 mediates ligand-dependent inhibition and ligand-independent promotion of cell migration and invasion via a reciprocal regulatory loop with akt. *Cancer Cell*, *16*(1), 9–20.

Mizoguchi, M., Nutt, C. L., Mohapatra, G., & Louis, D. N. (2004). Genetic alterations of phosphoinositide 3-kinase subunit genes in human glioblastomas. *Brain Pathology (Zurich, Switzerland)*, *14*(4), 372–377.

Molina, J. R., Agarwal, N. K., Morales, F. C., Hayashi, Y., Aldape, K. D., Cote, G., et al. (2012). PTEN, NHERF1 and PHLPP form a tumor suppressor network that is disabled in glioblastoma. *Oncogene*, *31*(10), 1264–1274.

Morford, L. A., Boghaert, E. R., Brooks, W. H., & Roszman, T. L. (1997). Insulin-like growth factors (IGF) enhance three-dimensional (3D) growth of human glioblastomas. *Cancer Letters*, *115*(1), 81–90.

Mourad, P. D., Farrell, L., Stamps, L. D., Santiago, P., Fillmore, H. L., Broaddus, W. C., et al. (2003). Quantitative assessment of glioblastoma invasion in vivo. *Cancer Letters, 192*(1), 97–107.

Murai, T., Miyazaki, Y., Nishinakamura, H., Sugahara, K. N., Miyauchi, T., Sako, Y., et al. (2004). Engagement of CD44 promotes rac activation and CD44 cleavage during tumor cell migration. *The Journal of Biological Chemistry, 279*(6), 4541–4550.

Nagane, M., Levitzki, A., Gazit, A., Cavenee, W. K., & Huang, H. J. (1998). Drug resistance of human glioblastoma cells conferred by a tumor-specific mutant epidermal growth factor receptor through modulation of bcl-XL and caspase-3-like proteases. *Proceedings of the National Academy of Sciences of the United States of America, 95*(10), 5724–5729.

Nagy, A., Gertsenstein, M., Vintersten, K., & Behringer, R. (2003). *Manipulating the mouse embryo: A laboratory manual.* Cold Spring Harbor, NY: Cold Spring Harbor Laboratory Press.

Nakada, M., Nakada, S., Demuth, T., Tran, N. L., Hoelzinger, D. B., & Berens, M. E. (2007). Molecular targets of glioma invasion. *Cellular and Molecular Life Sciences: CMLS, 64*(4), 458–478.

Narayana, A., Kunnakkat, S. D., Medabalmi, P., Golfinos, J., Parker, E., Knopp, E., et al. (2012). Change in pattern of relapse after antiangiogenic therapy in high-grade glioma. *International Journal of Radiation Oncology, Biology, Physics, 82*(1), 77–82.

Narita, Y., Nagane, M., Mishima, K., Huang, H. J., Furnari, F. B., & Cavenee, W. K. (2002). Mutant epidermal growth factor receptor signaling down-regulates p27 through activation of the phosphatidylinositol 3-kinase/akt pathway in glioblastomas. *Cancer Research, 62*(22), 6764–6769.

Natsume, A., Kato, T., Kinjo, S., Enomoto, A., Toda, H., Shimato, S., et al. (2012). Girdin maintains the stemness of glioblastoma stem cells. *Oncogene, 31*(22), 2715–2724.

Nazarenko, I., Hedren, A., Sjodin, H., Orrego, A., Andrae, J., Afink, G. B., et al. (2011). Brain abnormalities and glioma-like lesions in mice overexpressing the long isoform of PDGF-A in astrocytic cells. *PloS One, 6*(4), e18303.

Neely, G. G., Kuba, K., Cammarato, A., Isobe, K., Amann, S., Zhang, L., et al. (2010). A global in vivo drosophila RNAi screen identifies NOT3 as a conserved regulator of heart function. *Cell, 141*(1), 142–153.

Neumuller, R. A., & Knoblich, J. A. (2009). Dividing cellular asymmetry: Asymmetric cell division and its implications for stem cells and cancer. *Genes & Development, 23*(23), 2675–2699.

Nishikawa, R., Sugiyama, T., Narita, Y., Furnari, F., Cavenee, W. K., & Matsutani, M. (2004). Immunohistochemical analysis of the mutant epidermal growth factor, deltaEGFR, in glioblastoma. *Brain Tumor Pathology, 21*(2), 53–56.

Norden, A. D., Young, G. S., Setayesh, K., Muzikansky, A., Klufas, R., Ross, G. L., et al. (2008). Bevacizumab for recurrent malignant gliomas: Efficacy, toxicity, and patterns of recurrence. *Neurology, 70*(10), 779–787.

Noushmehr, H., Weisenberger, D. J., Diefes, K., Phillips, H. S., Pujara, K., Berman, B. P., et al. (2010). Identification of a CpG island methylator phenotype that defines a distinct subgroup of glioma. *Cancer Cell, 17*(5), 510–522.

Nutt, C. L., Zerillo, C. A., Kelly, G. M., & Hockfield, S. (2001). Brain enriched hyaluronan binding (BEHAB)/brevican increases aggressiveness of CNS-1 gliomas in Lewis rats. *Cancer Research, 61*(19), 7056–7059.

Nyberg, P., Xie, L., & Kalluri, R. (2005). Endogenous inhibitors of angiogenesis. *Cancer Research, 65*(10), 3967–3979.

Ohgaki, H., Kita, D., Favereaux, A., Huang, H., Homma, T., Dessen, P., et al. (2006). Brain tumors in S100beta-v-erbB transgenic rats. *Journal of Neuropathology and Experimental Neurology, 65*(12), 1111–1117.

Ohgaki, H., & Kleihues, P. (2005). Population-based studies on incidence, survival rates, and genetic alterations in astrocytic and oligodendroglial gliomas. *Journal of Neuropathology and Experimental Neurology*, *64*(6), 479–489.

Onishi, M., Ichikawa, T., Kurozumi, K., & Date, I. (2011). Angiogenesis and invasion in glioma. *Brain Tumor Pathology*, *28*(1), 13–24.

Orban, P. C., Chui, D., & Marth, J. D. (1992). Tissue- and site-specific DNA recombination in transgenic mice. *Proceedings of the National Academy of Sciences of the United States of America*, *89*(15), 6861–6865.

Owens, G. C., Orr, E. A., DeMasters, B. K., Muschel, R. J., Berens, M. E., & Kruse, C. A. (1998). Overexpression of a transmembrane isoform of neural cell adhesion molecule alters the invasiveness of rat CNS-1 glioma. *Cancer Research*, *58*(9), 2020–2028.

Ozawa, T., Brennan, C. W., Wang, L., Squatrito, M., Sasayama, T., Nakada, M., et al. (2010). PDGFRA gene rearrangements are frequent genetic events in PDGFRA-amplified glioblastomas. *Genes & Development*, *24*(19), 2205–2218.

Pandita, A., Aldape, K. D., Zadeh, G., Guha, A., & James, C. D. (2004). Contrasting in vivo and in vitro fates of glioblastoma cell subpopulations with amplified EGFR. *Genes, Chromosomes & Cancer*, *39*(1), 29–36.

Panner, A., James, C. D., Berger, M. S., & Pieper, R. O. (2005). mTOR controls FLIPS translation and TRAIL sensitivity in glioblastoma multiforme cells. *Molecular and Cellular Biology*, *25*(20), 8809–8823.

Parsa, A. T., Chakrabarti, I., Hurley, P. T., Chi, J. H., Hall, J. S., Kaiser, M. G., et al. (2000). Limitations of the C6/wistar rat intracerebral glioma model: Implications for evaluating immunotherapy. *Neurosurgery*, *47*(4), 993–999, discussion 999–1000.

Parsa, A. T., Wachhorst, S., Lamborn, K. R., Prados, M. D., McDermott, M. W., Berger, M. S., et al. (2005). Prognostic significance of intracranial dissemination of glioblastoma multiforme in adults. *Journal of Neurosurgery*, *102*(4), 622–628.

Parsons, D. W., Jones, S., Zhang, X., Lin, J., Leary, R., Angenendt, P., et al. (2008). An integrated genomic analysis of human glioblastoma multiforme. *Science*, *321*(5897), 1807–1812.

Patenaude, A., Parker, J., & Karsan, A. (2010). Involvement of endothelial progenitor cells in tumor vascularization. *Microvascular Research*, *79*(3), 217–223.

Paternot, S., & Roger, P. P. (2009). Combined inhibition of MEK and mammalian target of rapamycin abolishes phosphorylation of cyclin-dependent kinase 4 in glioblastoma cell lines and prevents their proliferation. *Cancer Research*, *69*(11), 4577–4581.

Persson, A. I., Petritsch, C., Swartling, F. J., Itsara, M., Sim, F. J., Auvergne, R., et al. (2010). Non-stem cell origin for oligodendroglioma. *Cancer Cell*, *18*(6), 669–682.

Phillips, H. S., Kharbanda, S., Chen, R., Forrest, W. F., Soriano, R. H., Wu, T. D., et al. (2006). Molecular subclasses of high-grade glioma predict prognosis, delineate a pattern of disease progression, and resemble stages in neurogenesis. *Cancer Cell*, *9*(3), 157–173.

Pillay, V., Allaf, L., Wilding, A. L., Donoghue, J. F., Court, N. W., Greenall, S. A., et al. (2009). The plasticity of oncogene addiction: Implications for targeted therapies directed to receptor tyrosine kinases. *Neoplasia (New York, NY)*, *11*(5), 448–458, 2 p following 458.

Pong, W. W., & Gutmann, D. H. (2011). The ecology of brain tumors: Lessons learned from neurofibromatosis-1. *Oncogene*, *30*(10), 1135–1146.

Pope, W. B., Prins, R. M., Albert Thomas, M., Nagarajan, R., Yen, K. E., Bittinger, M. A., et al. (2012). Non-invasive detection of 2-hydroxyglutarate and other metabolites in IDH1 mutant glioma patients using magnetic resonance spectroscopy. *Journal of Neuro-Oncology*, *107*(1), 197–205.

Quant, E. C., Norden, A. D., Drappatz, J., Muzikansky, A., Doherty, L., Lafrankie, D., et al. (2009). Role of a second chemotherapy in recurrent malignant glioma patients who progress on bevacizumab. *Neuro-Oncology*, *11*(5), 550–555.

Radaelli, E., Ceruti, R., Patton, V., Russo, M., Degrassi, A., Croci, V., et al. (2009). Immunohistopathological and neuroimaging characterization of murine orthotopic xenograft models of glioblastoma multiforme recapitulating the most salient features of human disease. *Histology and Histopathology*, *24*(7), 879–891.

Raimundo, N., Baysal, B. E., & Shadel, G. S. (2011). Revisiting the TCA cycle: Signaling to tumor formation. *Trends in Molecular Medicine*, *17*(11), 641–649.

Rankin, S. L., Zhu, G., & Baker, S. J. (2012). Review: Insights gained from modelling high-grade glioma in the mouse. *Neuropathology and Applied Neurobiology*, *38*(3), 254–270.

Read, R. D. (2011). Drosophila melanogaster as a model system for human brain cancers. *Glia*, *59*(9), 1364–1376.

Read, R. D., Cavenee, W. K., Furnari, F. B., & Thomas, J. B. (2009). A drosophila model for EGFR-ras and PI3K-dependent human glioma. *PLoS Genetics*, *5*(2), e1000374.

Read, R. D., Fenton, T. R., Gomez, G. G., Wykosky, J., Vandenberg, S. R., Babic, I., et al. (2013). A kinome-wide RNAi screen in drosophila glia reveals that the RIO kinases mediate cell proliferation and survival through TORC2-akt signaling in glioblastoma. *PLoS Genetics*, *9*(2), e1003253.

Read, R. D., Goodfellow, P. J., Mardis, E. R., Novak, N., Armstrong, J. R., & Cagan, R. L. (2005). A drosophila model of multiple endocrine neoplasia type 2. *Genetics*, *171*(3), 1057–1081.

Reardon, D. A., Neyns, B., Weller, M., Tonn, J. C., Nabors, L. B., & Stupp, R. (2011). Cilengitide: An RGD pentapeptide alphanubeta3 and alphanubeta5 integrin inhibitor in development for glioblastoma and other malignancies. *Future Oncology (London, England)*, 7(3), 339–354.

Reifenberger, G., Reifenberger, J., Ichimura, K., Meltzer, P. S., & Collins, V. P. (1994). Amplification of multiple genes from chromosomal region 12q13-14 in human malignant gliomas: Preliminary mapping of the amplicons shows preferential involvement of CDK4, SAS, and MDM2. *Cancer Research*, *54*(16), 4299–4303.

Reilly, K. M., Loisel, D. A., Bronson, R. T., McLaughlin, M. E., & Jacks, T. (2000). Nf1; Trp53 mutant mice develop glioblastoma with evidence of strain-specific effects. *Nature Genetics*, *26*(1), 109–113.

Reilly, K. M., Rubin, J. B., Gilbertson, R. J., Garbow, J. R., Roussel, M. F., & Gutmann, D. H. (2008). Rethinking brain tumors: The fourth mouse models of human cancers consortium nervous system tumors workshop. *Cancer Research*, *68*(14), 5508–5511.

Reiter, L. T., & Bier, E. (2002). Using drosophila melanogaster to uncover human disease gene function and potential drug target proteins. *Expert Opinion on Therapeutic Targets*, *6*(3), 387–399.

Reznik, T. E., Sang, Y., Ma, Y., Abounader, R., Rosen, E. M., Xia, S., et al. (2008). Transcription-dependent epidermal growth factor receptor activation by hepatocyte growth factor. *Molecular Cancer Research: MCR*, *6*(1), 139–150.

Ricci-Vitiani, L., Pallini, R., Biffoni, M., Todaro, M., Invernici, G., Cenci, T., et al. (2010). Tumour vascularization via endothelial differentiation of glioblastoma stem-like cells. *Nature*, *468*(7325), 824–828.

Rieger, J., Naumann, U., Glaser, T., Ashkenazi, A., & Weller, M. (1998). APO2 ligand: A novel lethal weapon against malignant glioma? *FEBS Letters*, *427*(1), 124–128.

Riemenschneider, M. J., Mueller, W., Betensky, R. A., Mohapatra, G., & Louis, D. N. (2005). In situ analysis of integrin and growth factor receptor signaling pathways in human glioblastomas suggests overlapping relationships with focal adhesion kinase activation. *The American Journal of Pathology*, *167*(5), 1379–1387.

Robinson, J. P., VanBrocklin, M. W., Guilbeault, A. R., Signorelli, D. L., Brandner, S., & Holmen, S. L. (2010). Activated BRAF induces gliomas in mice when combined with Ink4a/arf loss or akt activation. *Oncogene*, *29*(3), 335–344.

Robinson, J. P., Vanbrocklin, M. W., Lastwika, K. J., McKinney, A. J., Brandner, S., & Holmen, S. L. (2011). Activated MEK cooperates with Ink4a/arf loss or akt activation to induce gliomas in vivo. *Oncogene, 30*(11), 1341–1350.

Rutka, J. T., Apodaca, G., Stern, R., & Rosenblum, M. (1988). The extracellular matrix of the central and peripheral nervous systems: Structure and function. *Journal of Neurosurgery, 69*(2), 155–170.

Rutka, J. T., Muller, M., Hubbard, S. L., Forsdike, J., Dirks, P. B., Jung, S., et al. (1999). Astrocytoma adhesion to extracellular matrix: Functional significance of integrin and focal adhesion kinase expression. *Journal of Neuropathology and Experimental Neurology, 58*(2), 198–209.

Saini, M., Roser, F., Samii, M., & Bellinzona, M. (2004). A model for intratumoural chemotherapy in the rat brain. *Acta Neurochirurgica, 146*(7), 731–734.

Sandsmark, D. K., Zhang, H., Hegedus, B., Pelletier, C. L., Weber, J. D., & Gutmann, D. H. (2007). Nucleophosmin mediates mammalian target of rapamycin-dependent actin cytoskeleton dynamics and proliferation in neurofibromin-deficient astrocytes. *Cancer Research, 67*(10), 4790–4799.

Sandstrom, M., Johansson, M., Bergstrom, P., Bergenheim, A. T., & Henriksson, R. (2008). Effects of the VEGFR inhibitor ZD6474 in combination with radiotherapy and temozolomide in an orthotopic glioma model. *Journal of Neuro-Oncology, 88*(1), 1–9.

Sanson, M., Marie, Y., Paris, S., Idbaih, A., Laffaire, J., Ducray, F., et al. (2009). Isocitrate dehydrogenase 1 codon 132 mutation is an important prognostic biomarker in gliomas. *Journal of Clinical Oncology: Official Journal of the American Society of Clinical Oncology, 27*(25), 4150–4154.

Sarbassov, D. D., Guertin, D. A., Ali, S. M., & Sabatini, D. M. (2005). Phosphorylation and regulation of akt/PKB by the rictor-mTOR complex. *Science (New York, NY), 307*(5712), 1098–1101.

Sarkar, D., Su, Z. Z., Vozhilla, N., Park, E. S., Gupta, P., & Fisher, P. B. (2005). Dual cancer-specific targeting strategy cures primary and distant breast carcinomas in nude mice. *Proceedings of the National Academy of Sciences of the United States of America, 102*(39), 14034–14039.

Sasaki, A., Yokoo, H., Tanaka, Y., Homma, T., Nakazato, Y., & Ohgaki, H. (2013). Characterization of microglia/macrophages in gliomas developed in S-100beta-v-erbB transgenic rats. *Neuropathology: Official Journal of the Japanese Society of Neuropathology, 33*, 505–514.

Sauer, B., & Henderson, N. (1988). Site-specific DNA recombination in mammalian cells by the cre recombinase of bacteriophage P1. *Proceedings of the National Academy of Sciences of the United States of America, 85*(14), 5166–5170.

Scheffzek, K., Ahmadian, M. R., Wiesmuller, L., Kabsch, W., Stege, P., Schmitz, F., et al. (1998). Structural analysis of the GAP-related domain from neurofibromin and its implications. *The EMBO Journal, 17*(15), 4313–4327.

Scherer, H. J. (1940). The forms of growth in gliomas and their practical significance. *Brain, 63*(1), 1–35.

Schlegel, J., Piontek, G., Kersting, M., Schuermann, M., Kappler, R., Scherthan, H., et al. (1999). The p16/Cdkn2a/Ink4a gene is frequently deleted in nitrosourea-induced rat glial tumors. *Pathobiology: Journal of Immunopathology, Molecular and Cellular Biology, 67*(4), 202–206.

Schlegel, J., Stumm, G., Brandle, K., Merdes, A., Mechtersheimer, G., Hynes, N. E., et al. (1994). Amplification and differential expression of members of the erbB-gene family in human glioblastoma. *Journal of Neuro-Oncology, 22*(3), 201–207.

Schmidek, H. H., Nielsen, S. L., Schiller, A. L., & Messer, J. (1971). Morphological studies of rat brain tumors induced by N-nitrosomethylurea. *Journal of Neurosurgery, 34*(3), 335–340.

Schmidt, E. E., Ichimura, K., Reifenberger, G., & Collins, V. P. (1994). CDKN2 (p16/MTS1) gene deletion or CDK4 amplification occurs in the majority of glioblastomas. *Cancer Research*, *54*(24), 6321–6324.

Sequist, L. V., Heist, R. S., Shaw, A. T., Fidias, P., Rosovsky, R., Temel, J. S., et al. (2011). Implementing multiplexed genotyping of non-small-cell lung cancers into routine clinical practice. *Annals of Oncology*, *22*(12), 2616–2624.

Shah, M. R., & Ramsey, W. J. (2003). CD8+ T-cell mediated anti-tumor responses cross-reacting against 9L and RT2 rat glioma cell lines. *Cellular Immunology*, *225*(2), 113–121.

Shannon, P., Sabha, N., Lau, N., Kamnasaran, D., Gutmann, D. H., & Guha, A. (2005). Pathological and molecular progression of astrocytomas in a GFAP:12 V-ha-ras mouse astrocytoma model. *The American Journal of Pathology*, *167*(3), 859–867.

Shih, A. H., Dai, C., Hu, X., Rosenblum, M. K., Koutcher, J. A., & Holland, E. C. (2004). Dose-dependent effects of platelet-derived growth factor-B on glial tumorigenesis. *Cancer Research*, *64*(14), 4783–4789.

Shih, A. H., & Holland, E. C. (2006). Platelet-derived growth factor (PDGF) and glial tumorigenesis. *Cancer Letters*, *232*(2), 139–147.

Shin, J., Padmanabhan, A., de Groh, E. D., Lee, J. S., Haidar, S., Dahlberg, S., et al. (2012). Zebrafish neurofibromatosis type 1 genes have redundant functions in tumorigenesis and embryonic development. *Disease Models & Mechanisms*, *5*, 881–894. http://dx.doi.org/10.1242/dmm.009779.

Shinojima, N., Tada, K., Shiraishi, S., Kamiryo, T., Kochi, M., Nakamura, H., et al. (2003). Prognostic value of epidermal growth factor receptor in patients with glioblastoma multiforme. *Cancer Research*, *63*(20), 6962–6970.

Shu, Q., Wong, K. K., Su, J. M., Adesina, A. M., Yu, L. T., Tsang, Y. T., et al. (2008). Direct orthotopic transplantation of fresh surgical specimen preserves CD133+ tumor cells in clinically relevant mouse models of medulloblastoma and glioma. *Stem Cells (Dayton, Ohio)*, *26*(6), 1414–1424.

Sibenaller, Z. A., Etame, A. B., Ali, M. M., Barua, M., Braun, T. A., Casavant, T. L., et al. (2005). Genetic characterization of commonly used glioma cell lines in the rat animal model system. *Neurosurgical Focus*, *19*(4), E1.

Simon, M. A. (2000). Receptor tyrosine kinases: Specific outcomes from general signals. *Cell*, *103*(1), 13–15.

Singh, D., Chan, J. M., Zoppoli, P., Niola, F., Sullivan, R., Castano, A., et al. (2012). Transforming fusions of FGFR and TACC genes in human glioblastoma. *Science (New York, NY)*, *337*(6099), 1231–1235.

Sirko, S., von Holst, A., Wizenmann, A., Gotz, M., & Faissner, A. (2007). Chondroitin sulfate glycosaminoglycans control proliferation, radial glia cell differentiation and neurogenesis in neural stem/progenitor cells. *Development (Cambridge, England)*, *134*(15), 2727–2738.

Smilowitz, H. M., Weissenberger, J., Weis, J., Brown, J. D., O'Neill, R. J., & Laissue, J. A. (2007). Orthotopic transplantation of v-src-expressing glioma cell lines into immunocompetent mice: Establishment of a new transplantable in vivo model for malignant glioma. *Journal of Neurosurgery*, *106*(4), 652–659.

Smith, J. S., Wang, X. Y., Qian, J., Hosek, S. M., Scheithauer, B. W., Jenkins, R. B., et al. (2000). Amplification of the platelet-derived growth factor receptor-A (PDGFRA) gene occurs in oligodendrogliomas with grade IV anaplastic features. *Journal of Neuropathology and Experimental Neurology*, *59*(6), 495–503.

Snuderl, M., Eichler, A. F., Ligon, K. L., Vu, Q. U., Silver, M., Betensky, R. A., et al. (2009). Polysomy for chromosomes 1 and 19 predicts earlier recurrence in anaplastic oligodendrogliomas with concurrent 1p/19q loss. *Clinical Cancer Research*, *15*(20), 6430–6437.

Sohr, S., & Engeland, K. (2008). RHAMM is differentially expressed in the cell cycle and downregulated by the tumor suppressor p53. *Cell Cycle (Georgetown, TX)*, 7(21), 3448–3460.
Song, M. S., Carracedo, A., Salmena, L., Song, S. J., Egia, A., Malumbres, M., et al. (2011). Nuclear PTEN regulates the APC-CDH1 tumor-suppressive complex in a phosphatase-independent manner. *Cell, 144*(2), 187–199.
Song, H., Li, Y., Lee, J., Schwartz, A., & Bu, G. (2009). Low-density lipoprotein receptor-related protein 1 promotes cancer cell migration and invasion by inducing the expression of matrix metalloproteinases 2 and 9. *Cancer Research, 69*(3), 879–886.
Soroceanu, L., Kharbanda, S., Chen, R., Soriano, R. H., Aldape, K., Misra, A., et al. (2007). Identification of IGF2 signaling through phosphoinositide-3-kinase regulatory subunit 3 as a growth-promoting axis in glioblastoma. *Proceedings of the National Academy of Sciences of the United States of America, 104*(9), 3466–3471.
St Johnston, D. (2002). The art and design of genetic screens: Drosophila melanogaster. *Nature Reviews. Genetics, 3*(3), 176–188.
Stettner, M. R., Wang, W., Nabors, L. B., Bharara, S., Flynn, D. C., Grammer, J. R., et al. (2005). Lyn kinase activity is the predominant cellular SRC kinase activity in glioblastoma tumor cells. *Cancer Research, 65*(13), 5535–5543.
Stockhausen, M. T., Broholm, H., Villingshoj, M., Kirchhoff, M., Gerdes, T., Kristoffersen, K., et al. (2011). Maintenance of EGFR and EGFRvIII expressions in an in vivo and in vitro model of human glioblastoma multiforme. *Experimental Cell Research, 317*(11), 1513–1526.
Stoica, G., Lungu, G., Martini-Stoica, H., Waghela, S., Levine, J., & Smith, R., 3rd. (2009). Identification of cancer stem cells in dog glioblastoma. *Veterinary Pathology, 46*(3), 391–406.
Stojiljkovic, M., Piperski, V., Dacevic, M., Rakic, L., Ruzdijic, S., & Kanazir, S. (2003). Characterization of 9L glioma model of the wistar rat. *Journal of Neuro-Oncology, 63*(1), 1–7.
Stommel, J. M., Kimmelman, A. C., Ying, H., Nabioullin, R., Ponugoti, A. H., Wiedemeyer, R., et al. (2007). Coactivation of receptor tyrosine kinases affects the response of tumor cells to targeted therapies. *Science (New York, NY), 318*(5848), 287–290.
Stuhr, L. E., Raa, A., Oyan, A. M., Kalland, K. H., Sakariassen, P. O., Petersen, K., et al. (2007). Hyperoxia retards growth and induces apoptosis, changes in vascular density and gene expression in transplanted gliomas in nude rats. *Journal of Neuro-Oncology, 85*(2), 191–202.
Szatmari, T., Lumniczky, K., Desaknai, S., Trajcevski, S., Hidvegi, E. J., Hamada, H., et al. (2006). Detailed characterization of the mouse glioma 261 tumor model for experimental glioblastoma therapy. *Cancer Science, 97*(6), 546–553.
Szerlip, N. J., Pedraza, A., Chakravarty, D., Azim, M., McGuire, J., Fang, Y., et al. (2012). Intratumoral heterogeneity of receptor tyrosine kinases EGFR and PDGFRA amplification in glioblastoma defines subpopulations with distinct growth factor response. *Proceedings of the National Academy of Sciences of the United States of America, 109*(8), 3041–3046.
Tabatabai, G., Frank, B., Mohle, R., Weller, M., & Wick, W. (2006). Irradiation and hypoxia promote homing of haematopoietic progenitor cells towards gliomas by TGF-beta-dependent HIF-1alpha-mediated induction of CXCL12. *Brain: A Journal of Neurology, 129*(Pt 9), 2426–2435.
Tchougounova, E., Kastemar, M., Brasater, D., Holland, E. C., Westermark, B., & Uhrbom, L. (2007). Loss of arf causes tumor progression of PDGFB-induced oligodendroglioma. *Oncogene, 26*(43), 6289–6296.
Teodorczyk, M., & Martin-Villalba, A. (2010). Sensing invasion: Cell surface receptors driving spreading of glioblastoma. *Journal of Cellular Physiology, 222*(1), 1–10.

Turcan, S., Rohle, D., Goenka, A., Walsh, L. A., Fang, F., Yilmaz, E., et al. (2012). IDH1 mutation is sufficient to establish the glioma hypermethylator phenotype. *Nature, 483*(7390), 479–483.

Tzeng, J. J., Barth, R. F., Orosz, C. G., & James, S. M. (1991). Phenotype and functional activity of tumor-infiltrating lymphocytes isolated from immunogenic and non-immunogenic rat brain tumors. *Cancer Research, 51*(9), 2373–2378.

Ueki, K., Ono, Y., Henson, J. W., Efird, J. T., von Deimling, A., & Louis, D. N. (1996). CDKN2/p16 or RB alterations occur in the majority of glioblastomas and are inversely correlated. *Cancer Research, 56*(1), 150–153.

Uhrbom, L., Dai, C., Celestino, J. C., Rosenblum, M. K., Fuller, G. N., & Holland, E. C. (2002). Ink4a-arf loss cooperates with KRas activation in astrocytes and neural progenitors to generate glioblastomas of various morphologies depending on activated akt. *Cancer Research, 62*(19), 5551–5558.

Uhrbom, L., Hesselager, G., Nister, M., & Westermark, B. (1998). Induction of brain tumors in mice using a recombinant platelet-derived growth factor B-chain retrovirus. *Cancer Research, 58*(23), 5275–5279.

Valerie, K., Brust, D., Farnsworth, J., Amir, C., Taher, M. M., Hershey, C., et al. (2000). Improved radiosensitization of rat glioma cells with adenovirus-expressed mutant herpes simplex virus-thymidine kinase in combination with acyclovir. *Cancer Gene Therapy*, 7(6), 879–884.

Valerie, K., Hawkins, W., Farnsworth, J., Schmidt-Ullrich, R., Lin, P. S., Amir, C., et al. (2001). Substantially improved in vivo radiosensitization of rat glioma with mutant HSV-TK and acyclovir. *Cancer Gene Therapy, 8*(1), 3–8.

van Kempen, L. C., Ruiter, D. J., van Muijen, G. N., & Coussens, L. M. (2003). The tumor microenvironment: A critical determinant of neoplastic evolution. *European Journal of Cell Biology, 82*(11), 539–548.

Verhaak, R. G., Hoadley, K. A., Purdom, E., Wang, V., Qi, Y., Wilkerson, M. D., et al. (2010). Integrated genomic analysis identifies clinically relevant subtypes of glioblastoma characterized by abnormalities in PDGFRA, IDH1, EGFR, and NF1. *Cancer Cell, 17*(1), 98–110. http://dx.doi.org/10.1016/j.ccr.2009.12.020.

Voas, M. G., & Rebay, I. (2004). Signal integration during development: Insights from the drosophila eye. *Developmental Dynamics: An Official Publication of the American Association of Anatomists, 229*(1), 162–175.

von Deimling, A., Korshunov, A., & Hartmann, C. (2011). The next generation of glioma biomarkers: MGMT methylation, BRAF fusions and IDH1 mutations. *Brain Pathology (Zurich, Switzerland), 21*(1), 74–87.

Vredenburgh, J. J., Desjardins, A., Herndon, J. E., 2nd., Dowell, J. M., Reardon, D. A., Quinn, J. A., et al. (2007). Phase II trial of bevacizumab and irinotecan in recurrent malignant glioma. *Clinical Cancer Research: An Official Journal of the American Association for Cancer Research, 13*(4), 1253–1259.

Vredenburgh, J. J., Desjardins, A., Herndon, J. E., 2nd., Marcello, J., Reardon, D. A., Quinn, J. A., et al. (2007). Bevacizumab plus irinotecan in recurrent glioblastoma multiforme. *Journal of Clinical Oncology: Official Journal of the American Society of Clinical Oncology, 25*(30), 4722–4729.

Wakimoto, H., Mohapatra, G., Kanai, R., Curry, W. T., Jr., Yip, S., Nitta, M., et al. (2012). Maintenance of primary tumor phenotype and genotype in glioblastoma stem cells. *Neuro-Oncology, 14*(2), 132–144.

Wang, D. D., & Bordey, A. (2008). The astrocyte odyssey. *Progress in Neurobiology, 86*(4), 342–367.

Wang, R., Chadalavada, K., Wilshire, J., Kowalik, U., Hovinga, K. E., Geber, A., et al. (2010). Glioblastoma stem-like cells give rise to tumour endothelium. *Nature, 468*(7325), 829–833.

Wang, S. C., & Hung, M. C. (2009). Nuclear translocation of the epidermal growth factor receptor family membrane tyrosine kinase receptors. *Clinical Cancer Research: An Official Journal of the American Association for Cancer Research, 15*(21), 6484–6489.

Wang, X., Trotman, L. C., Koppie, T., Alimonti, A., Chen, Z., Gao, Z., et al. (2007). NEDD4-1 is a proto-oncogenic ubiquitin ligase for PTEN. *Cell, 128*(1), 129–139.

Wang, H., Wang, H., Zhang, W., Huang, H. J., Liao, W. S., & Fuller, G. N. (2004). Analysis of the activation status of akt, NFkappaB, and Stat3 in human diffuse gliomas. *Laboratory Investigation; a Journal of Technical Methods and Pathology, 84*(8), 941–951.

Waziri, A. (2010). Glioblastoma-derived mechanisms of systemic immunosuppression. *Neurosurgery Clinics of North America, 21*(1), 31–42.

Wei, Q., Clarke, L., Scheidenhelm, D. K., Qian, B., Tong, A., Sabha, N., et al. (2006). High-grade glioma formation results from postnatal pten loss or mutant epidermal growth factor receptor expression in a transgenic mouse glioma model. *Cancer Research, 66*(15), 7429–7437.

Weiler, M., Pfenning, P. N., Thiepold, A. L., Blaes, J., Jestaedt, L., Gronych, J., et al. (2013). Suppression of proinvasive RGS4 by mTOR inhibition optimizes glioma treatment. *Oncogene, 32*(9), 1099–1109.

Weis, S. M., & Cheresh, D. A. (2011). Tumor angiogenesis: Molecular pathways and therapeutic targets. *Nature Medicine, 17*(11), 1359–1370.

Weiss, W. A., Burns, M. J., Hackett, C., Aldape, K., Hill, J. R., Kuriyama, H., et al. (2003). Genetic determinants of malignancy in a mouse model for oligodendroglioma. *Cancer Research, 63*(7), 1589–1595.

Weissenberger, J., Steinbach, J. P., Malin, G., Spada, S., Rulicke, T., & Aguzzi, A. (1997). Development and malignant progression of astrocytomas in GFAP-v-src transgenic mice. *Oncogene, 14*(17), 2005–2013.

Weizsacker, M., Nagamune, A., Winkelstroter, R., Vieten, H., & Wechsler, W. (1982). Radiation and drug response of the rat glioma RG2. *European Journal of Cancer & Clinical Oncology, 18*(9), 891–895.

Wells, A. (1999). EGF receptor. *The International Journal of Biochemistry & Cell Biology, 31*(6), 637–643.

Wen, P. Y., Macdonald, D. R., Reardon, D. A., Cloughesy, T. F., Sorensen, A. G., Galanis, E., et al. (2010). Updated response assessment criteria for high-grade gliomas: Response assessment in neuro-oncology working group. *Journal of Clinical Oncology: Official Journal of the American Society of Clinical Oncology, 28*(11), 1963–1972.

Wild-Bode, C., Weller, M., Rimner, A., Dichgans, J., & Wick, W. (2001). Sublethal irradiation promotes migration and invasiveness of glioma cells: Implications for radiotherapy of human glioblastoma. *Cancer Research, 61*(6), 2744–2750.

Wilson, C. W., & Chuang, P. T. (2010). Mechanism and evolution of cytosolic hedgehog signal transduction. *Development (Cambridge, England), 137*(13), 2079–2094.

Witte, H. T., Jeibmann, A., Klambt, C., & Paulus, W. (2009). Modeling glioma growth and invasion in Drosophila melanogaster. *Neoplasia (New York, NY), 11*(9), 882–888.

Wykosky, J., Gibo, D. M., Stanton, C., & Debinski, W. (2005). EphA2 as a novel molecular marker and target in glioblastoma multiforme. *Molecular Cancer Research: MCR, 3*(10), 541–551.

Xiao, A., Wu, H., Pandolfi, P. P., Louis, D. N., & Van Dyke, T. (2002). Astrocyte inactivation of the pRb pathway predisposes mice to malignant astrocytoma development that is accelerated by PTEN mutation. *Cancer Cell, 1*(2), 157–168.

Xiao, A., Yin, C., Yang, C., Di Cristofano, A., Pandolfi, P. P., & Van Dyke, T. (2005). Somatic induction of pten loss in a preclinical astrocytoma model reveals major roles in disease progression and avenues for target discovery and validation. *Cancer Research, 65*(12), 5172–5180.

Xu, W., Yang, H., Liu, Y., Yang, Y., Wang, P., Kim, S. H., et al. (2011). Oncometabolite 2-hydroxyglutarate is a competitive inhibitor of alpha-ketoglutarate-dependent dioxygenases. *Cancer Cell*, *19*(1), 17–30.

Yacoub, A., Hamed, H., Emdad, L., Dos Santos, W., Gupta, P., Broaddus, W., et al. (2008). MDA-7/IL-24 plus radiation enhance survival in animals with intracranial primary human GBM tumors. *Cancer Biology Therapy*, 7(6), 917–933.

Yacoub, A., Mitchell, C., Hong, Y., Gopalkrishnan, R. V., Su, Z. Z., Gupta, P., et al. (2004). MDA-7 regulates cell growth and radiosensitivity in vitro of primary (non-established) human glioma cells. *Cancer Biology & Therapy*, *3*(8), 739–751.

Yan, H., Parsons, D. W., Jin, G., McLendon, R., Rasheed, B. A., Yuan, W., et al. (2009). IDH1 and IDH2 mutations in gliomas. *The New England Journal of Medicine*, *360*(8), 765–773.

Yang, X. J., Chen, G. L., Yu, S. C., Xu, C., Xin, Y. H., Li, T. T., et al. (2013). TGF-beta1 enhances tumor-induced angiogenesis via JNK pathway and macrophage infiltration in an improved zebrafish embryo/xenograft glioma model. *International Immunopharmacology*, *15*(2), 191–198.

Yang, X. J., Cui, W., Gu, A., Xu, C., Yu, S. C., Li, T. T., et al. (2013). A novel zebrafish xenotransplantation model for study of glioma stem cell invasion. *PLoS One*, *8*(4), e61801.

Yao, R., Natsume, Y., Saiki, Y., Shioya, H., Takeuchi, K., Yamori, T., et al. (2012). Disruption of Tacc3 function leads to in vivo tumor regression. *Oncogene*, *31*(2), 135–148.

Yi, D., Hua, T. X., & Lin, H. Y. (2011). EGFR gene overexpression retained in an invasive xenograft model by solid orthotopic transplantation of human glioblastoma multiforme into nude mice. *Cancer Investigation*, *29*(3), 229–239.

Yip, S., Butterfield, Y. S., Morozova, O., Chittaranjan, S., Blough, M. D., An, J., et al. (2012). Concurrent CIC mutations, IDH mutations, and 1p/19q loss distinguish oligodendrogliomas from other cancers. *The Journal of Pathology*, *226*(1), 7–16.

Yokoo, H., Tanaka, Y., Nobusawa, S., Nakazato, Y., & Ohgaki, H. (2008). Immunohistochemical and ultrastructural characterization of brain tumors in S100β-v-erbB transgenic rats. *Neuropathology*, *28*(6), 591–598. http://dx.doi.org/10.1111/j.1440-1789.2008.00923.x.

Yordanova, Y. N., Moritz-Gasser, S., & Duffau, H. (2011). Awake surgery for WHO grade II gliomas within "noneloquent" areas in the left dominant hemisphere: Toward a "supratotal" resection. Clinical article. *Journal of Neurosurgery*, *115*(2), 232–239.

Yoshikawa, S., McKinnon, R. D., Kokel, M., & Thomas, J. B. (2003). Wnt-mediated axon guidance via the drosophila derailed receptor. *Nature*, *422*(6932), 583–588.

Zhao, S., Lin, Y., Xu, W., Jiang, W., Zha, Z., Wang, P., et al. (2009). Glioma-derived mutations in IDH1 dominantly inhibit IDH1 catalytic activity and induce HIF-1alpha. *Science (New York, NY)*, *324*(5924), 261–265.

Zheng, H., Ying, H., Yan, H., Kimmelman, A. C., Hiller, D. J., Chen, A. J., et al. (2008). p53 and pten control neural and glioma stem/progenitor cell renewal and differentiation. *Nature*, *455*(7216), 1129–1133.

Zhu, H., Acquaviva, J., Ramachandran, P., Boskovitz, A., Woolfenden, S., Pfannl, R., et al. (2009). Oncogenic EGFR signaling cooperates with loss of tumor suppressor gene functions in gliomagenesis. *Proceedings of the National Academy of Sciences of the United States of America*, *106*(8), 2712–2716.

Zhu, Y., Guignard, F., Zhao, D., Liu, L., Burns, D. K., Mason, R. P., et al. (2005). Early inactivation of p53 tumor suppressor gene cooperating with NF1 loss induces malignant astrocytoma. *Cancer Cell*, *8*(2), 119–130.

Zimmerman, H. M., & Arnold, H. (1941). Experimental brain tumors. I. tumors produced with methylcholanthrene. *Cancer Research*, *1*(12), 919–938, ER.

CHAPTER EIGHT

Genetically Engineered Mice as Experimental Tools to Dissect the Critical Events in Breast Cancer

Mitchell E. Menezes*, Swadesh K. Das*,†, Luni Emdad*,†,‡, Jolene J. Windle*,†,‡, Xiang-Yang Wang*,†,‡, Devanand Sarkar*,†,‡, Paul B. Fisher*,†,‡,1

*Department of Human and Molecular Genetics, Virginia Commonwealth University, School of Medicine, Richmond, Virginia, USA
†VCU Institute of Molecular Medicine, Virginia Commonwealth University, School of Medicine, Richmond, Virginia, USA
‡VCU Massey Cancer Center, Virginia Commonwealth University, School of Medicine, Richmond, Virginia, USA
[1]Corresponding author: e-mail address: pbfisher@vcu.edu

Contents

Abstract

Elucidating the mechanism of pathogenesis of breast cancer has greatly benefited from breakthrough advances in both genetically engineered mouse (GEM) models and xenograft transplantation technologies. The vast array of breast cancer mouse models

Advances in Cancer Research, Volume 121
ISSN 0065-230X
http://dx.doi.org/10.1016/B978-0-12-800249-0.00008-1

currently available is testimony to the complexity of mammary tumorigenesis and attempts by investigators to accurately portray the heterogeneity and intricacies of this disease. Distinct molecular changes that drive various aspects of tumorigenesis, such as alterations in tumor cell proliferation and apoptosis, invasion and metastasis, angiogenesis, and drug resistance have been evaluated using the currently available GEM breast cancer models. GEM breast cancer models are also being exploited to evaluate and validate the efficacy of novel therapeutics, vaccines, and imaging modalities for potential use in the clinic. This review provides a synopsis of the various GEM models that are expanding our knowledge of the nuances of breast cancer development and progression and can be instrumental in the development of novel prevention and therapeutic approaches for this disease.

1. INTRODUCTION

Breast cancer is the most frequently diagnosed cancer among American women, excluding skin cancers, and is the second leading cause of cancer-related deaths among women. It is estimated that one in eight women in the United States will develop invasive breast cancer in their lifetime (http://www.cancer.org/cancer/breastcancer/detailedguide/breast-cancer-key-statistics). Understanding the molecular basis of the initiation and progression of breast cancer is essential to developing a cure for this disease as well as for producing vaccines or other approaches to prevent the disease. Although significant progress has been made with respect to understanding breast cancer progression and spread, treatment options are still limited once metastasis has occurred.

Animal models have been used for decades to help determine causes, and provide platforms for testing therapies and potential cures for human diseases. The first genetically engineered mouse (GEM) model, a transgenic mouse engineered to express HSV (herpes simplex virus) thymidine kinase, was developed in 1981 (Brinster et al., 1981), and multiple GEM models of breast cancer were generated shortly thereafter (e.g., Muller, Sinn, Pattengale, Wallace, & Leder, 1988; Sinn et al., 1987; Stewart, Pattengale, & Leder, 1984; Tsukamoto, Grosschedl, Guzman, Parslow, & Varmus, 1988). As our understanding of the various genetic alterations involved in cancer has progressed, GEM models, including both transgenic and knockout models, have become essential tools for defining the role of these genetic alterations in cancer initiation, progression, and metastasis. These models allow not only for determining the consequences of alterations in individual genes of potential relevance but also for determining the mechanisms of cooperation between multiple genes in tumorigenesis. Interestingly, many of the genes that are altered in human breast cancer are also found to promote the development of mammary tumors in GEMs (Cardiff, 2003), allowing for the creation of genetically accurate models of this disease. Such models are useful both for understanding the pathogenesis of breast cancer, as well as for evaluating novel therapeutic approaches that may target specific genetic alterations or pathways and strategies for the prevention of breast cancer.

Breast cancer is not a single disease, but is a heterogeneous family of diseases with diverse genetic and histopathologic characteristics and varying clinical outcomes. It would therefore be naïve to expect a single animal model to fully recapitulate all aspects of breast cancer. In fact, a wide variety of approaches have been utilized to model breast cancer in the mouse, including the use of spontaneous and carcinogen- or virally-induced models, xenograft models, and most recently GEMs (Cardiff & Kinney, 2007). The use of chemical carcinogens to induce mammary tumors in rodents has proven useful in the genetic analysis of initiation and progression of mammary tumors (Sukumar, McKenzie, & Chen, 1995) and for prevention studies, and xenograft models have been widely used for studying various aspects of disease progression and for evaluating therapeutic approaches. However, these approaches present drawbacks that can be addressed, at least in part, by the use of GEM models (Borowsky, 2011; Vargo-Gogola & Rosen, 2007). The development of GEMs provided a new avenue for understanding the events involved in cancer initiation and progression, and for creating genetically accurate models of human breast cancer. Vast arrays of GEM breast cancer models are currently available that represent different facets of this heterogeneous disease. Constitutive tissue-specific expression of oncogenes and global or tissue-specific knockout of tumor suppressor genes has produced tumor-prone mice that are useful for delineating the underlying molecular mechanisms of breast cancer, as well as for aiding in the design and investigation of treatment options for this disease. GEM models are also useful in the development of novel imaging modalities to detect tumors, particularly, during early stages, as well as metastases. Given the number of GEM models available today, it is important to select a model appropriate for the question being addressed or the breast cancer subtype one is endeavoring to understand. An informative set of reviews on the mammary gland as an experimental model (Bissell, Polyak, & Rosen, 2011) and a thorough history of the many experimental approaches to studying mammary tumor biology in mice (Cardiff & Kinney, 2007) have been published. Additional reviews have described and categorized many of the current transgenic models of breast cancer (e.g., Borowsky, 2011; Cardiff et al., 2000; Ottewell, Coleman, & Holen, 2006; Vargo-Gogola & Rosen, 2007). In this review, we will examine some of the prominent transgenic models that are proving invaluable to our understanding of the molecular processes of breast cancer, how these models have advanced the evaluation of experimental therapies, as well as current approaches in the application of these models in the development of small molecule inhibitors and other approaches for the treatment and prevention of breast cancer.

1.1. Development of the mouse mammary gland and human breast

The mouse mammary gland undergoes three distinct phases of proliferation: embryonic, at puberty and during pregnancy-lactation. Mouse mammary gland development begins shortly after mid-gestation. The milk line or the bilateral epidermal thickening, that runs from the neck to the inguinal regions, forms the base on which five placodes develop. These placodes later invaginate to form the nipples. As the mouse develops, limited branching occurs in these early structures to form the rudimental ductal tree (Figure 8.1) (Hens & Wysolmerski, 2005; Robinson, 2007). The mammary glands become more extensively branched at puberty when hormone-dependent events activate the formation of terminal end buds and more primary ducts in a radial pattern. Alveolar growth and differentiation occurs during pregnancy and lactation, and glandular regression follows lactation (Watson & Khaled, 2008). In contrast, humans have two breasts that develop from the milk buds on the anterior chest wall. Details of the human breast development from fetal to adult life have recently been reviewed (Gusterson & Stein, 2012; Russo & Russo, 2004). While the complexity of branching and molecular details of development is different in humans and mice, both species have mammary glands that grow mainly from the terminal end buds. These buds contain undifferentiated progenitor cells or stem cells and eventually give rise to milk-producing lobuloalveolar (LA) structures in mice, or the terminal ductal lobuloalveolar units (TDLU) in humans (Cardiff & Wellings, 1999). In most cases, breast cancer originates from these lobules or ducts of the mammary glands (Dimri, Band, & Band, 2005).

The mouse mammary gland offers several distinct advantages for modeling human breast cancer, despite many of the above-mentioned differences in their development. The mouse mammary gland has been extensively studied in the context of normal development, lactation changes, and in

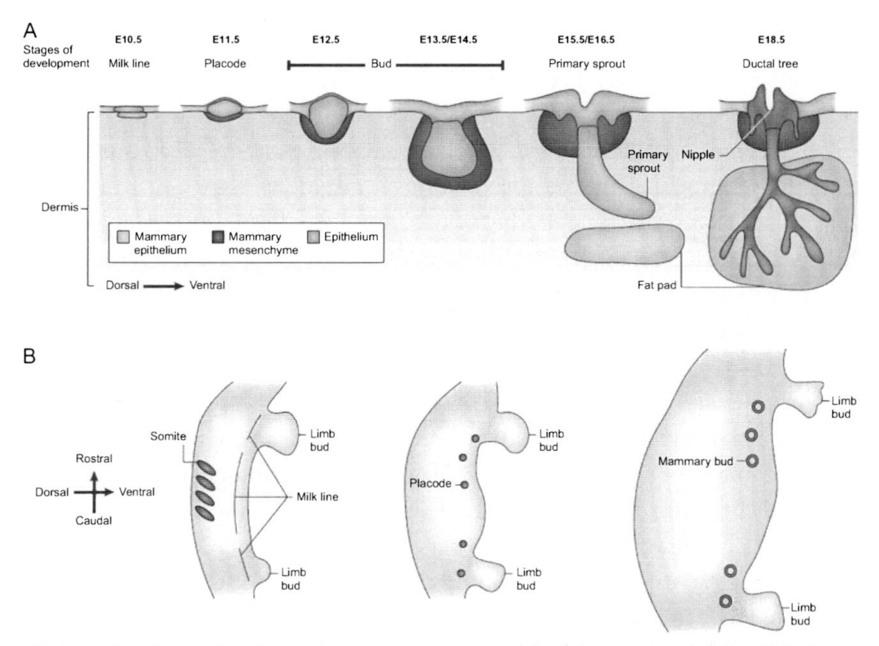

Figure 8.1 ***Schematic of embryonic mammary gland development in mice.*** (A) Cross-sectional view through the trunk of the mouse during mammary gland development. The milk line (orange) is formed by a slight thickening and stratification of the ectoderm (gray) at around embryonic day 10 (E10). The milk line breaks up into individual placodes (orange) and condensation of the underlying mammary mesenchyme (blue) begins on E11.5. Placodes sink deeper into the dermis and the mammary mesenchyme is organized in concentric layers around the mammary bud (orange) over the next few days. Proliferation of the mammary epithelium (orange) begins at the tip and the primary sprout pushes through the mammary mesenchyme toward the fat pad (green) starting on day E15.5. The elongating duct grows into the fat pad and branches into a small ductal system on E18.5. Cells of the mammary mesenchyme form the nipple consisting of specialized epidermal cells (purple). (B) Schematic of the position of the milk line, placodes, and mammary buds along the lateral body wall of early mouse embryos. *Reprinted with permission from MacMillan Publishers Ltd:* Nature Reviews Genetics *(Robinson, 2007), copyright 2007.* (See the color plate.)

development and characteristics of spontaneous tumors. The mouse mammary gland is also transplantable into syngeneic or immunodeficient athymic mice, providing an extra platform to separately study individual tumors formed from the multifocal neoplasia in mice (Caligiuri, Rizzolio, Boffo, Giordano, & Toffoli, 2012). Additionally, there are several mammary-specific gene promoters available that assist in engineering mammary-specific target gene expression.

1.2. Genetically engineering mouse models of breast cancer

Several approaches have been used to genetically engineer mouse models of breast cancer (Borowsky, 2003). The first approach is to create transgenic mice in which a gene of interest (e.g., an oncogene or reporter gene) is introduced into the germline of the mouse genome *via* random insertion, by directly injecting DNA into one of the two pronuclei of a fertilized egg (Cho, Haruyama, & Kulkarni, 2009; Voncken, 2011). This approach has been used to create transgenic mice expressing a wide range of oncogenes under the control of various mammary-specific or -selective promoters (discussed in Section 1.3), resulting in lines of mice that develop mammary tumors with varying latencies and properties. The second approach is to create knockout mice in which a gene of interest has been inactivated, an approach that is, particularly, suited for investigating the role of tumor suppressor genes in breast and other cancers. Knockout mice are generated by introducing a targeting vector encoding a modified, nonfunctional version of the gene of interest into mouse embryonic stem (ES) cells and selecting clones in which the targeting vector has integrated into the genome by homologous recombination, thereby replacing one of the wild-type alleles. The resulting ES cells are then used to generate chimeric mice, and eventually mice carrying the targeted allele in the germline in either a heterozygous or homozygous state (Hughes & Saunders, 2011). Global inactivation of tumor suppressor genes can in some cases model specific human hereditary tumor syndromes, for example, Li-Fraumeni syndrome, in which patients have inherited a mutant p53 allele and develop a range of tumor types, including breast cancer (Donehower et al., 1992; Malkin, 2011). However, most tumor-associated mutations arise somatically and are restricted to only the tumor cells in the patient. Further, global inactivation of some genes in mice results in embryonic lethality, thereby precluding analysis of the effects on tumorigenesis. Thus, in some cases it is preferable to create conditional knockout mice in which the gene of interest is modified in the germline, but becomes inactivated only in somatic cells in a fashion that can be both tissue specific and temporally regulated. The most well-established approach for generating tissue-specific conditional knockouts employs the Cre/loxP recombinase system (Lewandoski, 2001). LoxP sites are 34 bp elements from the P1 bacteriophage that are specifically recognized by the phage Cre recombinase enzyme. Using the gene targeting approach described above, "knock-in" mice can be generated in which two loxP sites have been inserted into intronic (or promoter) regions flanking critical exons in the gene of interest, creating a "floxed" allele, which remains functionally wild-type despite the presence of the two loxP sites. However, when these mice are crossed with transgenic mice expressing Cre recombinase under the transcriptional control of a tissue-specific promoter, Cre-mediated recombination results in deletion of the sequences between the two loxP sites, creating a knockout allele, but only in the tissues in which Cre is expressed. Tissue-specific inducible conditional knockout mice can be obtained by crossing mice carrying a floxed gene of interest with mice expressing Cre-ER, a fusion protein in which Cre is linked to a mutant estrogen receptor ligand binding domain. In these models, spatial control again comes from the promoter driving Cre-ER expression, while temporal control is mediated by administration of Tamoxifen, since the fusion protein can only translocate to the nucleus when bound to Tamoxifen (Friedel, Wurst, Wefers, & Kuhn, 2011). Similarly the tetracycline-inducible (Tet-on/Tet-off) systems can be utilized for inducible expression or silencing of a gene of interest (Fantozzi & Christofori, 2006; Saunders, 2011). More detailed information of various strategies for generating GEM models can be found elsewhere (Hofker & van Deursen, 2011; Pease & Saunders, 2011).

1.3. Mammary-specific promoters

Significant progress in the development of transgenic mouse models to investigate mammary tumorigenesis has been enabled by the identification of mammary-specific or -selective promoter elements (Table 8.1). The most widely used regulatory element for inducing mammary-selective transgene expression is the mouse mammary tumor virus (MMTV) long terminal

Table 8.1 Mammary gland-specific promoters

Promoter	Origin	Expression	Activation
MMTV-LTR	Mouse mammary tumor virus	Breast epithelial cells, several other tissues	Steroid hormones
WAP	Whey acidic protein	Secretory mammary epithelium	Lactogenic hormones
C3(1)	Rat prostate steroid-binding protein (PSBP)	Epithelial cells of prostate and mammary gland	Estrogen (ductal and alveolar mammary epithelium)
BLG	Bovine β-lactoglobulin	Mammary gland	Pregnancy and lactation
MT	Metallothionein	Most mammary cells	Zn^{2+}

The various mammary gland specific promoters used to drive mammary gland-specific expression of transgenes are shown in this table.

repeat (LTR). Several strains of *Mus musculus* (BR6, C3H, GR, RIII) develop spontaneous mammary tumors due to a high incidence of MMTV in the mice. MMTV is a nonacute transforming retrovirus that induces tumors by insertional mutagenesis in the DNA of the host genome, causing deregulated expression of adjacent cellular genes. This highly infectious virus is transmitted congenitally through the milk to the offspring of the mice of these strains (Asch, 1996). The MMTV LTR contains response elements for glucocorticoids, androgens, and progesterone, which are regulated by the estrous cycle in virgin mammary glands. The expression of genes adjacent to the LTR is highly upregulated during pregnancy, and peak during lactation. Transcription from the MMTV LTR also results in lower level of transgene expression in a wide range of other tissues, including the salivary glands and the male accessory glands (Wagner et al., 2001).

The promoter of the gene encoding the milk serum protein, whey acidic protein (WAP), is also commonly used for directing expression of transgenes to the mammary gland. The WAP promoter is hormonally regulated by prolactin, hydrocortisone, estrogen and insulin, and strongly activates mammary-specific expression of genes during mid-pregnancy and lactation, although it also directs low levels of expression in other tissues as well (Wen, Kawamata, Tojo, Tanaka, & Tachi, 1995).

The ovine BLG (beta-lactoglobulin) promoter is another mammary-specific promoter that has been used to generate transgenic mice (Webster, Wallace, Clark, & Whitelaw, 1995). BLG is hormonally regulated by prolactin, estrogen, hydrocortisone and insulin, and transgenes under the transcriptional regulation of the BLG promoter are induced during pregnancy and lactation earlier than transgenes expressed from the WAP promoter. However, transcription from the BLG promoter is not very tightly regulated, as a low level transgene expression is observed in virgin animals.

Additional promoters that have been used to direct oncogene expression in mammary tumor models include the 5′-flanking region of the C3(1) subunit of the rat prostatic steroid-binding protein, and the mouse metallothionein 1 (MT) promoter. The C3(1) promoter has been shown to target transgene expression to the epithelium of both the mammary and prostate glands (Green et al., 2000). Unlike the promoters described above, expression from this promoter is not influenced by pregnancy, and thus some of the artificial effects of pregnancy on MMTV- and WAP-driven transgenic models may be avoided with this promoter (Green et al., 2000; Yoshidome, Shibata, Couldrey, Korach, & Green, 2000). The mouse MT promoter directs expression to most epithelial cells, including those in the mammary gland, and is activated by heavy metals such as zinc and cadmium (Palmiter, Sandgren, Koeller, & Brinster, 1993).

Transgenic mouse tumor models in which oncogene expression is driven by the MMTV, WAP, or other pregnancy/lactation-dependent promoters are sometimes referred to as "artificial" models, and the extent to which they mimic the tumor biology of human breast cancer has been the center of considerable scientific debate (Borowsky, 2011; Vargo-Gogola & Rosen, 2007). In humans, pregnancy is often associated with a reduced risk of breast cancer (Britt, Ashworth, & Smalley, 2007; Russo, Moral, Balogh, Mailo, & Russo, 2005). In contrast, pregnancy generally promotes tumorigenesis in transgenic mouse models in which expression of an oncogene is under the control of the MMTV LTR or WAP promoter, due to direct activation of the promoters by pregnancy-associated hormones, resulting in increased expression of the oncogene. These models are thus not generally well suited for studies of the mechanisms mediating hormonal effects on breast tumor initiation or progression. Additional caveats with transgenic mouse models driven by any of the promoters mentioned above include potential differences in the specific subpopulations of mammary epithelial cells targeted by these promoters, as compared to the tumor-initiating cell types in human breast cancer, and artificially high levels of oncogene expression that may be driven by these promoters. "Knock-in" models, in which an oncogenic version of a gene replaces the wild-type protooncogene in its normal genomic locus and under the transcriptional regulation of its natural promoter, can potentially alleviate some of the concerns of MMTV- or WAP-driven breast tumor models (Andrechek et al., 2000). However, despite the potential artifacts and caveats of the conventional transgenic models, they have contributed substantially to a better understanding of the genetics and biology of breast cancer and provide powerful tools for the evaluation of targeted therapeutic approaches for the treatment of breast cancer.

2. MODELING VARIOUS ASPECTS OF HUMAN BREAST CANCER INITIATION AND PROGRESSION IN MICE

Breast cancer is a multifactorial disease (Figure 8.2). The hallmarks of cancer as defined by Hanahan and Weinberg describe the various biological traits acquired during the multistep development of tumors (Hanahan & Weinberg, 2000, 2011). Cancer cells gain attributes that allow them to sustain proliferative signaling, evade growth suppressors, activate invasion and metastasis, gain replicative immortality, induce angiogenesis, resist cell

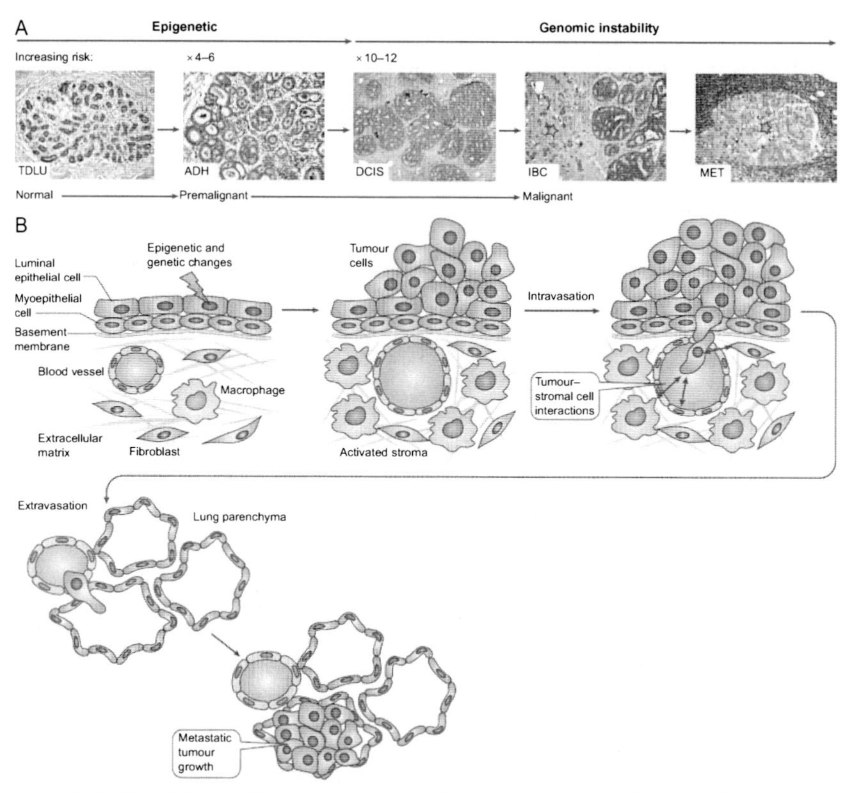

Figure 8.2 ***The biology of breast cancer.*** (a) Breast cancer is a multifaceted disease that is genetically and genomically heterogenous. The normal breast terminal ductal lobular unit (TDLU) is comprised of lobules and ducts that consist of bilayered epithelium of luminal and myoepithelial cells. Characterized by abnormal cells within the duct or lobule, atypical ductal hyperplasia (ADH) is a premalignant lesion. ADH is thought to be followed by ductal carcinoma *in situ* (DCIS), a noninvasive lesion with abnormal cells. Some of the noninvasive DCIS lesions might eventually give rise to invasive breast cancer (IBC). Finally cells that have gained invasive capabilities might metastasis to a distant area in the body. In breast cancer, the primary site of metastasis (MET) is the lymph nodes. (B) Schematic of the progression of breast cancer. Various factors play a role in the transformation of breast epithelial cells to eventually give rise to metastatic breast cancer. This multistep process involves loss of control of proliferation, gain in survival abilities, increased migration, and aberrant tumor–stromal cell interactions that facilitate transformation. Cells must invade through the basement membrane, intravasate into the blood circulation, survive, extravasate from the blood circulation, and establish a new tumor at a distant site in the body, for metastasis to develop. *Reprinted with permission from Macmillan Publishers Ltd:* Nature Reviews Cancer *(Vargo-Gogola & Rosen, 2007), copyright 2007.* (See the color plate.)

death, avoid immune destruction, and deregulate cellular energetics. Insight into the mechanisms that allow cancer cells to acquire such abilities can be teased out using GEM models.

Several genes have previously been identified as tumor suppressors or oncogenes in breast cancer. The cancer genome atlas network recently analyzed comprehensive molecular changes occurring in human breast tumors (Koboldt et al., 2012). Mutations in a number of genes such as *PIK3CA, PTEN, AKT1, TP53, GATA3, CDH1, RB1, MLL3, MAPK1, CDKN1B, RUNX1*, and *NF1* have been implicated to play a role in breast cancer. A number of germline variants such as *ATM, BRCA1, BRCA2*, and *PTEN* also predispose humans to breast cancer. Also contributing to the development of many breast cancers is the enhanced expression of protooncogenes such as human epidermal growth factor receptor (EGFR) family members (HER2 or NEU). Such altered gene expression results in an abnormal cellular signaling and protein signature of the tumor. In theory, any gene that might play a role in breast cancer can be evaluated using mammary-specific transgenic or knockout animals. As such, many of the current mouse models aim to target key molecular events that are critical mediators of the overall cancer phenotypes such as cell growth, cell cycle progression, survival, apoptosis, migration, invasion, metastasis, angiogenesis, and drug resistance. GEM models that target a single transgene such as an oncogene, tumor suppressor, growth factor and/or receptor, or combinations of the above, have been developed to gain a better understanding of the role of the transgene/s in breast tumorigenesis. Although the number of breast cancer GEM models continues to expand, no single model is currently able to completely replicate the complexity of the genetic and molecular deregulation of human breast cancer.

2.1. Modeling effects of oncogenes on cell proliferation, survival, and apoptosis in breast cancer

Cancer cells overexpress/downregulate molecules that regulate cell proliferation by functioning at multiple levels, including transcriptional, translational, and/or posttranslational, to activate the cascade of cyclins and cyclin-dependent kinases and entry into the cell cycle. The tumorigenic potential of cancer cells is frequently enhanced through the synchronized expression of molecules that promote survival and permit evasion from apoptotic death. The following transgenic models will be discussed in this section: the protooncogenes c-myc and v-Ha-Ras, the cell cycle regulator cyclin D1, the growth factor TGFα, and the viral SV40 T antigen.

2.1.1 c-Myc transgenic models

The protooncogene *c-myc* encodes a 62-kDa multifunctional, nuclear phosphoprotein, and is located on chromosome 8q24.21 in humans. c-Myc functions as a transcription factor that activates expression of a number of genes required for cell cycle progression from G1 to S phase. Paradoxically, c-Myc expression can also strongly promote apoptotic cell death. c-Myc expression increases during pregnancy-related proliferation in the normal mammary gland. However, it is absent in differentiated alveolar cells during lactation, but increases again during the normal apoptotic mammary involution process (Liao & Dickson, 2000). In human tumors, deregulation of c-myc expression may occur *via* multiple mechanisms. The *c-myc* gene is amplified in about 16% and rearranged in about 5% of human breast cancers, while the c-Myc protein is overexpressed in 45–70% of human breast cancers (Blackshear, 2001; Deming, Nass, Dickson, & Trock, 2000; Spandidos, Field, Agnantis, Evan, & Moore, 1989). Amplification of *c-myc* correlates with poor patient prognosis (Deming et al., 2000).

Activation of *c-myc* has been implicated as a critical step in the development of mammary carcinoma in both humans and mice. The Leder laboratory developed MMTV-c-myc transgenic mice that express c-myc specifically in the mammary glands at high levels in female mice, and in the salivary glands of both males and females at lower levels (Stewart et al., 1984). These mice develop mammary adenocarcinomas in multiparous animals with a 100% incidence. Given the hormone responsiveness of the MMTV promoter, it is not surprising that tumor incidence and latency depends on the number of pregnancies. In multiparous females, adenocarcinomas normally develop at 5- to 6-months of age and correspond with the second or third pregnancy. Metastatic tumors are rare, but have been observed in the lung. No tumors are observed in male mice (Stewart et al., 1984).

Transgenic mice expressing c-myc under the control of the murine WAP promoter have also been generated by two-independent groups (Sandgren et al., 1995; Schoenenberger et al., 1988). The frequency of tumors in WAP-c-myc mice was much higher than that of their MMTV-driven counterpart. WAP-c-myc female mice begin to develop mammary tumors within 2-months of initiation of transgene expression (pregnancy). Approximately 80% of females develop palpable tumors after undergoing two rounds of pregnancy, and individual mice often develop multiple tumors affecting a single or multiple glands (Schoenenberger et al., 1988). Virgin mice do not develop tumors (monitored up to

14 months). WAP-c-myc mice form mammary tumors that are well-differentiated and acquire constitutive expression and secretion of β-casein and transgenic myc.

While these studies validate the potential of c-myc to drive mammary tumorigenesis, further observations revealed the requirement for additional genetic events in MMTV-c-myc and WAP-c-myc derived mammary carcinoma development. For example, tumors arose focally, even though c-Myc expression was observed throughout the normal mammary epithelium of both MMTV-c-myc and WAP-c-myc female mice during the latter part of pregnancy and throughout lactation. In addition, although random chromosomal integration of c-myc in MMTV-c-myc mice caused the expression of the transgene in multiple tissues, c-myc only elicited neoplasms in mammary tissue. Additionally, the long latency of tumors and dependence on pregnancy suggests that c-myc itself might not be sufficient for the induction of mammary tumors.

The observation that c-myc on its own was insufficient to drive tumorigenesis in the entire mammary glands instructed follow-up studies using double transgenic (or bitransgenic) mouse models. Double transgenic mice carrying both the c-myc and another gene, such as v-Ha-ras (Sinn et al., 1987), TGF-α (transforming growth factor α) (Amundadottir, Johnson, Merlino, Smith, & Dickson, 1995), bcl-2, or neu, develop mammary tumors at higher frequencies and with shorter latency. MMTV-c-myc/MMTV-v-Ha-ras and MMTV-c-myc/MT-TGFα double transgenic mice helped elucidate that c-myc could synergize with v-Ha-ras and TGFα to decrease tumor latency and eliminate the requirement for pregnancy for tumor initiation. In particular, the double transgenic MMTV-c-myc/MT-TGFα mice do not develop normal mammary tissue, and mammary tissue from 3-week-old mice can form tumors when transplanted into athymic mice. This indicates that c-myc and TGFα can synergize in transforming mammary epithelium, requiring few or no additional genetic alterations (Amundadottir et al., 1995; Amundadottir, Nass, Berchem, Johnson, & Dickson, 1996).

The Chodosh laboratory developed a transgenic mouse model that allowed for temporal control of mammary-specific expression of c-myc upon treatment with tetracycline derivates (Boxer, Jang, Sintasath, & Chodosh, 2004). In this model, the MMTV promoter drives expression of reverse tetracycline-dependent transcriptional activator, rtTA, and in the presence of doxycycline, induces expression of c-myc fused to a tetracycline-dependent promoter (MMTV-rtTA/TetO-MYC transgenic mice). Chronic induction of c-myc in these transgenic mice results in solitary mammary adenocarcinomas after about 22-weeks. This model is of importance as it allows for the assessment of tumor dependence upon continued c-myc expression once the tumor has formed, or "oncogene dependence." Once mammary tumors were observed, the mice were taken off doxycycline to assess whether the tumors would regress. Interestingly, only about half of the tumors regressed after inhibition of c-myc expression by removal of doxycycline. Thus, c-myc inactivation did not lead to a reversal of tumors and the tumors were able to grow in the absence of c-myc overexpression. Among the tumors that failed to regress following c-myc inactivation, nearly all the tumors resumed growth even in the absence of c-myc expression. This suggests that the majority of primary mammary tumors that arose from c-myc activation already acquired the ability to grow in the absence of c-myc overexpression. Among the tumors that had regressed upon absence of c-myc, a majority of the tumors spontaneously recurred at the initial site within 1 year following doxycycline withdrawal. Mammary tumors were found to recur due to the presence of residual neoplastic cells that persisted in the mammary glands of nearly all animals with fully regressed tumors. Reactivation of c-myc in the residual neoplastic cells resulted in a rapid regrowth of tumors. Additionally, successive cycles of c-myc expression resulted in progression of nearly all tumors to a doxycycline-independent state.

2.1.2 v-Ha-ras transgenic models

The protooncogene *Ha-ras* (Harvey rat sarcoma viral oncogene homolog) is a small, guanine nucleotide binding protein located on chromosome 11p15.5 in humans. *Ha-ras*, one of three *ras* oncogenes (the others being *K-ras* and *N-ras*), encodes a 21-kDa protein. These evolutionarily conserved GTPases play pivotal roles in many cellular processes. In response to growth factor stimulation, GTP-bound or activated Ras interacts with multiple effector molecules, particularly, the well-studied Ras/MEK/ERK or MAPK (mitogen-activated protein kinase) signaling cascade that ultimately promotes cell proliferation. In many human tumors, one or another of the *ras* genes is found to have acquired a point mutation that renders it constitutively activated, such that downstream signaling is activated even in the absence of growth factors. Although ras mutations have been observed in breast cancer patients (Bos, 1989), they do not appear to play a pivotal role in the etiology of breast cancer (Spandidos, 1987). Elevated levels of Ras, both at the transcript and protein level, have also been documented in human breast cancer patients (DeBortoli, Abou-Issa, Haley, & Cho-Chung, 1985; Hand et al., 1984; Slamon, deKernion, Verma, & Cline, 1984), although amplification and rearrangement of *ras* genes were not observed in breast cancer patients (Schondorf et al., 1999). Interestingly, in studies using the chemical carcinogen *N*-nitroso-*N*-methylurea to induce mammary tumors in rodents, activation of *Ha-Ras* was identified as the earliest genetic event observed in the mammary gland (Sukumar et al., 1995).

Transgenic mice expressing v-Ha-ras (a virally derived *Ha-ras* gene carrying an activating point mutation) under the control of the MMTV promoter were developed by the Leder laboratory (Sinn et al., 1987) and the Jolicoeur laboratory (Tremblay et al., 1989). Mice from multiple-independent lines from both groups were found to express the transgene in the mammary and salivary glands, the Harderian gland (an intraorbital lacrimal gland present in rodents), and at lower levels, the spleen, thymus, and lung. Expression in seminal vesicles of male mice was also reported by Sinn et al. Further, mice from both groups developed mammary and salivary adenocarcinomas and Harderian gland hyperplasias, as well as occasional lung tumors, and lymphomas (Sinn et al., 1987) or splenomegaly (Tremblay et al., 1989). Although Sinn et al. reported detecting mammary adenocarcinomas as early as 32 days of age in two females, the mean age for detection of mammary adenocarcinomas in females was 168 days of age in their best-characterized model. Similarly, Tremblay et al. reported mammary adenocarcinomas arising between 4 and 10 months of age in females from multiple lines. In both cases, mammary tumors were less common in males than females. Histologically, the mammary tumors in these mice were locally invasive with some metastasis to the liver and/or lung. Transgenic mice expressing v-Ha-ras under the control of the WAP promoter were developed by the Gerlinger laboratory (Andres et al., 1987). In this model, less than 1% of transgenic mice developed mammary tumors (and/or salivary gland tumors) with a tumor latency of about 1-year.

Double transgenic mice expressing MMTV-c-myc and MMTV-v-Ha-ras were also developed by the Leder laboratory (Sinn et al., 1987). The two oncogenes interacted synergistically, leading to markedly accelerated tumor kinetics in both female and male mice. However, uniform malignant transformation of the mammary gland did not occur even upon expression of both oncogenes, as evidenced by the development of focal, though multiple, stochastic mammary tumors. Similarly, double transgenic mice expressing WAP-c-myc and WAP-v-Ha-ras showed a synergistic effect on mammary tumorigenesis, although the tumors that arose were still focal in nature (Andres et al., 1988). Thus, it could be concluded from these studies that even two "hits" are insufficient to induce mammary tumorigenesis.

The tumor growth characteristics of MMTV-c-myc (Stewart et al., 1984), MMTV-v-Ha-ras (Sinn et al., 1987), and the double transgenic tumors were evaluated to better understand the potential cooperative role of the myc and ras oncogenes in mammary tumorigenesis (Hundley et al., 1997). MMTV-c-myc tumors showed the highest level of spontaneous apoptosis, followed by MMTV-c-myc/MMTV-v-Ha-ras, while MMTV-v-Ha-ras tumors had very low levels of spontaneous apoptosis. The cell cycle characteristics of tumors obtained from the three transgenic mice were also significantly different, with MMTV-c-myc tumors showing the highest rates of proliferation (higher S-phase fractions), MMTV-v-Ha-ras tumors having very low rates of proliferation, and MMTV-c-myc/MMTV-v-Ha-ras having intermediate levels. Despite these differences in rates of apoptosis and proliferation, the tumor growth rates in these three transgenic mouse models were similar, indicating the importance of the balance between cell cycle regulation and cell death in determining kinetics of tumor growth and the fact that different oncogenes had a profound influence on that balance (Hundley, Koester, Troyer, Hilsenbeck, Barrington, et al., 1997).

The influence of another cell cycle regulator, a cell cycle-stimulating phosphatase CDC25A, on tumorigenesis in MMTV-v-Ha-ras, MMTV-neu, and MMTV-c-myc mice was also evaluated (Ray et al., 2007). CDC25A is known to be significantly overexpressed in human cancers including breast tumors. CDC25A might contribute to some of the tumorigenic effects mediated by Ha-ras and neu oncogenes. Cdc25A$^{+/-}$ mice develop normally. Mating Cdc25A$^{+/-}$ mice with MMTV-Ha-ras or MMTV-neu transgenic mice caused development of mammary and salivary tumors after about 18-weeks, however Cdc25A$^{+/+}$/MMTV-Ha-ras mice develop tumors after about 60-weeks. In fact, about 10% of Cdc25A$^{+/-}$/MMTV-Ha-ras mice did not develop tumors even at 2-years of age. Similar results were obtained with Cdc25A$^{+/-}$/MMTV-neu mice. Cdc25A$^{+/-}$/MMTV-neu mice developed tumors at about 32-weeks of age compared to Cdc25A$^{+/+}$/MMTV-neu mice, which formed tumors at about 23-weeks of age. Thus, hemizygous loss of CDC25A delays or inhibits ras- and neu-induced tumorigenicity. In contrast to ras- and neu-induced tumorigenesis, tumor latency was not statistically different between Cdc25A$^{+/-}$/MMTV-c-myc and Cdc25A$^{+/+}$/MMTV-c-myc mice

(Ray et al., 2007). Thus, CDC25A might be a critical factor in neu- and ras-mediated tumorigenesis. These studies also suggest that both c-myc and ras oncogenes have profound influence on the necessary balance between cell cycle regulation and cell death in determining kinetics of tumor growth.

2.1.3 Cyclin D1 transgenic and knockout models

Cyclin *D1* is a 36-kDa protein located on chromosome 11q13, a region commonly amplified in breast cancer (Fantl, Smith, Brookes, Dickson, & Peters, 1993). Cyclin D1 functions in regulation of progression of the cell cycle through the G1 to S phase and in cellular metabolism (Fu, Wang, Li, Sakamaki, & Pestell, 2004). Cyclin D1 plays an essential role in mammary gland development, where it is required for normal lobuloalveolar development. In fact, cyclin D1 null mice have impaired mammary development with undeveloped alveoli during pregnancy and lactation does not occur (Fantl, Stamp, Andrews, Rosewell, & Dickson, 1995; Sicinski et al., 1995). Cyclin *D1* is amplified in 15–20% of human breast cancers (Lammie & Peters, 1991) and is associated with poor prognosis of estrogen receptor positive patients. Cyclin D1 is overexpressed in about 45–50% of breast carcinomas (Bartkova et al., 1994; Buckley et al., 1993; Gillett et al., 1994). Overexpression of cyclin D1 was observed in all histological types of breast cancer (Bartkova et al., 1994), although a majority of human breast tumors that overexpress cyclin D1 tend to be well differentiated and estrogen receptor positive (Barnes & Gillett, 1998). Cyclin D1 can be detected at the various stages of breast cancer from ductal carcinoma in situ all the way to metastatic lesions (Bartkova et al., 1994; Gillett et al., 1996)

The Schmidt laboratory developed MMTV-driven cyclin D1 transgenic mice (MMTV-cyclin D1) (Wang et al., 1994). MMTV-cyclin D1 virgin mice show increased proliferation and precocious alveolar development in the mammary gland, mimicking pregnancy. MMTV-cyclin D1 mice developed proliferative abnormalities in the mammary gland that increased with age, number of pregnancies, and level of cyclin D1 expression (Wang et al., 1994). Female transgenic mice develop focal mammary adenocarcinomas after a latency of more than 1-year (with mean tumor onset age of 551 ± 26 days). Male mice also develop tumors. Random nonspecific testicular and uterine wall tumors with liver metastasis were observed in one strain. Given the long latency and focal nature of the tumors, cyclin D1 appears to promote mammary tumorigenesis, however, additional genetic changes are necessary for development of mammary tumors.

Another transgenic mouse that expressed constitutively nuclear and nondegradable cyclin D1 (phosphorylation-deficient cyclin D1 T286A) under the transcriptional control of the MMTV promoter was generated by the Diehl laboratory (Lin et al., 2008). The mutated cyclin D1 transgenic mice developed multifocal adenocarcinomas at 18-months of age while the normal cyclin D1 transgenic took 22-months to develop adenocarcinomas with less than 100% incidence, in both cases (Lin et al., 2008). As seen in human tumors, tumors from both the cyclin D1 transgenic mice were estrogen receptor positive and exhibited estrogen-dependent growth. This suggests that temporal control of cyclin D1 localization and proteolysis are essential for maintaining mammary epithelium homeostasis.

The Law laboratory generated transgenic mice with a cyclin D1-Cdk2 fusion protein under the transcriptional control of the MMTV promoter (MMTV-cyclin D1-Cdk2) that showed resemblance to human basal-like subtype of breast cancer (also known as triple negative), which is highly invasive, has a high postsurgical recurrence rate and poor prognosis (Corsino et al., 2008). Previous mouse models of basal-like breast cancer involved deletion of BRCA1 and p53. However BRCA1-inactivated tumors express progesterone receptor and overexpress HER2 and are not representative of the human basal-like breast cancer. The MMTV-cyclin D1-Cdk2 mice develop mammary fibrosis and hyperplasia that ultimately leads to mammary tumors with desmoplasia. This is a useful model for testing CDK2 inhibitors as well as for the development and testing of novel therapeutic agents targeting tumor cell/fibroblast paracrine loops.

To determine the role of p53 in MMTV-cyclin D1-derived mammary tumors, MMTV-cyclin D1 mice were crossed with p53 null mice ($p53^{-/-}$) (Hosokawa et al., 2001). Mammary tumors were not observed in MMTV-cyclin D1/$p53^{+/-}$ and MMTV-cyclin D1/$p53^{-/-}$ mice suggesting that p53 inactivation did not cooperate with cyclin D1 in mammary tumorigenesis.

The Sicinski lab, that had previously developed cyclin D1 null mice (Sicinski et al., 1995), crossed these mice to four mammary tumor-prone transgenic mice to further elucidate the role of cyclin D1 in breast cancer. Cyclin D1 null mice (cyclin $D1^{-/-}$) were crossed with MMTV-c-myc, MMTV-v-Ha-ras, MMTV-c-neu, and MMTV-Wnt-1 transgenic mice (Yu, Geng, & Sicinski, 2001). Loss of cyclin D1 expression inhibited mammary tumor formation in MMTV-c-neu and MMTV-v-Ha-ras mice but not in MMTV-c-myc and MMTV-Wnt-1 mice. Mammary tumor inhibition in MMTV-v-Ha-ras appears to be cyclin D1 specific, as mice generated by crossing cyclin D2 and cyclin D3 null mice with MMTV-v-Ha-ras mice developed mammary tumors (Yu et al., 2001). Additionally mice lacking the interacting partner of cyclin D1, CDK4, are resistant to mammary tumors in MMTV-c-neu mice (Yu et al., 2006). Cyclin D1 was not only able to inhibit mammary tumors arising in MMTV-c-neu mice, but also suppressed tumorigenesis in mice carrying a mutant constitutively active c-neu oncogene (Bowe, Kenney, Adereth, & Maroulakou, 2002). Interestingly, compensation by cyclin E1 resulted in recovered tumor potential in 35% of MMTV- mutant c-neu/cyclin D1$^{-/-}$ mice.

2.1.4 TGFα transgenic models

Transforming growth factor α (TGFα) is a member of the epidermal growth factor (EGF) family of cytokines. *TGFα* encodes a 17-kDa protein and is located on chromosome 2p13. It was originally observed in culture media of certain retrovirus transformed fibroblasts. TGFα binds to the EGFR and activates its endogenous tyrosine kinase activity and further EGFR downstream signaling pathways. TGFα is overexpressed in breast carcinomas and stimulates epithelial cell migration and promotes angiogenesis (Schreiber, Winkler, & Derynck, 1986). TGFα has been directly implicated in modulation of tumor transformation.

The Coffey laboratory developed MMTV-TGFα transgenic mice (human TGFα being expressed under the transcriptional control of the MMTV promoter/enhancer) (Matsui, Halter, Holt, Hogan, & Coffey, 1990). High expression of TGFα was observed in the normal mammary glands in both virgin and pregnant, multiparous females, and in neoplastic mammary tissue. While the TGFα transgene was expressed solely in the mammary glands in the females, transgene expression was also detected in the testes, seminal vesicles, salivary glands, and lungs in the males. Significant alveolar hyperplasia was evident in virgin transgenic females at approximately 100-days of age but not at 4-weeks of age, suggesting a role of sexual maturity and hormonal influences for hyperplastic changes. Among the multiparous females, hyperplastic alveolar nodules and cysts were observed, with an increase in hyperplastic and dysplastic characteristics upon increasing number of pregnancies. Only one multiparous female developed adenocarcinoma but no metastases were observed. Interestingly, no histological abnormalities were observed in the males up to 6-months of age despite expression of TGFα. A follow-up study by the same group evaluated the generation of hyperplasia and tumors in greater detail (Halter et al., 1992). By 12-months of age, hyperplasia was observed in 65% of multiparous and 45% of virgin females. By 16-months of age, 40% of multiparous and 30% of virgin females developed adenocarcinoma.

Another TGFα transgenic mouse was generated by the Lee laboratory in which transgenic mice expressed rat TGFα under the transcriptional control of the WAP promoter (Sandgren et al., 1995). Expression of TGFα was not detected in virgin females and these mice remained tumor free. TGFα was expressed at high levels during pregnancy and lactation, and parous WAP-TGFα females developed tumors with a 100% incidence but variable tumor latency. Tumor incidence and tumor latency was increased in the WAP-TGFα mice as compared to the MMTV-TGFα model. The majority of the tumors observed were well-differentiated adenomas and adenocarcinomas (53%) and fibroadenomas (35%) with characteristics of late pregnant mammary epithelium. Most of the tumors that formed were transplantable. Some of the high TGFα expressing females were unable to lactate. Following cessation of the lactogenic stimulus, TGFα overexpression also delayed or inhibited normal mammary involution. This delay in involution might predispose the proliferating epithelial cells to transformation, further aiding in the malignant transformation of the mammary gland. Cyclin D1 expression was found to be elevated in the WAP-TGFα induced tumors, suggesting cyclin D1 might act in concert with TGFα during mammary tumorigenesis. Mating of WAP-c-myc mice with WAP-TGFα resulted in double transgenic WAP-c-myc/WAP-TGFα mice that had further decreased tumor latency, increased tumor growth (more than five large mammary tumors in females) and also induced mammary tumors in virgin female and male mice. Double transgenic females had an average life span of less than 4-months as compared to WAP-c-myc mice (about 8-months) and WAP-TGFα mice (about 9.5-months). This suggests that the two transgenes cooperate during mammary tumorigenesis.

2.1.5 SV40 T antigen transgenic models

Simian virus 40 (SV40) is a papovavirus that causes sarcomas when injected into newborn hamsters. Monkeys are the natural host for the SV40 virus. Although the SV40 T antigen is not known to initiate human breast cancer, it deregulates cellular pathways that are known to be important in human breast cancer. Most importantly, SV40 T antigen binds and functionally inactivates both the p53 and Retinoblastoma (Rb) tumor suppressor proteins (Ahuja, Saenz-Robles, & Pipas, 2005).

Transgenic mice expressing SV40 T antigen under the transcriptional control of the WAP promoter were developed by the Graessmann laboratory (Tzeng, Guhl, Graessmann, & Graessmann, 1993). Eight-independent WAP-SV40 T antigen transgenic mouse lines were generated. Female mice from three of the eight lines developed mammary tumors with high frequency, usually observed during lactation. Transgenic females were not capable of producing milk, as T antigen expression induced apoptosis during late pregnancy, causing premature mammary gland involution before birth of the litters. Despite the mammary gland apoptosis, female transgenic mice developed mammary tumors following the first lactation period (Tzeng, Gottlob, Santarelli, & Graessmann, 1996). Interestingly, 5–10% of the tumors showed a reduction in expression of T antigen during tumor progression. Both T antigen positive and negative tumor cells maintained expression of estrogen and progesterone receptors at comparable levels, indicating that T antigen expression was independent of hormone receptor level. Approximately 70% of virgin mice developed T antigen positive breast tumors. However, ovariectomized animals had reduced tumor rates, indicating a critical role of ovary-associated hormones in breast cancer formation (Santarelli, Tzeng, Zimmermann, Guhl, & Graessmann, 1996).

The Furth laboratory performed further studies using one of the eight previously generated WAP-SV40 T antigen mice to understand the molecular changes induced by expression of SV40 T antigen in the mammary gland (Li, Lewis, Capuco, Laucirica, & Furth, 2000). Mammary adenocarcinomas, ranging from well- to poorly-differentiated, formed in the female mice with a 100% tumor incidence by 8- to 9-months of age. On average, palpable tumors were observed in these mice at 6-months of age following at least a single pregnancy (Li, Lewis, Capuco, Laucirica, & Furth, 2000). The mechanisms underlying the inability of the transgenic females to produce milk were investigated. With the onset of the normal mammary involution process, there is a cessation in milk protein gene expression and alveolar cells undergo apoptosis. Both p53-dependent and p53-independent mechanisms of apoptosis have been identified *in vitro* using mammary epithelial cells. To determine whether p53 was involved in apoptosis during mammary involution, involution was examined in the presence and absence of functional p53. Mammary involution and remodeling was determined to be independent of p53. The mammary cells that escaped apoptosis continued to differentiate; however milk production, processing, and secretion were impaired and hence lactation failed (Li, Hu, Heermeier, Hennighausen, & Furth, 1996a,1996b).

The tumorigenic process in these mice appears to be influenced by competition between proliferation and apoptosis. Expression of T antigen during the first pregnancy triggers both proliferation as well as p53-independent apoptosis. While cell proliferation was maintained throughout tumorigenesis, apoptosis was decreased as cells transitioned to hyperplasia and adenocarcinoma (Li, Lewis, Capuco, Laucirica, & Furth, 2000). As mammary cells transitioned from normal to adenocarcinoma, the expression of several cell survival and cell cycle control proteins were elevated. For example, proteins involved in cell cycle regulation such as cyclin D1, CDK2 and E2F-1, and cell survival proteins Bcl-xl, Bax, Bag-1, and Bad were increased in the adenocarcinomas. About 10% of the tumors also showed *K-Ras* amplification. Hormonal changes also appeared to play a role in tumor progression, with estrogen mainly acting during the early stages of tumor progression. Additionally, transient exposure to glucocorticoids resulted in tetraploidy, premature appearance of irreversible hyperplasia and early tumor development. Overexpression of the cell survival protein, Bcl-2 caused acceleration of tumor appearance. Bcl-2 reduced apoptosis and decreased cell proliferation rate. However, Bcl-2 inhibition in cell proliferation was lost once the cells transition to adenocarcinoma (Li, Lewis, Capuco, Laucirica, & Furth, 2000). This WAP-SV40 T antigen model can be used to investigate the molecular mechanisms underlying cell proliferation, apoptosis, DNA mutation, and DNA repair as modulated by extrinsic and intrinsic signals during mammary tumorigenesis (Li, Lewis, Capuco, Laucirica, & Furth, 2000).

The Green laboratory developed transgenic mice that placed the SV40 large T antigen under the transcriptional control of the C3(1) promoter (Maroulakou, Anver, Garrett, & Green, 1994). Similar to the WAP-SV40 T antigen transgenic mice, the C3(1)-SV40 T antigen female transgenic mice developed mammary hyperplasia by 3-months of age and mammary adenocarcinomas developed by 6-months of age with a 100% tumor incidence, with occasional metastasis to the lung. As would be expected, onset of mammary tumors correlated with expression of T antigen. Of note, the male transgenic mice developed prostate adenomas or adenocarcinomas that were capable of metastasizing to the lung, and these mice serve as a model for understanding the mechanisms underlying development of prostate cancer. As observed in the WAP-SV40 T antigen model, apoptosis occurred by p53- and Bcl-2-independent pathways, and was suppressed during the transition from preneoplasia to carcinoma. Thus, suppression of apoptosis appears to play a role in the tumorigenic process (Shibata et al., 1996). Since Bax was one of the proteins elevated during the preneoplastic stages where apoptosis was increased, the role of Bax in tumor progression was assessed using C3(1)-SV40 T antigen crossed with Bax knockout mice. C3(1)-SV40 T antigen/Bax$^{-/-}$ mice developed increased tumor numbers with larger tumor mass, accelerated tumor growth rates, and decreased survival as compared to mice with wild-type Bax. Loss of a single *Bax* allele was sufficient to accelerate tumorigenesis. Thus, Bax appears to be a critical suppressor of mammary tumor progression at the preneoplastic stage *via* upregulation of apoptosis, but this protective effect is lost during the transition from preneoplastic to invasive carcinoma (Shibata et al., 1999).

Double transgenic animals were developed to determine the significance of other genes in tumorigenesis in WAP-SV40-T antigen transgenic mice. The ability of Maspin, a unique serpin, in inhibiting tumor progression was evaluated. Transgenic animals expressing Maspin under the transcriptional control of WAP promoter were generated. These mice were then crossed with the WAP-SV40 T antigen mice to generate double transgenic WAP-SV40 T antigen/WAP-Maspin mice. Introduction of Maspin reduced tumor growth and pulmonary metastases. Maspin appears to slow tumor progression by increasing apoptosis, decreasing angiogenesis, and inhibiting tumor cell migration (Zhang, Shi, Magit, Furth, & Sager, 2000). Some of the mammary adenocarcinomas that developed in the WAP-SV40 T antigen transgenic mice displayed constitutively active Stat5a/b and Stat3. The significance of Stat5a in WAP-SV40 T antigen tumorigenesis was therefore evaluated by crossing WAP-SV40 T antigen mice with mice carrying germ line deletion of Stat5a (Ren, Cai, Li, & Furth, 2002). Reducing Stat5a expression in the WAP-SV40 T antigen mice resulted in delayed tumor onset, smaller tumor size, and increased apoptosis, indicating that Stat5a plays a role in WAP-SV40 T antigen-mediated tumorigenesis.

2.2. Modeling breast tumor metastasis

Metastasis is a complex, multistep process that involves local invasion and dissemination of tumor cells from the primary tumor site, intravasation into the vasculature, survival, and extravasation from the vasculature, followed by colonization at a distant organ in the body (Figure 8.2b) (Chambers, Groom, & MacDonald, 2002; Nguyen, Bos, & Massague, 2009). Understanding the molecular events that regulate tumor cell migration and metastasis from the primary tumor site to distant organs will greatly aid in developing novel therapeutic targets to inhibit tumor metastasis. The following transgenic models have been demonstrated to display significant rates of metastasis and will be discussed in this section: the growth factor receptors ERBB2 and EGFR, the viral polyoma middle T antigen and the Wnt protooncogenes.

2.2.1 ERBB2/NEU transgenic models

An obvious strategy for tumor cells to gain growth advantage is by overexpression of one or more of its growth-related receptors, which results in constitutive receptor activation and subsequent aberrant downstream signaling. One such growth-related receptor is the protooncogene *ERBB2*. ERBB2 (also known as HER2 and the rat homologue, Neu) belongs to the EGFR family and encodes a receptor tyrosine kinase. *ERBB2* is located on human chromosome 17q12. ERBB2 signals downstream through the Ras/MAPK cascade, activating c-myc, c-jun and c-fos, and activated protein-1 (AP-1). By heterodimerization with ERBB3, ERBB2 can also activate the phosphatidylinositol 3 kinase (PI3K)/AKT signaling pathway. Both the Ras/MAPK and the PI3K/AKT pathways drive cell proliferation, migration, transformation, and survival through the inhibition of apoptosis. Frequently amplified and overexpressed in breast cancer patients, ERBB2 is associated with an aggressive tumor phenotype, with reduced responsiveness to hormonal therapy and poor patient outcome. Fifteen to twenty percent of human breast cancers are associated with *ERBB2* gene amplification, while 30% of breast cancers show increased expression of activated ERBB2. ERBB2 overexpression is used as a prognostic marker of breast cancer in the clinic (Taneja et al., 2009). The role of ERBB2 in mammary tumorigenesis was elucidated by several laboratories using transgenic mice that expressed ErbB2 under the transcriptional control of mammary-specific promoters. A review by Ursini-Siegel et al. provided a timeline of all ErbB2 mouse models developed up to 2005 (Ursini-Siegel, Schade, Cardiff, & Muller, 2007). The list has grown and continues to grow since the publication of that timeline.

Transgenic mice expressing an activated form of rat Neu under the control of the MMTV promoter were developed by both the Leder laboratory (Muller et al., 1988) and the Jolicoeur laboratory (Bouchard, Lamarre, Tremblay, & Jolicoeur, 1989) and helped elucidate the role of ErbB2 overexpression in tumor initiation in the mammary gland. In the first MMTV-neu transgenic model, mice developed tumors in multiple mammary glands with a 3-month latency (Muller et al., 1988). The involvement of multiple mammary glands with a short tumor latency suggested that few additional genetic changes were necessary to transform the mammary epithelium and promote tumorigenesis when an activated form of Neu was overexpressed (Muller et al., 1988). Increased Src tyrosine kinase activity was observed in mammary tumors from the MMTV-neu mice, implying that Src may be activated by Neu. In fact, c-Src might be activated through direct interaction with activated Neu during mammary tumorigenesis (Muthuswamy, Siegel, Dankort, Webster, & Muller, 1994). In the second MMTV-neu transgenic model, however, mice developed focal, stochastic mammary tumors with longer latency (Bouchard et al., 1989) than the mice developed by the Leder laboratory. This discrepancy might have been due to differences in integration site and copy number of the transgene, resulting in differences in transgene expression levels achieved in the mammary epithelium. Subsequent studies with additional transgenic mice showed multifocal mammary tumors with short latencies (Guy, Cardiff, & Muller, 1996) suggesting that overexpression of activated Neu may be sufficient to efficiently transform mammary epithelial cells in transgenic mice (Ursini-Siegel et al., 2007).

Constitutive activation of Neu is observed in rats that have a valine to glutamic acid substitution (at amino acid 664) in the transmembrane domain. This mutation stabilizes dimerization of the receptor, resulting in constitutive tyrosine kinase activity. Of note, comparable activating mutations have not been detected in ERBB2 positive breast cancer in humans, where wild-type ERBB2 is overexpressed primarily through gene amplification. Thus, transgenic models that more closely mimic human breast cancer were generated using wild-type neu. MMTV-neu transgenic mice develop focal mammary tumors with an 8- to 12-month latency period and subsequent lung metastasis (Guy et al., 1992). However analysis of the tumors from these mice showed sporadic in-frame deletions, point mutations, or cysteine residue insertions within the NEU extracellular domain. These cysteine insertions lead to formation of intermolecular disulphide bridges that promote receptor dimerization, causing receptor activation, a process that is normally triggered by ligand binding (Guy et al., 1996; Siegel, Dankort, Hardy, & Muller, 1994). These studies suggest that activating mutations of NEU are essential for mammary tumor initiation in these transgenic mice. Interestingly, an alternatively spliced, constitutively active ERBB2 isoform with a 16-amino acid in-frame deletion has been detected in humans (Siegel, Ryan, Cardiff, & Muller, 1999). This ERBB2 splice variant has only been observed at a 5% level with the wild-type receptor, in both normal and cancer breast tissue, and hence further validation of the role of this ERBB2 splice variant is required to clearly understand its role in tumorigenesis.

The Chodosh laboratory generated mice that conditionally expressed activated Neu under the control of both the MMTV promoter and a tetracycline regulatory element (MMTV-rtTA/TetO-NeuNT mice) (Moody et al., 2002). Administration of doxycycline induced the expression of activated Neu in the mammary gland and resulted in multiple invasive and metastatic mammary carcinomas similar to tumors arising in MMTV-activated neu mice. Removal of doxycycline resulted in a loss of Neu expression and complete regression of mammary tumors and lung metastasis. However, many of the mice later developed Neu-independent mammary tumors, which might have resulted from the reactivation of dormant tumor cells that persisted in the primary tumor. Of note, a recent study has carried out extensive proteome and transcriptome profiling of the Her2/Neu-driven transgenic mammary model (MMTV-rtTA/TetO-NeuNT), and all the data including biospecimens are available to the research community (Schoenherr et al., 2011).

Transgenic mice expressing ErbB2 under the transcriptional control of the MMTV or WAP promoters illustrated the importance of ErbB2 in induction of mammary tumors. However, the use of these strong promoters induced nonphysiologically elevated levels of ErbB2 expression. Therefore, to determine the role of Neu in mammary tumor induction in a model that was more relevant to that observed in humans, the Muller laboratory developed a knock-in transgenic mouse using the endogenous Neu promoter to drive transcription of activated Neu (Andrechek et al., 2000). In this modified locus, expression of activated Neu was transcriptionally silenced by an upstream loxP-flanked neomycin cassette. However, crossing these mice with MMTV-Cre transgenic mice resulted in the mammary-specific excision of the neo cassette and expression of activated Neu from the endogenous mouse ErbB2 promoter. Expression of activated Neu accelerated lobuloalveolar development, and focal mammary tumors were observed after a long latency. Interestingly, in all the mammary tumors, Neu expression was increased through gene amplification, and this therefore represents a model that more closely mimics human ERBB2 positive breast tumors (Andrechek et al., 2000).

As mentioned previously, abrogating CDK4 was able to suppress tumor growth in MMTV-Neu mice. CDK2 was also tested for its role in MMTV-Neu-mediated tumorigenesis (Ray, Terao, Christov, Kaldis, & Kiyokawa, 2011). Only 13% of MMTV-Neu/Cdk2$^{-/-}$ mice developed mammary tumors with a longer latency than MMTV-Neu/Cdk2$^{+/-}$ and

MMTV-Neu/Cdk2$^{+/+}$ mice. The rest of the mice remained tumor free up to 16-months, indicating a role of Cdk2 in ERBB2-mediated tumorigenesis.

2.2.2 EGFR transgenic models

EGFR is another member of the receptor tyrosine kinase family that is located on chromosome 7p12 in humans. EGFR plays a key role in normal mammary development and function and is present in the ductal epithelial cells of normal breast (Moller et al., 1989; Wiesen, Young, Werb, & Cunha, 1999). EGFR is overexpressed in about 30–50% of human breast cancers, and elevated EGFR expression correlates with poor prognosis and increased invasiveness (Koenders, Beex, Kienhuis, Kloppenborg, & Benraad, 1993).

To evaluate the role of EGFR in mammary tumor development, transgenic mice expressing human EGFR under the transcriptional regulation of both the MMTV and BLG promoters were developed by the Theuring laboratory (Brandt et al., 2000). As would be expected from our understanding of the respective promoters, the BLG-EGFR mice expressed the EGFR transgene exclusively in female mammary glands, while in the MMTV-EGFR mice, the transgene was expressed in other organs such as salivary glands, ovary, and testis as well. Mammary gland development was affected in both virgin and lactating female transgenic MMTV-EGFR and BLG-EGFR animals. Mammary epithelial hyperplasia was observed in the virgin females, while dysplasia and tubular adenocarcinomas were observed in lactating female transgenic mice. With increasing number of pregnancies an increase in the number of dysplasias were observed. Small tubular adenocarcinomas were observed in one out of six multiparous BLG-EGFR female transgenic mice and in two out of six multiparous MMTV-EGFR female transgenic mice. Transgenic mice expressing human EGFR under the transcriptional regulation of the MMTV promoter were also generated in the Parsons laboratory to determine the role of EGFR on multiple steroid hormone-responsive tissues (Marozkina, Stiefel, Frierson, & Parsons, 2008). Although the EGFR transgene was expressed in the mammary gland, only a few visible tumors developed over 2-years. Fifty-five percent of the female transgenic mice developed hyperplasia, hypertrophy, or slight dysplasia in the mammary glands, as was observed in the earlier transgenic models. Interestingly, hyperplasia, hypertrophy, or slight dysplasia was also observed in 89% of the female transgenics in the uterus or uterine horn and in 100% female transgenics in the ovaries or oviduct. Thus, although EGFR overexpression caused impaired mammary gland development and induced epithelial cell transformation, it was not able to singly transform the mammary gland.

2.2.3 PyVmT transgenic models

Polyoma virus, like SV40, is a DNA tumor virus that encodes a potent transforming protein, Polyoma virus middle tumor antigen (PyVmT). Infection of newborn or athymic mice with the Polyoma virus results in formation of mammary adenocarcinomas along with multiple other tumors types (Asch, 1996), and it was demonstrated that the middle T antigen is required for tumor induction (Mes & Hassell, 1982). PyVmT binds to and activates a number of signaling proteins, including c-Src, Shc, protein phosphatase 2A (PP2A), and PI3K, leading to activation of multiple downstream signaling pathways, including the Ras/MAPK pathway (Dilworth, 2002).

The Muller laboratory generated transgenic mice with mammary-specific expression of PyVmT under the transcriptional control of the MMTV promoter/enhancer (Guy, Cardiff, & Muller, 1992). Along with mammary expression of PyVmT, PyVmT was also observed at lower levels in salivary glands and ovaries in females, and in salivary glands, epididymis, and seminal vesicles in males but tumors were only observed in the mammary gland. Interestingly, in older animals, PyVmT was also detected in the lungs in female and male transgenic mice and correlated with the appearance of multifocal lung metastases. Synchronous multifocal mammary adenocarcinomas involving all the mammary glands were observed in MMTV-PyVmT mice. The mice develop tumors with a short latency, 100% incidence and majority of tumor-bearing mice develop secondary metastatic tumors in the lungs by 3-months of age. Transplanting the primary mammary tumor into the fat pads of normal syngeneic mice resulted in metastatic foci in the lymphatics or the lungs, further validating the metastatic capabilities of this PyVmT model. Male mice also develop mammary tumors although tumor onset was delayed compared to females. Virgin female transgenic mice developed multifocal mammary tumors indicating that tumor initiation was independent of pregnancy. The simultaneous appearance of multifocal tumors, the short tumor latency, and short latency between transgene expression and malignant transformation suggests PyVmT is sufficient for transforming the mammary epithelium with requirements for few, if any, additional genetic events. Important insights regarding metastatic progression might be obtained using this model since expression of PyVmT was closely associated with pulmonary metastases.

The MMTV-PyVmT model has a far greater metastasis incidence as compared to other MMTV-oncogene bearing transgenic mice, where metastasis is only rarely observed. Since metastatic tumors retain the expression of mammary markers such as β-casein, they likely originated from the primary tumors. One of the advantages of the MMTV-PyVmT transgenic mouse is that the mice develop focal tumors at the various mammary glands. In contrast, in the MMTV-ERBB2 the entire mammary gland coalesces into a large tumor. Although the MMTV-PyVmT model mimics breast cancer observed in patients, certain aspects of the model do not correlate with the human disease. Most of the tumors derived from MMTV-PyVmT mice are estrogen receptor negative and metastasis occurs primarily in the lungs, with some metastasis observed in the lymphatics. In contrast, human tumors show varying estrogen receptor expression and metastasize most commonly to the bone, lungs, and liver. Although another study did show that MMTV-ERBB2 mice develop polyclonal mammary tumors independent of a second event. Of note, both oncogenes ErbB2 and PyVmT are associated with constitutive tyrosine kinase activities that are not subject to normal cellular regulation (Guy, Cardiff, & Muller, 1992). The role of the tyrosine kinase c-Src in PyVmT-mediated tumorigenesis was evaluated by crossing MMTV-PyVmT mice with mice with a disrupted c-Src. In the absence of functional c-Scr in MMTV-PyVmT transgenics, mice rarely developed mammary tumors. Following a long latency, abnormal hyperplastic mammary tissue was observed. Thus, c-Scr might be required for PyVmT induced mammary tumorigenesis (Guy, Muthuswamy, Cardiff, Soriano, & Muller, 1994). Additionally the role of various genes in the inhibition of metastasis can be evaluated using MMTV-PyVmT. For example, silencing the expression of metastasis-stimulating protein, S100A4 ($S100A4^{-/-}$/PyVmT mice), resulted in a reduction of the metastatic burden in lungs (Grum-Schwensen et al., 2010).

2.2.4 Wnt transgenic models

Wnt-1 is a secreted protein that is cysteine-rich, glycosylated, and located on chromosome 12q13 in humans. Wnt-1 functions primarily by stabilizing and increasing the levels of cytosolic β-catenin. Wnt-1 was first identified as a protooncogene activated by viral insertion in mouse mammary tumors. Wnt-1 is normally absent in the mammary gland and it has not been directly implicated in human breast cancer. However, other members of the Wnt family are expressed in the breast and are overexpressed in breast cancer (Li, Hively, & Varmus, 2000).

MMTV-Wnt-1 transgenic mice were developed in the Varmus laboratory (Tsukamoto et al., 1988). MMTV-Wnt-1 mice show extensive ductal hyperplasia early in life and mammary adenocarcinomas were observed in about 50% of the female mice by 6-months of age. Potent mitogenic effects were evident in the mammary epithelium upon ectopic expression of Wnt-1 and extensive ductal hyperplasia resulted in an inability of female transgenic mice to feed their pups. Female transgenic mice develop mammary adenocarcinomas with pregnancy causing a slight increase in tumor onset. Although metastasis was not observed at the time tumors were detected, lymph node and/or lung metastasis were observed after removal of primary tumors in these mice. About 15% of male mice also develop mammary gland tumors with extensive hyperplasia by 12-months of age. The genetic background of the transgenic mice does not seem to influence the latency of Wnt-1-mediated tumor development. The long latency between ductal hyperplasia in the mammary gland and the appearance of mammary tumors suggests that MMTV-Wnt-1 mice might require additional cooperative genetic events for tumor formation. Studies conducted on ovarectomized MMTV-Wnt-1 mice demonstrated that Wnt-1 functions independently of estrogen signaling during mammary tumorigenesis (Bocchinfuso, Hively, Couse, Varmus, & Korach, 1999). Tumors were still observed although with delayed onset. Additionally, tumors formed after an increased latency in estrogen receptor α-null mice.

In cooperation with another oncogene, fibroblast growth factor 3 (FGF3), Wnt-1 showed increased tumorigenic abilities. MMTV-FGF3/MMTV-Wnt-1 double transgenic female mice developed tumors faster than either of the individual transgene bearing mice. The effect was more pronounced in double transgenic males (Kwan et al., 1992). To determine whether Wnt-1 cooperates with TGF-β during the tumorigenic process, MMTV-Wnt-1/MMTV-TGF-β double transgenic mice were generated. MMTV-Wnt-1/MMTV-TGF-β double transgenic mice did not show an increase in tumor incidence or latency, indicating TGF-β cannot inhibit Wnt-1-mediated mammary tumorigenesis (Li, Hively, & Varmus, 2000). MMTV-Wnt-1 mice were also crossed with p53 null mice to determine the influence of p53-deficiency on mammary tumorigenesis (Donehower et al., 1995). The MMTV-Wnt-1/p53$^{-/-}$ mice developed tumors earlier than MMTV-Wnt-1/p53$^{+/-}$ and MMTV-Wnt-1/p53$^{+/+}$ mice suggesting that inactivation of p53 is important in Wnt-1-mediated tumorigenesis.

Like *Wnt-1*, *Wnt-10b* is also located on chromosome 12q13 in humans. Wnt-10b, like Wnt-1, was also identified as a protooncogene activated by viral insertion in mouse mammary tumors. In mice, Wnt-10b is expressed during the formation of the mammary rudiment in embryos and its expression continues through puberty. The Leder laboratory generated Wnt-10b transgenic mice with Wnt-10b under the transcriptional control of the MMTV promoter/enhancer (MMTV-Wnt-10b) (Lane & Leder, 1997). Overexpression of Wnt-10b resulted in profound developmental alterations in the mammary gland of both virgin females and males, with increased susceptibility to mammary adenocarcinomas. The mammary adenocarcinomas that formed in the MMTV-Wnt-10b mice were solitary indicating Wnt10b was necessary but not sufficient to bring about tumorigenesis. The FGF family of transcription factors tends to collaborate with Wnts, however FGF expression was not elevated in MMTV-Wnt-10b tumors. Crossing MMTV-Wnt-10b mice with MMTV-FGF3/int-2 mice produced sterile offspring with highly disorganized mammary epithelium, suggesting potent interaction between Wnt-10b and FGF3 signaling pathways.

2.3. Modeling angiogenesis through the vascular endothelial growth factor family

Angiogenesis is the development of tumor-associated vasculature that aids tumor growth and maintenance by supplying nutrients and oxygen and eliminating metabolic wastes and carbon dioxide. The growing tumor causes normal vasculature to sprout new vessels to sustain tumor growth and the process also aids in metastasis (Folkman, 1992; Hanahan & Weinberg, 2011). Transgenic mouse models based upon expression of vascular endothelial growth factor (VEGF) will be discussed in this section.

2.3.1 VEGF transgenic models

VEGF is a glycosylated protein that is located on chromosome 6p12. VEGF primarily affects endothelial cells and induces changes such as increased vascular permeability, promotes cell migration, inhibits apoptosis and induces angiogenesis, vasculogenesis and endothelial cell growth. Breast cancer patients show overexpression of VEGF (Yoshiji, Gomez, Shibuya, & Thorgeirsson, 1996) and VEGF is a prognostic indicator of relapse free and overall survival in node-negative breast cancer (Obermair et al., 1997).

Transgenic mice expressing human VEGF under the transcriptional control of the MMTV promoter (MMTV-VEGF) were generated by the Thorgeirsson laboratory (Schoeffner et al., 2005). VEGF expression was observed at high levels in the mammary gland and at lower levels in the salivary glands, lungs, and spleen in female transgenic animals. Only female transgenic mice were used in this study. Another laboratory reported that, in addition to mammary gland expression, MMTV-VEGF transgenic males expressed VEGF in the testis and epididymis that resulted in infertility (Korpelainen et al., 1998). Preliminary studies using heterozygous VEGF expressing females observed over a 2-year period showed no significant mammary gland abnormalities despite high levels of VEGF expression (Schoeffner et al., 2005). However, when MMTV-VEGF mice were crossed with MMTV-PyVmT mice in an attempt to induce metastatic mammary carcinomas in the double transgenic animals (MMTV-VEGF/MMTV-PyVmT), mammary tumor latency was decreased as compared to the MMTV-PyVmT mice, and an increase in vasodilation in the tumors was observed at 4-weeks of age. By 6- to 8-weeks of age, mammary tumor incidence, multiplicity (number of tumors/mouse), and tumor weight was increased in the double transgenic mice compared to that of MMTV-PyVmT mice. As would be expected, double transgenic mice showed a dramatic increase in metastatic abilities as compared to MMTV-PyVmT mice. Thirty-three percent of tumor-bearing double transgenic mice had macroscopic and microscopic lung metastases by 6-weeks of age and by 8-weeks of age this increased to 100%. An increase in microvascular density and an increase in tumor cell proliferation and survival were responsible for enhanced tumor growth. The study helped elucidate that VEGF contributed to mammary tumor growth by a combination of increasing neovascularization, autocrine stimulation of growth, and inhibition of apoptosis.

2.4. Modeling tumor suppression mediated by tumor suppressor genes

Tumor suppressor genes play a crucial role in regulating and limiting cell growth and proliferation. Various tumor suppressor genes have been identified which function by several mechanisms including regulation of cell cycle progression, inhibition of uncontrolled growth factor receptor signaling, promoting apoptosis, and regulating DNA repair mechanisms. In general, the loss or inactivation of both alleles of a tumor suppressor gene promotes tumor development. The following transgenic models will be discussed in this section: tumor suppressors BRCA1 and p53.

2.4.1 BRCA1 knockout models

BRCA1 (breast cancer 1) is a nuclear phosphoprotein involved in maintaining genomic stability. Located on human chromosome 17q21,

BRCA1 acts as a tumor suppressor. Individuals carrying germline mutations in the BRCA1 gene have a dramatically increased susceptibility to both breast and ovarian cancers. Germline mutations of *BRCA1* are detected in about 50% of hereditary breast cancer patients and about 90% of hereditary breast and ovarian cancer patients (Casey, 1997; Hill, Doyle, McDermott, & O'Higgins, 1997). Brca1 expression is elevated in mammary epithelial cells during pregnancy and downregulated during lactation in mice (Marquis et al., 1995). Homozygous Brca1 knockout results in early embryonic lethality in mice, indicating the importance of Brca1 in normal development (Gowen, Johnson, Latour, Sulik, & Koller, 1996; Hakem et al., 1996; Liu, Flesken-Nikitin, Li, Zeng, & Lee, 1996).

In order to investigate the role of Brca1 loss in mammary tumorigenesis in mice, the Deng laboratory developed mammary gland specific conditional Brca1 knockout mice using the Cre–loxP system (Xu et al., 1999). These mice carried one Brca1-null allele and one Brca1 conditional knockout (floxed) allele, as well as either an MMTV-Cre or WAP-Cre transgene. Increased apoptosis and abnormal ductal development was observed in mammary epithelial cells in these animals. Mammary tumors were observed at a low frequency and following a long latency. The tumors exhibited genomic instability, including both aneuploidy and chromosomal rearrangements, resulting in genomic imbalances similar to those observed in human breast cancer (Weaver et al., 2002). It has been reported that BRCA1 familial tumors frequently contain p53 mutations (Deng & Scott, 2000). Thus, to investigate the potential role of p53 in Brca1-associated tumorigenesis, the Brca1 conditional knockout mice described above were bred to p53 heterozygous knockout mice to generate mice lacking one p53 allele, which resulted in a significant increase in both tumor frequency and kinetics of onset (Xu et al., 1999).

2.4.2 p53 Transgenic and knockout models

The *p53* tumor suppressor gene, located on human chromosome 17p13.1, is the gene found to be most frequently mutated in human tumors overall, and is mutated in approximately 25% of breast tumors (Olivier, Hollstein, & Hainaut, 2010; Petitjean, Achatz, Borresen-Dale, Hainaut, & Olivier, 2007). In response to a wide range of cellular stressors, p53 normally becomes activated and subsequently promotes several possible cellular responses, including cell cycle arrest, apoptosis, senescence, and DNA damage repair. Tumor-associated *p53* mutations lead to loss of these protective mechanisms, resulting in genomic instability and a dramatically increased susceptibility to cellular transformation. Patients with Li-Fraumeni syndrome, who carry a germline mutation in one *p53* allele, have a strong predisposition to development of a wide range of tumor types, the most frequent of which is breast cancer (Olivier et al., 2010; Petitjean et al., 2007). Additionally, patients with mutations in *BRCA1* and *BRCA2* also tend to show somatic *p53* abnormalities, illustrating that p53 plays an important role in suppression of breast tumors (Blackburn & Jerry, 2002).

When p53 knockout mice were orginally generated, it was expected that complete p53 deficiency would result in embryonic lethality, as is the case for many other tumor suppressor genes. It was therefore surprising to find that mice homozygous for the p53 null allele did not display developmental abnormalities. However, both $p53^{-/-}$ and $p53^{+/-}$ mice developed a variety of tumor types, most notably lymphomas, and to a lesser extent, sarcomas, with a mean age of 6 months in $p53^{-/-}$ mice (Donehower et al., 1992). However no mammary tumors were observed. Thus, although these mice model the overall tumor susceptibility seen in Li-Fraumeni patients, the tumor spectrum is quite different. Several approaches have therefore been taken to create mouse models for examining the role of p53 mutation in mammary tumorigenesis. One approach was to backcross the p53 null allele onto a BALB/c mouse background, a strain that exhibits a higher rate of radiation-induced mammary tumorigenesis than many other strains (Ullrich, Bowles, Satterfield, & Davis, 1996). As with the original p53 null mice, both BALB/c-$p53^{-/-}$ and BALB/c-$p53^{+/-}$ mice developed multiple tumor types, with a shorter latency in the homozygous knockout mice (Kuperwasser et al., 2000). BALB/c-$p53^{+/-}$ mice had a longer life span (54-weeks) as compared to BALB/c-$p53^{-/-}$ mice (15.4-weeks). While the tumor spectrum in the BALB/c-$p53^{-/-}$ mice was similar to that seen in the original $p53^{-/-}$ mice, including primarily lymphomas and sarcomas, 55% of female BALB/c-$p53^{+/-}$ mice developed mammary carcinomas, which was the most common tumor type seen in these mice. Further, the mammary tumors from the BALB/c-$p53^{+/-}$ mice showed frequent loss of the remaining wild-type p53 allele, as do tumors arising in Li-Fraumeni patients. Thus, the BALB/c-$p53^{+/-}$ mice represent a relatively faithful model of human Li-Fraumeni syndrome.

Another widely used approach for investigating the role of p53 mutation in mammary tumorigenesis has been to cross the $p53^{-/-}$ mice with various transgenic mouse mammary tumor models to examine the effects of p53 deficiency on tumor properties in these lines of mice. For example, Hundley et al. crossed MMTV-v-Ha-ras transgenic mice with p53 null mice to generate MMTV-v-Ha-ras/p53$^{+/+}$, MMTV-v-Ha-ras/p53$^{+/-}$ and MMTV-v-Ha-ras/p53$^{-/-}$ mice (Hundley et al., 1997). As expected, the MMTV-v-Ha-ras/p53$^{-/-}$ mice developed tumors more rapidly than mice of the other two genotypes. However, there was a surprising shift in the tumor distribution in the p53-deficient mice. While MMTV-v-Ha-ras/p53$^{+/+}$ and MMTV-v-Ha-ras/p53$^{+/-}$ mice developed primarily mammary tumors and few salivary tumors, the MMTV-v-Ha-ras/p53$^{-/-}$ mice developed primarily salivary tumors, indicating that the acceleration of tumor onset resulting from p53 deficiency was more dramatic in the salivary gland than the mammary gland. Tumors arising in MMTV-v-Ha-ras/p53$^{-/-}$ were also of a higher histologic grade and displayed increased tumor growth rates and increased genomic instability and heterogeneity, as compared to MMTV-v-Ha-ras/p53$^{+/+}$ mice. The increased tumor growth rates in the MMTV-v-Ha-ras/p53$^{-/-}$ mice were found to result from increased tumor cell proliferation rather than decreased apoptosis, indicating that p53-mediated tumor suppression can occurs independent of its role in promoting apoptosis (Hundley, Koester, Troyer, Hilsenbeck, Subler, et al., 1997). Numerous other transgenic mouse tumor models have similarly been crossed with p53 null mice, as with the MMTV-Wnt-1 mice described above, generally resulting in accelerated tumor kinetics, and altered tumor properties consistent with the poorer prognosis seen in patients whose tumors carry p53 mutations.

The p53 models described above all employed p53 null alleles, from which no p53 protein is produced. While the majority of tumor-associated mutations seen in most tumor suppressor genes are null alleles, p53 is unique in that the majority of human tumor-associated mutations, including those in breast tumors, are missense mutations that result in the production of a full-length protein carrying a single amino acid substitution (Petitjean et al., 2007). Almost all of the missense mutations occur in the central half of the protein encoding the DNA binding domain, particularly, at certain "hot-spot" residues. These mutations all result in impaired DNA binding and transcriptional activation of target genes, and can thus be considered loss-of-function mutations. However, many of these missense mutants also appear to possess gain-of-function properties, that is, they have acquired oncogenic properties that promote tumorigenesis and tumor progression (Muller & Vousden, 2013). To investigate the functions of p53 gain-of-function mutants, several lines of transgenic mice expressing mutant p53 genes under the control of cell type-specific promoters have been generated. Of relevance to this review is a line of transgenic mice developed by the

Rosen laboratory in which mouse p53-R172H (carrying an arginine to histidine substitution at amino acid residue 172) was expressed under the transcriptional control of the WAP promoter (Li et al., 1998). This mutation in the mouse *p53* gene is equivalent to p53-R175H in the human *p53* gene, corresponding to one of the most common "hot-spot" mutations. The transgene did not affect normal mammary gland development and spontaneous tumors were not observed in these mice. However, the mice were more susceptible to mammary tumor development upon administration of the chemical carcinogen dimethylbenz(a)anthracene than wild-type littermates, and the tumors showed increased genomic instability.

3. DEVELOPING NOVEL THERAPEUTICS AND IMAGING TECHNIQUES USING TRANSGENIC ANIMALS

3.1. Evaluating the use of novel therapeutics using transgenic animals

The *in vivo* evaluation of potential anticancer therapies has traditionally been carried out almost exclusively using human tumor xenografts, mouse allografts, or tumor tissue explants grown in immunocompromised mice. There are numerous advantages to the use of such models, including the ability to study effects of investigational agents on actual human tumor cells carrying all of the genetic alterations present in the primary tumors from which they are derived (Neve et al., 2006), the availability of a large number of tumor cell lines representing a broad array of tumor types, and the ability to reproducibly generate a large cohort of tumor-bearing animals at one time. However, there are also several drawbacks to xenograft models, most notably that the host animals must be immunocompromised to permit engraftment of human cells, thus creating an artificial setting that does not model the effects of the immune system on tumor cell growth or response to treatment. In addition, for most xenograft approaches, the microenvironment surrounding the inoculated tumor cells very poorly models the normal microenvironment of a developing tumor. The use of GEM models for drug development can potentially provide a complementary approach that eliminates some of the drawbacks of xenograft models, since tumors arise in immunocompetent animals, and arise stochastically in the presence of a normal microenvironment. In addition, GEM models afford the unique opportunity to tailor the genetics of the tumor to specific targeted therapies being investigated. Because they have an intact immune system, GEM models are also well suited for evaluating vaccine approaches for the prevention of cancer.

Transgenic animals have been instrumental in testing the efficacy and mode of action of chemotherapeutics that are predicted to target the driving oncogene, either directly or indirectly. For example, cyclooxygenase 2 (COX-2) is overexpressed in HER-2/neu-positive human breast cancers and in mammary tumors from MMTV-neu mice. Hence celecoxib, a selective COX-2 inhibitor, was evaluated for efficacy of tumor reduction in MMTV-neu transgenic mice (Howe et al., 2002). Treatment of MMTV-neu mice with celecoxib resulted in a decrease in the incidence of mammary tumors along with a 50% reduction in mammary prostaglandin E_2 levels. Similar results were obtained by another group that tested both COX-1 and COX-2 inhibitors (SC560 and celecoxib, respectively) in MMTV-neu mice (Lanza-Jacoby et al., 2003). Only celecoxib was able to significantly lower mammary tumor incidence, number of tumors, and increase tumor latency. As in the earlier study, the protective effect of celecoxib was thought to be mediated by reduction in prostacyclin and prostaglandin E_2.

Another group of chemotherapeutic agents, farnesyltransferase inhibitors (FTIs), have been tested in the MMTV-v-Ha-ras transgenic mouse model. Farnesyltransferase is an enzyme that mediates a posttranslational modification of ras required for its activity. Because of the prevalence of ras mutations in human cancer, considerable effort has been directed toward developing FTIs that could potentially block the function of mutant ras in tumor cells. The efficacy of the FTI L-744,832, developed by investigators at Merck Research Laboratories, was investigated in tumor-bearing MMTV-v-Ha-ras mice as a proof-of-principle that tumors carrying Ha-ras mutations would be sensitive to farnesyltransferase inhibition (Barrington et al., 1998). L-744,832 indeed promoted dramatic tumor regression in MMTV-v-Ha-ras mice, primarily *via* induction of apoptosis. This response was largely p53 independent, as tumors from MMTV-v-Ha-ras/p53$^{-/-}$ mice also rapidly regressed in response to L-744,832. Tumors from MMTV-v-Ha-ras/MMTV-c-myc mice also initially regressed in response to L-744,832, although the response was less durable than in the MMTV-v-Ha-ras mice. In contrast to each of the models carrying an activated Ha-ras transgene, no tumor regression was observed upon treatment of MMTV-c-neu mice with L-744,832. These studies demonstrated the utility of GEM models for evaluating novel agents that target a genetic component of the model.

Additionally, both novel and clinically utilized chemotherapeutics can be evaluated alone and in combination with other therapeutics in GEM models to determine treatments that would synergize and allow for enhanced tumor reduction. The tyrosine kinase inhibitor SU11248 was assessed both as a single agent and in combination with "standard of care" therapeutics in MMTV-v-Ha-ras transgenic mice and xenograft models (Abrams et al., 2003). SU11248 is an oral multitargeted tyrosine kinase inhibitor that has antitumor and antiangiogenic properties. Treatment of tumor-bearing MMTV-v-Ha-ras mice with SU11248 caused a dramatic regression of mammary tumors. In the xenograft models, treatment with SU11248 in combination with docetaxel, 5-fluorouracil or doxorubicin caused a 50% reduction in tumor growth.

To evaluate therapeutics for potential use in estrogen receptor-negative breast cancer patients, treatment with an EGFR inhibitor was evaluated in MMTV-neu (unactivated neu) mice, since these mice form estrogen receptor-negative mammary tumors. Gefitinib (Iressa or ZD1839) is an oral EGFR tyrosine kinase inhibitor. Treatment of MMTV-neu mice with Gefitinib, starting at 6 months of age, resulted in a suppression of mammary tumors with increased tumor latency. Gefitinib reduced proliferation of tumor cells and caused an increase in expression of the cell cycle regulator p27 (Lu et al., 2003).

Activities of drugs for chemoprevention, treatment, and inhibition of metastasis have also been evaluated in transgenic animals. For example, dipyridamole was tested as a chemoprevention strategy in MMTV-PyVmT transgenic mice. Dipyridamole treatment beginning at 4 weeks of age delayed onset of palpable tumors, delayed tumor progression, and suppressed lung metastasis (Wang, Schwab, Fan, Seagroves, & Buolamwini, 2013). The ability of an antimatrix metalloproteinase-9 DNAzyme to decrease tumor growth of mammary tumors in the MMTV-PyVmT mouse model has also been evaluated. This DNAzyme was able to reduce tumor growth and final tumor load when administered weekly through intratumoral injection (Hallett et al., 2013).

Transgenic animals have also been utilized to assess the effectiveness of cancer vaccines. The HER2/NEU protein was assessed for its potential as a tumor antigen for vaccination and a DNA vaccination strategy against Neu was developed and tested in transgenic animals. The neu gene was inserted into a novel plasmid ELVIS containing a Sindbis virus replicon that reproduces multiple copies of mRNA (Lachman et al., 2001). Immunization of MMTV-neu transgenic mice intramuscularly using this ELVIS-neu vaccine was protective against development of spontaneous mammary tumors. The vaccine also showed effectiveness in reducing incidence and number of lung metastasis in BALB/c mice following tumor cell injection in the mammary fat pad or tail vein, respectively. This vaccine was suggested to have potential use as a therapeutic vaccination for breast cancer patients with HER2/neu-expressing tumors to help reduce metastasis.

The efficacy of a novel tripartite cancer vaccine was also tested in transgenic animals. The tripartite vaccine consisted of an immunoadjuvant Pam_3CysSK_4, a peptide T_{helper} epitope, and an aberrantly glycosylated MUC1 peptide, designed to elicit both recognition by cytotoxic T lymphocytes (CTL) and antibody-dependent cell-mediated cytotoxicity (ADCC) responses against aberrantly glycosylated mucin MUC1 commonly found in breast cancer (Lakshminarayanan et al., 2012). MUC1 transgenic mice that express human MUC1 at physiological levels were first immunized using the tripartite vaccine. The mice were then injected with mammary tumor cells (obtained from tumors derived from MUC1 transgenic mice crossed with MMTV-PyVmT mice). A week after injection with tumor cells, a final immunization was administered to the mice. Interestingly, the vaccine was able to elicit both CTL and ADCC response and resulted in a significant reduction in tumor burden.

3.2. Imaging using transgenic animals

Various imaging techniques have been developed that aid in detecting tumors and metastasis, and noninvasively monitoring efficacy of therapeutic interventions in human breast cancer patients. In fact the field of imaging is rapidly growing. Using computed tomography (CT), magnetic resonance imaging (MRI), and ultrasound (US), anatomical and physiological information can be obtained and these techniques are widely used both in the clinic and in preclinical applications. Imaging at the molecular level can be obtained using positron emission tomography (PET), single-photon emission computed tomography (SPECT), which are used in the clinic as well as fluorescence reflectance imaging (FRI), fluorescence-mediated tomography (FMT), bioluminescence imaging (BLI), laser-scanning confocal microscopy (LCSM), and mulitphoton microscopy (MPM) which are being optimized for future use in the clinic. Details on these imaging modalities have been reviewed by Condeelis and Weissleder (2010).

MMTV-Neu (activated neu) transgenic mice were utilized to follow the progression of mammary tumors using MRI and ultrasonography (Galie et al., 2004). US imaging was found useful for obtaining basic information such as the size of developing tumors as well as identifying necrotic areas. On the other hand, advanced analysis of morphological aspects was possible with

MRI allowing high resolution of differentiation of details of necrotic area such as coagulation, liquefaction, biphasic splitting of cysts, and fibrotic and lipidic infiltration (Galie et al., 2004).

The Segall laboratory developed transgenic animals that specifically express green fluorescent protein (GFP) in mammary tumor cells to evaluate the behavior of tumor cells in mammary tumors (Ahmed et al., 2002). Mice with GFP expression under the transcriptional control of the MMTV promoter were generated. These mice expressed low levels of fluorescence in the mammary and salivary glands of transgenic animals. Next the MMTV-GFP mice were crossed with MMTV-PyVmT mice generating double transgenic MMTV-PyVmT/MMTV-GFP mice. These double transgenic mice developed GFP expressing tumors that showed high metastatic rates. Given the short tumor latency, high penetrance and reproducibility of tumor progression, this model is ideal for monitoring tumor progression and efficacy of therapeutic interventions using imaging. Another transgenic model was also developed by the same laboratory that used a Cre-activatable promoter, CAG-CAT-EGFP in combination with WAP-Cre or MMTV-Cre mice (Ahmed et al., 2002). Even more intense fluorescence was obtained in the mammary glands of virgin female mice. Of note, virgin MMTV-Cre/CAG-CAT-EGFP transgenic females showed more uniform GFP expression than virgin WAP-Cre/CAG-CAT-EGFP transgenic mice. To evaluate the potential of these transgenic mice in following mammary tumors the mice were crossed with MMTV-PyVmT mice and MMTV-neu mice. The mammary tumors that formed showed bright fluorescence. Interestingly, motile cells could be visualized within the tumor using multiphoton microscope and time lapse images showed a cell moving toward a blood vessel followed by intravasation (Ahmed et al., 2002). This study showed that *in vivo* imaging could be used to visualize metastatic properties in living tumors.

The ability of mammary tumor cells to intravasate into blood vessels and the role of macrophages in aiding this process was also evaluated using transgenic animals (Wyckoff et al., 2007). Interaction between tumor cell and macrophages was observed using two-color fluorescent mice with mammary tumors, MMTV-PyVmT/MMTV-iCre-CAG-CAC-ECFP/c-fms-GFP (Mammary tumors were labeled with CFP (cyan fluorescent protein) and macrophages were labeled with GFP) (Wyckoff et al., 2004, 2007).

Although GFP-based imaging modalities are valuable, autofluorescence, and low tissue penetration of light emitted by GFP can lead to significant drawbacks in using this imaging modality. Bioluminescence based imaging modalities are more sensitive for broader *in vivo* applications. A transgenic mouse model was developed that utilized luciferase and bioluminescent imaging to monitor tumor progression *in vivo* (Zagozdzon et al., 2012). The transgenic mice expressed luciferase (Luc2, the next generation firefly luciferase, optimized for high expression, and reduced anomalous transcription) under the transcriptional control of the MMTV promoter (MMTV-Luc2). Homozygous MMTV-Luc2 mice were crossed with MMTV- PyVmT mice to generate double transgenic mice with mammary tumors expressing luciferase, allowing visualization of tumor progression using bioluminescent imaging. Of note, metastasis, for example to the lung, was not easily visualizable in these transgenic mice. This double transgenic model is ideal for early detection and monitoring of mammary tumor progression.

Near-infrared non invasive optical imaging was also used for detection of microscopic mammary cancer in ErbB2 transgenic mice (Conti et al., 2013). DA364, a near-infrared fluorescence arginine-glycine-aspartic acid cyclin probe that binds the integrin $\alpha_v\beta_3$ and is then taken up specifically by cancer cells was utilized for imaging. This probe was, particularly, useful as it was not taken up by healthy mammary glands of both virgin and lactating wild-type mice indicating tumor specific uptake. DA364 was useful for non-invasive visualization of atypical hyperplasia and microscopic foci before mammary lesions were palpable. Since DA364 allowed visualization of early microscopic preneoplastic changes in the mammary gland, it was also useful to assess the efficacy of DNA vaccination against ErbB2.

Thus, transgenic animals are very useful models to optimize and develop the various imaging modalities. Future efforts are and will be directed towards developing higher resolution, cancer-specific imaging techniques that will be invaluable tools in the clinic.

4. CONCLUSIONS AND FUTURE PERSPECTIVES

Transgenic mouse models have in the past and continue to provide invaluable insights into our understanding of the biology and molecular mechanisms underlying the progression of breast cancer (Figure 8.3). The ability to specifically overexpress/silence a single transgene has allowed us to determine the role of that transgene in the initiation, progression, and metastasis of breast cancer. Overexpressing multiple transgenes in the mammary gland have helped understand whether the genes act in concert to enhance mammary tumor progression or whether the two genes play differing roles in breast cancer. The influence of hormones and changes

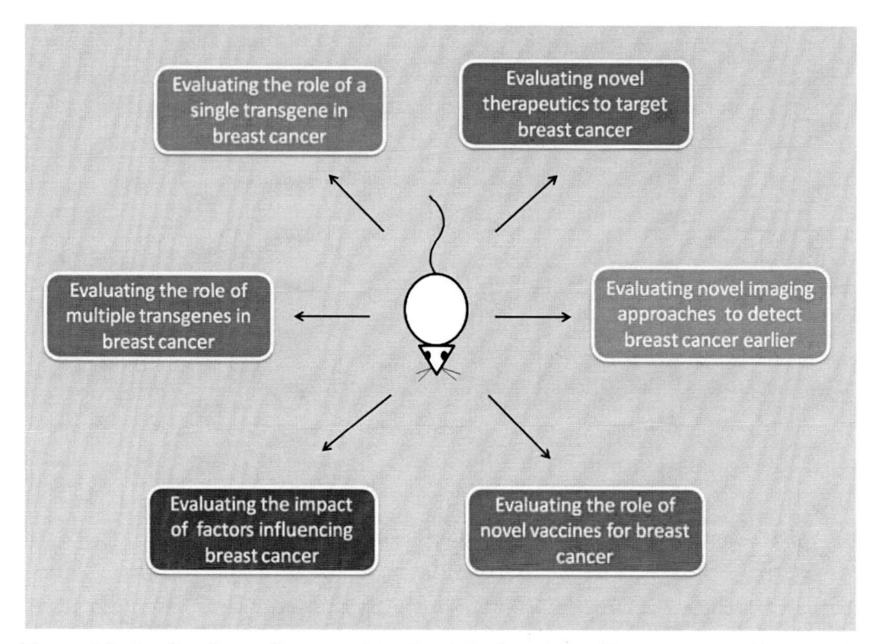

Figure 8.3 ***Applications of transgenic animals in the study of breast cancer.*** Transgenic animals have greatly advanced our understanding of the biology of breast cancer. They have been invaluable tools in evaluating the role of single or multiple transgenes in the progression of breast cancer and evaluating the impact of factors that might influence the initiation and/or progression of breast cancer. Of particular significance, transgenic animals have been used to evaluate the role of novel therapeutics to target breast cancer, to evaluate novel imaging techniques, and to evaluate the efficacy of newly developed vaccines against breast cancer. (See the color plate.)

in genomic instability upon expression of transgene(s) can also be evaluated. Furthermore, one can evaluate the molecular mechanism of action of the transgene(s), potentially aiding in the design of better, more targeted therapeutics downstream of the transgene for treatment of human breast cancer. Generation of transgenic animals that express alternatively spliced isoforms (e.g., ErbB2) might help identify the biological importance of specific isoforms in breast cancer. Importantly, identification of a specific isoform that might be sufficient to transform the mammary gland will aid in therapeutic targeting by generating antibodies or drugs that specifically target the particular isoform, allowing for rationally targeted therapy of breast cancer. Transgenic animals also aid in our understanding of mechanisms underlying tumor resistance to chemotherapy and tumor dormancy and recurrence. Novel therapeutic interventions can be evaluated for efficacy in transgenic animals. In particular, therapeutic interventions that prove useful in transgenic mice that overexpress ErbB2, for example, will more likely be useful in patients that show ERBB2 overexpression. Consequently, we can target novel therapeutics to patients that are most likely to respond to the treatment by first testing the drug in a genetically similar transgenic animal. Similarly, cell lines can be generated from the established transgenic mice that can be useful for preliminary screening of drugs and other novel therapeutics *in vitro* for future *in vivo* efficacy studies. Transgenic mice can also be used to evaluate novel vaccines that can prevent breast cancer. Further studies might also help elucidate the mechanisms involved in familial or hereditary breast cancer and will pave the way for therapeutics and early prevention strategies. Finally, development of novel imaging modalities evaluated in transgenic animals that aid in early detection of cancer will be extremely useful in detecting early tumors in patients so that therapeutic intervention is possible before metastatic spread.

In these contexts, while no transgenic animal is fully able to recapitulate the different breast tumors that arise in humans, they have enabled an enhanced understanding of at least some of the complex molecular events involved in the pathogenesis of breast cancer. Current and newer models will help in developing improved methods for identifying, treating, and potentially preventing breast cancer formation and progression to metastasis.

ACKNOWLEDGMENTS

We apologize to those authors whose work we could not cite directly due to the large number of publications available on transgenic mouse mammary models. The authors acknowledge insight on the topic from former laboratory members, Regina A. Oyesanya and Santanu Dasgupta. Support for our laboratories was provided in part by National Institutes of Health grants P01 CA104177 (P. B. F.), R01 CA097318 (P. B. F.), R01 CA127641 (P. B. F.), R01 CA168517 (Maurizio Pellecchia and P. B. F.) and R01 CA154708 (X. Y. W.); Department of Defense synergy grant W81XWH-10-PCRP-SIDA (P. B. F., X. Y. W.); the National Foundation for Cancer Research (P. B. F.); the Samuel Waxman Cancer Research Foundation (P. B. F. and D. S.); an A. David Mazzone Prostate Cancer Foundation Challenge Award (Martin G. Pomper, P.B.F., and George Sguoros); NCI Cancer Center Support Grant to VCU Massey Cancer Center (P. B. F., D. S., X. Y. W., and J. J. W.); and VCU Massey Cancer Center developmental funds (P. B. F.); The James S. McDonnell Foundation and National Cancer Institute Grant R01 CA138540-01A1 (D. S.). X. Y. W. and D. S. are Harrison Scholars in the VCU Massey Cancer Center. P. B. F and D. S. are SWCRF Investigators. P. B. F. holds the Thelma Newmeyer Corman Chair in Cancer Research in the VCU Massey Cancer Center.

REFERENCES

Abrams, T. J., Murray, L. J., Pesenti, E., Holway, V. W., Colombo, T., Lee, L. B., et al. (2003). Preclinical evaluation of the tyrosine kinase inhibitor SU11248 as a single agent and in combination with "standard of care" therapeutic agents for the treatment of breast cancer. *Molecular Cancer Therapeutics*, *2*, 1011–1021.

Ahmed, F., Wyckoff, J., Lin, E. Y., Wang, W., Wang, Y., Hennighausen, L., et al. (2002). GFP expression in the mammary gland for imaging of mammary tumor cells in transgenic mice. *Cancer Research*, *62*, 7166–7169.

Ahuja, D., Saenz-Robles, M. T., & Pipas, J. M. (2005). SV40 large T antigen targets multiple cellular pathways to elicit cellular transformation. *Oncogene*, *24*, 7729–7745.

Amundadottir, L. T., Johnson, M. D., Merlino, G., Smith, G. H., & Dickson, R. B. (1995). Synergistic interaction of transforming growth factor alpha and c-myc in mouse mammary and salivary gland tumorigenesis. *Cell Growth & Differentiation: The molecular biology journal of the American Association for Cancer Research*, *6*, 737–748.

Amundadottir, L. T., Nass, S. J., Berchem, G. J., Johnson, M. D., & Dickson, R. B. (1996). Cooperation of TGF alpha and c-Myc in mouse mammary tumorigenesis: Coordinated stimulation of growth and suppression of apoptosis. *Oncogene*, *13*, 757–765.

Andrechek, E. R., Hardy, W. R., Siegel, P. M., Rudnicki, M. A., Cardiff, R. D., & Muller, W. J. (2000). Amplification of the neu/erbB-2 oncogene in a mouse model of mammary tumorigenesis. *Proceedings of the National Academy of Sciences of the United States of America*, *97*, 3444–3449.

Andres, A. C., Schonenberger, C. A., Groner, B., Hennighausen, L., LeMeur, M., & Gerlinger, P. (1987). Ha-ras oncogene expression directed by a milk protein gene promoter: Tissue specificity, hormonal regulation, and tumor induction in transgenic mice. *Proceedings of the National Academy of Sciences of the United States of America*, *84*, 1299–1303.

Andres, A. C., van der Valk, M. A., Schonenberger, C. A., Fluckiger, F., LeMeur, M., Gerlinger, P., et al. (1988). Ha-ras and c-myc oncogene expression interferes with morphological and functional differentiation of mammary epithelial cells in single and double transgenic mice. *Genes & Development*, *2*, 1486–1495.

Asch, B. B. (1996). Tumor viruses and endogenous retrotransposons in mammary tumorigenesis. *Journal of Mammary Gland Biology and Neoplasia*, *1*, 49–60.

Barnes, D. M., & Gillett, C. E. (1998). Cyclin D1 in breast cancer. *Breast Cancer Research and Treatment*, *52*, 1–15.

Barrington, R. E., Subler, M. A., Rands, E., Omer, C. A., Miller, P. J., Hundley, J. E., et al. (1998). A farnesyltransferase inhibitor induces tumor regression in transgenic mice harboring multiple oncogenic mutations by mediating alterations in both cell cycle control and apoptosis. *Molecular and Cellular Biology*, *18*, 85–92.

Bartkova, J., Lukas, J., Muller, H., Lutzhoft, D., Strauss, M., & Bartek, J. (1994). Cyclin D1 protein expression and function in human breast cancer. *International Journal of Cancer*, *57*, 353–361.

Bissell, M. J., Polyak, K., & Rosen, J. M. (Eds.), (2011). *The mammary gland as an experimental model. Cold Spring Harbor perspectives in biology*. Cold Spring Harbor Laboratory Press, Online ISSN: 1943-0264.

Blackburn, A. C., & Jerry, D. J. (2002). Knockout and transgenic mice of Trp53: What have we learned about p53 in breast cancer? *Breast Cancer Research: BCR*, *4*, 101–111.

Blackshear, P. E. (2001). Genetically engineered rodent models of mammary gland carcinogenesis: An overview. *Toxicologic Pathology*, *29*, 105–116.

Bocchinfuso, W. P., Hively, W. P., Couse, J. F., Varmus, H. E., & Korach, K. S. (1999). A mouse mammary tumor virus-Wnt-1 transgene induces mammary gland hyperplasia and tumorigenesis in mice lacking estrogen receptor-alpha. *Cancer Research*, *59*, 1869–1876.

Borowsky, A. D. (2003). Genetically engineering a mouse. *Comparative Medicine*, *53*, 249–250.

Borowsky, A. D. (2011). Choosing a mouse model: Experimental biology in context - the utility and limitations of mouse models of breast cancer. *Cold Spring Harbor Perspectives in Biology*, *3*, a009670.

Bos, J. L. (1989). ras oncogenes in human cancer: A review. *Cancer Research*, *49*, 4682–4689.

Bouchard, L., Lamarre, L., Tremblay, P. J., & Jolicoeur, P. (1989). Stochastic appearance of mammary tumors in transgenic mice carrying the MMTV/c-neu oncogene. *Cell*, *57*, 931–936.

Bowe, D. B., Kenney, N. J., Adereth, Y., & Maroulakou, I. G. (2002). Suppression of Neu-induced mammary tumor growth in cyclin D1 deficient mice is compensated for by cyclin E. *Oncogene*, *21*, 291–298.

Boxer, R. B., Jang, J. W., Sintasath, L., & Chodosh, L. A. (2004). Lack of sustained regression of c-MYC-induced mammary adenocarcinomas following brief or prolonged MYC inactivation. *Cancer Cell*, *6*, 577–586.

Brandt, R., Eisenbrandt, R., Leenders, F., Zschiesche, W., Binas, B., Juergensen, C., et al. (2000). Mammary gland specific hEGF receptor transgene expression induces neoplasia and inhibits differentiation. *Oncogene*, *19*, 2129–2137.

Brinster, R. L., Chen, H. Y., Trumbauer, M., Senear, A. W., Warren, R., & Palmiter, R. D. (1981). Somatic expression of herpes thymidine kinase in mice following injection of a fusion gene into eggs. *Cell*, *27*, 223–231.

Britt, K., Ashworth, A., & Smalley, M. (2007). Pregnancy and the risk of breast cancer. *Endocrine-Related Cancer*, *14*, 907–933.

Buckley, M. F., Sweeney, K. J., Hamilton, J. A., Sini, R. L., Manning, D. L., Nicholson, R. I., et al. (1993). Expression and amplification of cyclin genes in human breast cancer. *Oncogene*, *8*, 2127–2133.

Caligiuri, I., Rizzolio, F., Boffo, S., Giordano, A., & Toffoli, G. (2012). Critical choices for modeling breast cancer in transgenic mouse models. *Journal of Cellular Physiology*, *227*, 2988–2991.

Cardiff, R. D. (2003). Mouse models of human breast cancer. *Comparative Medicine*, *53*, 250–253.

Cardiff, R. D., Anver, M. R., Gusterson, B. A., Hennighausen, L., Jensen, R. A., Merino, M. J., et al. (2000). The mammary pathology of genetically engineered mice: The consensus report and recommendations from the Annapolis meeting. *Oncogene*, *19*, 968–988.

Cardiff, R. D., & Kinney, N. (2007). Mouse mammary tumor biology: A short history. *Advances in Cancer Research*, *98*, 53–116.

Cardiff, R. D., & Wellings, S. R. (1999). The comparative pathology of human and mouse mammary glands. *Journal of Mammary Gland Biology and Neoplasia*, *4*, 105–122.

Casey, G. (1997). The BRCA1 and BRCA2 breast cancer genes. *Current Opinion in Oncology*, *9*, 88–93.

Chambers, A. F., Groom, A. C., & MacDonald, I. C. (2002). Dissemination and growth of cancer cells in metastatic sites. *Nature Reviews. Cancer*, *2*, 563–572.

Cho, A., Haruyama, N., & Kulkarni, A. B. (2009). Generation of transgenic mice. *Current Protocols in Cell Biology*. Chapter 19, Unit 19 11.

Condeelis, J., & Weissleder, R. (2010). In vivo imaging in cancer. *Cold Spring Harbor Perspectives in Biology*, *2*, a003848.

Conti, L., Lanzardo, S., Iezzi, M., Montone, M., Bolli, E., Brioschi, C., et al. (2013). Optical imaging detection of microscopic mammary cancer in ErbB-2 transgenic mice through the DA364 probe binding alphav beta3 integrins. *Contrast Media & Molecular Imaging*, *8*, 350–360.

Corsino, P. E., Davis, B. J., Norgaard, P. H., Parker, N. N., Law, M., Dunn, W., et al. (2008). Mammary tumors initiated by constitutive Cdk2 activation contain an invasive basal-like component. *Neoplasia*, *10*, 1240–1252.

DeBortoli, M. E., Abou-Issa, H., Haley, B. E., & Cho-Chung, Y. S. (1985). Amplified expression of p21 ras protein in hormone-dependent mammary carcinomas of humans and rodents. *Biochemical and Biophysical Research Communications*, *127*, 699–706.

Deming, S. L., Nass, S. J., Dickson, R. B., & Trock, B. J. (2000). C-myc amplification in breast cancer: A meta-analysis of its occurrence and prognostic relevance. *British Journal of Cancer*, *83*, 1688–1695.

Deng, C. X., & Scott, F. (2000). Role of the tumor suppressor gene Brca1 in genetic stability and mammary gland tumor formation. *Oncogene*, *19*, 1059–1064.

Dilworth, S. M. (2002). Polyoma virus middle T antigen and its role in identifying cancer-related molecules. *Nature Reviews. Cancer*, *2*, 951–956.

Dimri, G., Band, H., & Band, V. (2005). Mammary epithelial cell transformation: Insights from cell culture and mouse models. *Breast Cancer Research: BCR*, 7, 171–179.

Donehower, L. A., Godley, L. A., Aldaz, C. M., Pyle, R., Shi, Y. P., Pinkel, D., et al. (1995). Deficiency of p53 accelerates mammary tumorigenesis in Wnt-1 transgenic mice and promotes chromosomal instability. *Genes & Development*, *9*, 882–895.

Donehower, L. A., Harvey, M., Slagle, B. L., McArthur, M. J., Montgomery, C. A., Jr., Butel, J. S., et al. (1992). Mice deficient for p53 are developmentally normal but susceptible to spontaneous tumours. *Nature*, *356*, 215–221.

Fantl, V., Smith, R., Brookes, S., Dickson, C., & Peters, G. (1993). Chromosome 11q13 abnormalities in human breast cancer. *Cancer Surveys*, *18*, 77–94.

Fantl, V., Stamp, G., Andrews, A., Rosewell, I., & Dickson, C. (1995). Mice lacking cyclin D1 are small and show defects in eye and mammary gland development. *Genes & Development*, *9*, 2364–2372.

Fantozzi, A., & Christofori, G. (2006). Mouse models of breast cancer metastasis. *Breast Cancer Research*, *8*, 212.

Folkman, J. (1992). The role of angiogenesis in tumor growth. *Seminars in Cancer Biology*, *3*, 65–71.

Friedel, R. H., Wurst, W., Wefers, B., & Kuhn, R. (2011). Generating conditional knockout mice. *Methods in Molecular Biology*, *693*, 205–231.

Fu, M., Wang, C., Li, Z., Sakamaki, T., & Pestell, R. G. (2004). Minireview: Cyclin D1: Normal and abnormal functions. *Endocrinology*, *145*, 5439–5447.

Galie, M., D'Onofrio, M., Calderan, L., Nicolato, E., Amici, A., Crescimanno, C., et al. (2004). In vivo mapping of spontaneous mammary tumors in transgenic mice using MRI and ultrasonography. *Journal of Magnetic Resonance Imaging*, *19*, 570–579.

Gillett, C., Fantl, V., Smith, R., Fisher, C., Bartek, J., Dickson, C., et al. (1994). Amplification and overexpression of cyclin D1 in breast cancer detected by immunohistochemical staining. *Cancer Research*, *54*, 1812–1817.

Gillett, C., Smith, P., Gregory, W., Richards, M., Millis, R., Peters, G., et al. (1996). Cyclin D1 and prognosis in human breast cancer. *International Journal of Cancer*, *69*, 92–99.

Gowen, L. C., Johnson, B. L., Latour, A. M., Sulik, K. K., & Koller, B. H. (1996). Brca1 deficiency results in early embryonic lethality characterized by neuroepithelial abnormalities. *Nature Genetics*, *12*, 191–194.

Green, J. E., Shibata, M. A., Yoshidome, K., Liu, M. L., Jorcyk, C., Anver, M. R., et al. (2000). The C3(1)/SV40 T-antigen transgenic mouse model of mammary cancer: Ductal epithelial cell targeting with multistage progression to carcinoma. *Oncogene*, *19*, 1020–1027.

Grum-Schwensen, B., Klingelhofer, J., Grigorian, M., Almholt, K., Nielsen, B. S., Lukanidin, E., et al. (2010). Lung metastasis fails in MMTV-PyMT oncomice lacking S100A4 due to a T-cell deficiency in primary tumors. *Cancer Research*, *70*, 936–947.

Gusterson, B. A., & Stein, T. (2012). Human breast development. *Seminars in Cell & Developmental Biology*, *23*, 567–573.

Guy, C. T., Cardiff, R. D., & Muller, W. J. (1992). Induction of mammary tumors by expression of polyomavirus middle T oncogene: A transgenic mouse model for metastatic disease. *Molecular and Cellular Biology*, *12*, 954–961.

Guy, C. T., Cardiff, R. D., & Muller, W. J. (1996). Activated neu induces rapid tumor progression. *The Journal of Biological Chemistry*, *271*, 7673–7678.

Guy, C. T., Muthuswamy, S. K., Cardiff, R. D., Soriano, P., & Muller, W. J. (1994). Activation of the c-Src tyrosine kinase is required for the induction of mammary tumors in transgenic mice. *Genes & Development*, *8*, 23–32.

Guy, C. T., Webster, M. A., Schaller, M., Parsons, T. J., Cardiff, R. D., & Muller, W. J. (1992). Expression of the neu protooncogene in the mammary epithelium of transgenic mice induces metastatic disease. *Proceedings of the National Academy of Sciences of the United States of America*, *89*, 10578–10582.

Hakem, R., de la Pompa, J. L., Sirard, C., Mo, R., Woo, M., Hakem, A., et al. (1996). The tumor suppressor gene Brca1 is required for embryonic cellular proliferation in the mouse. *Cell*, *85*, 1009–1023.

Hallett, M. A., Teng, B., Hasegawa, H., Schwab, L. P., Seagroves, T. N., & Pourmotabbed, T. (2013). Anti-matrix metalloproteinase-9 DNAzyme decreases tumor growth in the MMTV-PyMT mouse model of breast cancer. *Breast Cancer Research: BCR*, *15*, R12.

Halter, S. A., Dempsey, P., Matsui, Y., Stokes, M. K., Graves-Deal, R., Hogan, B. L., et al. (1992). Distinctive patterns of hyperplasia in transgenic mice with mouse mammary tumor virus transforming growth factor-alpha. Characterization of mammary gland and skin proliferations. *The American Journal of Pathology*, *140*, 1131–1146.

Hanahan, D., & Weinberg, R. A. (2000). The hallmarks of cancer. *Cell*, *100*, 57–70.

Hanahan, D., & Weinberg, R. A. (2011). Hallmarks of cancer: The next generation. *Cell*, *144*, 646–674.

Hand, P. H., Thor, A., Wunderlich, D., Muraro, R., Caruso, A., & Schlom, J. (1984). Monoclonal antibodies of predefined specificity detect activated ras gene expression in human mammary and colon carcinomas. *Proceedings of the National Academy of Sciences of the United States of America*, *81*, 5227–5231.

Hens, J. R., & Wysolmerski, J. J. (2005). Key stages of mammary gland development: Molecular mechanisms involved in the formation of the embryonic mammary gland. *Breast Cancer Research: BCR*, *7*, 220–224.

Hill, A. D., Doyle, J. M., McDermott, E. W., & O'Higgins, N. J. (1997). Hereditary breast cancer. *The British Journal of Surgery*, *84*, 1334–1339.

Hofker, M. H., & van Deursen, J. (Eds.), (2011). *Transgenic mouse methods and protocols*. (2nd ed.). *Methods in molecular biology*: *Vol. 693*. ISBN: 978-1-60761-973-4.

Hosokawa, Y., Papanikolaou, A., Cardiff, R. D., Yoshimoto, K., Bernstein, M., Wang, T. C., et al. (2001). In vivo analysis of mammary and non-mammary tumorigenesis in MMTV-cyclin D1 transgenic mice deficient in p53. *Transgenic Research*, *10*, 471–478.

Howe, L. R., Subbaramaiah, K., Patel, J., Masferrer, J. L., Deora, A., Hudis, C., et al. (2002). Celecoxib, a selective cyclooxygenase 2 inhibitor, protects against human epidermal growth factor receptor 2 (HER-2)/neu-induced breast cancer. *Cancer Research*, *62*, 5405–5407.

Hughes, E. D., & Saunders, T. L. (2011). Gene targeting in embryonic stem cells. In S. Pease & T. L. Saunders (Eds.), *Advanced protocols for animal transgenesis: An ISTT manual*. Berlin: Springer Berlin Heidelberg. ISBN: 978-3-642-20791-4.

Hundley, J. E., Koester, S. K., Troyer, D. A., Hilsenbeck, S. G., Barrington, R. E., & Windle, J. J. (1997). Differential regulation of cell cycle characteristics and apoptosis in MMTV-myc and MMTV-ras mouse mammary tumors. *Cancer Research*, *57*, 600–603.

Hundley, J. E., Koester, S. K., Troyer, D. A., Hilsenbeck, S. G., Subler, M. A., & Windle, J. J. (1997). Increased tumor proliferation and genomic instability without decreased apoptosis in MMTV-ras mice deficient in p53. *Molecular and Cellular Biology*, *17*, 723–731.

Koboldt, D. C., Fulton, R. S., McLellan, M. D., Schmidt, H., Kalicki-Veizer, J., McMichael, J. F., et al. (2012). Comprehensive molecular portraits of human breast tumors. *Nature*, *490*, 61–70.

Koenders, P. G., Beex, L. V., Kienhuis, C. B., Kloppenborg, P. W., & Benraad, T. J. (1993). Epidermal growth factor receptor and prognosis in human breast cancer: A prospective study. *Breast Cancer Research and Treatment*, *25*, 21–27.

Korpelainen, E. I., Karkkainen, M. J., Tenhunen, A., Lakso, M., Rauvala, H., Vierula, M., et al. (1998). Overexpression of VEGF in testis and epididymis causes infertility in transgenic mice: Evidence for nonendothelial targets for VEGF. *The Journal of Cell Biology*, *143*, 1705–1712.

Kuperwasser, C., Hurlbut, G. D., Kittrell, F. S., Dickinson, E. S., Laucirica, R., Medina, D., et al. (2000). Development of spontaneous mammary tumors in BALB/c p53 heterozygous mice. A model for Li-Fraumeni syndrome. *The American Journal of Pathology*, *157*, 2151–2159.

Kwan, H., Pecenka, V., Tsukamoto, A., Parslow, T. G., Guzman, R., Lin, T. P., et al. (1992). Transgenes expressing the Wnt-1 and int-2 proto-oncogenes cooperate during mammary carcinogenesis in doubly transgenic mice. *Molecular and Cellular Biology*, *12*, 147–154.

Lachman, L. B., Rao, X. M., Kremer, R. H., Ozpolat, B., Kiriakova, G., & Price, J. E. (2001). DNA vaccination against neu reduces breast cancer incidence and metastasis in mice. *Cancer Gene Therapy*, *8*, 259–268.

Lakshminarayanan, V., Thompson, P., Wolfert, M. A., Buskas, T., Bradley, J. M., Pathangey, L. B., et al. (2012). Immune recognition of tumor-associated mucin MUC1 is achieved by a fully synthetic aberrantly glycosylated MUC1 tripartite vaccine. *Proceedings of the National Academy of Sciences of the United States of America*, *109*, 261–266.

Lammie, G. A., & Peters, G. (1991). Chromosome 11q13 abnormalities in human cancer. *Cancer Cells*, *3*, 413–420.

Lane, T. F., & Leder, P. (1997). Wnt-10b directs hypermorphic development and transformation in mammary glands of male and female mice. *Oncogene*, *15*, 2133–2144.

Lanza-Jacoby, S., Miller, S., Flynn, J., Gallatig, K., Daskalakis, C., Masferrer, J. L., et al. (2003). The cyclooxygenase-2 inhibitor, celecoxib, prevents the development of mammary tumors in Her-2/neu mice. *Cancer Epidemiology, Biomarkers & Prevention*, *12*, 1486–1491.

Lewandoski, M. (2001). Conditional control of gene expression in the mouse. *Nature Reviews. Genetics*, *2*, 743–755.

Li, Y., Hively, W. P., & Varmus, H. E. (2000). Use of MMTV-Wnt-1 transgenic mice for studying the genetic basis of breast cancer. *Oncogene*, *19*, 1002–1009.

Li, M., Hu, J., Heermeier, K., Hennighausen, L., & Furth, P. A. (1996a). Expression of a viral oncoprotein during mammary gland development alters cell fate and function: Induction of p53-independent apoptosis is followed by impaired milk protein production in surviving cells. *Cell Growth & Differentiation*, 7(1), 3–11.

Li, M., Hu, J., Heermeier, K., Hennighausen, L., & Furth, P. A. (1996b). Apoptosis and remodeling of mammary gland tissue during involution proceeds through p53-independent pathways. *Cell Growth & Differentiation: The molecular biology journal of the American Association for Cancer Research*, 7(1), 13–20.

Li, M., Lewis, B., Capuco, A. V., Laucirica, R., & Furth, P. A. (2000). WAP-TAg transgenic mice and the study of dysregulated cell survival, proliferation, and mutation during breast carcinogenesis. *Oncogene*, *19*, 1010–1019.

Li, B., Murphy, K. L., Laucirica, R., Kittrell, F., Medina, D., & Rosen, J. M. (1998). A transgenic mouse model for mammary carcinogenesis. *Oncogene*, *16*, 997–1007.

Liao, D. J., & Dickson, R. B. (2000). c-Myc in breast cancer. *Endocrine-related Cancer*, 7, 143–164.

Lin, D. I., Lessie, M. D., Gladden, A. B., Bassing, C. H., Wagner, K. U., & Diehl, J. A. (2008). Disruption of cyclin D1 nuclear export and proteolysis accelerates mammary carcinogenesis. *Oncogene*, *27*, 1231–1242.

Liu, C. Y., Flesken-Nikitin, A., Li, S., Zeng, Y., & Lee, W. H. (1996). Inactivation of the mouse Brca1 gene leads to failure in the morphogenesis of the egg cylinder in early postimplantation development. *Genes & Development*, *10*, 1835–1843.

Lu, C., Speers, C., Zhang, Y., Xu, X., Hill, J., Steinbis, E., et al. (2003). Effect of epidermal growth factor receptor inhibitor on development of estrogen receptor-negative mammary tumors. *Journal of the National Cancer Institute*, *95*(24), 1825–1833.

Malkin, D. (2011). Li-fraumeni syndrome. *Genes & Cancer*, *2*, 475–484.

Maroulakou, I. G., Anver, M., Garrett, L., & Green, J. E. (1994). Prostate and mammary adenocarcinoma in transgenic mice carrying a rat C3(1) simian virus 40 large tumor antigen fusion gene. *Proceedings of the National Academy of Sciences of the United States of America*, *91*, 11236–11240.

Marozkina, N. V., Stiefel, S. M., Frierson, H. F., Jr., & Parsons, S. J. (2008). MMTV-EGF receptor transgene promotes preneoplastic conversion of multiple steroid hormone-responsive tissues. *Journal of Cellular Biochemistry*, *103*, 2010–2018.

Marquis, S. T., Rajan, J. V., Wynshaw-Boris, A., Xu, J., Yin, G. Y., Abel, K. J., et al. (1995). The developmental pattern of Brca1 expression implies a role in differentiation of the breast and other tissues. *Nature Genetics*, *11*, 17–26.

Matsui, Y., Halter, S. A., Holt, J. T., Hogan, B. L., & Coffey, R. J. (1990). Development of mammary hyperplasia and neoplasia in MMTV-TGF alpha transgenic mice. *Cell*, *61*, 1147–1155.

Mes, A. M., & Hassell, J. A. (1982). Polyoma viral middle T-antigen is required for transformation. *Journal of Virology*, *42*, 621–629.

Moller, P., Mechtersheimer, G., Kaufmann, M., Moldenhauer, G., Momburg, F., Mattfeldt, T., et al. (1989). Expression of epidermal growth factor receptor in benign and malignant primary tumours of the breast. *Virchows Archiv. A, Pathological Anatomy and Histopathology*, *414*, 157–164.

Moody, S. E., Sarkisian, C. J., Hahn, K. T., Gunther, E. J., Pickup, S., Dugan, K. D., et al. (2002). Conditional activation of Neu in the mammary epithelium of transgenic mice results in reversible pulmonary metastasis. *Cancer Cell*, *2*, 451–461.

Muller, W. J., Sinn, E., Pattengale, P. K., Wallace, R., & Leder, P. (1988). Single-step induction of mammary adenocarcinoma in transgenic mice bearing the activated c-neu oncogene. *Cell*, *54*, 105–115.

Muller, P. A. J., & Vousden, K. H. (2013). p53 mutations in cancer. *Nature Cell Biology*, *15*, 2–8.

Muthuswamy, S. K., Siegel, P. M., Dankort, D. L., Webster, M. A., & Muller, W. J. (1994). Mammary tumors expressing the neu proto-oncogene possess elevated c-Src tyrosine kinase activity. *Molecular and Cellular Biology*, *14*, 735–743.

Neve, R. M., Chin, K., Fridlyand, J., Yeh, J., Baehner, F. L., Fevr, T., et al. (2006). A collection of breast cancer cell lines for the study of functionally distinct cancer subtypes. *Cancer Cell*, *10*(6), 515–527.

Nguyen, D. X., Bos, P. D., & Massague, J. (2009). Metastasis: From dissemination to organ-specific colonization. *Nature Reviews. Cancer*, *9*, 274–284.

Obermair, A., Kucera, E., Mayerhofer, K., Speiser, P., Seifert, M., Czerwenka, K., et al. (1997). Vascular endothelial growth factor (VEGF) in human breast cancer: Correlation with disease-free survival. *International Journal of Cancer*, *74*, 455–458.

Olivier, M., Hollstein, M., & Hainaut, P. (2010). TP53 mutations in human cancers: Origins, consequences, and clinical use. *Cold Spring Harbor Perspectives in Biology*, *2*, a001008.

Ottewell, P. D., Coleman, R. E., & Holen, I. (2006). From genetic abnormality to metastases: Murine models of breast cancer and their use in the development of anticancer therapies. *Breast Cancer Research and Treatment, 96*, 101–113.

Palmiter, R. D., Sandgren, E. P., Koeller, D. M., & Brinster, R. L. (1993). Distal regulatory elements from the mouse metallothionein locus stimulate gene expression in transgenic mice. *Molecular and Cellular Biology, 13*, 5266–5275.

Pease, S., & Saunders, T. L. (Eds.), (2011). *Advanced protocols for animal transgenesis. An ISTT manual.* ISBN: 978-3-642-20791-4.

Petitjean, A., Achatz, M. I., Borresen-Dale, A. L., Hainaut, P., & Olivier, M. (2007). TP53 mutations in human cancers: Functional selection and impact on cancer prognosis and outcomes. *Oncogene, 26*(15), 2157–2165.

Ray, D., Terao, Y., Christov, K., Kaldis, P., & Kiyokawa, H. (2011). Cdk2-null mice are resistant to ErbB-2-induced mammary tumorigenesis. *Neoplasia, 13*, 439–444.

Ray, D., Terao, Y., Nimbalkar, D., Hirai, H., Osmundson, E. C., Zou, X., et al. (2007). Hemizygous disruption of Cdc25A inhibits cellular transformation and mammary tumorigenesis in mice. *Cancer Research, 67*, 6605–6611.

Ren, S., Cai, H. R., Li, M., & Furth, P. A. (2002). Loss of Stat5a delays mammary cancer progression in a mouse model. *Oncogene, 21*, 4335–4339.

Robinson, G. W. (2007). Cooperation of signalling pathways in embryonic mammary gland development. *Nature Reviews. Genetics, 8*, 963–972.

Russo, J., Moral, R., Balogh, G. A., Mailo, D., & Russo, I. H. (2005). The protective role of pregnancy in breast cancer. *Breast Cancer Research*, 7, 131–142.

Russo, J., & Russo, I. H. (2004). Development of the human breast. *Maturitas, 49*, 2–15.

Sandgren, E. P., Schroeder, J. A., Qui, T. H., Palmiter, R. D., Brinster, R. L., & Lee, D. C. (1995). Inhibition of mammary gland involution is associated with transforming growth factor alpha but not c-myc-induced tumorigenesis in transgenic mice. *Cancer Research, 55*, 3915–3927.

Santarelli, R., Tzeng, Y. J., Zimmermann, C., Guhl, E., & Graessmann, A. (1996). SV40 T-antigen induces breast cancer formation with a high efficiency in lactating and virgin WAP-SV-T transgenic animals but with a low efficiency in ovariectomized animals. *Oncogene, 12*, 495–505.

Saunders, T. L. (2011). Inducible transgenic mouse models. *Methods in Molecular Biology, 693*, 103–115.

Schoeffner, D. J., Matheny, S. L., Akahane, T., Factor, V., Berry, A., Merlino, G., et al. (2005). VEGF contributes to mammary tumor growth in transgenic mice through paracrine and autocrine mechanisms. *Laboratory Investigation; a journal of technical methods and pathology, 85*, 608–623.

Schoenenberger, C. A., Andres, A. C., Groner, B., van der Valk, M., LeMeur, M., & Gerlinger, P. (1988). Targeted c-myc gene expression in mammary glands of transgenic mice induces mammary tumours with constitutive milk protein gene transcription. *The EMBO Journal*, 7, 169–175.

Schoenherr, R. M., Kelly-Spratt, K. S., Lin, C., Whiteaker, J. R., Liu, T., Holzman, T., et al. (2011). Proteome and transcriptome profiles of a Her2/Neu-driven mouse model of breast cancer. *Proteomics. Clinical Applications, 5*, 179–188.

Schondorf, T., Andrack, A., Niederacher, D., Scharl, A., Becker, M., Engel, H., et al. (1999). H-ras gene amplification or mutation is not common in human primary breast cancer. *Oncology Reports, 6*, 1029–1033.

Schreiber, A. B., Winkler, M. E., & Derynck, R. (1986). Transforming growth factor-alpha: A more potent angiogenic mediator than epidermal growth factor. *Science, 232*, 1250–1253.

Shibata, M. A., Liu, M. L., Knudson, M. C., Shibata, E., Yoshidome, K., Bandey, T., et al. (1999). Haploid loss of bax leads to accelerated mammary tumor development in

C3(1)/SV40-TAg transgenic mice: Reduction in protective apoptotic response at the preneoplastic stage. *The EMBO Journal, 18*, 2692–2701.

Shibata, M. A., Maroulakou, I. G., Jorcyk, C. L., Gold, L. G., Ward, J. M., & Green, J. E. (1996). p53-independent apoptosis during mammary tumor progression in C3(1)/SV40 large T antigen transgenic mice: Suppression of apoptosis during the transition from preneoplasia to carcinoma. *Cancer Research, 56*, 2998–3003.

Sicinski, P., Donaher, J. L., Parker, S. B., Li, T., Fazeli, A., Gardner, H., et al. (1995). Cyclin D1 provides a link between development and oncogenesis in the retina and breast. *Cell, 82*, 621–630.

Siegel, P. M., Dankort, D. L., Hardy, W. R., & Muller, W. J. (1994). Novel activating mutations in the neu proto-oncogene involved in induction of mammary tumors. *Molecular and Cellular Biology, 14*, 7068–7077.

Siegel, P. M., Ryan, E. D., Cardiff, R. D., & Muller, W. J. (1999). Elevated expression of activated forms of Neu/ErbB-2 and ErbB-3 are involved in the induction of mammary tumors in transgenic mice: Implications for human breast cancer. *The EMBO Journal, 18*, 2149–2164.

Sinn, E., Muller, W., Pattengale, P., Tepler, I., Wallace, R., & Leder, P. (1987). Coexpression of MMTV/v-Ha-ras and MMTV/c-myc genes in transgenic mice: Synergistic action of oncogenes in vivo. *Cell, 49*, 465–475.

Slamon, D. J., deKernion, J. B., Verma, I. M., & Cline, M. J. (1984). Expression of cellular oncogenes in human malignancies. *Science, 224*, 256–262.

Spandidos, D. A. (1987). Oncogene activation in malignant transformation: A study of H-ras in human breast cancer. *Anticancer Research, 7*, 991–996.

Spandidos, D. A., Field, J. K., Agnantis, N. J., Evan, G. I., & Moore, J. P. (1989). High levels of c-myc protein in human breast tumours determined by a sensitive ELISA technique. *Anticancer Research, 9*, 821–826.

Stewart, T. A., Pattengale, P. K., & Leder, P. (1984). Spontaneous mammary adenocarcinomas in transgenic mice that carry and express MTV/myc fusion genes. *Cell, 38*, 627–637.

Sukumar, S., McKenzie, K., & Chen, Y. (1995). Animal models for breast cancer. *Mutation Research, 333*, 37–44.

Taneja, P., Frazier, D. P., Kendig, R. D., Maglic, D., Sugiyama, T., Kai, F., et al. (2009). MMTV mouse models and the diagnostic values of MMTV-like sequences in human breast cancer. *Expert Review of Molecular Diagnostics, 9*, 423–440.

Tremblay, P. J., Pothier, F., Hoang, T., Tremblay, G., Brownstein, S., Liszauer, A., et al. (1989). Transgenic mice carrying the mouse mammary tumor virus ras fusion gene: Distinct effects in various tissues. *Molecular and Cellular Biology, 9*, 854–859.

Tsukamoto, A. S., Grosschedl, R., Guzman, R. C., Parslow, T., & Varmus, H. E. (1988). Expression of the int-1 gene in transgenic mice is associated with mammary gland hyperplasia and adenocarcinomas in male and female mice. *Cell, 55*, 619–625.

Tzeng, Y. J., Gottlob, K., Santarelli, R., & Graessmann, A. (1996). The SV40 T-antigen induces premature apoptotic mammary gland involution during late pregnancy in transgenic mice. *FEBS Letters, 380*, 215–218.

Tzeng, Y. J., Guhl, E., Graessmann, M., & Graessmann, A. (1993). Breast cancer formation in transgenic animals induced by the whey acidic protein SV40 T antigen (WAP-SV-T) hybrid gene. *Oncogene, 8*, 1965–1971.

Ullrich, R. L., Bowles, N. D., Satterfield, L. C., & Davis, C. M. (1996). Strain-dependent susceptibility to radiation-induced mammary cancer is a result of differences in epithelial cell sensitivity to transformation. *Radiation Research, 146*, 353–355.

Ursini-Siegel, J., Schade, B., Cardiff, R. D., & Muller, W. J. (2007). Insights from transgenic mouse models of ERBB2-induced breast cancer. *Nature Reviews. Cancer, 7*, 389–397.

Vargo-Gogola, T., & Rosen, J. M. (2007). Modelling breast cancer: One size does not fit all. *Nature Reviews Cancer, 7*, 659–672.

Voncken, J. W. (2011). Genetic modification of the mouse: General technology - pronuclear and blastocyst injection. *Methods in Molecular Biology*, *693*, 11–36.

Wagner, K. U., McAllister, K., Ward, T., Davis, B., Wiseman, R., & Hennignausen, L. (2001). Spatial and temporal expression of the Cre gene under the control of the MMTV-LTR in different lines of transgenic mice. *Transgenic Research*, *10*, 545–553.

Wang, T. C., Cardiff, R. D., Zukerberg, L., Lees, E., Arnold, A., & Schmidt, E. V. (1994). Mammary hyperplasia and carcinoma in MMTV-cyclin D1 transgenic mice. *Nature*, *369*, 669–671.

Wang, C., Schwab, L. P., Fan, M., Seagroves, T. N., & Buolamwini, J. K. (2013). Chemoprevention activity of dipyridamole in the MMTV-PyMT transgenic mouse model of breast cancer. *Cancer Prevention Research (Phila)*, *6*, 437–447.

Watson, C. J., & Khaled, W. T. (2008). Mammary development in the embryo and adult: A journey of morphogenesis and commitment. *Development*, *135*, 995–1003.

Weaver, Z., Montagna, C., Xu, X., Howard, T., Gadina, M., Brodie, S. G., et al. (2002). Mammary tumors in mice conditionally mutant for Brca1 exhibit gross genomic instability and centrosome amplification yet display a recurring distribution of genomic imbalances that is similar to human breast cancer. *Oncogene*, *21*, 5097–5107.

Webster, J., Wallace, R. M., Clark, A. J., & Whitelaw, C. B. (1995). Tissue-specific, temporally regulated expression mediated by the proximal ovine beta-lactoglobulin promoter in transgenic mice. *Cellular & Molecular Biology Research*, *41*, 11–15.

Wen, J., Kawamata, Y., Tojo, H., Tanaka, S., & Tachi, C. (1995). Expression of whey acidic protein (WAP) genes in tissues other than the mammary gland in normal and transgenic mice expressing mWAP/hGH fusion gene. *Molecular and Reproductive Development*, *41*, 399–406.

Wiesen, J. F., Young, P., Werb, Z., & Cunha, G. R. (1999). Signaling through the stromal epidermal growth factor receptor is necessary for mammary ductal development. *Development*, *126*, 335–344.

Wyckoff, J. B., Wang, Y., Lin, E. Y., Li, J. F., Goswami, S., Stanley, E. R., et al. (2007). Direct visualization of macrophage-assisted tumor cell intravasation in mammary tumors. *Cancer Research*, *67*, 2649–2656.

Wyckoff, J., Wang, W., Lin, E. Y., Wang, Y., Pixley, F., Stanley, E. R., et al. (2004). A paracrine loop between tumor cells and macrophages is required for tumor cell migration in mammary tumors. *Cancer Research*, *64*, 7022–7029.

Xu, X., Wagner, K. U., Larson, D., Weaver, Z., Li, C., Ried, T., et al. (1999). Conditional mutation of Brca1 in mammary epithelial cells results in blunted ductal morphogenesis and tumour formation. *Nature Genetics*, *22*, 37–43.

Yoshidome, K., Shibata, M. A., Couldrey, C., Korach, K. S., & Green, J. E. (2000). Estrogen promotes mammary tumor development in C3(1)/SV40 large T-antigen transgenic mice: Paradoxical loss of estrogen receptoralpha expression during tumor progression. *Cancer Research*, *60*, 6901–6910.

Yoshiji, H., Gomez, D. E., Shibuya, M., & Thorgeirsson, U. P. (1996). Expression of vascular endothelial growth factor, its receptor, and other angiogenic factors in human breast cancer. *Cancer Research*, *56*, 2013–2016.

Yu, Q., Geng, Y., & Sicinski, P. (2001). Specific protection against breast cancers by cyclin D1 ablation. *Nature*, *411*, 1017–1021.

Yu, Q., Sicinska, E., Geng, Y., Ahnstrom, M., Zagozdzon, A., Kong, Y., et al. (2006). Requirement for CDK4 kinase function in breast cancer. *Cancer Cell*, *9*, 23–32.

Zagozdzon, A. M., O'Leary, P., Callanan, J. J., Crown, J., Gallagher, W. M., & Zagozdzon, R. (2012). Generation of a new bioluminescent model for visualisation of mammary tumour development in transgenic mice. *BMC Cancer*, *12*, 209.

Zhang, M., Shi, Y., Magit, D., Furth, P. A., & Sager, R. (2000). Reduced mammary tumor progression in WAP-TAg/WAP-maspin bitransgenic mice. *Oncogene*, *19*, 6053–6058.

CHAPTER NINE

Life is Three Dimensional—As *In Vitro* Cancer Cultures Should Be

I. Levinger, Y. Ventura, R. Vago[1]

Avram and Stella Goldstein-Goren Department of Biotechnology Engineering, Ben-Gurion University, Beer-Sheva, Israel

[1]Corresponding author: e-mail address: rvago@bgu.ac.il

Contents

Abstract

For many decades, fundamental cancer research has relied on two-dimensional *in vitro* cell culture models. However, these provide a poor representation of the complex three-dimensional (3D) architecture of living tissues. The more recent 3D culture systems, which range from ridged scaffolds to semiliquid gels, resemble their natural counterparts more closely. The arrangement of the cells in 3D systems allows better cell–cell interaction and facilitates extracellular matrix secretion, with concomitant effects on gene and protein expression and cellular behavior. Many studies have reported differences between 3D and 2D systems as regards responses to therapeutic agents and pivotal cellular processes such as cell differentiation, morphology, and signaling pathways, demonstrating the importance of 3D culturing for various cancer cell lines.

Cells cultured in 2D systems are forced to settle onto a flat surface, which means that nearly 50% of their surface area is dedicated to adhesion, leaving only a limited portion available for cell–cell interactions (Fig. 9.1A). This enforced cell

Advances in Cancer Research, Volume 121
ISSN 0065-230X
http://dx.doi.org/10.1016/B978-0-12-800249-0.00009-3

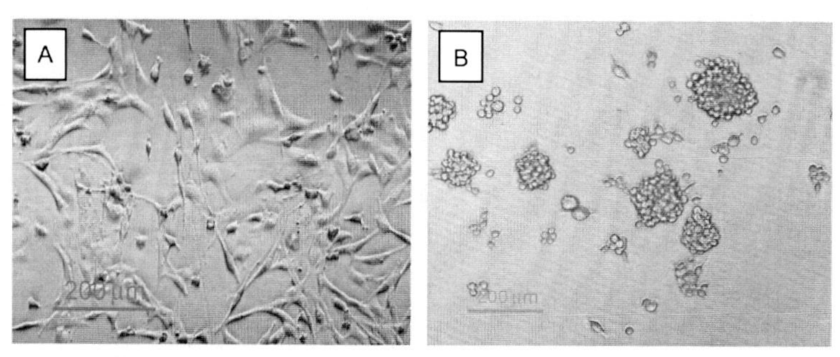

Figure 9.1 ***Cell morphology.*** (A) 2D traditional monolayer. (B) Fibrosarcoma cells aggregates 5 days post seeding.

configuration influences cell population properties, including cell arrangement and morphology (Ghosh et al., 2005), and disrupts normal cellular processes such as cell–cell communications and cell–matrix interactions (Kim, Stein, & O'Hare, 2005). In cancer research, however, accurate mimicry of native cancer tissues in three-dimensional (3D) cultures is crucial for a reliable understanding of the characteristics of cancer cells at the molecular and cellular levels in *in vivo* processes (Ghosh et al., 2005; Kim, 2005). The development of suitable 3D models for cancer research is thus of the utmost importance.

It has recently become apparent that the cell population in the cancer microenvironment is heterogeneous and comprises tumor-supporting stromal cell types such as fibroblasts, endothelial cells, and, probably, dynamic stem cells. Therefore 3D culture systems, if they are to mimic living tissues, must also enable culturing of heterogeneous cell populations and support interactions between different cell types. In fact, even if seeded with only one cell type, 3D microenvironments in variably display heterogeneity—unlike 2D cultures. This is because the 3D matrix provides an assortment of conditions for cell development, leading to the creation of diverse subpopulations—proliferating cells at the periphery and nonproliferating, or even necrotic, cells in portions of the matrix farther removed from the supplemented medium. Such a culture mimics the features of solid tumors in a more realistic way than homogeneous 2D cultures (Figs. 9.2 and 9.3A–C).

1. 3D CELL CULTURE METHODS AND SCAFFOLDING MATERIALS

There are three types of 3D cell culture: (1) Cultures relying on induction of spontaneous cell aggregation, (2) cultures using natural and synthetic hydrogels, and (3) cultures on natural and synthetic solid materials.

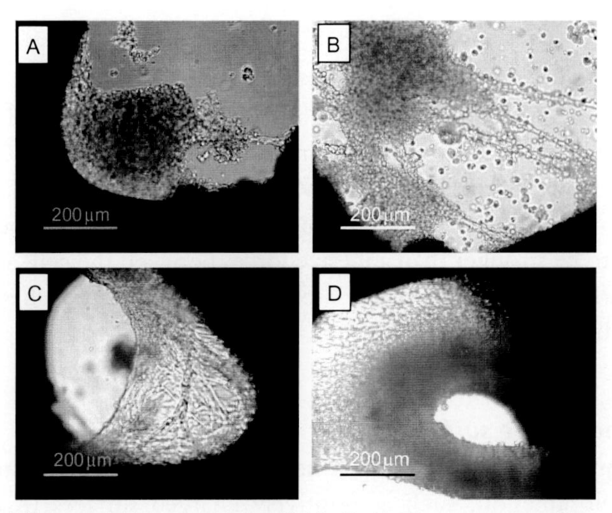

Figure 9.2 ***Cells morphology on 3D scaffold.*** (A–C) Fibrosarcoma cells seeded on 3D scaffold from *Tetraclita rifotincta* forming a tissue-like structure, pictures were taken 7, 9, 12, 16 days post seeding. (See the color plate.)

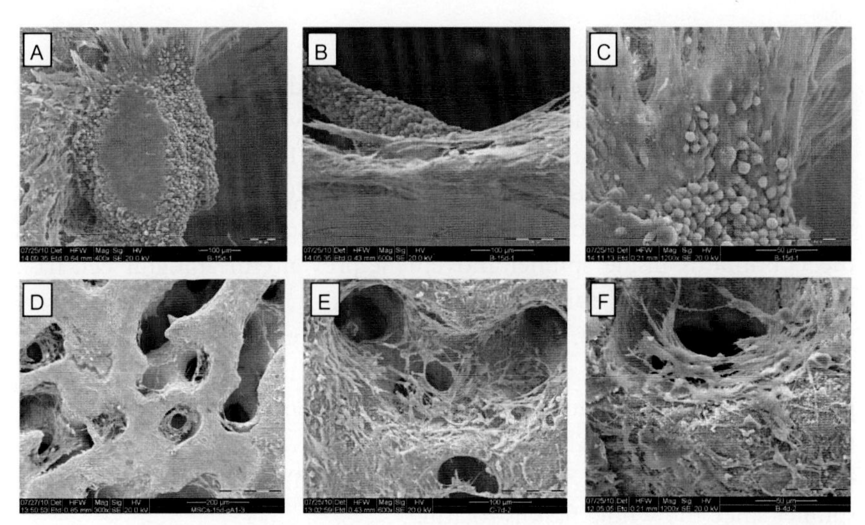

Figure 9.3 ***Cells morphology by SEM on 3D scaffolds.*** (A–C) Fibrosarcoma cells seeded on 3D scaffold from *Tetraclita rifotincta* forming cells aggregate with round shape cells, magnification ×400, 600, 1200, respectively. (D–F) Fibrosarcoma cells seeded on *Porites lutea* aragonite scaffold forming tissue-like structures, magnification ×300, 600, 1200, respectively. (C and F) Cells secrete ECM.

1.1. Spontaneous cell aggregation into multicellular spheroids

Cultures of cells aggregated into spheroids have been shown to successfully mimic solid tumor heterogeneity. Already in 1970, Sutherland et al. proposed a multispheroid model as a bridge between traditional 2D culture and living tissues (Sutherland, Inch, McCredie, & Kruuv, 1970). However, in order for cells to aggregate into spheroids, adhesive forces between the cells need to be stronger than adhesive forces between the cells and the surface (Fig. 9.1B; Kim, Stein, & O'Hare, 2004). Several approaches have been used to achieve this.

1.1.1 Nonadhesive surface

In one approach, spheroids are formed by seeding cells onto surfaces that prevent cell adhesion, such as positively charged surfaces (Koide et al., 1990; Lin & Chang, 2008), hydrophobic polymers, polyethylene glycol, albumin coating (Koide et al., 1990), agarose thin films (Emfietzoglou et al., 2005; Yuhas, Li, Martinez, & Ladman, 1977), or poly-2-hydroxyethyl methacrylate (HEMA; Bae et al., 1999). Being unable to attach to the surface, cells tend to attach to each other and form spheroids.

1.1.2 External force

Alternatively, spheroids can form spontaneously due to physical induction by an external force. The spinner flask is the device most widely used to induce spheroid formation by this method (Rodday, Hirschhaeuser, Walenta, & Mueller-Klieser, 2011; Wang et al., 2013). In this kind of bioreactor, cells are grown in a liquid suspension in a static cell culture flask, and the continuous inside spinner movement prevents meaningful cell-flask contact, allowing the formation of spheroids of uniform shape and size (Bartholoma et al., 2005; Goodman, Olive, & Pun, 2007).

Beside spinner flasks, there are other methods based on a similar principle, such as gyratory rotators, where the cell suspension is rotated in a gyratory rotation incubator till spheroids of the required size are obtained (Kim et al., 2004; Pan, Zhou, Zhang, Gao, & Xu, 2013); here, there is no inside spinner since the whole flask is rotating.

There are also microgravity systems, where cells remain suspended in a weightless environment and aggregate spontaneously to form 3D microstructures (Infanger et al., 2006; Ingram et al., 1997; Jessup et al., 2000; Marrero, Messina, & Heller, 2009).

1.1.3 Liquid overlay

Cell adherence to surfaces can also be prevented using liquid overlay techniques. As a first step, tumor cells are cultured in dishes with a nonadhesive surface; the cells form small aggregates that grow within a few days into multicellular spheroids. The latter can then be transferred to agar-covered surfaces and cultured for many days (Metzger et al., 2011; Nagelkerke, Bussink, Sweep, & Span, 2013; Wu et al., 2010).

1.2. Natural and synthetic hydrogels

Hydrogels are hydrophilic networks of polymer chains, natural or synthetic, in a water-based dispersion medium. They are ideal materials for 3D tissue platforms as their easily controlled density and flexibility facilitate the mimicking of extracellular matrices (ECMs). Hydrogels can contain structural proteins such as laminin, entactin, and collagen (Hughes, Postovit, & Lajoie, 2010; Jin et al., 2013; Rao et al., 2013; Szot, Buchanan, Freeman, & Rylander, 2011; Yamada, Hozumi, Katagiri, Kikkawa, & Nomizu, 2013), and they facilitate the creation of models of various thicknesses to meet the researcher's needs (Fig. 9.4A and C). 3D gel-based models are important in cancer biology as a tool for studying cell migration and invasiveness (David et al., 2004; Kumar et al., 2010; Olsen, Moreira, Lukanidin, & Ambartsumian, 2010; Sarig-Nadir & Seliktar, 2010; Zeng, Cai, & Wu, 2011; Fig. 9.4B) and for evaluation of anticancer therapies (Chen, Nguyen, & Shively, 2010; Choudhary et al., 2010; Fromigue et al., 2003; Gurski, Jha, Zhang, Jia, & Farach-Carson, 2009; Horning et al., 2008). Hydrogels can also serve as a platform for cocultures designed to reproduce the natural heterogenic architecture of cells (Litzenburger et al., 2009; Olsen et al., 2010). Many types of hydrogels are used in cancer research: Matrigel™ (Hughes et al., 2010), Puramatrix™ (Abu-Yousif, Rizvi, Evans, Celli, & Hasan, 2009; Yang & Zhao, 2011), hyaluronan hydrogel (David et al., 2008; Gurski et al., 2009), fibrin hydrogels (Ahearne & Kelly, 2013; Carriel et al., 2013; Gao et al., 2013), and alginate polymer platforms are some examples (Lee et al., 2008; Sung & Shuler, 2009; Zhang et al., 2005).

1.3. Natural and synthetic solid materials

Scaffolds may be prepared from natural or synthesized substances. Natural origin scaffolds are source-variable: they may be tissue based matrices (from bone orcartilage), or they may consist of proteins (collagen, fibronectin), proteoglycans (chitosan), and other substances.

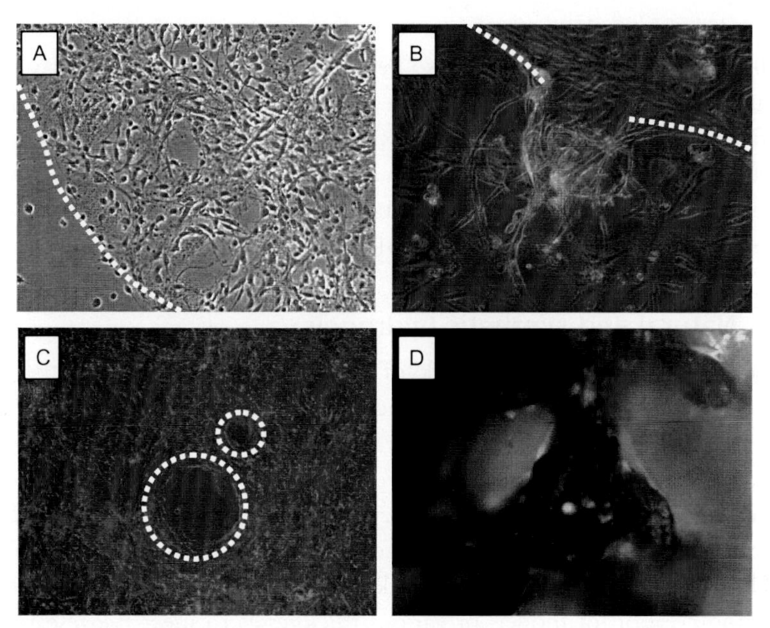

Figure 9.4 ***Different 3D microenvironments for cancer culture and coculture models.*** (A) Matrigel™ droplet with fibrosarcoma cells embedded. (B) Invasion assay in coculture: cancer cells invade to Matrigel™ contains MSCs (red). (C) MSCs (red) and cancer cells (green) in coculture in Matrigel™, tube formation occurs (circles). (D) MSCs (red) and cancer cells (green) in coculture on 3D solid aragonite POR scaffold. (See the color plate.)

1.3.1 Natural solid materials

a. Chitosan, a biodegradable, biocompatible, and nontoxic material shown to be suitable for natural tissue imitation (Elder, Gottipati, Zelenka, & Bumgardner, 2013; Liu et al., 2013; Park et al., 2013), has been used as a scaffold for culturing breast cancer cells (MCF-7; Dhiman, Ray, & Panda, 2004, 2005; Horning et al., 2008); the compatibility of chitosan for this specific cell culture was proved, as one example, by enhanced lactate production by the cancerous breast cells as compared with 2D or with other 3D models (Dhiman et al., 2005).

b. Another example of a successful choice of a 3D model for a particular cell type was reported by Gotoh et al. These researchers cultured hepatocellular carcinoma on a lactose-silk fibroin scaffold (Gotoh, Ishizuka, Matsuura, & Niimi, 2011) and were able to demonstrate elevated expression of genes related to liver-specific functions as compared with another 3D model / other 3D models (Gotoh et al., 2011). The same silk scaffold was used in a other studies (Kundu, Saha, Datta, & Kundu, 2013; Li, Wang, Liu, Xie, & Wei, 2013; Tan, Aung, Toh, Goh, & Nathan,

2011), for example, study with prostate cancer cells; its porous solid structure was shown to be well suited mechanically to supporting the cells of the latter tissue (Kwon et al., 2010).

1.3.2 Synthetic solid materials

a. In a bladder carcinoma study, polycarbonate scaffolds exhibited parameters consistent with native gall bladder and offered a convenient new research tool (Purdum, Ulissi, Hylemon, Shiffman, & Moore, 1993). PLGA/ poly scaffold proved to be adapted to 3D culturing of several cell types (Chang et al., 2013; Kang & Bae, 2009; Sadiasa, Kim, & Lee, 2013), including MCF-7 breast cancer cells (Kang & Bae, 2009; Yang, Basu, Tomasko, Lee, & Yang, 2005).

b. Porous ceramic nonmineralized or mineralized polymeric scaffolds proved useful for evaluating underlying cellular and molecular mechanisms in mineralized and nonmineralized matrices (Gao et al., 2013; Kido et al., 2013; Zhang, Jiang, Zhang, Wang, & Zhen, 2013), as in a study of breast cancer bone metastasis (Pathi, Kowalczewski, Tadipatri, & Fischbach, 2010).

As can be seen, a wide spectrum of models differing in origin (natural, synthetic), material (ceramic, carbon, metal), physical properties (flexibility, density, porosity), and form (gel/hydrogel, fibers/mesh, porous solid) are available for research in the 3D cancer culture field. To assure optimum mirroring of the *in vivo* reality, it is important to select the 3D model most compatible with the cell line under investigation. 3D platforms details are summarized in Table 9.1, while their possible applications and related assays are given in Table 9.2.

The choice of an optimal 3D model should be guided by the particular research goals or questions addressed the methods to be used, and the cell lines. Thus, if researchers wish to investigate cellular properties, say cell line response to a certain anticancer treatment, they should strive to select the model that most closely mimics the natural tumor environment of every cell line included in the study. However, if the aim is to explore tumor-microenvironment coinfluence, the best choice would be a model amenable to modification of characteristics such as matrix rigidity, which may affect cancer cell tumorgenicity.

In studies focusing on the effects of specific molecules on cancer cells, researchers should examine the possibility of using solid models that can be coated with the molecule of interest, as well as hydrogels, in which the molecules can be embedded. Bioactive solid scaffolds should be

Table 9.1 Summary of 3D culturing methods

Method	Description	Advantages	Disadvantages	References
Spontaneous cell aggregation	Wide range of production methods			
a. Multicellular spheroids formed on nonadhesive surfaces	On nonadhesive surfaces, cell–cell adhesive forces are stronger than cell–surface forces	• Suitable for many cell lines • Mimics cells' heterogeneity with high degree of similarity to *in vivo* • Suitable for coculture experiments and studies of interactions between different cells • Inexpensive • Easy to produce	• Variability in size/shape/cell number • Necrotic core forms to due to blocking of nutrients • Limited mass produced • Each spheroid contains a relatively small number of cells	Lin and Chang (2008), Koide et al. (1990), Yuhas et al. (1977), Emfietzoglou et al. (2005), and Bae et al. (1999)
b. Multicellular spheroids formed by application of external force: **(1)** Spinner flask **(2)** Gyratory rotator **(3)** Microgravity system	Cell are induced to aggregate by external forces	• Simple to use • Inexpensive • Massive production • Spheroids obtained are uniform in shape and size • Long-lasting culture • Easy controlled physicochemical parameters	• Requires specialized equipment • The external forces influence the cells, may alter gene expression	Bartholoma et al. (2005), Goodman et al. (2007), Ingram et al. (1997), Jessup et al. (2000), and Infanger et al. (2006)

c. Multicellular spheroids formed by liquid overlay method	Adhesion is prevented by use of nonadhesive surface. After small aggregates grow into multicellular spheroids, latter are transferred to agar-covered surfaces	• Homological spheroids obtained can provide suitable platform for drug scanning • Can be use in adhesion properties study • Inexpensive • Easy to set up and rapid production • Suitable for a variety of cell lines	• Produces only a small amount of spheroids	Metzger et al. (2011) and Wu et al. (2010)
Natural and synthetic hydrogels	Networks of polymer/protein chains in water-based media with controlled density and flexibility	• Biocompatible • Flexibility, density, pore size can be regulated and adapted to the studied cell culture • Suitable for study of 3D cell invasion characteristics and migration mechanisms	• Often expensive • Hydrogels matrix proteins may disrupt study of cell cultures at the protein level	Hughes et al. (2010), Zeng et al. (2011), Kumar et al. (2010), Olsen et al. (2010), Chen et al. (2010), Fromigue et al. (2003), Gurski et al. (2009), Litzenburger et al. (2009), Yang and Zhao (2011), Abu-Yousif et al. (2009), David et al. (2008), Zhang et al. (2005), Sung and Shuler (2009), and Lee et al. (2008)

Continued

Table 9.1 Summary of 3D culturing methods—cont'd

Method	Description	Advantages	Disadvantages	References
Natural and synthetic solid materials	Solid scaffold serves as habitat of seeded cells. Cells are attached to scaffold surface and allowed to produce 3D architecture	• Provide good 3D extracellular support more closely approximating *in vivo* form • Cell morphology and differentiation more physiological than in 2D culture • Variable levels of flexibility, density, and porosity, allowing wide choice of physical properties for platforms depending on research needs	• Scaffold fabrication is often expensive • Optimal seeding concentrations need to be calibrated separately for every platform • Variable ability to mimic *in vivo* conditions, so the same experiment may yield different results on different 3D models	*Natural materials:* Horning et al. (2008), Dhiman et al. (2005), Dhiman et al. (2004), Gotoh et al. (2011), and Kwon et al. (2010) *Synthetic materials:* Purdum et al. (1993), Kang and Bae (2009), Yang et al. (2005), and Pathi et al. (2010)

Table 9.2 3D culturing platforms—products, properties, and possible applications

3D culturing model classification	Diagrammatic representation	Type	Available products	Properties and remarks	Possible applications
Cell aggregates		Multispheroids	• Devices for external force induction, such as gyratory systems, spinner flasks e c.t. • Nonadhesive surface culturing systems	• Suitable for any adherent cells, and in particularly for cells from neurolines	• Gene assays such as screening models • Drug efficiency evaluation • Nano particles drug delivery evaluation • Cancer cell lines growth patterns evaluation • Cells sensitivity to radiation
Soft platforms		Hydrogels	• Natural hydrogels: collagen I based (as Matrigel ™) (Desai, Kisaalita, Keith, & Wu, 2006; Hughes et al., 2010; Szot et al., 2011), alginate (Boxberger & Meyer, 1994), and hyaluronic hydrogels (Gurski et al., 2009)	• Suitable for all cancer cell lines and primary cultures • Tissue-like flexibility • Viscoelasticity • Diffusive transport of media • HA hydrogels may serve as bioactive ECM for cancer cells due to HA-CD44 activation pathways	• Tumor–stromal interactions: cancer cells cocultures with variety of cell types: MSCs, fibroblasts, endothelial cells e t.c. • Microenvironment mechanic influences evaluation (various thicknesses). • Invasion assays • Cancer cells encapsulation patterns

Continued

Table 9.2 3D culturing platforms—products, properties, and possible applications—cont'd

3D culturing model classification	Diagrammatic representation	Type	Available products	Properties and remarks	Possible applications
			• Synthetic peptides hydrogels, such as Puramix™ (Abu-Yousif et al., 2009).	• Puramix™ was reported as preferable platform for ovarian cancer cell line (Yang & Zhao, 2011)	• Suitable platform for neurocell lines culturing • *In vivo* vascularization assays of transplanted tumor models • Drug efficiency evaluation
Solid 3D platforms		Solid synthetic and natural scaffolds (porous and fibrous)	Natural derived scaffolds based on: • chitosan • fibroin • collagen (OPLA) (Tepliashin et al., 2007) • silk (Park, Gil, Kim, Lee, & Kaplan, 2010; Tan et al., 2011)	• Biodegradable • Mechanically robust • Easily available • Bioactive, interact with cells • Fiber-based scaffolds lack control over porosity and pore size	• Microenvironment effects: biomechanic influence on tumor cells • Anticancer treatments efficiency evaluation • Specific tumor-microenvironment interactions • Drug delivery assays • Bioactive molecules effect on cancer cells, as in case of scaffold coated with peptides/ cell ligands e c.t.

		Synthetic scaffolds such as: • PLGA (Kang & Bae, 2009; Yang et al., 2005) • 45S5 bioactive glass (BG) (Rivadeneira, Carina Audisio, Boccaccini, & Gorustovich, 2013)	• Inert • High porosity • Interconnected structure, supporting tissue-like formation • Long shelf-life • Easily designed for desired porosity, mechanical properties, and degradation time • Well-defined architecture, predictable, and reproducible structure • May be cryopresevable	
	Synthetic and natural calcium providing scaffolds	Aragonite scaffolds (*Porites lutea*), calcite scaffolds (*Tetraclita rifotincta*), fibrous matrix (Chen, Chen, & Yang, 2003), calcium phosphate scaffolds.	• Connective tissue cancer lines • Osteosarcoma cells • Cell lines capable of bone metastasis	• Cells sensitivity to anticancer treatments • Bone metastasis, cancer cells coculturing with bone microenvironment cell lines such as osteoclasts, cancer bone metastasis-related processes such as vicious cycle • CSCs niche studies

considered for studies focusing on tumor-microenvironment corelations, as in the of bone-marrow niche representation by a calcium providing micro-environment; in such studies the authors recommend seeding bone-related cell types in the solid matrix in order to provide improved resemblance to the bone microenvironment. Another advantage of solid scaffolds is the possibility they offer of embedding different cell types in the desired order, resulting in a close representation of the natural tumor microenvironment.

Hydrogels, on the other hand, are more suitable for invasion assays and for vascularization assays of transplanted matrix in *in vivo* studies, due to their soft texture which allows cells to move more easily inside the matrix. Moreover, hydrogels can be used as tools for measuring the ability of outside-seeded cancer cells to invade the matrix in response to an attractant inside the hydrogel (a chemical compound or other cell type, as demonstrated in Fig. 9.4B). Lastly, owing to the translucency of the matrix, hydrogels facilitate observation and staining of the cell culture under investigation. This property of hydrogels makes them a good choice for researchers wishing to investigate cellular morphology in 3D culture and coculture as well as to perform various cell staining/immune staining operations on the cell culture.

In conclusion, existing models provide cancer researchers with a wide range of choices for 3D culturing in an environment that closely mimics the native surroundings of the tumor.

The influence of the diverse 3D models on the properties of cancer cell cultures is discussed below.

2. CELL MORPHOLOGY

Under *in vivo* conditions, both normal and malignant cells are tightly packed, have a volumetric shape, are supported by an ECM, and communicate with their neighbors and with the ECM via various cell–cell/cell–matrix connecting molecules. Almost two decades ago, Boxberger and Meyer (Boxberger & Meyer, 1994) demonstrated differences in morphology between bladder carcinoma cells grown in a 3D alginate model and those grown in 2D culture: cells in monolayers had a flat shape and irregular organization, whereas cells on 3D scaffolds were arranged in tightly packed spheroids and showed well-developed cell–cell contacts and tissue-like organization. Later studies seeking to reproduce these findings for other cancers also found differences in cell shape, size, and polarity (Chen et al., 2003; Luo & Yang, 2004; Sainz, TenCate, & Uprichard, 2009). For instance, human hybridoma cells grown in 3D fibrous matrix formed large, tightly packed aggregates with a small cell size relative to

2D culture and hence a greater surface area, allowing better nutrient transport and more cell–cell contacts (Luo & Yang, 2004). Thus large cell aggregates tend to exhibit "economical" behavior: they consume less oxygen and require lower concentrations of nutrients, enabling cells to reach high densities and maintain long-term stability. Chen et al. (2003) showed that culturing osteosarcoma cells on a fibrous bed bioreactor resulted in decrease in apoptosis by enabling retention of viable cells and disposal of apoptotic cells. These studies demonstrate how 3D fibrous models affect cell morphology and promote long-term culturing by enabling removal of nonviable cells and preventing healthy cells from being affected by the secretion of apoptotic factors to the media.

A study of human hepatoma also revealed differences in cell morphology between monolayers and a rotating wall vessel system (Sainz et al., 2009). The 3D arrangement made it possible to examine the cells' response to Hepatitis C virus infection in more natural cell culture architecture.

Recently we showed that a novel 3D scaffolds produced from the marine invertebrate *Porites lutea* (POR) and sea barnacle *Tetraclita rifotincta*, which consists of $CaCO_3$ in its crystalline form of aragonite and calcite, respectively, provide a suitable microenvironment for cells culturing (Astachov, Nevo, Brosh, & Vago, 2011) and cancer cells in particularly (Figs. 9.2 and 9.3). Both scaffolds combine high porosity with stiff surfaces (Abramovitch-Gottlib, Geresh, & Vago, 2006; Astachov et al., 2011; Gross-Aviv & Vago, 2009), and the Ca^{2+} released from the scaffold to the media can be consumed by the cultured cells. Cancer cells exploit the unique architecture to adhere to the pores' perimeters (Figs. 9.2 and 9.3 a,b,d). The proliferation rate of the cultivated cancer cells, as well as other tumorigenic-related attributes such as migration and invasion, are also affected by the POR microenvironment.

The studies described above used 3D culturing to reproduce native cell arrangement and morphology, and consequently their findings are more reliable than data obtained by culturing cells in a traditional monolayer. However, the influence of the 3D microenvironment extends well beyond cell morphology. More specifically, 3D cell studies can reveal evidence of extensive changes influencing downstream processes, particularly at the molecular level.

3. CELL PROLIFERATION

Many properties of cancer cells—including their proliferation rate—are affected by the *in vitro* 3D microenvironment, in ways that depend on both cell type and the particular 3D model. In a study of cervical tumor epithelial cells, it was shown that both the proliferation rate and the length of the exponential growth period were lower in 2D culture than in a rotating wall vessel culture (Chopra, Dinh, & Hannigan, 1997). However, a low proliferation rate in colon cancer cultures on multicellular spheroids versus monolayers was demonstrated by Nakamura et al., and similar results were reported for human breast cancer grown on collagen matrices by Torisawa et al. (Nakamura et al., 2003; Torisawa, Shiku, Yasukawa, Nishizawa, & Matsue, 2005). These contradictory findings were not resolved by a study of Desai et al. (2006) in which human neuroblastoma cells were grown on collagen hydrogel; the findings showed no significant differences between control monolayers and 3D cultures.

The effect of the microenvironment on cancer cell proliferation can be mechanical—expressed, for instance, in limited growth area or matrix stiffness—or chemobiological, for example, leading to release of molecules to the media by the matrix and their consumption by the cells. Since no single trendline has emerged regarding cancer cell proliferation on 3D platforms versus monolayers, it must be concluded that each particular model exerts its own distinct effect on the given cell type. The goal being to match the *in vivo* reality as closely as possible, the choice of 3D platform should depend on the specific *in vivo* microenvironment the researcher is seeking to mimic.

Since cell proliferation rate and survival ability are believed to be interrelated (Iyoda & Fukai, 2012; Rasmuson et al., 2012), further studies of cell viability in 3D models are required.

4. CELL VIABILITY AND DRUG METABOLISM EFFECTS

It has been shown that, compared with 2D cultures, the cross-talk between cells in 3D systems is superior in that the cells pass on and receive more information, which confers an advantage in dealing with stress (Luo & Yang, 2004). Most studies report better survival of cancer cell lines in 3D versus 2D culture and more especially under various stress situations, such as exposure to drugs, nutrient deficiency in the media, hypoxia, etc. (Chen et al., 2003; Li et al., 2008; Loessner et al., 2010). It is important to note that comparisons between 3D aggregate models and 2D monolayers are relevant only for cancer cell spheroids smaller than 200 μm in diameter or in 3D microenvironments that allow media flow; otherwise, necrotic processes may develop at the center of the growing spheroid due to lack of nutrients for the "core" cells.

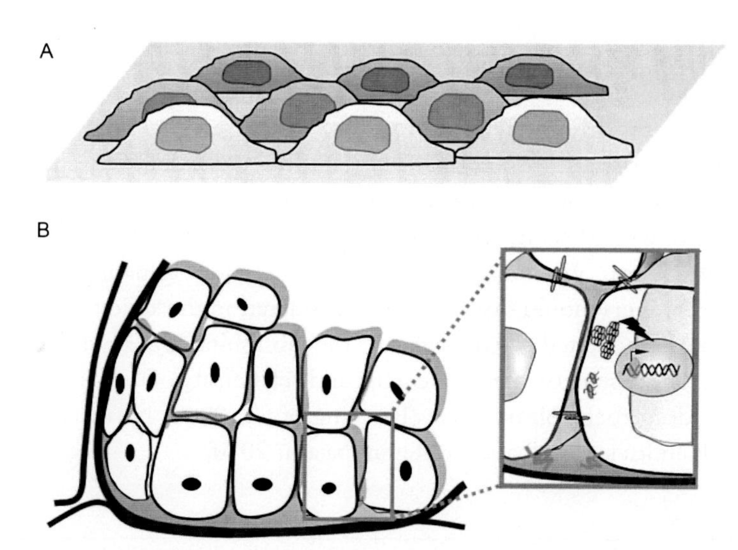

Figure 9.5 ***2D and 3D culture models compared.*** (A) 2D culture. Cell connectivity is limited due to the disposition of the cells as a flat monolayer on the 2D surface. (B) 3D culture. Each cell has many neighbors and cell connectivity is high. Cells interconnect with each other and with the 3D platform, which resembles an ECM. (See the color plate.)

Cancer cell monolayers spread over a plastic surface come into direct contact with the medium over the major part of their surface area, which can lead to inaccurate predictions regarding the efficacy of anticancer agents (Horning et al., 2008). It has also been suggested that many important signals and key regulators, such as cell–cell connectivity molecules and receptors, are down-regulated in 2D cultures due to loss of the structural and molecular variations present in 3D cell cultures (Lee, Kenny, Lee, & Bissell, 2007). It is therefore reasonable to expect that signaling pathways and cellular responses to a specific drug in 3D microenvironments would differ from those observed in monolayers. Human MCF-7 cells, for example, exhibited decreased chemosensitivity to a number of anticancer agents when grown on large and porous biodegradable polymeric 3D scaffolds, as reflected by a 12- to 23-fold difference in the cells' IC_{50} values as compared with cells cultured in 2D (Horning et al., 2008); this divergence was attributed to the low percentage (26%) of cells exposed to toxic drug concentrations in the 3D model vis-à-vis the cells in 2D culture (Horning et al., 2008). Similarly, in the studies of Bokhari et al. (Bokhari, Carnachan, Cameron, & Przyborski, 2007a,2007b) and Li et al. (Bokhari et al., 2007a; Li et al., 2008), 3D cancer cell cultures exhibited elevated ability to resist drug-induced apoptosis and maintain viability, again due to lower exposure of the inner cell to the drug; another possible factor is better communication between the cells in 3D culturing, which may improve the cells' ability to prepare for stress caused by drug toxicity.

In all these studies monolayer cell cultures were demonstrated to be physiologically unsuitable environments for studies of drug efficacy and cell metabolism, with 3D cultures being shown to be a better platform for testing drug effectiveness. When using a 3D environment for evaluating anticancer drugs, researchers should first take into consideration the action mechanism of the drug being tested and then choose the most suitable 3D system. Moreover, the heterogeneity, metabolic state, and cell density of the culture influence the drug metabolism effect (Horning et al., 2008; Nirmalanandhan, Duren, Hendricks, Vielhauer, & Sittampalam, 2010).

5. CELL RESPONSE TO EXTERNAL STIMULI

Cell shape and architecture have a profound impact on cellular performance, including response to external (as opposed to drug induced) stimuli such as tissue-specific effects, radiation therapy, or certain physical forces (Huang & Ingber, 2005; Tang et al., 2010; Wu et al., 2008). The study of Desai et al. (2006) on human neuroblastoma cells, in which resting membrane potential in 2D and on collagen hydrogel 3D model cultured tissues was investigated, provides a good example of tissue-specific external stimuli research. The authors found that neuroblastoma cells cultured on 3D were less responsive to increase of intracellular Ca^{2+} than the 2D control, demonstrating how 2D culturing may falsify cellular response to calcium dynamics.

Another external stimulus of relevance for cancer therapy is radiation. For reliable *in vitro* evaluation of cell survival following radiotherapy, a number of relevant parameters should be taken into consideration—cell morphology, physical forces between cell membranes, the actin cytoskeleton, the nuclear matrix. 3D culturing modifies these parameters. The tighter cell arrangement and more volumetric cell morphology characteristic of 3D models are examples of features that promote triggering of signaling pathways, leading to [possible] changes in chromatin organization and alterations in the expression of radiosensitivity-related genes, such as tumor suppressor genes. Storch et al. (2010) showed that cells in 3D culture exhibited increased survival and a reduced number of DNA breaks after exposure to radiation, underlining the importance of such models for the successful evaluation of radiotherapy.

3D models endowed with certain physical and biological properties make it possible to study tumor features that cannot be investigated using traditional 2D culture. An example of such a feature is the mechanobiological stimulus exerted by a solid tumor growing against adjacent tissues, creating nonfluid related pressure, which is considered to promote cancer spread (Suresh, 2007). Recently it has been shown that mechanobiological stimulation that alters cell morphology regulates metastasis-associated genes (*caveolin-1*, *integrinb1*, *Rac1*; Demou, 2010), thus indicating that these genes are involved in decreasing cell–cell contact, increasing ECM degradation, and promoting tumor progression (Rath, Nam, Knobloch, Lannutti, & Agarwal, 2008; Suresh, 2007).

6. DIFFERENTIATION

Though they share the same DNA code, cells in an organism may express different genes, a fact that explains the presence of diverse cell types and behaviors. Epigenetics influences several cellular processes, including cellular differentiation and cancer progression (Bissell, Radisky, Rizki, Weaver, & Petersen, 2002; Katto & Mahlknecht, 2011). Epigenetic processes may alter cell–cell communication characteristics and cell invasion ability, with cellular adhesion molecules playing an essential role in regulating these parameters (Katto & Mahlknecht, 2011). Because of the importance of cell adhesion molecules in cell migration and invasiveness (Zeromski, Nyczak, & Dyszkiewicz, 2001), alterations in their expression level may influence the differentiation stage of cancer cells. Downregulation of adhesion genes enables cancer cells to break free from the primary tumor and move downstream toward secondary growing sites, causing metastasis. These mobile cancer cells are considered to be less differentiated than their tightly packed counterparts in the primary tumor (Asano et al., 2004; Bringuier et al., 1993; Umbas et al., 1992).

Recent studies have shown that tumor cells–microenvironment interactions play a pivotal role in tumor progression and development and should therefore not be ignored in cancer *in vitro* research (Albini & Sporn, 2007). Stromal–tumor interactions have been intensively studied in the last few years in various 3D microenvironments that mimic stromal rigidity and elasticity features, such as collagen (Kenny, Krausz, Yamada, & Lengyel, 2007; Pageau, Sazonova, Wong, Soto, & Sonnenschein, 2011; Yang et al., 2010) and polyacrylamide gels (Tilghman et al., 2010).

In this context, it is important to note that the expression of adhesion molecules is highly dependent on the features of the cells' microenvironment (Huang & Ingber, 2005). For example, Tilghman et al. (2010) showed that cellular levels of epithelial and mesenchymal factors are influenced by hydrogel matrix stiffness. Their study demonstrated that upregulation of adhesion molecule E-cadherin together with downregulation of mesenchymal transcription factors may be considered as the first step in the chain of events leading to epithelial differentiation of lung carcinoma (Tilghman et al., 2010).

E-cadherin levels seem to be dependent on 3D cell arrangement and not necessarily on matrix properties. Li et al. (2008) having observed that human hepatoma cells cultured in multicellular spheroids exhibited tighter cellular arrangement and elevated E-cadherin expression, attributed the higher resistance to apoptosis of the hepatoma cells to the improved cell–cell adhesion afforded by 3D culturing. Since tightly packed cancer cells are characterized by decreased stemness and mobility, cell cultures on 3D platforms may be more representative of the native differentiation state than cells produced in a traditional 2D environment.

7. CANCER STEM CELLS

At present, most therapeutic regimes for cancer rely on the notion that all the highly proliferative cancer cells within a particular tumor are equal in their ability to drive tumorigenesis and metastasis (Yi, Hao, Nan, & Fan, 2013), and thus conventional chemotherapies and/or radiation therapies are designed to target highly proliferative cells (Karamboulas & Ailles, 2013; Yi et al., 2013). If, however, we accept the premise that cancer is a disease of stem cells and that any given tumor cell population includes a small, distinct subset of cancer cells with stem-like abilities, then it becomes possible to understand why some tumor cells are not affected by the conventional therapies that target rapidly dividing cells (Sugihara & Saya, 2013; Yi et al., 2013). This new and developing insight into cancer biology has pointed up the need to better understand the heterogeneity of the cancer cell population and of the processes that may lead to the establishment of a new subpopulation of cells with stem cell-like abilities.

Diverse 3D microenvironments may exercise contrary effects on cancer tissues: some may induce cancer cell differentiation while others may cause cells to acquire the characteristics of cancer stem cells (CSCs; Campbell, Davidenko, Caffarel, Cameron, & Watson, 2011; Chen et al., 2012). Our group results demonstrate calcium providing POR scaffolds ability to enhance fibrosarcoma tumorigenic potential at the same time as they elevate markers related to CSCs, demonstrating that the microenvironment may be capable of shifting cancer cell differentiation toward a more stem-like state (paper in process).

A decline in adhesion molecule levels could indicate that cancer cells are transiting to a less differentiated state. In a study of human breast cancer cells cultured on a 3D collagen scaffold, Chen et al. (2012) demonstrated simultaneous downregulation of the epithelial cell marker and upregulation of breast CSCs markers. The gel-like 3D scaffold investigated by Chen et al. (2012) exhibited an ability to induce expression of CSCs phenotypes via the process of epithelial-mesenchymal transition (EMT). Another model linking EMT to the acquisition of CSCs capacities was proposed (Debeb et al., 2010). The results of the latter study showed simultaneous elevation of mesenchymal gene expression and growth of a cancer stem-like $CD44^{+}/CD24^{-}$ subpopulationin a 293T cancer culture (Debeb et al., 2010). Intriguingly, stem-like cancer cells grown in 3D spheres were found to be more resistant to radiation, due to upregulation of stem cell survival signals such as Notch1 and Survivin (Debeb et al., 2010).

Our group has found that cancer cells cultured on the novel 3D platform prepared from *Porites lutea*, demonstrated increased tumorigenic potential in migration and invasion assays as compared with a 2D control, as well as elevated expression of tumor-related markers. It is possible that this calcium providing bone-like microenvironment (Figs. 9.3D–F and 9.4D), which is composed of $CaCO_3$, harbors the cancer cells in a stem cell niche-like habitat, influencing their tumorigenic potential and modifying their position on the stemness-differentiation axis.

The ability of 3D microenvironments to induce formation of a CSCs-like subpopulation among cancer cells illustrates the importance of exercising great care in selecting 3D culturing platforms for *in vitro* cancer studies. An appropriate choice of 3D microenvironment should facilitate accurate assessment of the tumorigenic potential of the cancer cell culture, a parameter that is probably underestimated in classical 2D studies.

8. GENE AND PROTEIN EXPRESSION

When grown in 3D systems—as opposed to 2D models—cells tend to acquire tissue-like properties, such as cell polarization (Bokhari et al., 2007b) and tissue-specific function, as reflected in the expression of unique markers (Bokhari et al., 2007a). Many studies have reported higher survival rates of cancer cells in 3D models, which may be attributed to a greater cell connectivity approaching that characterizing *in vivo* conditions. Such cell capabilities are mediated by the altered expression of genes and subconsequential proteins, which can be classified based on their biological roles, such as cell–cell and cell–ECM adhesion, metabolic factors, cell structure, and signal transduction factors.

Molecular changes associated with culturing in 3D systems have also been shown in several microarray studies performed on human cancer cells (neuroblastoma, breast cancer, melanoma, and others), with a 1.5- to 2-fold increase over 2D control levels being reported for many genes in the categories of protein binding (Horning et al., 2008; Kumar et al., 2008; Nakamura et al., 2003), cell structure (Horning et al., 2008; Li, Livi, Gourd, Deweerd, & Hoffman-Kim, 2007), and signal transduction (Fischbach et al., 2009; Horning et al., 2008; Kumar et al., 2008; Li et al., 2007; Quiros et al., 2008).Since these elevated gene levels were detected on diverse 3D models—multispheroids (Kumar et al., 2008; Nakamura et al., 2003), soft platforms such as hydrogels (Fischbach et al., 2009; Li et al., 2007; Quiros et al., 2008), and solid microenvironments (Horning et al., 2008)—it appears that there is no relation between the particular type of 3Dplatformadopted and the observed alteration of gene level. We may therefore infer that the elevated gene expression reported for 3D microenvironments is due to the cellular arrangement and cell–cell interactions characteristic of 3D systems in general rather than to interactions between the cells and the 3D model itself.

As we saw earlier, adhesion molecules and most particularly E-cadherin have received special attention in recent analyses of gene and protein expression in 3D versus 2D culture. Several studies of breast cancer (MCF-7) cells grown as multicellular spheroids and compared with 2D monolayers have demonstrated increased expression of E-cadherin and p27 (an important factor in cell division cycle regulation) in the 3D model (do Amaral, Rezende-Teixeira, Freitas, & Machado-Santelli, 2011; Kim et al., 2004; Nakamura et al., 2003).

The importance of E-cadherin for regulating cell proliferation, differentiation, and chemosensitivity was brought out by the finding that exposure to anti-E-cadherin neutralizing antibody led to decreasedp27 expression (Nakamura et al., 2003). As noted, enhanced expression of E-cadherin in cells cultured in a 3D microenvironment has been shown to cause cancer cell differentiation and metastasis inhibition (Bringuier et al., 1993; Kim et al., 2004; Umbas et al., 1992); it would be instructive to study this gene alteration in 3D versus 2D cultures. However, parallel to elevated E-cadherin expression—and a shift toward a more differentiated state—increased expression of other genes vital to keeping cells undifferentiated (such as MEIS1) has also been reported (Horning et al., 2008), testifying to the complexity of cellular decision-making pathways and possible compensation processes.

Aside from E-cadherin, 3D microenvironments are associated with differences in the expression of other important molecules as well. In human hepatoma cells grown in multicellular spheroids, an increase was registered in the levels of CD44v6, VEGF, VEGFR2, endostatin, and cytochrome-c in 3D culturing, followed by enhanced cell resistance to drug-induced apoptosis and upregulated angiogenic potential (Fischbach et al., 2009; Li et al., 2008). Increases in angiogenic potential were also demonstrated in a study with oral squamous cell carcinoma cells (Fischbach et al., 2009) grown on alginate hydrogel, where cell–ECM interactions via integrins led to enhanced interleukin 8 secretion. That study was followed by an *in vivo* experiment in mice, where the interleukin 8-delivery system caused increased tumor vascularization; finding, which clearly demonstrates the importance of 3D culturing in reaching in *in vitro* experiments an adequate assessment of cancer potential and cancer related processes occurring *in vivo*.

To sum up the influence of 3D culturing on cancer cell gene expression, we can say that 3D cultures differ significantly in their gene expression profiles from the traditional 2D models. Based on the tissue-like cell arrangement and morphology of 3D microenvironments, which can affect such attributes as cell sensitivity to anticancer agents, it may be argued that these expression profiles are indeed representative of *in vivo* conditions. The tissue-like character of 3D *in vitro* models suggests that they have a role to play in future *in vivo* evaluations of anticancer agents (in animals and in clinical trials), perhaps helping to improve the trials' reliability. However, to our knowledge, no serious attempt has been made to compare gene expression in cells cultured *in vitro* (2D/3D) and tissues *in vivo*. If it emerges that 3D cultured cells resemble animal tissues over a broad range of parameters—cell proliferation, differentiation, morphology, density, cell/matrix arrangement—then we would have solid evidence for the superiority of 3D models over 2D in cancer research.

9. FUTURE PERSPECTIVES

3D culturing is a rapidly developing field. There is an impressive assortment of available 3D models and culturing methods, including induction of spheroid formation, soft structure hydrogel models, and natural or synthetic scaffolds.

As shown in Fig. 9.1, cells in 3D culture, unlike 2D, are arranged in volumetric clusters, have many neighbors on each side, and exhibit multiple cell–cell interactions involving gap junctions, cadherin molecules, desmosomes, and adhesion belts. Cells in 3D microenvironments interact with the 3D biolattice as well, via cell–ECM adhesion molecules such as integrins, and utilize the 3D platform to form tissue-like constructs. Increased cell connectivity causes subconsequential signaling between the cells, one result being a gene expression pattern distinct from 2D. As mentioned above, the outcomes of this altered gene expression may include better adjustment of the cancer cells to stresses such as radiotherapy and anticancer drugs, as well as cell differentiation, or, on the contrary, acquisition of CSC properties.

However, several limitations of 3D culturing cannot be ignored:

- Currently, 3D microenvironments lack vasculature and hence both nutrient supply and normal transport of small molecules
- Unlike continuous *in vivo* models, they mimic static or short-term conditions
- They do not model interactions with other cell types or the influence of the latter upon the culture

One of the major weaknesses of 3D cancer models today is that, in contradistinction to natural tumors, cell cultures grown in such microenvironments are in most cases homogenic. While 3D models may be seen as a bridge between 2D cultures and *in vivo* reality, it is important to recall that ECM secretion *in vivo* originates from surrounding noncancerous cells, such as stroma cells. In 3D models, on the other hand, the ECM is produced by the cancer cells themselves (Kim et al., 2004), entraining differences in gene/protein expression between 3D cultured cells and *in vivo* tissues. Such differences could influence research points of interest and distort predictions regarding the efficacy of anticancer agents. To overcome this obstacle, cancer biologists should consider coculturing several cell types in suitably chosen 3D models coculturing tumor and connective tissue cell types should enable production of ECM by the stromal cells. This more natural ECM secretion will mimic living tissue more correctly and will allow better imitation of paracrine signals, cell–cell interactions, and response of tumor cells to treatments. However, success in this enterprise has to await the development of simple and reproducible models of the tumor that can accommodate cell types such as fibroblasts, endothelial cells, and mesenchymal stem cells.

Despite having these cons, 3D culturing platforms overcome those by offering many advantages for cancer *in vitro* research. Here, we summarize the main ones:

- Cell morphology and arrangement are more tissue-like than in traditional 2D culture
- Gene and protein expressions are closer to the *in vivo* pattern
- Cell–ECM/cell–cell interactions and adhesion properties mimic tissue conditions more closely
- 3D models provide a suitable tool for evaluating cell response to external stimuli and for testing the efficacy of drugs

The cell heterogeneity characteristic of 3D cultures is likely to cause regional differences in marker expression and cell signaling, and these in turn could point to similar processes *in vivo*, yielding possible implications for the targeted therapy of solid tumors.

In addition, as we saw, the characteristics of the particular 3D microenvironment chosen for cell culturing may affect cellular behavior. For this reason, in cancer research involving 3D culturing, matrix chemical composition and physical properties should be considered with reference to natural tissue properties (Table 9.2). Future planning of novel 3D systems should strive to reach maximum resemblance to the investigated tissue, in order to reduce the gap between *in vitro* studies and *in vivo* reality.

Another reason to prefer 3D culturing is the fact that animal studies often fail to offer an adequate imitation of human tissue metabolism and cell behavior. Studies of human cells *in vitro* in a correctly chosen 3D model can provide a good platform for evaluating the efficacy of anticancer agents, discovering potential target genes for therapy, and revealing signal pathways relevant for tumor progression to be explored in future studies.

To sum up, 3D culture systems offer a wide variety of parameters and features that can be manipulated to reveal important information about cell interaction, behavior, and possible differentiation, adhesion, and morphogenesis regulation, tumorigenic potential, and the cross-influences between these diverse processes. Without a doubt, 3D models have become crucial scientific tools for cancer cell research.

REFERENCES

Abramovitch-Gottlib, L., Geresh, S., & Vago, R. (2006). Biofabricated marine hydrozoan: A bioactive crystalline material promoting ossification of mesenchymal stem cells. *Tissue Engineering*, *12*(4), 729–739.

Abu-Yousif, A. O., Rizvi, I., Evans, C. L., Celli, J. P., & Hasan, T. (2009). PuraMatrix encapsulation of cancer cells. *Journal of Visualized Experiments*. (34). http://dx.doi.org/10.3791/1692, pii: 1692.

Ahearne, M., & Kelly, D. J. (2013). A comparison of fibrin, agarose and gellan gum hydrogels as carriers of stem cells and growth factor delivery microspheres for cartilage regeneration. *Biomedical Materials (Bristol, England)*, *8*(3), 035004, 035004-6041/8/3/035004.

Albini, A., & Sporn, M. B. (2007). The tumour microenvironment as a target for chemoprevention. *Nature Reviews. Cancer, 7*(2), 139–147.

Asano, K., Duntsch, C. D., Zhou, Q., Weimar, J. D., Bordelon, D., Robertson, J. H., et al. (2004). Correlation of N-cadherin expression in high grade gliomas with tissue invasion. *Journal of Neuro-oncology, 70*(1), 3–15.

Astachov, L., Nevo, Z., Brosh, T., & Vago, R. (2011). The structural, compositional and mechanical features of the calcite shell of the barnacle Tetraclita rufotincta. *Journal of Structural Biology, 175*(3), 311–318.

Bae, S. I., Kang, G. H., Kim, Y. I., Lee, B. L., Kleinman, H. K., & Kim, W. H. (1999). Development of intracytoplasmic lumens in a colon cancer cell line cultured on a non-adhesive surface. *Cancer Biochemistry Biophysics, 17*(1–2), 35–47.

Bartholoma, P., Impidjati, Reininger-Mack, A., Zhang, Z., Thielecke, H., & Robitzki, A. (2005). A more aggressive breast cancer spheroid model coupled to an electronic capillary sensor system for a high-content screening of cytotoxic agents in cancer therapy: 3-Dimensional in vitro tumor spheroids as a screening model. *Journal of Biomolecular Screening, 10*(7), 705–714.

Bissell, M. J., Radisky, D. C., Rizki, A., Weaver, V. M., & Petersen, O. W. (2002). The organizing principle: Microenvironmental influences in the normal and malignant breast. *Differentiation; Research in Biological Diversity, 70*(9–10), 537–546.

Bokhari, M., Carnachan, R. J., Cameron, N. R., & Przyborski, S. A. (2007a). Culture of HepG2 liver cells on three dimensional polystyrene scaffolds enhances cell structure and function during toxicological challenge. *Journal of Anatomy, 211*(4), 567–576.

Bokhari, M., Carnachan, R. J., Cameron, N. R., & Przyborski, S. A. (2007b). Novel cell culture device enabling three-dimensional cell growth and improved cell function. *Biochemical and Biophysical Research Communications, 354*(4), 1095–1100.

Boxberger, H. J., & Meyer, T. F. (1994). A new method for the 3-D in vitro growth of human RT112 bladder carcinoma cells using the alginate culture technique. *Biology of the Cell/Under the Auspices of the European Cell Biology Organization, 82*(2–3), 109–119.

Bringuier, P. P., Umbas, R., Schaafsma, H. E., Karthaus, H. F., Debruyne, F. M., & Schalken, J. A. (1993). Decreased E-cadherin immunoreactivity correlates with poor survival in patients with bladder tumors. *Cancer Research, 53*(14), 3241–3245.

Campbell, J. J., Davidenko, N., Caffarel, M. M., Cameron, R. E., & Watson, C. J. (2011). A multifunctional 3D co-culture system for studies of mammary tissue morphogenesis and stem cell biology. *PloS One, 6*(9), e25661.

Carriel, V., Garrido-Gomez, J., Hernandez-Cortes, P., Garzon, I., Garcia-Garcia, S., Saez-Moreno, J. A., et al. (2013). Combination of fibrin-agarose hydrogels and adipose-derived mesenchymal stem cells for peripheral nerve regeneration. *Journal of Neural Engineering, 10*(2), 026022, 026022-2560/10/2/026022.

Chang, N. J., Lam, C. F., Lin, C. C., Chen, W. L., Li, C. F., Lin, Y. T., et al. (2013). Transplantation of autologous endothelial progenitor cells in porous PLGA scaffolds create a microenvironment for the regeneration of hyaline cartilage in rabbits. *Osteoarthritis and Cartilage/OARS, Osteoarthritis Research Society, 21*, 1613–1622.

Chen, C., Chen, K., & Yang, S. T. (2003). Effects of three-dimensional culturing on osteosarcoma cells grown in a fibrous matrix: Analyses of cell morphology, cell cycle, and apoptosis. *Biotechnology Progress, 19*(5), 1574–1582.

Chen, C. J., Nguyen, T., & Shively, J. E. (2010). Role of calpain-9 and PKC-delta in the apoptotic mechanism of lumen formation in CEACAM1 transfected breast epithelial cells. *Experimental Cell Research, 316*(4), 638–648.

Chen, L., Xiao, Z., Meng, Y., Zhao, Y., Han, J., Su, G., et al. (2012). The enhancement of cancer stem cell properties of MCF-7 cells in 3D collagen scaffolds for modeling of cancer and anti-cancer drugs. *Biomaterials, 33*(5), 1437–1444.

Chopra, V., Dinh, T. V., & Hannigan, E. V. (1997). Three-dimensional endothelial-tumor epithelial cell interactions in human cervical cancers. *In Vitro Cellular & Developmental Biology. Animal, 33*(6), 432–442.

Choudhary, R., Li, H., Winn, R. A., Sorenson, A. L., Weiser-Evans, M. C., & Nemenoff, R. A. (2010). Peroxisome proliferator-activated receptor-gamma inhibits transformed growth of non-small cell lung cancer cells through selective suppression of Snail. *Neoplasia (New York, N.Y.), 12*(3), 224–234.

David, L., Dulong, V., Le Cerf, D., Cazin, L., Lamacz, M., & Vannier, J. P. (2008). Hyaluronan hydrogel: An appropriate three-dimensional model for evaluation of anticancer drug sensitivity. *Acta Biomaterialia, 4*(2), 256–263.

David, L., Dulong, V., Le Cerf, D., Chauzy, C., Norris, V., Delpech, B., et al. (2004). Reticulated hyaluronan hydrogels: A model for examining cancer cell invasion in 3D. *Matrix Biology: Journal of the International Society for Matrix Biology, 23*(3), 183–193.

Debeb, B. G., Zhang, X., Krishnamurthy, S., Gao, H., Cohen, E., Li, L., et al. (2010). Characterizing cancer cells with cancer stem cell-like features in 293T human embryonic kidney cells. *Molecular Cancer, 9*, 180.

Demou, Z. N. (2010). Gene expression profiles in 3D tumor analogs indicate compressive strain differentially enhances metastatic potential. *Annals of Biomedical Engineering, 38*(11), 3509–3520.

Desai, A., Kisaalita, W. S., Keith, C., & Wu, Z. Z. (2006). Human neuroblastoma (SH-SY5Y) cell culture and differentiation in 3-D collagen hydrogels for cell-based biosensing. *Biosensors & Bioelectronics, 21*(8), 1483–1492.

Dhiman, H. K., Ray, A. R., & Panda, A. K. (2004). Characterization and evaluation of chitosan matrix for in vitro growth of MCF-7 breast cancer cell lines. *Biomaterials, 25*(21), 5147–5154.

Dhiman, H. K., Ray, A. R., & Panda, A. K. (2005). Three-dimensional chitosan scaffold-based MCF-7 cell culture for the determination of the cytotoxicity of tamoxifen. *Biomaterials, 26*(9), 979–986.

do Amaral, J. B., Rezende-Teixeira, P., Freitas, V. M., & Machado-Santelli, G. M. (2011). MCF-7 cells as a three-dimensional model for the study of human breast cancer. *Tissue Engineering. Part C, Methods, 17*, 1097–1107.

Elder, S., Gottipati, A., Zelenka, H., & Bumgardner, J. (2013). Attachment, proliferation, and chondroinduction of mesenchymal stem cells on porous chitosan-calcium phosphate scaffolds. *The Open Orthopaedics Journal, 7*, 275–281.

Emfietzoglou, D., Kostarelos, K., Papakostas, A., Yang, W. H., Ballangrud, A., Song, H., et al. (2005). Liposome-mediated radiotherapeutics within avascular tumor spheroids: Comparative dosimetry study for various radionuclides, liposome systems, and a targeting antibody. *Journal of Nuclear Medicine: Official Publication Society of Nuclear Medicine, 46*(1), 89–97.

Fischbach, C., Kong, H. J., Hsiong, S. X., Evangelista, M. B., Yuen, W., & Mooney, D. J. (2009). Cancer cell angiogenic capability is regulated by 3D culture and integrin engagement. *Proceedings of the National Academy of Sciences of the United States of America, 106*(2), 399–404.

Fromigue, O., Louis, K., Wu, E., Belhacene, N., Loubat, A., Shipp, M., et al. (2003). Active stromelysin-3 (MMP-11) increases MCF-7 survival in three-dimensional Matrigel culture via activation of p42/p44 MAP-kinase. *International Journal of Cancer/Journal International du Cancer, 106*(3), 355–363.

Gao, J. M., Yan, J., Li, R., Li, M., Yan, L. Y., Wang, T. R., et al. (2013). Improvement in the quality of heterotopic allotransplanted mouse ovarian tissues with basic fibroblast growth factor and fibrin hydrogel. *Human Reproduction (Oxford, England), 28*, 2784–2793.

Gao, C., Yang, B., Hu, H., Liu, J., Shuai, C., & Peng, S. (2013). Enhanced sintering ability of biphasic calcium phosphate by polymers used for bone scaffold fabrication. *Materials Science & Engineering, C: Materials for Biological Applications*, *33*(7), 3802–3810.

Ghosh, S., Spagnoli, G. C., Martin, I., Ploegert, S., Demougin, P., Heberer, M., et al. (2005). Three-dimensional culture of melanoma cells profoundly affects gene expression profile: A high density oligonucleotide array study. *Journal of Cellular Physiology*, *204*(2), 522–531.

Goodman, T. T., Olive, P. L., & Pun, S. H. (2007). Increased nanoparticle penetration in collagenase-treated multicellular spheroids. *International Journal of Nanomedicine*, *2*(2), 265–274.

Gotoh, Y., Ishizuka, Y., Matsuura, T., & Niimi, S. (2011). Spheroid formation and expression of liver-specific functions of human hepatocellular carcinoma-derived FLC-4 cells cultured in lactose-silk fibroin conjugate sponges. *Biomacromolecules*, *12*(5), 1532–1539.

Gross-Aviv, T., & Vago, R. (2009). The role of aragonite matrix surface chemistry on the chondrogenic differentiation of mesenchymal stem cells. *Biomaterials*, *30*(5), 770–779.

Gurski, L. A., Jha, A. K., Zhang, C., Jia, X., & Farach-Carson, M. C. (2009). Hyaluronic acid-based hydrogels as 3D matrices for in vitro evaluation of chemotherapeutic drugs using poorly adherent prostate cancer cells. *Biomaterials*, *30*(30), 6076–6085.

Horning, J. L., Sahoo, S. K., Vijayaraghavalu, S., Dimitrijevic, S., Vasir, J. K., Jain, T. K., et al. (2008). 3-D tumor model for in vitro evaluation of anticancer drugs. *Molecular Pharmaceutics*, *5*(5), 849–862.

Huang, S., & Ingber, D. E. (2005). Cell tension, matrix mechanics, and cancer development. *Cancer cell*, *8*(3), 175–176.

Hughes, C. S., Postovit, L. M., & Lajoie, G. A. (2010). Matrigel: A complex protein mixture required for optimal growth of cell culture. *Proteomics*, *10*(9), 1886–1890.

Infanger, M., Kossmehl, P., Shakibaei, M., Bauer, J., Kossmehl-Zorn, S., Cogoli, A., et al. (2006). Simulated weightlessness changes the cytoskeleton and extracellular matrix proteins in papillary thyroid carcinoma cells. *Cell and Tissue Research*, *324*(2), 267–277.

Ingram, M., Techy, G. B., Saroufeem, R., Yazan, O., Narayan, K. S., Goodwin, T. J., et al. (1997). Three-dimensional growth patterns of various human tumor cell lines in simulated microgravity of a NASA bioreactor. *In Vitro Cellular & Developmental Biology. Animal*, *33*(6), 459–466.

Iyoda, T., & Fukai, F. (2012). Modulation of tumor cell survival, proliferation, and differentiation by the peptide derived from Tenascin-C: Implication of beta1-integrin activation. *International Journal of Cell Biology*, *2012*, 647594.

Jessup, J. M., Frantz, M., Sonmez-Alpan, E., Locker, J., Skena, K., Waller, H., et al. (2000). Microgravity culture reduces apoptosis and increases the differentiation of a human colorectal carcinoma cell line. *In Vitro Cellular & Developmental Biology. Animal*, *36*(6), 367–373.

Jin, L., Feng, T., Shih, H. P., Zerda, R., Luo, A., Hsu, J., et al. (2013). Colony-forming cells in the adult mouse pancreas are expandable in Matrigel and form endocrine/acinar colonies in laminin hydrogel. *Proceedings of the National Academy of Sciences of the United States of America*, *110*(10), 3907–3912.

Kang, S. W., & Bae, Y. H. (2009). Cryopreservable and tumorigenic three-dimensional tumor culture in porous poly(lactic-co-glycolic acid) microsphere. *Biomaterials*, *30*(25), 4227–4232.

Karamboulas, C., & Ailles, L. (2013). Developmental signaling pathways in cancer stem cells of solid tumors. *Biochimica et Biophysica Acta*, *1830*, 2481–2495.

Katto, J., & Mahlknecht, U. (2011). Epigenetic regulation of cellular adhesion in cancer. *Carcinogenesis*, *32*(10), 1414–1418.

Kenny, H. A., Krausz, T., Yamada, S. D., & Lengyel, E. (2007). Use of a novel 3D culture model to elucidate the role of mesothelial cells, fibroblasts and extra-cellular matrices on adhesion and invasion of ovarian cancer cells to the omentum. *International Journal of Cancer/Journal international du Cancer, 121*(7), 1463–1472.

Kido, H. W., Ribeiro, D. A., de Oliveira, P., Parizotto, N. A., Camilo, C. C., Fortulan, C. A., et al. (2013). Biocompatibility of a porous alumina ceramic scaffold coated with hydroxyapatite and bioglass. *Journal of Biomedical Materials Research. Part A*. http://dx.doi.org/10.1002/jbm.a.34877.

Kim, J. B. (2005). Three-dimensional tissue culture models in cancer biology. *Seminars in Cancer Biology, 15*(5), 365–377.

Kim, J. B., Stein, R., & O'Hare, M. J. (2004). Three-dimensional in vitro tissue culture models of breast cancer—A review. *Breast Cancer Research and Treatment, 85*(3), 281–291.

Kim, J. B., Stein, R., & O'Hare, M. J. (2005). Tumour-stromal interactions in breast cancer: The role of stroma in tumourigenesis. *Tumour Biology: The Journal of the International Society for Oncodevelopmental Biology and Medicine, 26*(4), 173–185.

Koide, N., Sakaguchi, K., Koide, Y., Asano, K., Kawaguchi, M., Matsushima, H., et al. (1990). Formation of multicellular spheroids composed of adult rat hepatocytes in dishes with positively charged surfaces and under other nonadherent environments. *Experimental Cell Research, 186*(2), 227–235.

Kumar, A., Xu, J., Brady, S., Gao, H., Yu, D., Reuben, J., et al. (2010). Tissue transglutaminase promotes drug resistance and invasion by inducing mesenchymal transition in mammary epithelial cells. *PloS One, 5*(10), e13390.

Kumar, H. R., Zhong, X., Hoelz, D. J., Rescorla, F. J., Hickey, R. J., Malkas, L. H., et al. (2008). Three-dimensional neuroblastoma cell culture: Proteomic analysis between monolayer and multicellular tumor spheroids. *Pediatric Surgery International, 24*(11), 1229–1234.

Kundu, B., Saha, P., Datta, K., & Kundu, S. C. (2013). A silk fibroin based hepatocarcinoma model and the assessment of the drug response in hyaluronan-binding protein 1 overexpressed HepG2 cells. *Biomaterials, 34*, 9462–9474.

Kwon, H., Kim, H. J., Rice, W. L., Subramanian, B., Park, S. H., Georgakoudi, I., et al. (2010). Development of an in vitro model to study the impact of BMP-2 on metastasis to bone. *Journal of Tissue Engineering and Regenerative Medicine, 4*(8), 590–599.

Lee, G. Y., Kenny, P. A., Lee, E. H., & Bissell, M. J. (2007). Three-dimensional culture models of normal and malignant breast epithelial cells. *Nature Methods, 4*(4), 359–365.

Lee, M. Y., Kumar, R. A., Sukumaran, S. M., Hogg, M. G., Clark, D. S., & Dordick, J. S. (2008). Three-dimensional cellular microarray for high-throughput toxicology assays. *Proceedings of the National Academy of Sciences of the United States of America, 105*(1), 59–63.

Li, G. N., Livi, L. L., Gourd, C. M., Deweerd, E. S., & Hoffman-Kim, D. (2007). Genomic and morphological changes of neuroblastoma cells in response to three-dimensional matrices. *Tissue Engineering, 13*(5), 1035–1047.

Li, C. L., Tian, T., Nan, K. J., Zhao, N., Guo, Y. H., Cui, J., et al. (2008). Survival advantages of multicellular spheroids vs. monolayers of HepG2 cells in vitro. *Oncology Reports, 20*(6), 1465–1471.

Li, Q., Wang, J., Liu, H., Xie, B., & Wei, L. (2013). Tissue-engineered mesh for pelvic floor reconstruction fabricated from silk fibroin scaffold with adipose-derived mesenchymal stem cells. *Cell and Tissue Research, 354*, 471–480.

Lin, R. Z., & Chang, H. Y. (2008). Recent advances in three-dimensional multicellular spheroid culture for biomedical research. *Biotechnology Journal, 3*(9–10), 1172–1184.

Litzenburger, B. C., Kim, H. J., Kuiatse, I., Carboni, J. M., Attar, R. M., Gottardis, M. M., et al. (2009). BMS-536924 reverses IGF-IR-induced transformation of mammary epithelial cells and causes growth inhibition and polarization of MCF7 cells. *Clinical Cancer Research: An Official Journal of the American Association for Cancer Research, 15*(1), 226–237.

Liu, B. J., Ma, L. N., Su, J., Jing, W. W., Wei, M. J., & Sha, X. Z. (2013). Biocompatibility assessment of porous chitosan-Nafion and chitosan-PTFE composites in vivo. *Journal of Biomedical Materials Research. Part A*. http://dx.doi.org/10.1002/jbm.a.34830.

Loessner, D., Stok, K. S., Lutolf, M. P., Hutmacher, D. W., Clements, J. A., & Rizzi, S. C. (2010). Bioengineered 3D platform to explore cell-ECM interactions and drug resistance of epithelial ovarian cancer cells. *Biomaterials*, *31*(32), 8494–8506.

Luo, J., & Yang, S. T. (2004). Effects of three-dimensional culturing in a fibrous matrix on cell cycle, apoptosis, and MAb production by hybridoma cells. *Biotechnology Progress*, *20*(1), 306–315.

Marrero, B., Messina, J. L., & Heller, R. (2009). Generation of a tumor spheroid in a microgravity environment as a 3D model of melanoma. *In Vitro Cellular & Developmental Biology. Animal*, *45*(9), 523–534.

Metzger, W., Sossong, D., Bachle, A., Putz, N., Wennemuth, G., Pohlemann, T., et al. (2011). The liquid overlay technique is the key to formation of co-culture spheroids consisting of primary osteoblasts, fibroblasts and endothelial cells. *Cytotherapy*, *13*(8), 1000–1012.

Nagelkerke, A., Bussink, J., Sweep, F. C., & Span, P. N. (2013). Generation of multicellular tumor spheroids of breast cancer cells: How to go three-dimensional. *Analytical Biochemistry*, *437*(1), 17–19.

Nakamura, T., Kato, Y., Fuji, H., Horiuchi, T., Chiba, Y., & Tanaka, K. (2003). E-cadherin-dependent intercellular adhesion enhances chemoresistance. *International Journal of Molecular Medicine*, *12*(5), 693–700.

Nirmalanandhan, V. S., Duren, A., Hendricks, P., Vielhauer, G., & Sittampalam, G. S. (2010). Activity of anticancer agents in a three-dimensional cell culture model. *Assay and Drug Development Technologies*, *8*(5), 581–590.

Olsen, C. J., Moreira, J., Lukanidin, E. M., & Ambartsumian, N. S. (2010). Human mammary fibroblasts stimulate invasion of breast cancer cells in a three-dimensional culture and increase stroma development in mouse xenografts. *BMC Cancer*, *10*, 444.

Pageau, S. C., Sazonova, O. V., Wong, J. Y., Soto, A. M., & Sonnenschein, C. (2011). The effect of stromal components on the modulation of the phenotype of human bronchial epithelial cells in 3D culture. *Biomaterials*, *32*(29), 7169–7180.

Pan, K., Zhou, H., Zhang, Z., Gao, Y., & Xu, X. (2013). Evaluation of co-cultured CL-1 hepatocytes and hepatic stellate cells in rotatory cell culture system. *Nan fang yi ke da xue xue bao*, *33*(6), 902–905.

Park, H., Choi, B., Nguyen, J., Fan, J., Shafi, S., Klokkevold, P., et al. (2013). Anionic carbohydrate-containing chitosan scaffolds for bone regeneration. *Carbohydrate Polymers*, *97*(2), 587–596.

Park, S. H., Gil, E. S., Kim, H. J., Lee, K., & Kaplan, D. L. (2010). Relationships between degradability of silk scaffolds and osteogenesis. *Biomaterials*, *31*(24), 6162–6172.

Pathi, S. P., Kowalczewski, C., Tadipatri, R., & Fischbach, C. (2010). A novel 3-D mineralized tumor model to study breast cancer bone metastasis. *PloS One*, *5*(1), e8849.

Purdum, P. P., 3rd., Ulissi, A., Hylemon, P. B., Shiffman, M. L., & Moore, E. W. (1993). Cultured human gallbladder epithelia. Methods and partial characterization of a carcinoma-derived model. *Laboratory Investigation; A Journal of Technical Methods and Pathology*, *68*(3), 345–353.

Quiros, R. M., Valianou, M., Kwon, Y., Brown, K. M., Godwin, A. K., & Cukierman, E. (2008). Ovarian normal and tumor-associated fibroblasts retain in vivo stromal characteristics in a 3-D matrix-dependent manner. *Gynecologic Oncology*, *110*(1), 99–109.

Rao, S. S., Dejesus, J., Short, A. R., Otero, J. J., Sarkar, A., & Winter, J. O. (2013). Glioblastoma behaviors in three-dimensional collagen-hyaluronan composite hydrogels. *ACS Applied Materials & Interfaces*, *5*, 9276–9284.

Rasmuson, A., Kock, A., Fuskevag, O. M., Kruspig, B., Simon-Santamaria, J., Gogvadze, V., et al. (2012). Autocrine prostaglandin E2 signaling promotes tumor cell survival and proliferation in childhood neuroblastoma. *PloS One*, 7(1), e29331.

Rath, B., Nam, J., Knobloch, T. J., Lannutti, J. J., & Agarwal, S. (2008). Compressive forces induce osteogenic gene expression in calvarial osteoblasts. *Journal of Biomechanics*, *41*(5), 1095–1103.

Rivadeneira, J., Carina Audisio, M., Boccaccini, A. R., & Gorustovich, A. A. (2013). In vitro antistaphylococcal effects of a novel 45S5 bioglass/agar-gelatin biocomposite films. *Journal of Applied Microbiology*, *115*(2), 604–612.

Rodday, B., Hirschhaeuser, F., Walenta, S., & Mueller-Klieser, W. (2011). Semiautomatic growth analysis of multicellular tumor spheroids. *Journal of Biomolecular Screening*, *16*(9), 1119–1124.

Sadiasa, A., Kim, M. S., & Lee, B. T. (2013). Poly(lactide-co-glycolide acid)/biphasic calcium phosphate composite coating on a porous scaffold to deliver simvastatin for bone tissue engineering. *Journal of Drug Targeting*, *21*(8), 719–729.

Sainz, B., Jr., TenCate, V., & Uprichard, S. L. (2009). Three-dimensional Huh7 cell culture system for the study of Hepatitis C virus infection. *Virology Journal*, *6*, 103.

Sarig-Nadir, O., & Seliktar, D. (2010). The role of matrix metalloproteinases in regulating neuronal and nonneuronal cell invasion into PEGylated fibrinogen hydrogels. *Biomaterials*, *31*(25), 6411–6416.

Storch, K., Eke, I., Borgmann, K., Krause, M., Richter, C., Becker, K., et al. (2010). Three-dimensional cell growth confers radioresistance by chromatin density modification. *Cancer Research*, *70*(10), 3925–3934.

Sugihara, E., & Saya, H. (2013). Complexity of cancer stem cells. *International Journal of Cancer/Journal International du Cancer*, *132*, 1249–1259.

Sung, J. H., & Shuler, M. L. (2009). A micro cell culture analog (microCCA) with 3-D hydrogel culture of multiple cell lines to assess metabolism-dependent cytotoxicity of anti-cancer drugs. *Lab On a Chip*, *9*(10), 1385–1394.

Suresh, S. (2007). Biomechanics and biophysics of cancer cells. *Acta Biomaterialia*, *3*(4), 413–438.

Sutherland, R. M., Inch, W. R., McCredie, J. A., & Kruuv, J. (1970). A multi-component radiation survival curve using an in vitro tumour model. *International Journal of Radiation Biology and Related Studies in Physics, Chemistry, and Medicine*, *18*(5), 491–495.

Szot, C. S., Buchanan, C. F., Freeman, J. W., & Rylander, M. N. (2011). 3D in vitro bioengineered tumors based on collagen I hydrogels. *Biomaterials*, *32*(31), 7905–7912.

Tan, P. H., Aung, K. Z., Toh, S. L., Goh, J. C., & Nathan, S. S. (2011). Three-dimensional porous silk tumor constructs in the approximation of in vivo osteosarcoma physiology. *Biomaterials*, *32*(26), 6131–6137.

Tang, X., Kuhlenschmidt, T. B., Zhou, J., Bell, P., Wang, F., Kuhlenschmidt, M. S., et al. (2010). Mechanical force affects expression of an in vitro metastasis-like phenotype in HCT-8 cells. *Biophysical Journal*, *99*(8), 2460–2469.

Tepliashin, A. S., Korzhikova, S. V., Sharifullina, S. Z., Rostovskaia, M. S., Chupikova, N. I., Vasiunina, N. I., et al. (2007). Differentiation of multipotent mesenchymal stromal cells of bone marrow into cells of cartilage tissue by culturing in three-dementia OPLA scaffolds. *Tsitologiia*, *49*(7), 544–551.

Tilghman, R. W., Cowan, C. R., Mih, J. D., Koryakina, Y., Gioeli, D., Slack-Davis, J. K., et al. (2010). Matrix rigidity regulates cancer cell growth and cellular phenotype. *PloS One*, *5*(9), e12905.

Torisawa, Y. S., Shiku, H., Yasukawa, T., Nishizawa, M., & Matsue, T. (2005). Multi-channel 3-D cell culture device integrated on a silicon chip for anticancer drug sensitivity test. *Biomaterials*, *26*(14), 2165–2172.

Umbas, R., Schalken, J. A., Aalders, T. W., Carter, B. S., Karthaus, H. F., Schaafsma, H. E., et al. (1992). Expression of the cellular adhesion molecule E-cadherin is reduced or absent in high-grade prostate cancer. *Cancer Research, 52*(18), 5104–5109.

Wang, Y., Chou, B. K., Dowey, S., He, C., Gerecht, S., & Cheng, L. (2013). Scalable expansion of human induced pluripotent stem cells in the defined xeno-free E8 medium under adherent and suspension culture conditions. *Stem Cell Research, 11*(3), 1103–1116.

Wu, C. C., Chao, Y. C., Chen, C. N., Chien, S., Chen, Y. C., Chien, C. C., et al. (2008). Synergism of biochemical and mechanical stimuli in the differentiation of human placenta-derived multipotent cells into endothelial cells. *Journal of Biomechanics, 41*(4), 813–821.

Wu, Y. L., Yang, H. J., Wang, K. K., Tao, G. S., Liu, Y. Z., & Hu, Y. (2010). Reversion of resistance to cisplatin induced by MG132 in cervical cancer line HCE1 multicellular spheroid. *Zhonghua fu Chan ke za zhi, 45*(4), 287–291.

Yamada, Y., Hozumi, K., Katagiri, F., Kikkawa, Y., & Nomizu, M. (2013). Laminin-111-derived peptide-hyaluronate hydrogels as a synthetic basement membrane. *Biomaterials, 34*(28), 6539–6547.

Yang, Y., Basu, S., Tomasko, D. L., Lee, L. J., & Yang, S. T. (2005). Fabrication of well-defined PLGA scaffolds using novel microembossing and carbon dioxide bonding. *Biomaterials, 26*(15), 2585–2594.

Yang, S., Guo, L. J., Gao, Q. H., Xuan, M., Tan, K., Zhang, Q., et al. (2010). Derived vascular endothelial cells induced by mucoepidermoid carcinoma cells: 3-Dimensional collagen matrix model. *Journal of Zhejiang University. Science. B, 11*(10), 745–753.

Yang, Z., & Zhao, X. (2011). A 3D model of ovarian cancer cell lines on peptide nanofiber scaffold to explore the cell-scaffold interaction and chemotherapeutic resistance of anticancer drugs. *International Journal of Nanomedicine, 6*, 303–310.

Yi, S. Y., Hao, Y. B., Nan, K. J., & Fan, T. L. (2013). Cancer stem cells niche: A target for novel cancer therapeutics. *Cancer Treatment Reviews, 39*, 290–296.

Yuhas, J. M., Li, A. P., Martinez, A. O., & Ladman, A. J. (1977). A simplified method for production and growth of multicellular tumor spheroids. *Cancer Research, 37*(10), 3639–3643.

Zeng, G. F., Cai, S. X., & Wu, G. J. (2011). Up-regulation of METCAM/MUC18 promotes motility, invasion, and tumorigenesis of human breast cancer cells. *BMC Cancer, 11*, 113.

Zeromski, J., Nyczak, E., & Dyszkiewicz, W. (2001). Significance of cell adhesion molecules, CD56/NCAM in particular, in human tumor growth and spreading. *Folia Histochemica et Cytobiologica/Polish Academy of Sciences, Polish Histochemical and Cytochemical Society, 39*(Suppl. 2), 36–37.

Zhang, Y., Jiang, X., Zhang, X., Wang, D., & Zhen, L. (2013). Cytocompatibility of two porous bioactive glass-ceramic in vitro. *West Chin J. Stomatol., 31*(3), 294–299.

Zhang, X., Wang, W., Yu, W., Xie, Y., Zhang, X., Zhang, Y., et al. (2005). Development of an in vitro multicellular tumor spheroid model using microencapsulation and its application in anticancer drug screening and testing. *Biotechnology Progress, 21*(4), 1289–1296.

INDEX

Note: Page numbers followed by "*f*" indicate figures and "*t*" indicate tables.

C

S

T

V

W

Z

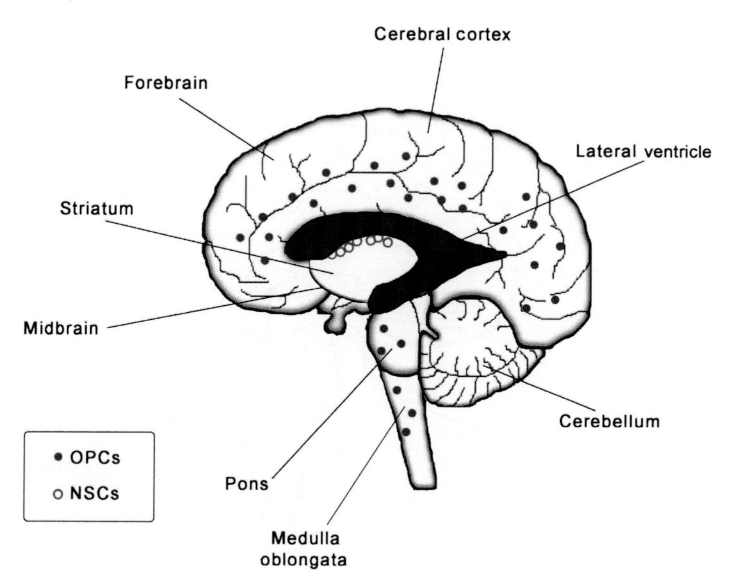

Figure 1.1, Shirin Ilkanizadeh *et al.* (See Page 4 of this volume.)

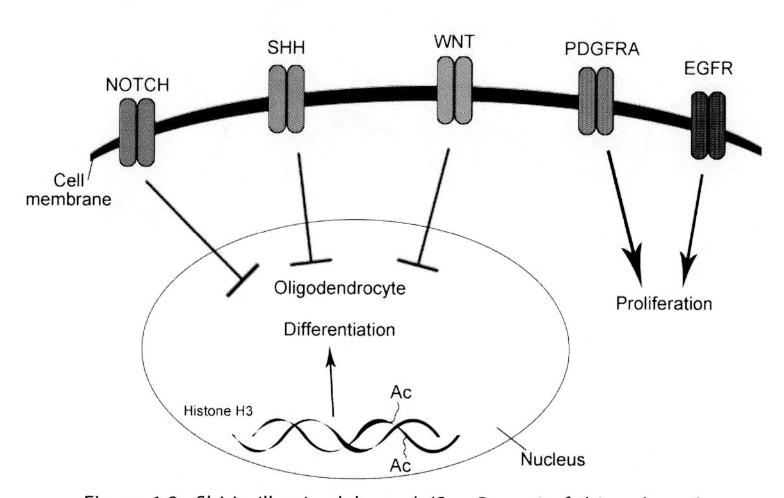

Figure 1.2, Shirin Ilkanizadeh *et al.* (See Page 6 of this volume.)

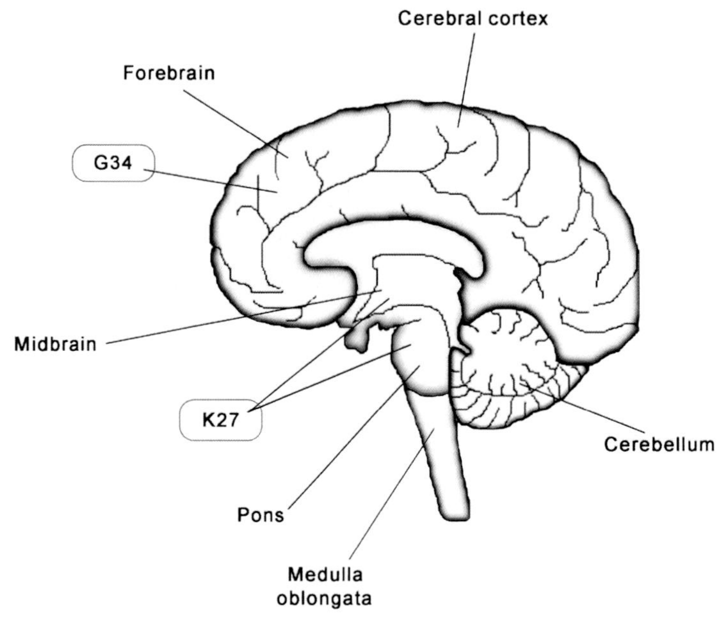

Figure 1.3, Shirin Ilkanizadeh *et al.* (See Page 11 of this volume.)

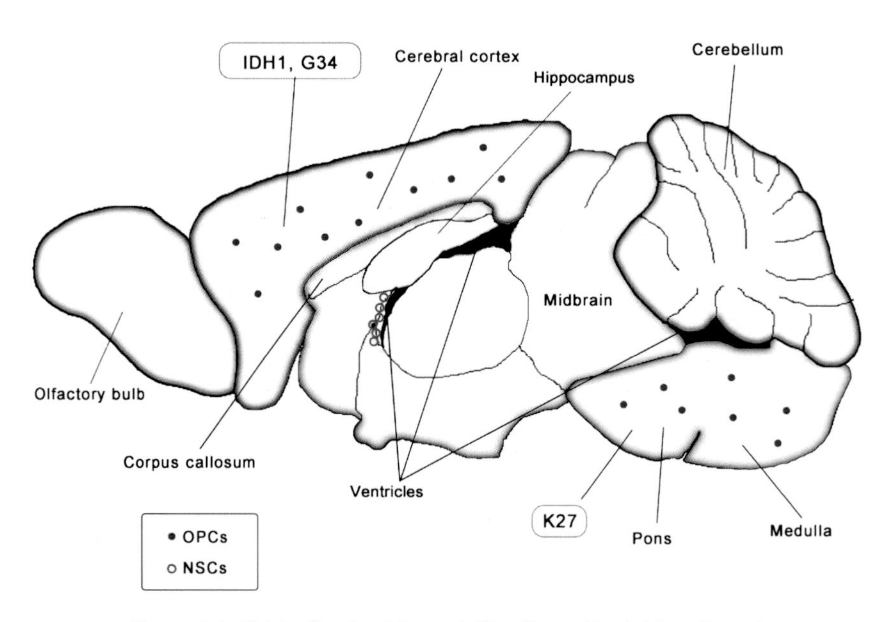

Figure 1.6, Shirin Ilkanizadeh *et al.* (See Page 41 of this volume.)

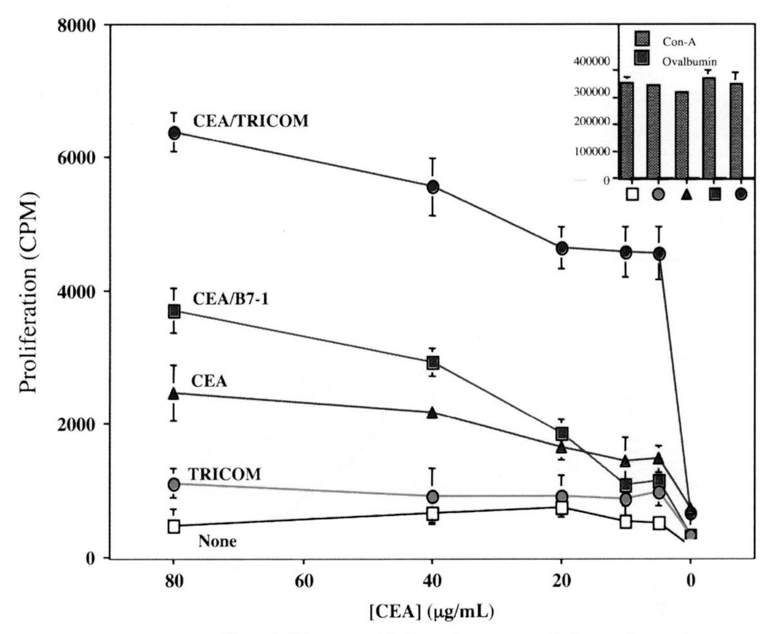

Figure 2.2, Jeffrey Schlom *et al.* (See Page 80 of this volume.)

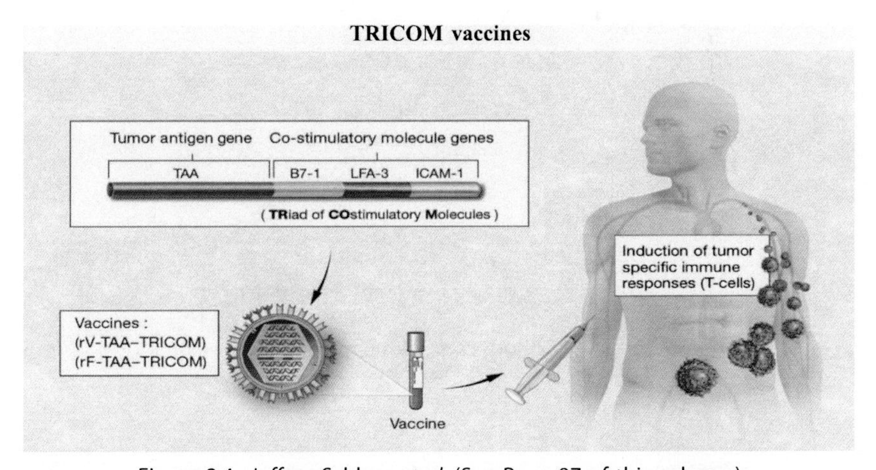

Figure 2.4, Jeffrey Schlom *et al.* (See Page 87 of this volume.)

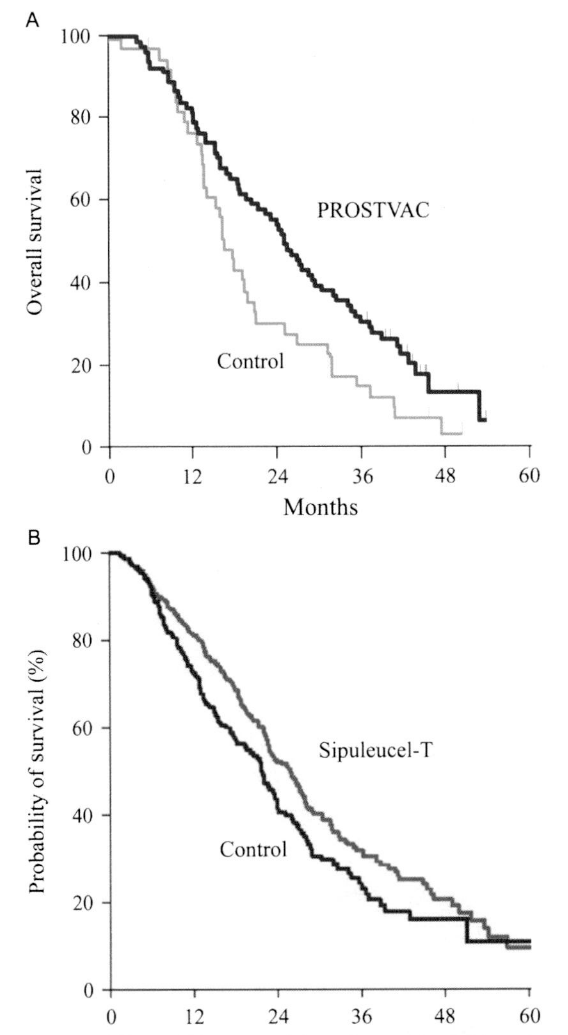

Figure 2.5, Jeffrey Schlom *et al.* (See Page 90 of this volume.)

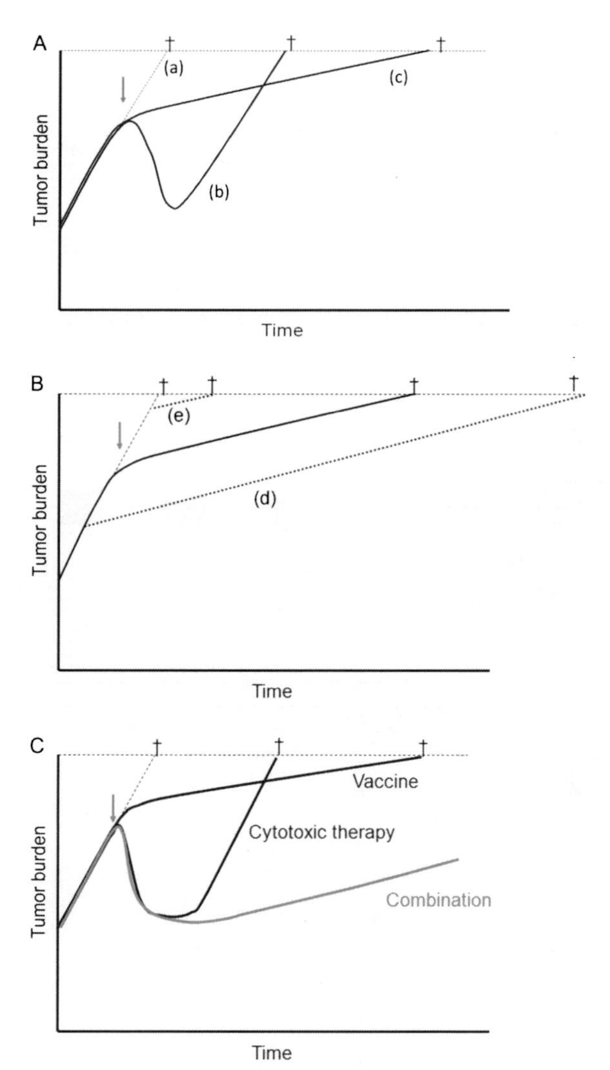

Figure 2.7, Jeffrey Schlom *et al.* (See Page 93 of this volume.)

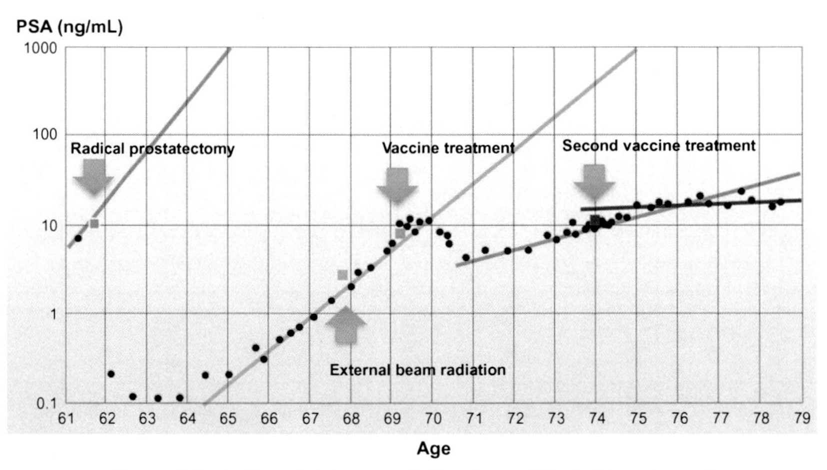

Figure 2.8, Jeffrey Schlom *et al.* (See Page 95 of this volume.)

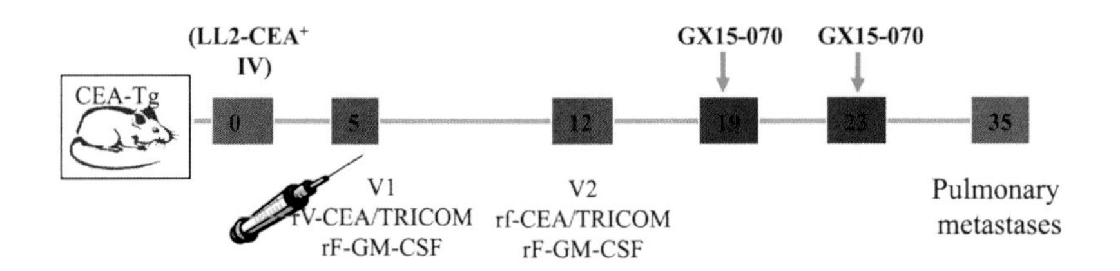

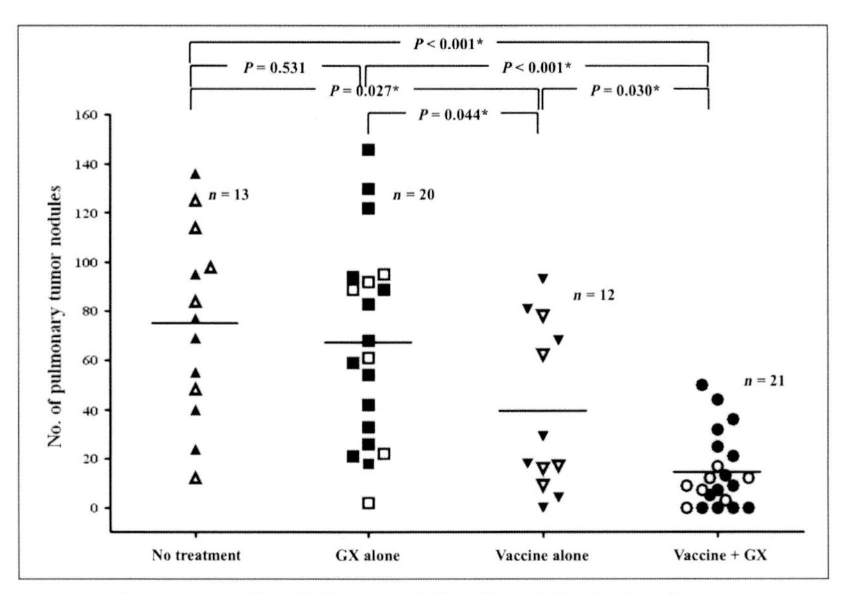

Figure 2.9, Jeffrey Schlom *et al.* (See Page 100 of this volume.)

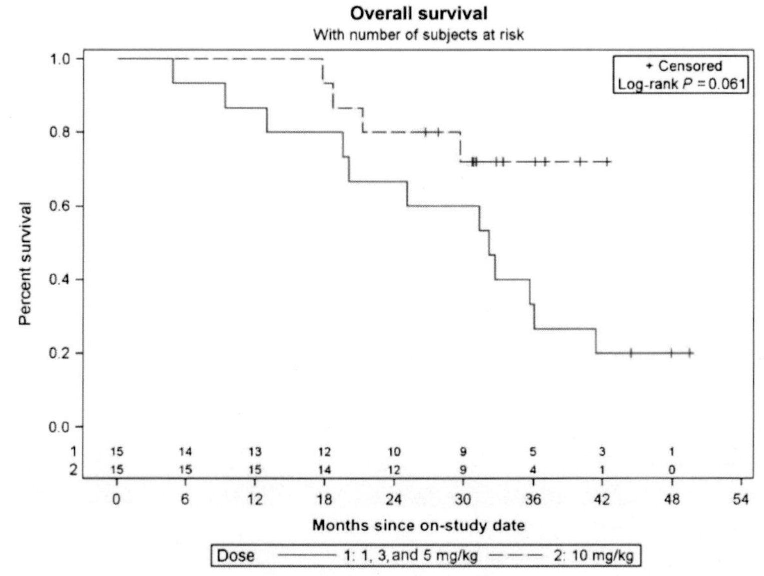

Figure 2.12, Jeffrey Schlom *et al.* (See Page 105 of this volume.)

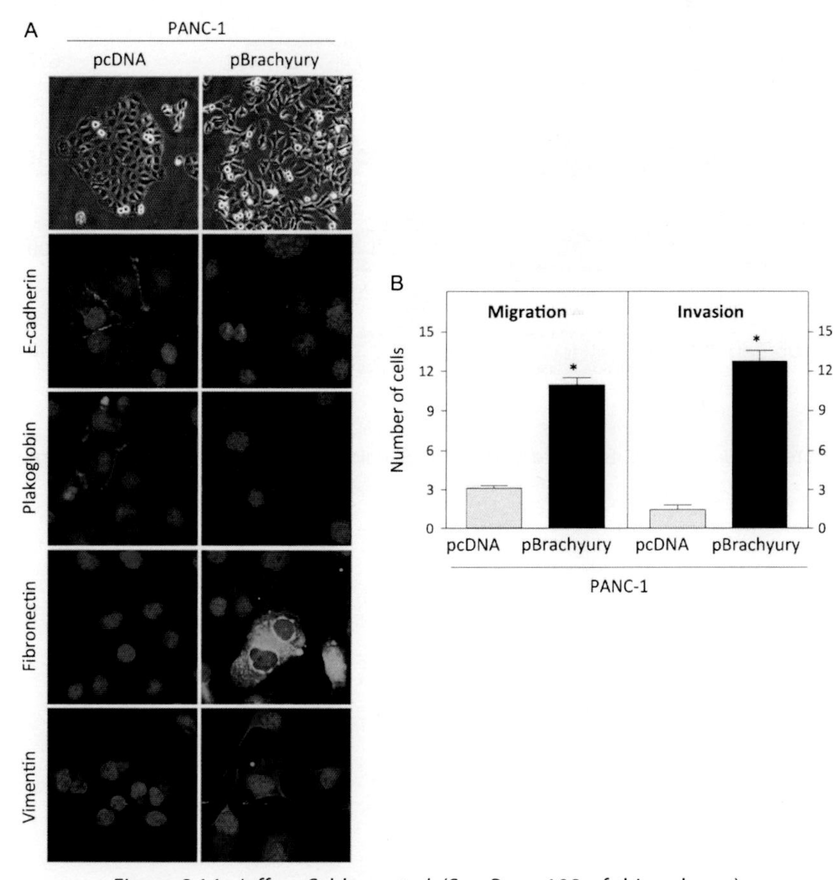

Figure 2.14, Jeffrey Schlom *et al.* (See Page 108 of this volume.)

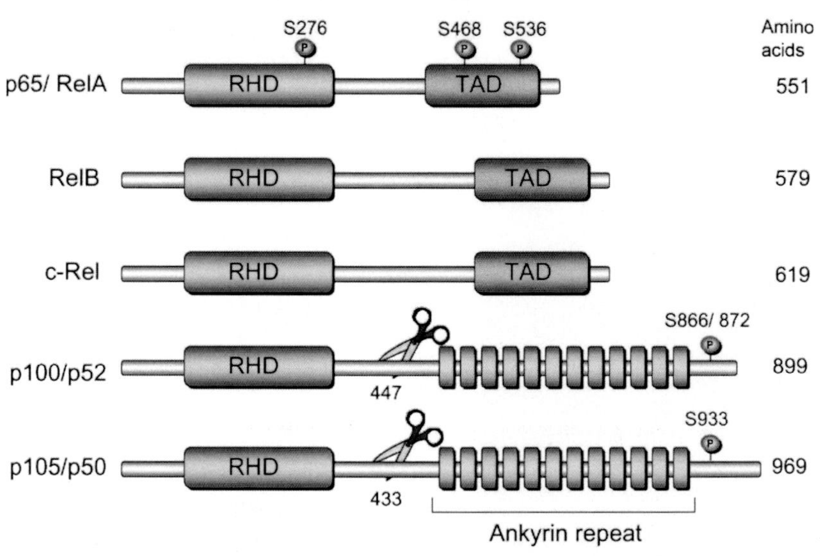

Figure 3.1, Jennifer W. Bradford and Albert S. Baldwin (See Page 126 of this volume.)

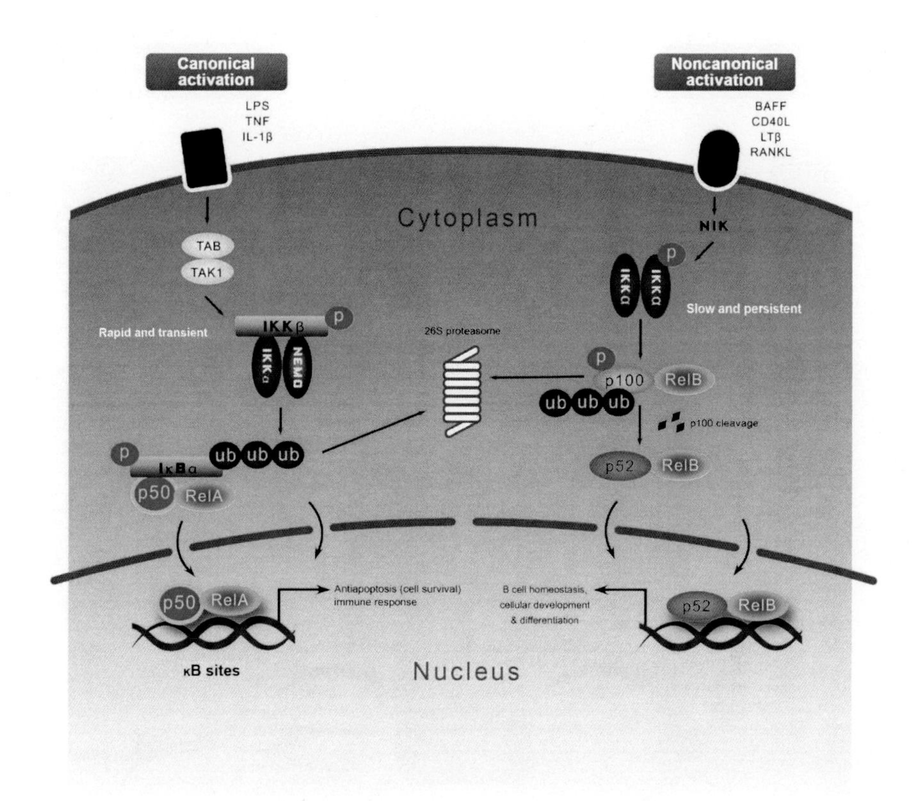

Figure 3.2, Jennifer W. Bradford and Albert S. Baldwin (See Page 128 of this volume.)

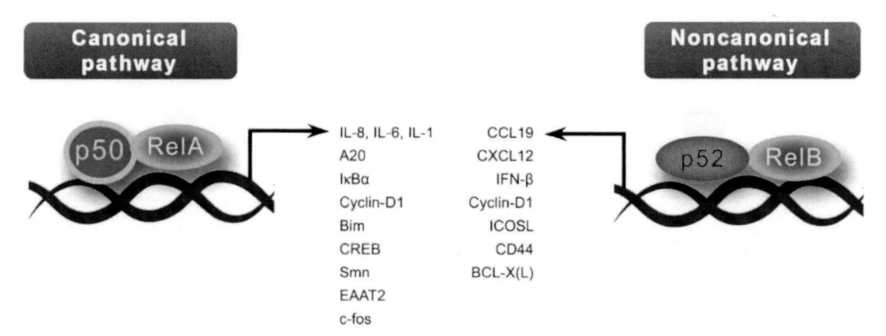

Figure 3.3, Jennifer W. Bradford and Albert S. Baldwin (See Page 128 of this volume.)

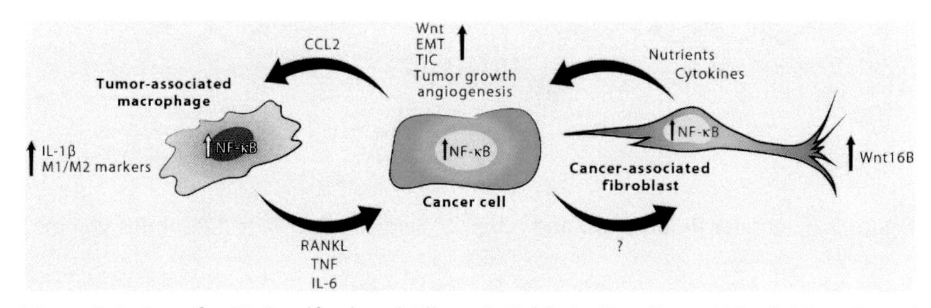

Figure 3.4, Jennifer W. Bradford and Albert S. Baldwin (See Page 132 of this volume.)

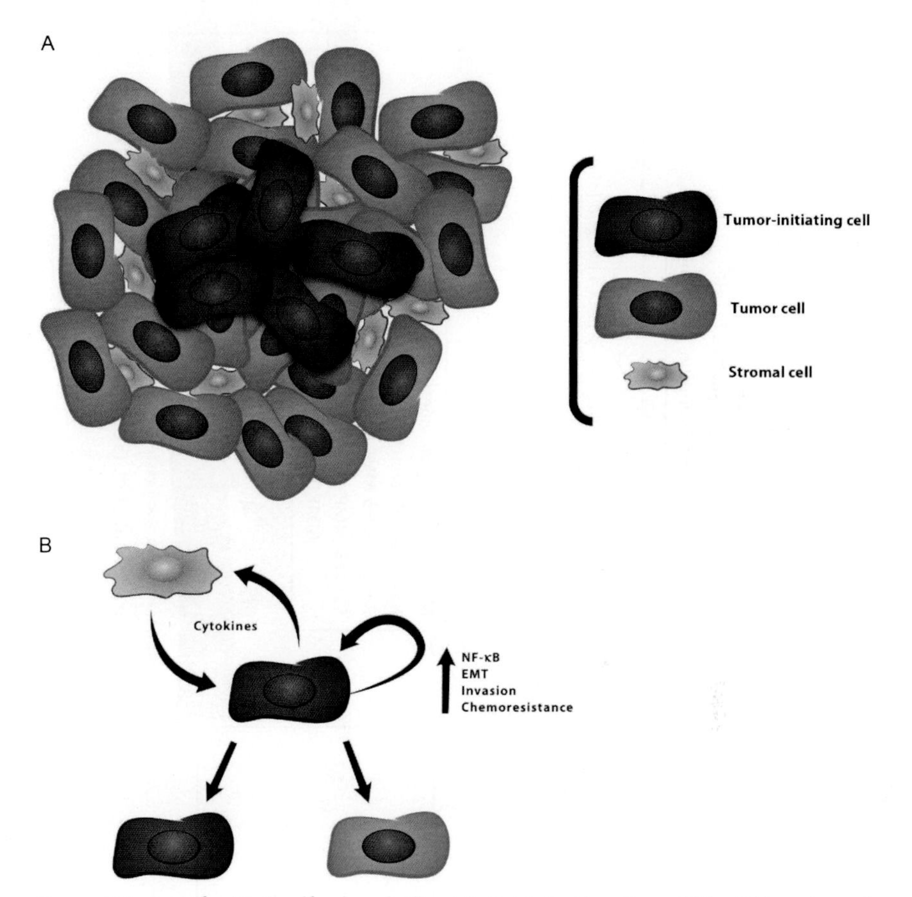

Figure 3.6, Jennifer W. Bradford and Albert S. Baldwin (See Page 138 of this volume.)

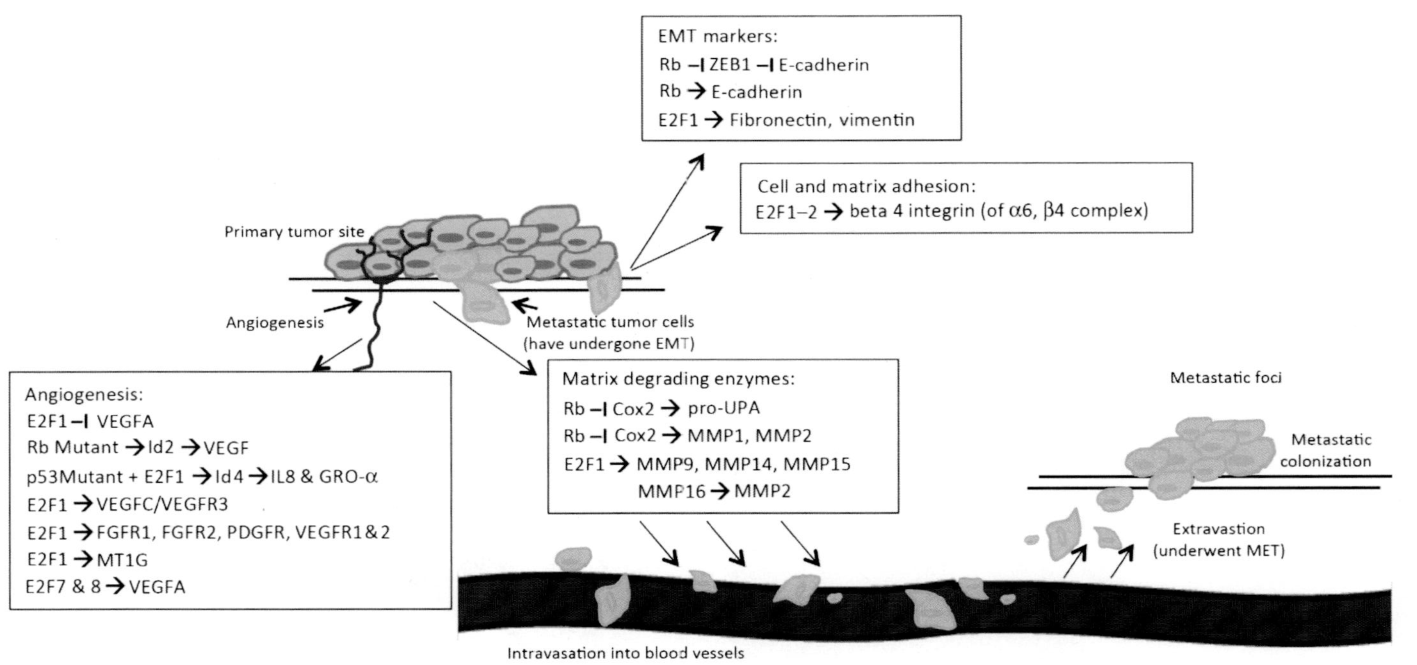

Figure 4.1, Courtney Schaal *et al.* (See Page 166 of this volume.)

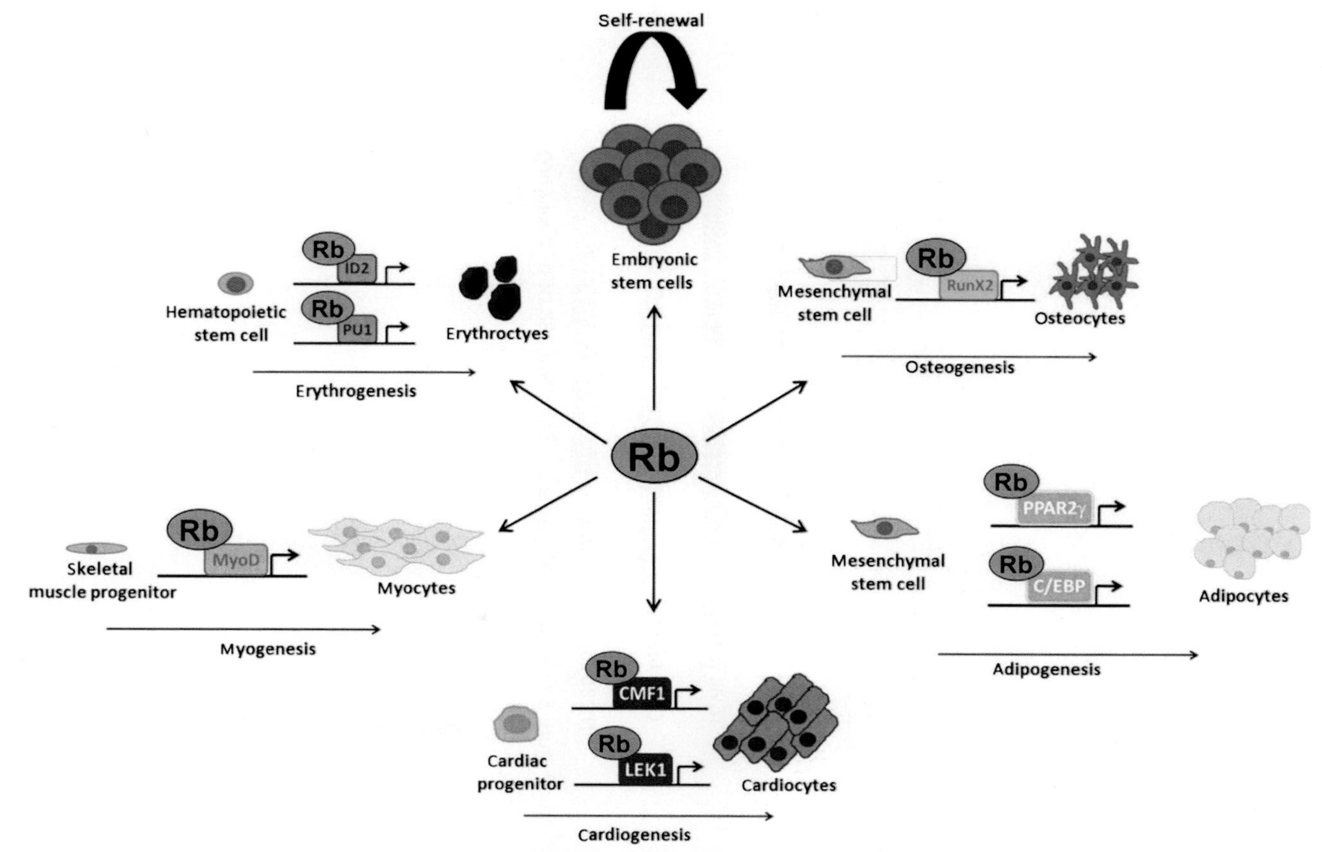

Figure 4.2, Courtney Schaal *et al.* (See Page 170 of this volume.)

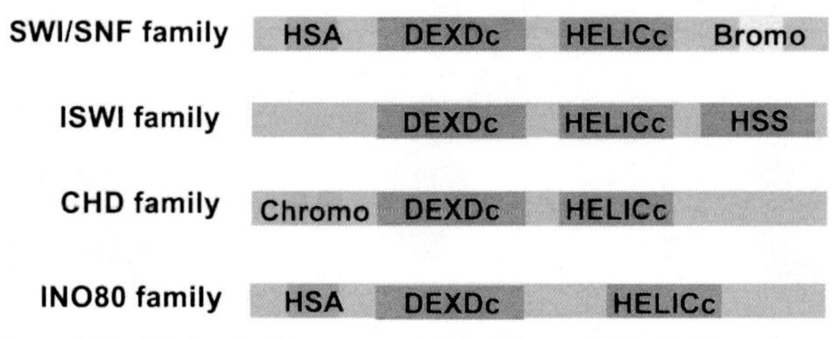

Figure 5.1, Kimberly Mayes *et al.* (See Page 188 of this volume.)

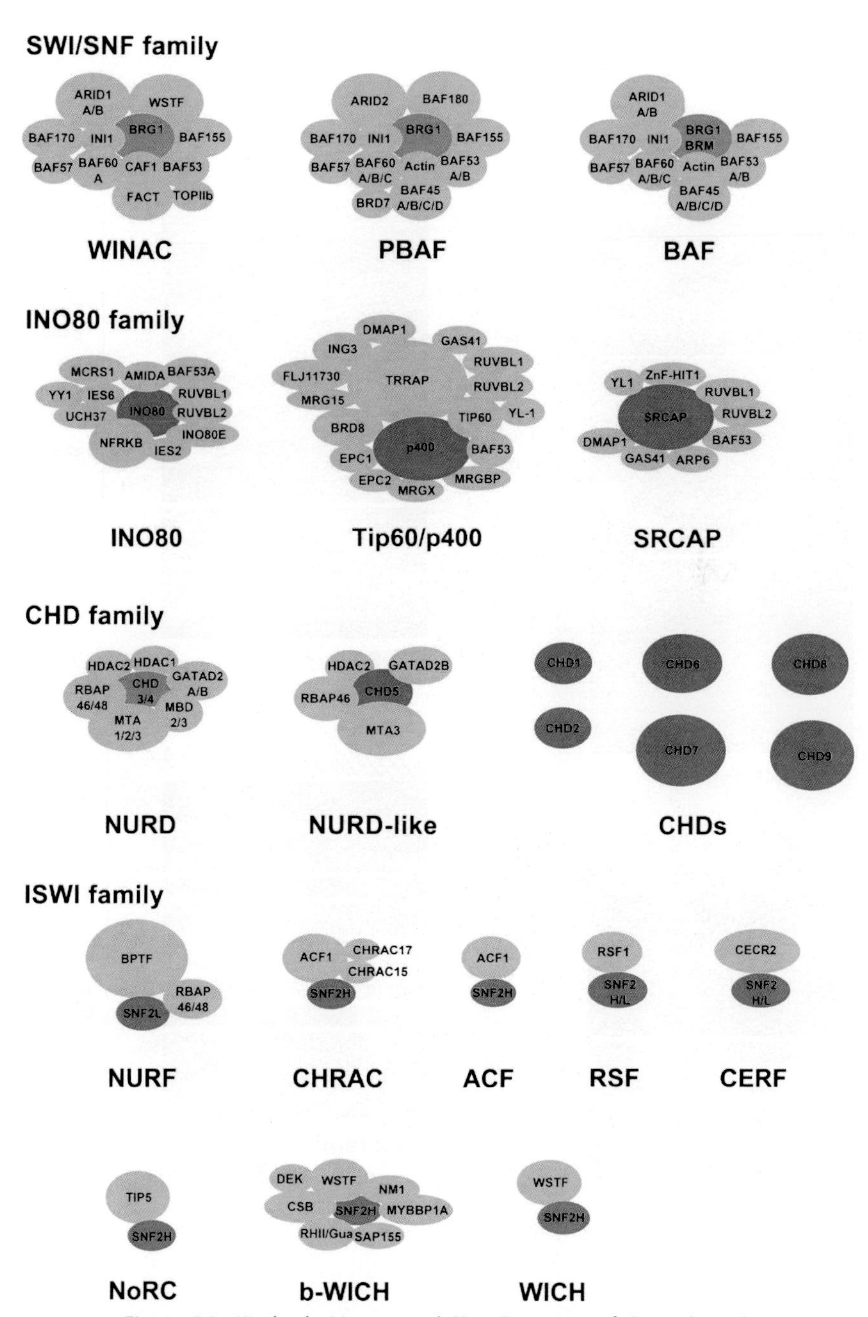

Figure 5.2, Kimberly Mayes *et al.* (See Page 189 of this volume.)

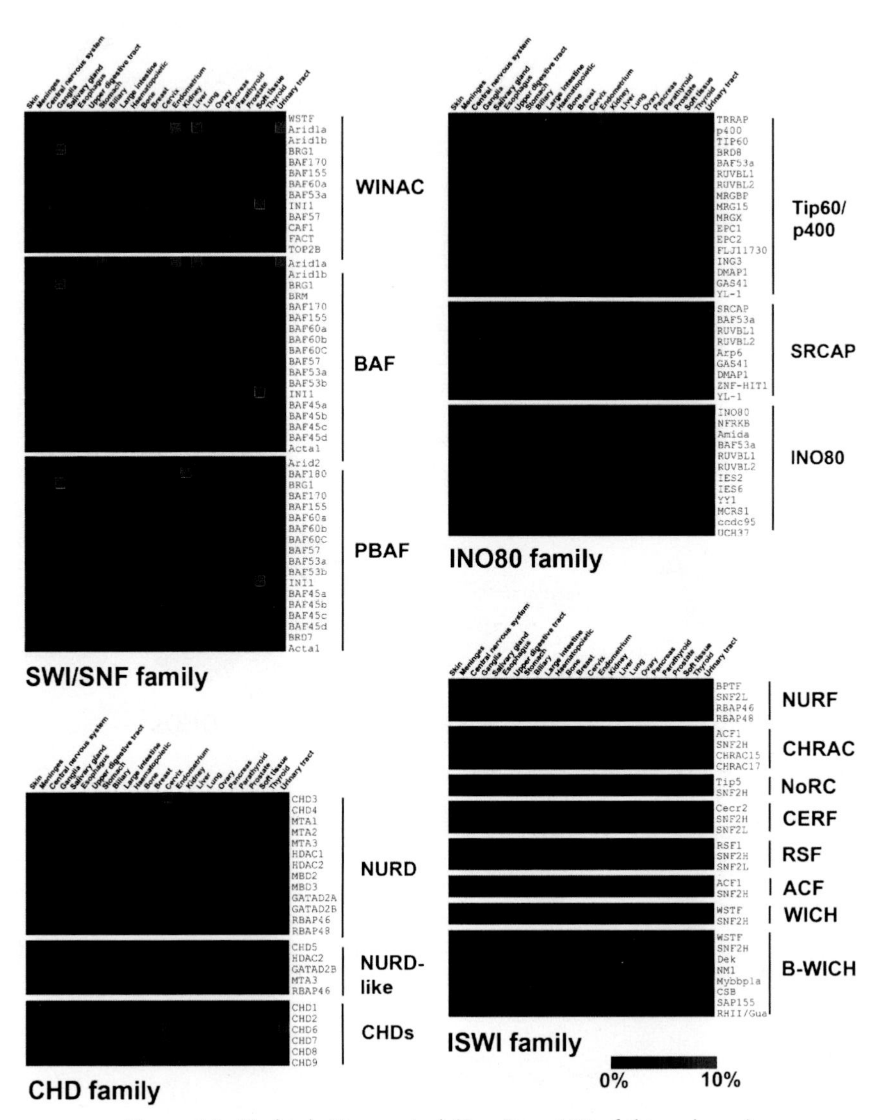

Figure 5.3, Kimberly Mayes *et al.* (See Page 193 of this volume.)

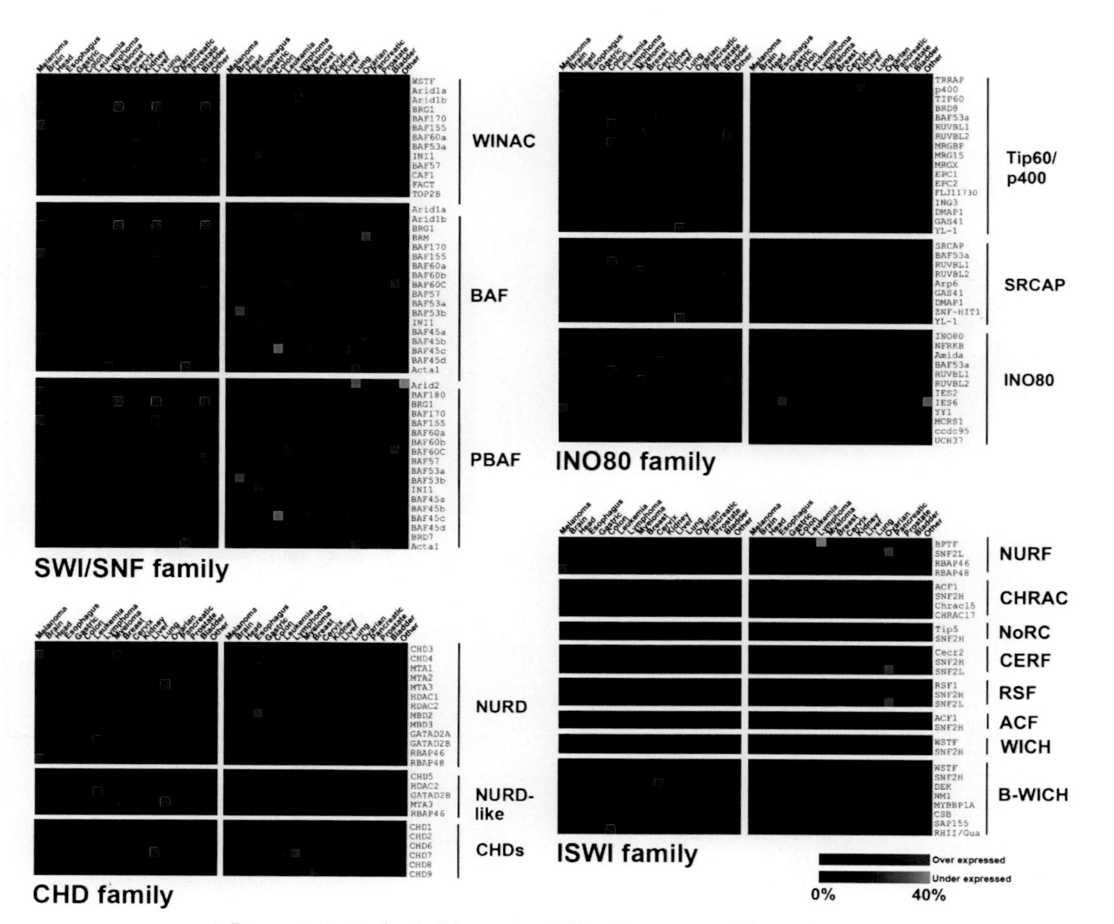

Figure 5.4, Kimberly Mayes *et al.* (See Page 196 of this volume.)

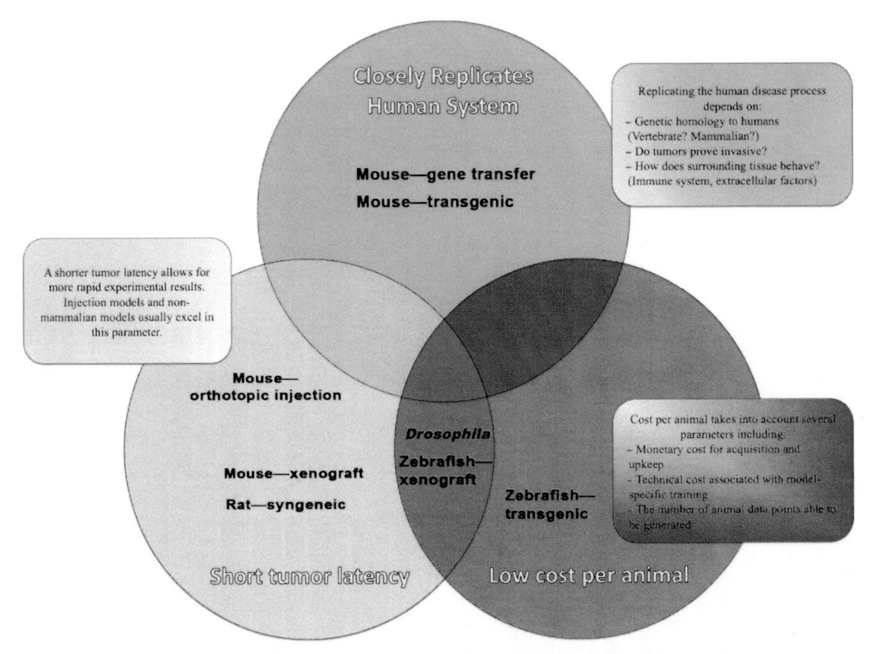

Figure 7.1, Timothy P. Kegelman *et al.* (See Page 278 of this volume.)

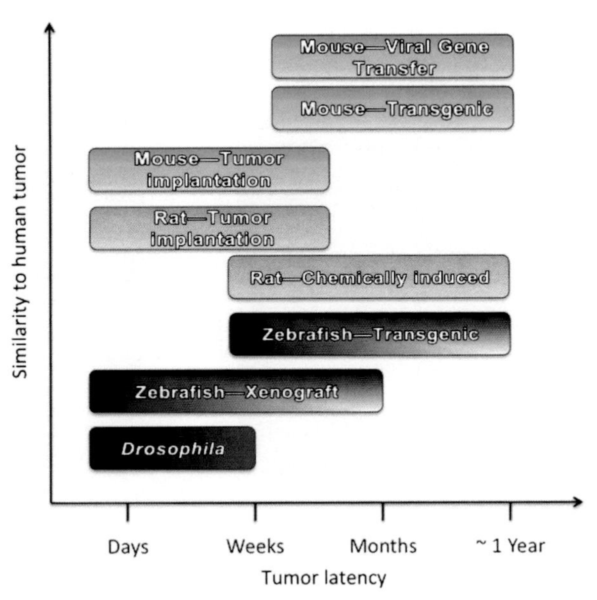

Figure 7.2, Timothy P. Kegelman *et al.* (See Page 279 of this volume.)

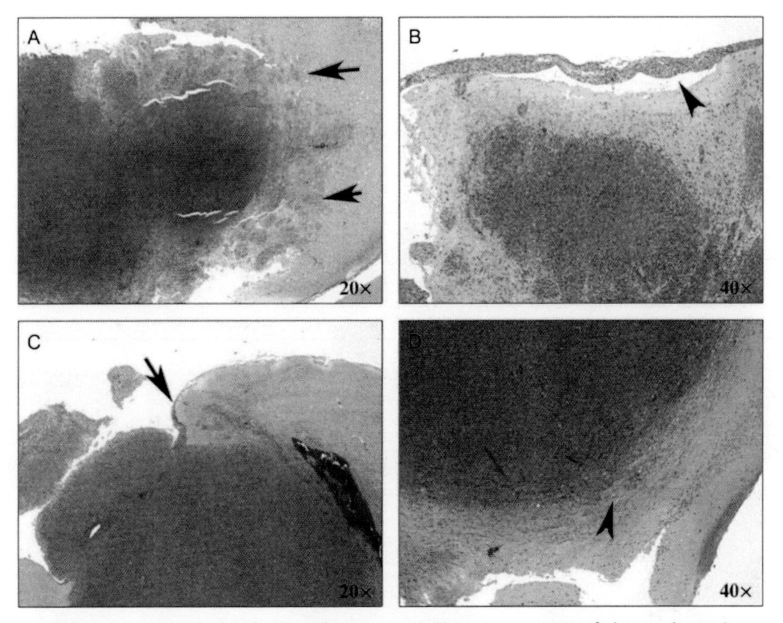

Figure 7.3, Timothy P. Kegelman *et al.* (See Page 289 of this volume.)

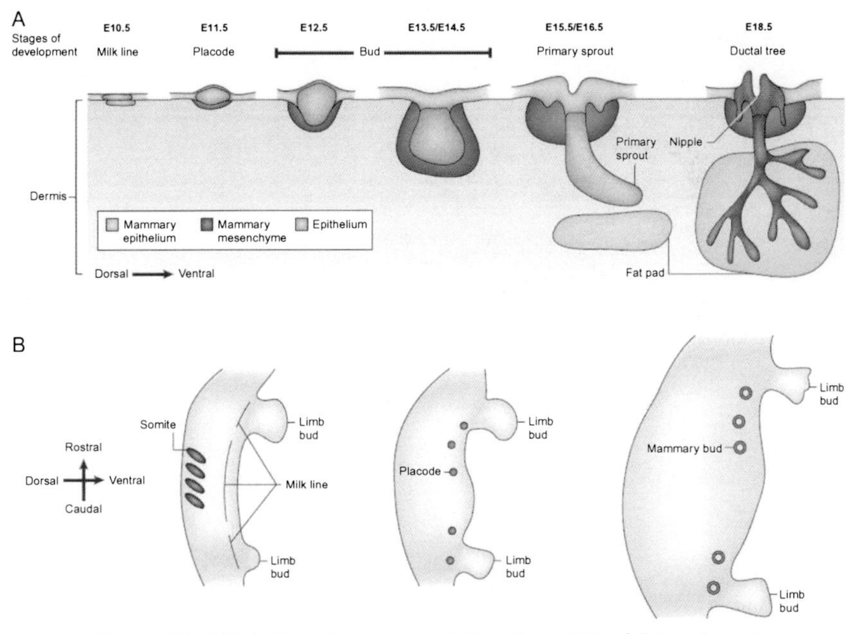

Figure 8.1, Mitchell E. Menezes *et al.* (See Page 335 of this volume.)

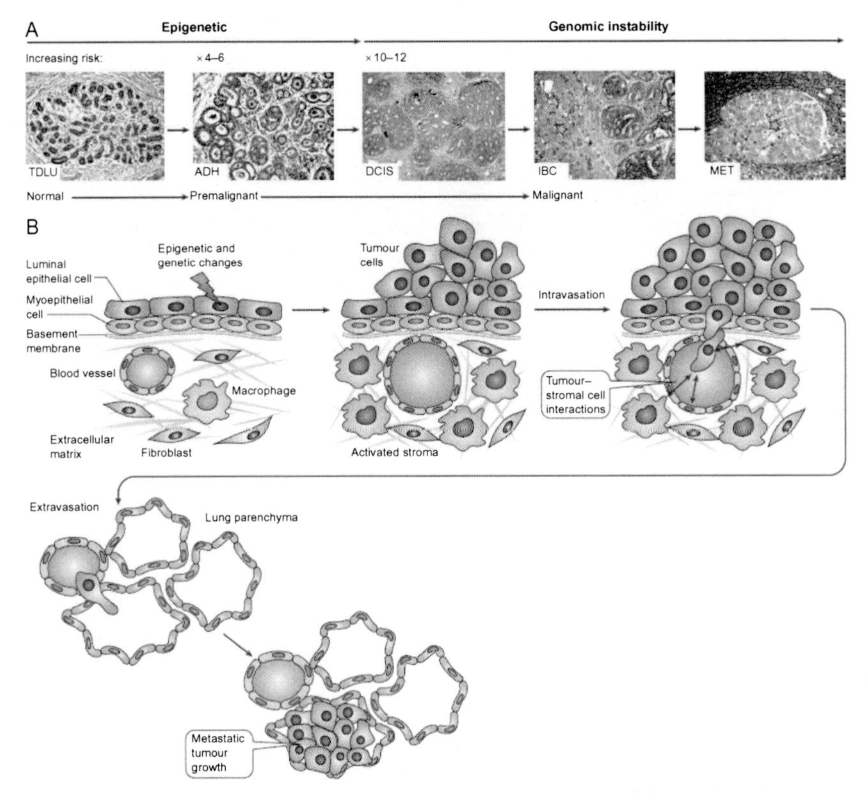

Figure 8.2, Mitchell E. Menezes *et al.* (See Page 340 of this volume.)

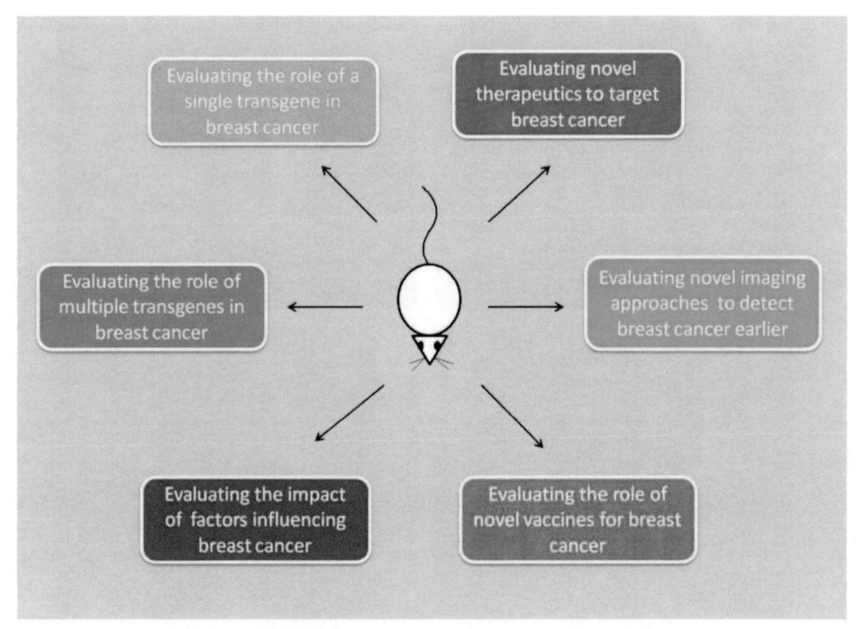

Figure 8.3, Mitchell E. Menezes *et al.* (See Page 372 of this volume.)

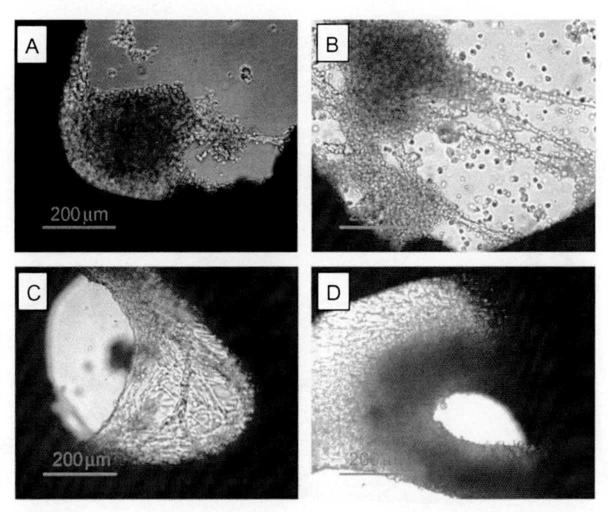

Figure 9.2, I. Levinger *et al.* (See Page 385 of this volume.)

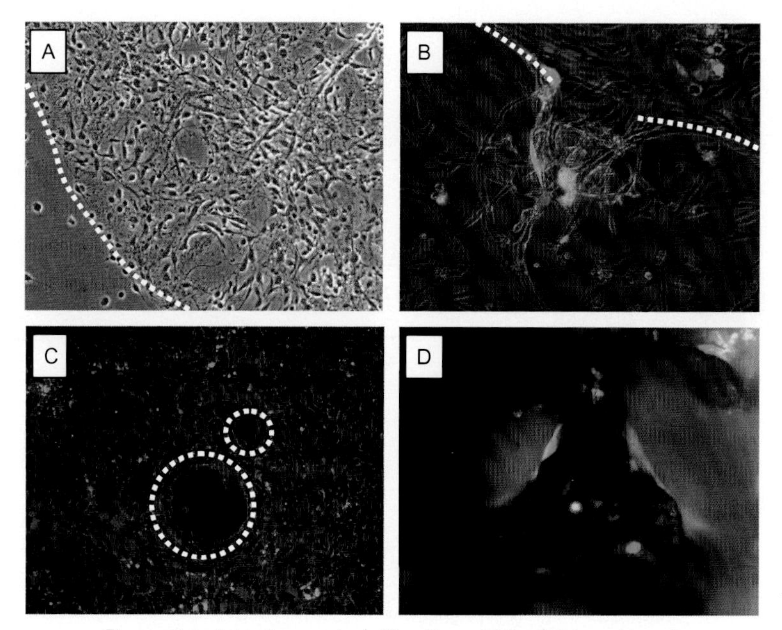

Figure 9.4, I. Levinger *et al.* (See Page 388 of this volume.)

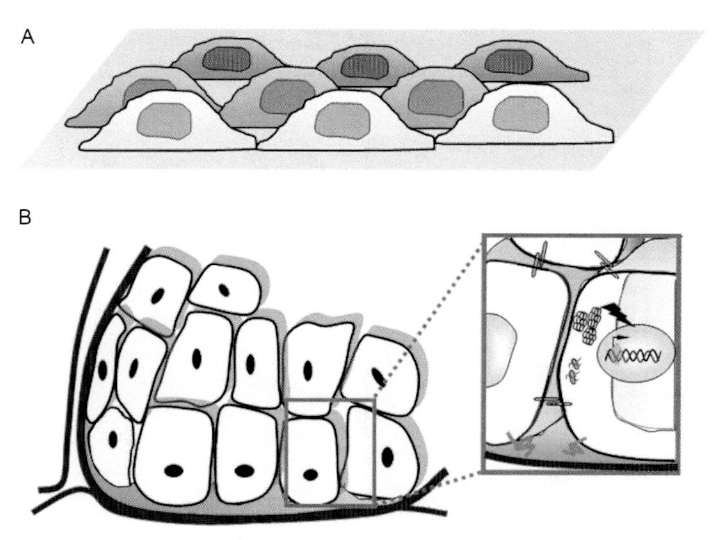

Figure 9.5, I. Levinger *et al.* (See Page 399 of this volume.)

CPI Antony Rowe
Eastbourne, UK
July 15, 2014